New Perspectives and Challenges in Traffic and Transportation Engineering Supporting Energy Saving in Smart Cities

New Perspectives and Challenges in Traffic and Transportation Engineering Supporting Energy Saving in Smart Cities

Editors

Elżbieta Macioszek
Anna Granà
Margarida Coelho
Paulo Fernandes

MDPI • Basel • Beijing • Wuhan • Barcelona • Belgrade • Manchester • Tokyo • Cluj • Tianjin

Editors
Elżbieta Macioszek
Silesian University of Technology
Poland

Anna Granà
University of Palermo
Italy

Margarida Coelho
University of Aveiro
Portugal

Paulo Fernandes
University of Aveiro
Portugal

Editorial Office
MDPI
St. Alban-Anlage 66
4052 Basel, Switzerland

This is a reprint of articles from the Special Issue published online in the open access journal *Energies* (ISSN 1996-1073) (available at: https://www.mdpi.com/journal/energies/special_issues/traffic_transportation_smart_cities).

For citation purposes, cite each article independently as indicated on the article page online and as indicated below:

LastName, A.A.; LastName, B.B.; LastName, C.C. Article Title. *Journal Name* **Year**, *Volume Number*, Page Range.

ISBN 978-3-0365-5523-2 (Hbk)
ISBN 978-3-0365-5524-9 (PDF)

Contents

About the Editors

Elżbieta Macioszek

DSc. PhD. Eng. Elżbieta Macioszek, Professor at the Silesian University of Technology. Works in the Department of Transport Systems, Traffic Engineering and Logistics, Faculty of Transport and Aviation Engineering, Silesian University of Technology, Poland. She is deputy head of the department, and is a member of the scientific board for civil engineering and transportation on the Faculty of Transport and Aviation Engineering, Silesian University of Technology. She has received eight awards from the Rector of the Silesian University of Technology: four for scientific achievements and four for organizational achievements.

Her professional experience includes more than ten national and international contract works in the field of transportation. Her scientific interests combine the problems of road traffic engineering (including traffic analysis and prognosis, transport systems modeling, and optimization of the transport networks) with the development of commuting behaviors within cities. She is an expert in performing assessments of projects' applications in Poland. Moreover, she is an author and coauthor of more than 230 papers and chapters in books; a member of numerous organizational, program, technical, and scientific committees of scientific conferences organized in various countries around the world; as well a reviewer and member of editorial boards in journals connected with transport, traffic, and civil engineering. She has completed scientific internships, the first in 2019 at the Department of Civil Engineering, Faculty of Science and Technology in Tokyo University of Science in Japan; and the second in 2021 in the Department of Engineering at the University of Palermo. In addition, she took part in and represented the Silesian University of Technology in many European Programs.

Other information can be found here:

https://orcid.org/0000-0002-1345-0022

https://www.scopus.com/authid/detail.uri?authorId=36662343300

https://publons.com/researcher/1672098/elzbieta-macioszek/

Anna Granà

Anna Granà works in the Department of Engineering, University of Palermo, Italy. She received her PhD in Road, Airport and Railway Infrastructure Engineering from the University of Palermo, Italy, and currently she is a professor in Road, Railway and Airport Construction at the same university.

She participates in the editorial committees of journals and editorial series: since April 2022, she has been Associate Editor of *Transport* (Proceedings of the ICE); since 2012, she has been the Editor-in-Chief of the *Journal of Sustainable Development*; and she is the Review Editor of the international journal *Transportation and Transit Systems* (Frontiers in Built Environment). Since 2003, she has been a member of the Italian Society of Road Railway and Airport Infrastructures (SIIV).

She is a member of the International Scientific Advisory Board for the Priority Research Area: Smart cities and future mobility (Silesian University of Technology, Poland); she was invited by the Rector of the Silesian University of Technology, Professor Arkadiusz Mężyk PhD, DSc in January 2021.

She is a member of several research units in National Relevant Research Programs. Her research activity is demonstrated by more than 90 articles in journals, book chapters, or proceedings of national and international conferences. Anna Granà has participated in several congresses; in many cases, she was called for the presentation of her own papers. She won the SIIV Research Award for the best paper presented at the 2008 SIIV annual meeting.

Her fields of scientific interest include: the analysis of the risk associated with road traffic; the functional design of road geometry, with particular reference to non-signalized intersections and roundabouts; capacity modelling and environmental sustainability of transport infrastructures; and traffic microsimulation modelling. The themes of the analysis of the risk associated with road traffic cover a wide range of interests including: the preventive analysis of road safety; the safety and reliability of transport infrastructures (roads and intersections); crash analysis; and microscopic traffic simulation modelling.

Other information can be found here:

https://orcid.org/0000-0001-6976-0807

Margarida Coelho

Prof. Margarida Coelho is Assistant Professor with Habilitation ("Agregação") of the Department of Mechanical Engineering and Vice-Director of the Centre for Mechanical Technology and Automation, University of Aveiro, Portugal. She is the Scientific Coordinator of the research team on Smart Mobility at the Department of Mechanical Engineering. Her research interests are: impacts of transportation systems (namely, traffic congestion, energy consumption, pollutant emissions, and road safety); connected and automated mobility; life cycle assessment; active modes; and decarbonization of ports. Margarida Coelho has more than 90 scientific papers published (or in press) in international journals (such as the *Sustainable Cities & Society, Applied Energy, International Journal of Hydrogen Energy, Science for Total Environment, Transportation Research Part D: Transport and Environment, International Journal on Sustainable Transportation, Atmospheric Environment, and Transportation Research Record*, etc.), besides more than 150 other publications in book chapters and proceedings of scientific conferences. Margarida Coelho has had extensive participation in transport-related projects. She is the PI of R&D Projects funded by the Portuguese Science and Technology Foundation (FCT) and the Luso-American Foundation / United States National Science Foundation. She is also a participating member in several projects funded by FCT, SUDOE, and INTERREG Europe Programmes; under the Interreg Projects CISMOB and PriMaaS, she has been involved in stakeholder community engagement regarding the penetration of ICT in transport (to tackle decarbonization objectives) as well as low-carbon mobility-as-service measures. She has already supervised 5 PhD students (6 other students are developing their PhD under her supervision) and more than 50 Master students. She was the Chairwoman of EWGT2021—EURO Working Group on Transportation Annual Meeting. She is a recognized Expert on Energy and Transportation by the Portuguese Association of Engineering, where she also serves as Vice-Coordinator of the Energy Commission.

Paulo Fernandes

Paulo Fernandes received his PhD in the Doctoral Programme in Mechanical Engineering at the University of Aveiro in May 2017. His current research interests are: (1) road traffic emissions monitoring; (2) emissions, noise, and safety modelling; and (3) simulation and analysis of transport systems, active modes, and new mobility forms. Between 2017 and 2020, he worked as a post-doc researcher and further as an assistant researcher in the MobiWise project at the Centre for Mechanical Technology and Automation (TEMA). His main role consisted of supervising young research fellows with respect to traffic, noise, emissions, and safety modelling. He also worked at the DICA-VE project as junior researcher at TEMA in the management of activities from research fellows, namely emissions monitoring and modelling (2020 and 2021). After that, he was junior researcher at TEMA in the Driving2Driverless project in activities related to the cost–benefit analysis of shared, autonomous, and electric vehicles services (2021). Currently, he is a First-Level Doctorate Researcher at TEMA with his own scientific project entitled "IDRIVE: Incorporating Driving Volatility Information into a Rating system to Inform drivers about Vehicle Emission rates", which is centred on developing rating systems for NOx and PM emissions from Euro 6 vehicles. In addition to previous works and being a team member of several scientific projects such as @CRUiSE or InFLOWence, he collaborated in 2019 as a technical consultant in an audit of the energy used of the transport sector in Bartica, Guyana with the Instituto do Ambiente e Desenvolvimento in activities related to experimental survey design, energy analysis, and policy recommendations. In the context of cooperation with society, he participated as a member in the Mobility Plan of the University of Aveiro (concluded in January 2020) in the establishment and evaluation of measures for promoting soft modes and improving quality of life within the university campus. Additionally, he joined to the Organizing Committee of the EWGT2021:24th Euro Working Group on Transportation in September 2021. Between 2017 and 2021, he supervised four master's students in the Mechanical Engineering Integrated Master Course. He has published research works with 47 coauthors, demonstrating a solid disposition to establish inter-institutional collaborations. He has published 31 articles and 12 conference papers in SCOPUS-indexed journals (17 as the main corresponding author, a total of 328 citations, and h-index of 11), with some of them in the top 20 most-cited Transport (Transportation Research Part-D, Sustainable Cities and Society, Transportation Research Part-A), Environmental Science (Science of the Total Environment, Atmospheric Environment) and Engineering (Applied Energy) journals according to the Cite Score 2020 rank; 3 book chapters; and 16 papers in international transport conferences (IEEE, EWGT, World Conference on Transport Research and Transportation Research Board).

energies

MDPI

Editorial

New Perspectives and Challenges in Traffic and Transportation Engineering Supporting Energy Saving in Smart Cities—A Multidisciplinary Approach to a Global Problem

Elżbieta Macioszek [1,*], Anna Granà [2], Paulo Fernandes [3] and Margarida C. Coelho [3]

1 Department of Transport Systems, Traffic Engineering and Logistics, Faculty of Transport and Aviation Engineering, Silesian University of Technology, 8 Krasińskiego St., 40-019 Katowice, Poland
2 Department of Engineering, University of Palermo, Viale delle Scienze ed 8, 90128 Palermo, Italy; anna.grana@unipa.it
3 Centre for Mechanical Technology and Automation (TEMA), Department of Mechanical Engineering, Campus Universitario de Santiago, University of Aveiro, 3810-193 Aveiro, Portugal; paulo.fernandes@ua.pt (P.F.); margarida.coelho@ua.pt (M.C.C.)
* Correspondence: elzbieta.macioszek@polsl.pl

Citation: Macioszek, E.; Granà, A.; Fernandes, P.; Coelho, M.C. New Perspectives and Challenges in Traffic and Transportation Engineering Supporting Energy Saving in Smart Cities—A Multidisciplinary Approach to a Global Problem. *Energies* **2022**, *15*, 4191. https://doi.org/10.3390/en15124191

Received: 10 May 2022
Accepted: 2 June 2022
Published: 7 June 2022

1. Introduction

Transportation, like other spheres of human activity, is constantly changing due to economic development. People are constantly improving the ways of moving using various energy sources, expanding infrastructures, and adapting cities to increasing traffic volumes. An efficient, modern, and demand-driven transportation infrastructure is the strength of a growing economy. The development of transportation improves the accessibility of regions and labor markets and reduces the nuisance caused by congestion. All these elements translate into economic benefits, competitiveness, increased productivity of enterprises and regions, as well as social benefits. However, transport consumes enormous energy resources; hence, all solutions in the scope of improving transportation infrastructure, vehicle construction, as well as modeling users' and drivers' behavior may constitute an element contributing to broadly saving energy.

This Special Issue will consist of papers describing the state-of-the-art in methods and solutions in traffic and transportation engineering supporting energy saving in smart cities as well as outlining trends already under way and future developments in this sector. Topics of interest include but are not limited to the following:

- Road traffic measurements, data analyzing;
- Road traffic (micro-, meso-, macro) modelling, simulation models;
- Road and intersection capacity;
- Optimization, route choice;
- Human factor in road traffic and transportation engineering;
- Road safety;
- Pedestrian and bicycle traffic and infrastructure;
- Public transportation solutions, parking;
- Contemporary problems of road traffic engineering and sustainable transportation;
- Intelligent transportation systems (ITS), traffic control and management;
- Smart grid services;
- Electric mobility;
- Environmental impacts of transportation systems;
- Life cycle analysis (LCA) of alternative energy vectors for road vehicles;
- Transportation systems and process modeling;
- Sustainable transportation development;
- Life cycle impact, fuel consumption, and emissions.

In the next section, we provide a brief review of the papers published in this Special Issue. These papers are included in the previously outlined thematic areas.

2. The Review of the Contributions in This Special Issue

This Special Issue presents original research and review articles addressing all aspects related to new perspectives and challenges in traffic and transportation engineering supporting energy saving in smart cities. Many of them are characterized by a multidisciplinary approach to a global problem. This collection spans a body of research papers that represents the efforts of 16 research papers.

H. Vasiutina et al. [1] examined the environmental impact of using cargo bikes in cities. It is a comprehensive review of existing approaches. The impact of the use of cargo bicycles for delivery processes on the environment is undeniably positive: it leads to a reduction in pollutant emissions, noise, and vibrations caused by traditional vehicles; decreases traffic jams; causes more effective use of public space; and others. But how should such an effect be measured? What tools should be used to justify the necessity for change to more sustainable means of transportation? How can we improve the state of the environment considering the interests of logistics service providers? There is a large amount of scientific literature dedicated to this problem: by using different modeling approaches, authors attempt to address the issue of sustainable transportation. This paper conducted a literature review in the field of green cargo deliveries, investigated the benefits and drawbacks of integrating cargo bikes in urban logistics schemes, and examined methodologies and techniques for evaluating the impact of using cargo bicycles on the environment. By providing an opportunity to become acquainted with the situation in the sphere of green deliveries, the authors aimed to encourage a breakthrough in the field of sustainable transportation that may be achieved by using cargo bikes in modern cities. The authors present a review of the existing approaches and tools for modeling transportation emissions and state the significant positive environmental consequences. The main approaches were the estimation of emissions based on analytical models, including the following: traffic performance characteristics and emissions using an integrated model consisting of the Transport Systems Modeling software (TSM); emissions and dispersion using an integrated model consisting of TSM software; simplified air pollution using an integrated model consisting of TSM software; and vehicle emission computation using an integrated model consisted from traffic emission and dispersion. Moreover, as the main tools used to the estimation of the environmental impact of transport, they indicated analytical models, ArcGIS Visual Basic and Moves, Mainsim Scipuf and CyberGIS, FlexSim and regression models, Paramics MS Access and IVE, and Emisens.

In the next paper, prepared by A. Sołowczuk, and D. Kacprzak [2], the problem of the identification of determinants of the effectiveness of on-road chicanes in transition zones to villages subject to a 70 km/h speed limit have been presented. In recent decades, traffic calming, especially in villages characterized by relevant through traffic, has become an urgent issue. Various schemes are applied in the transition zones to reduce the inbound traffic speeds and thus improve traffic safety. The studies conducted in several countries point to different determinants of the speed reduction obtained in this way. This article dealt with schemes including a central island horizontally deflecting one lane located in transition zones to villages with 70 km/h speed restriction on two-lane roads (6 m carriageway width). To identify the speed reduction determinants, the speeds before and after chicanes were measured, and the effects of the three criteria were investigated, characterizing the traffic management scheme, road design parameters, landscape elements present in the surroundings of the transition zone, and visibility conditions. Based on the confirmation of the logical tautology of many pre-selected factors, one aggregate parameter was proposed for the assessment of the practicable level of speed reduction, combining the effect of the selected factors in the above-mentioned criteria. Statistical analysis of the obtained results confirmed a statistically significant relationship between both the speed reduction value, the speed reduction index, and the aggregate parameter proposed by the authors. Factors related to the surrounding landscape and visibility conditions were found to have the greatest direct effect on speed reduction. The chicanes chosen in the final step of the proposed design process should be enhanced by additional solar-powered elements, ensuring their improved visibility. These devices should not, however, require any additional energy supply and should not increase the construction or maintenance costs. The further development of this topic has been included in the works [3–5].

The main purpose of the study [3] was to identify the speed reduction determinants in traffic calming schemes in village transition zones based on a central island horizontally deflecting one lane of a two-lane, two-way road with a 50 km/h speed restriction. As part of the study, vehicle speeds were measured just before and after the chicanes under analysis. Furthermore, the inbound lane traffic volumes were measured in field, and a number of factors were identified, including the applied traffic management scheme, road parameters, view of the road ahead and of the village skyline, isolated buildings, road infrastructure, and adjacent roadside developments. The obtained data were analyzed with a method employing the tautologies of the 32 selected factors affecting the drivers' perception. A single aggregate parameter was proposed for assessing the coincidence of the influence of selected factors on speed reduction. The analysis of the existing schemes and the results of statistical analyses carried out in this study confirmed the authors' hypothesis that the combined selected factors produce a desirable effect and that they should be additionally enhanced by the application of solar-powered devices.

The purpose of the study [4] was to investigate whether the applied traffic-calming measures had a considerable bearing on the reduction in speed to the desired level, as assumed in the traffic-calming plan. Three street sections starting and ending with different intersection types were chosen to examine the synergy of the applied traffic calming measures. The numbers and speeds of vehicles were measured in three-day-long continuous surveys. As expected, the amount of speed reduction depended on the hourly traffic volume on a one-way street and various other traffic engineering aspects. The obtained results may be used to modify the existing speed profile models and can guide traffic engineers in choosing the most effective traffic calming measures.

Meanwhile, in [5], the authors put forward a hypothesis of there being some determinants which, in combination, influence speed reduction in village transition zones. To corroborate their hypothesis on the combined impact of the transition zone features on speed reduction and in order to validate the established relationships, the authors conducted a verification study in transition zones containing chicanes or central islands. The Authors studied twenty transition zones and managed to confirm the hypothesis at a 95% confidence level. The contribution of this study is a further investigation of the synergy effect of various relevant factors, and the findings can assist in planning new transition zones or suggest additional measures to achieve the desired speed reduction in existing transition zones.

In turn, O. Pietrzak and K. Pietrzak [6] presented the economic effects of electromobility in Sustainable Urban Public Transportation. This paper focused on the effects of implementing zero-emission buses in public transportation fleets in urban areas in the context of electromobility assumptions. It filled the literature gap in the area of research on the impact of the energy mix of a given country on the issues raised in this article. The main purpose of this paper was to identify and analyze the economic effects of implementing zero-emission buses in public transportation in cities. The research area was the city of Szczecin, Poland. The research study was completed using the following research methods: a literature review, a document analysis (legal acts and internal documents), a case study, a ratio analysis, and a comparative analysis of the selected variants (investment variant and base variant). The conducted research study has shown that economic benefits resulting from implementing zero-emission buses in an urban transportation fleet are limited by the current energy mix structure of the given country. An unfavorable energy mix may lead to increased emissions of sulfur dioxide (SO_2) and carbon dioxide (CO_2) resulting from the operation of this vehicle mode. Therefore, achieving full effects in the field of electromobility in the given country depends on taking concurrent actions to diversify the sources of power generation and on increasing the share of Renewable Energy Sources (RES).

Furthermore, in the paper [7], a new back-of-queue model of a signal-controlled intersection approach development based on an analysis of vehicle driver behavior has been presented. In smart cities, it is expected that transportation, communication, as well as the movement of people and goods will take place in the shortest possible time while maintaining a high level of safety. In recent years, due to the significant increase in the number of passengers and vehicles on the road and the capacity limitations of transportation networks, it has become necessary to use innovative technologies for intelligent control and traffic management. Intelligent transportation systems use advanced technologies in the field of data gathering, information processing, and traffic control to meet current transportation needs. To be able to effectively control and manage road traffic, it is essential to have reliable mathematical models that allow for a faithful representation of the real traffic conditions. Models of this type are usually the basis of complex algorithms used in practice in road traffic control. The application of appropriate models reflecting the behavior of road users contributes to the reduction in congestion, the vehicles' travel time on the transportation network, fuel consumption, and emissions, which in turn support understood energy savings. The article proposes a model that allows for the estimation of the maximum queue size at the signal-controlled intersection approach (the so-called maximum back-of-queue). This model considers the most important traffic characteristics of the vehicles forming this queue. These traffic characteristics include the volume-to-capacity ratio, the capacity-to-starts ratio, the initial queue length on a given time interval, the green effective period ratio, the cycle length of a given interval, and the arrival flow rate of a given interval. The verification allowed the authors to conclude that the proposed model is characterized by high compliance with the actual traffic and road conditions at the intersections with signal controllers located in built-up areas in Poland. The obtained compliance confirms the possibility of using the model for practical applications in calculating the maximum back-of-queue at signal-controlled intersections located in built-up areas in Poland.

In [8], the authors presented their investigation of the impact of dynamic travel time information on drivers' route choice in an urban area—a case study based on the city of Białystok. Increasing traffic volumes in cities lead to common traffic congestions building up, especially during peak hours. To protect city dwellers from excessive fuel exhaust and traffic noise and to prevent drivers from time loss due to overloaded routes, it is important to inform them about real-time traffic conditions and possible delays in advance. Effectively influencing drivers' decisions to divert from an original route choice in case of traffic hinderance is essential, and the application of dynamic travel information in the form of variable message signs (VMSs) is believed to be effective in these terms. The paper examined drivers' willingness to divert from an initial route choice due to the information provided on VMS boards. Their behavior was analyzed in terms of their response to everyday and artificially elongated travel times displayed on the VMSs. Maximum simulated elongation reached 200% and 300% of the initial state, depending on the characteristics of the pre-peak conditions. To assess the effectiveness of VMSs, the changes in traffic intensities were statistically analyzed. In general, apart from few significant differences, the results revealed drivers' ignorance of the travel time information provided on the VMS, regardless of the extension of the original times.

J. Paszkowski et al., in their paper [9], present the problem of modelling the effects of traffic-calming introduction to volume–delay functions and traffic assignment. Traffic calming is introduced to minimize the negative results of motor vehicle use, for example, lower safety levels or quality of life, high noise levels, and pollution. It can be implemented through the introduction of road infrastructure reducing the velocity and the traffic volume. In this paper, they studied how traffic-calming influences the traffic assignment. For the research, a traffic-calming measure of speed cushions on Stachiewicza street in Krakow was taken. A method of extracting trajectories from aerial footage was shown, which was further used to build a model. For a given example, through driving characteristics research and microscopic modelling, volume–delay functions given by the Bureau of Public Roads (USA) were estimated for a street with and without traffic calming. Later, a toy network of two roads of the same length, connecting the same origin and destination, was simulated using an equilibrium traffic assignment method. Simulations were conducted both with the use of PTV Vissim and Visum software and through individual calculations. According to the results of this paper, there was a difference in traffic volumes according to the equilibrium traffic assignment in the aforementioned toy network as a function of the network's total traffic volume.

V. Naumov and M. Pawluś, in the paper [10], identified the optimal packing and routing to improve last-mile delivery using cargo bicycles. Efficient vehicle routing is a major concern for any supply chain, especially when dealing with last-mile deliveries in highly urbanized areas. In this paper problems considering last-mile delivery in areas with the restrictions of motorized traffic are described, and distinct types of cargo bikes were reviewed. The paper described methods for solving a combination of problems in what concerns the cargo bicycle logistics, including efficient packing, routing, and load-dependent speed constraints. Proposed models apply mathematical descriptions of problems, including the Knapsack Problem, the Traveling Salesman Problem, and the Traveling Thief Problem. Based on synthetically generated data, they examined the efficiency of the proposed algorithms. Models described in this paper were implemented in Python programming language and can be further developed and used for solving the problems of electric cargo bikes' routing under real-world conditions.

Moreover, A.A. Drabicki et al. [11] evaluated the impact of public transportation service disruptions upon passenger travel behavior in Krakow City. Public transportation (PT) service disruptions are common and unexpected events which often result in major impediments to passengers' typical travel routines. However, attitudes and behavioral responses to unexpected PT disruptions are still not fully examined in state-of-the-art research. The objective of this study was to understand how PT users adapt their travel choices and what travel information sources they utilize once they encounter sudden PT service disruptions. To this end, the authors conducted a passenger survey among PT users in the city of Kraków (Poland), consisting of a series of stated- and revealed-preference questions. The results showed that passengers' reported choices during past PT disruptions mostly involved adjusting the current PT travel routine, exposing a certain bias in their stated choices (which tend to overestimate the probability of modal shifts). Factors influencing travel behavior shifts included the frequency and recency of PT disruption experience, as well as the propensity to arrive on-time. With regards to travel information sources, staff announcements and personal experience appeared to play a key role in recognizing the emerging disruption, but real-time information (RTI) sources were showed to be the most useful in planning the onward journey after such events. Based on these results, the study highlighted the implications for future RTI policy during PT service disruptions: in particular, the provision of a reliable time estimate until normal service conditions are resumed. Such RTI content could foster passengers' tendencies to use PT services in uncertain conditions, especially as their stated wait time tolerance often matches the actual duration of PT disruptions.

In the next paper, prepared by J. Oskarbski et al. [12], a bicycle traffic model for sustainable urban mobility planning has been proposed. Modelling tools and transportation models are required to assess the impact of measures for the effective planning of cycling routes in cities. This paper introduced the methodology for developing a four-stage macroscopic model of bicycle traffic for the city of Gdynia (Poland) and its use in planning new bicycle routes, considering a modal shift. The model presented in this paper allows for the evaluation of the influence of the characteristics of the cycling infrastructure, along with the development of the cycling network based on the choice of cycling as an alternative to other modes of transportation, by considering the modal shift. The model incorporated the influence of the longitudinal gradients, links, and surface types of cycling routes on the distribution and demand for bicycle traffic. The results of this research allowed the authors to assess the impact of planned cycling routes on the reduction in the volume of car traffic, which is crucial for reducing energy consumption and negative environmental impacts. Experiences from the application of the model in Gdynia suggested that the model provides a strong basis to support mobility planning and monitoring processes in cities worldwide.

D. Biernacki and Ch. Lis in [13] studied the sustainable effects of urban-port road system reconstruction. The aim of the research was to identify and quantify the direct sustainable effects resulting from the improved road infrastructure in the local urban port system. This case study considered the city port of Szczecin (Poland). The effects were identified for the local road transportation system by comparing freight road transportation performance in two options: (i) with investment and (ii) without investment. The sustainable effects were quantified in terms of money and physical units. Sustainable economic, social, and environmental effects concerned generalized freight road transportation costs, i.e., truck operating costs and costs of truck drivers' working time, as well as freight transit time, energy consumption, greenhouse gas emissions, and environmental savings. To capture these effects, the forecasted truck traffic demand, unit vehicle operating costs, values of time, and air pollution and climate change values were elaborated and revealed in freight road transportation. The findings revealed that the primary effect of investment is the reduced traffic congestion, which enhances the velocity of trucks in the transportation system.

H. Fu-Shiung [14] compared three ridesharing cost-saving allocation schemes based on the number of acceptable shared rides. Shared mobility based on cars refers to a transportation mode in which travelers/drivers share vehicles to reduce the cost of the journey, emissions, air pollution, and parking demands. Cost savings provide a strong incentive for the shared mobility mode. Since the cost savings are due to the cooperation of the stakeholders in shared mobility systems, they should be properly divided and allocated to relevant participants. The improper allocation of cost savings will lead to the dissatisfaction of drivers/passengers and hinder the acceptance of the shared mobility mode. In practice, several schemes based on proportional methods to allocate cost savings have been proposed in shared mobility systems. Neither a guideline for selecting these proportional methods has been prepared nor a comparative study on effectiveness of these proportional methods. Although shared mobility has attracted much attention in the research community, there is still a lack of research of the influence of cost-saving allocation schemes on the performance of shared mobility systems. Motivated by the deficiencies in existing studies, this paper examined three proportional cost-saving allocation schemes by analyzing their performance in terms of the numbers of acceptable rides under different schemes. The authors focused on ridesharing based on cars in this study. The main study contribution relied on the development of a theory based on analysis to characterize the performance under different schemes to provide a guideline for selecting these proportional methods. The developed theory was verified by conducting experiments based on real geographical data.

In the next paper [15], an evaluation of the use of radar speed cameras and panels indicating vehicle speeds as traffic-calming measures (TCM) in short-length urban areas located along rural roads was performed. TCMs are typically implemented in urban areas to reduce vehicles' speeds. However, speed is still a problem in rural roads crossing small villages without a bypass and in short-length urban areas, since drivers do not normally reduce their speed for that short segment. Hence, various TCMs can be installed. It is necessary to maintain a calm area in these short segments to improve road safety, especially for pedestrians aiming to cross the road, and to save combustibles by avoiding a constant increase–decrease of speed. Four villages were selected to evaluate the efficiency of radar speed cameras and panels indicating vehicle speed. The results showed that the presence of radar speed cameras reduces the speed in the direction they can fine speeding cars, but with a lower effect in the non-fining direction. Additionally, a positive effect was observed in the fining direction in other points, such as pedestrian crossings. Nevertheless, the effect does not last long, and speed cameras may be considered as punctual measures. If the TCMs are placed far from the start of the village, they are not respected. Hence, it is recommended to place them near the real start of the build-up area. Lastly, it was verified that longer urban areas make the overall speed decrease. However, when drivers feel that they are arriving to the end of the urban area, signaled by the lack of buildings, they start to speed up.

The last paper [16] is devoted to assessing the environmental performances of urban roundabouts using the Vehicle Specific Power (VSP) methodology and AIMSUN. In line with globally shared environmental sustainability goals, the shift towards citizen-friendly mobility is changing the way people move through cities and road-user behavior. Building a sustainable road transportation requires design knowledge to develop increasingly green road infrastructures and monitoring the environmental impacts from mobile crowdsourced data. In this view, the paper presents an empirically based methodology that integrates the VSP emission model and AIMSUN microscopic traffic simulation to estimate second-by-second vehicle emissions at urban roundabouts. The distributions of time spent in each VSP mode from instantaneous vehicle trajectory data gathered in the field via smartphone were the starting points of the analysis. The versatility of AIMSUN in calibrating the model parameters to better reflect the field-observed speed-time trajectories and to enhance the estimation accuracy was assessed. The conversion of an existing roundabout within the sample into a turbo counterpart was also made as an attempt to confirm the reproducibility of the proposed procedure. The results shed light on new opportunities in the environmental performance evaluation of road units when changes in design or operation should be considered within traffic management strategies and highlighted the potential of the smart approach in collecting massive amounts of data through digital communities.

3. Conclusions

Based on the collection of articles in this Special Issue, it can be concluded that the topics discussed in the articles most often concerned such issues as the use of traffic-calming measures and their effectiveness, human factors in road traffic and transportation engineering, bicycle traffic and infrastructure, the environmental impacts of transportation systems, as well as sustainable urban public transport.

In conclusion, we believe that this research topic presents a broad range of conceptual and practical research and reviews and that the papers included in this Special Issue clearly contribute to expanding the knowledge in this field. Moreover, we hope that the presented collection of articles will find a wide audience among scientists dealing with new perspectives and challenges in traffic and transportation engineering in support of energy saving in smart cities, as well as among practitioners dealing with this subject. Taking advantage of this opportunity, we wish all of you a fruitful reading.

Author Contributions: Conceptualization, E.M., A.G., M.C.C. and P.F.; methodology, E.M., A.G., M.C.C. and P.F.; software, E.M., A.G., M.C.C. and P.F.; validation, E.M., A.G., M.C.C. and P.F.; formal analysis, E.M., A.G., M.C.C. and P.F.; investigation, E.M., A.G., M.C.C. and P.F.; resources, E.M., A.G., M.C.C. and P.F.; data curation, E.M., A.G., M.C.C. and P.F.; writing—original draft preparation, E.M., A.G., M.C.C. and P.F.; writing—review and editing, E.M., A.G., M.C.C. and P.F.; visualization, E.M., A.G., M.C.C. and P.F.; supervision, E.M., A.G., M.C.C. and P.F.; project administration, E.M., A.G., M.C.C. and P.F.; funding acquisition, E.M., A.G., M.C.C. and P.F. All authors have read and agreed to the published version of the manuscript.

Funding: This research was partially funded from UIDB/00481/2020 and UIDP/00481/2020-FCT-Fundação para a Ciência e a Tecnologia; and CENTRO-01-0145-FEDER-022083-Centro Portugal Regional Operational Programme (Centro2020), under the PORTUGAL 2020 Partnership Agreement, through the European Regional Development Fund.

Institutional Review Board Statement: Not applicable.

Informed Consent Statement: Not applicable.

Data Availability Statement: Not applicable.

Acknowledgments: The *Energies* Special Issue Editors would like to thank the anonymous Reviewers for their profound and valuable comments, which have contributed to enhancing the standard of each paper in this Special Issue and to enhancing the authors' future research in their research.

Conflicts of Interest: The authors declare no conflict of interest.

References

1. Vasiutina, H.; Szarata, A.; Rybicki, S. Evaluating the Environmental Impact of Using Cargo Bikes in Citie: A Comprehensive Review of Existing Approaches. *Energies* **2021**, *14*, 6462. [CrossRef]
2. Sołowczuk, A.; Kacprzak, D. Identification of the Determinants of the Effectiveness of On-Road Chicanes in Transition Zones to Villages Subject to a 70 km/h Speed Limit. *Energies* **2020**, *13*, 5244. [CrossRef]
3. Sołowczuk, A.; Kacprzak, D. Identification of the Determinants of the Effectiveness of On-Road Chicanes in the Village Transition Zones Subject to a 50 km/h speed limit. *Energies* **2021**, *14*, 4002. [CrossRef]
4. Sołowczuk, A. Effect of traffic calming in a downtown district of Szczecin, Poland. *Energies* **2021**, *14*, 5838. [CrossRef]
5. Sołowczuk, A.; Kacprzak, D. Synergy Effect of Factors Characterising Village Transition Zones on Speed Reduction. *Energies* **2021**, *14*, 8474. [CrossRef]
6. Pietrzak, O.; Pietrzak, K. The Economic Effects of Electromobility in Sustainable Urban Public Transport. *Energies* **2021**, *14*, 878. [CrossRef]
7. Macioszek, E.; Iwanowicz, D. A Back-Queue Model of a Signal-Controlled Intersection Approach Developed Based on analysis of Vehicle Driver Bahavior. *Energies* **2021**, *14*, 1204. [CrossRef]
8. Ziółkowski, R.; Dziejma, Z. Investigations of the Dynamic Travel Time Information Impact on Drivers' Route Choice in an Urban Area—A Case Study Based on the City of Bialystok. *Energies* **2021**, *14*, 1645. [CrossRef]
9. Paszkowski, J.; Herrmann, M.; Richter, M.; Szarata, A. Modelling the effects of traffic-calming introduction to volume-delay functions and traffic assignment. *Energies* **2021**, *14*, 3726. [CrossRef]
10. Naumov, V.; Pawluś, M. Identifying the Optimal Packing and Routing to Improve Last-Mile Delivery using Cargo Bicycles. *Energies* **2021**, *14*, 4132. [CrossRef]
11. Drabicki, A.A.; Md Islam, F.; Szarata, A. Investigating the impact of public transport service distruptions upon passenger travel behaviour-results from Krakow City. *Energies* **2021**, *14*, 4889. [CrossRef]
12. Oskarbski, J.; Birr, K.; Żarski, K. Bicycle traffic model for sustainable urban mobility planning. *Energies* **2021**, *14*, 5970. [CrossRef]
13. Biernacki, D.; Lis, C. Exploring the sustainable effects of urban-port road system recosntruction. *Energies* **2021**, *14*, 6512. [CrossRef]
14. Fu-Shiung, H. A comparison of three ridersharing cost saving allocation schemes based on the number of acceptable shared rides. *Energies* **2021**, *14*, 6931. [CrossRef]
15. Pérez-Acebo, H.; Ziółkowski, R.; Gonzalo-Orden, H. Evaluation of the radar speed cameras and panels indicating the vehicles' speed as traffic calming measures (TCM) in short length urban areas located along rural roads. *Energies* **2021**, *14*, 8146. [CrossRef]
16. Acuto, F.; Coelho, M.C.; Fernandes, P.; Giuffrè, T.; Macioszek, E.; Granà, A. Assessing the Environmental Performaces of Urban Roundabouts using the VSP Methodology and AIMSUN. *Energies* **2022**, *15*, 1371. [CrossRef]

energies

MDPI

Article

Identification of the Determinants of the Effectiveness of On-Road Chicanes in Transition Zones to Villages Subject to a 70 km/h Speed Limit

Alicja Sołowczuk and Dominik Kacprzak *

Road and Bridge Department, West Pomeranian University of Technology Szczecin, 71-311 Szczecin, Poland; alicja.solowczuk@zut.edu.pl

* Correspondence: kdim.zut@gmail.com or dominik.kacprzak@zut.edu.pl; Tel.: +48-(091)-449-40-36

Received: 7 September 2020; Accepted: 29 September 2020; Published: 9 October 2020

Abstract: In recent decades traffic calming, especially in villages situated on through roads, has become an urgent issue. Various schemes are applied in the transition zones to reduce the inbound traffic speeds and thus improve the traffic safety. The studies conducted in several countries point to different determinants of the speed reduction obtained in this way. This article deals with the schemes including a central island horizontally deflecting one lane, located in transition zones to villages with 70 km/h speed restriction on two-lane roads (6 m carriageway width). In order to identify the speed reduction determinants, the speeds before and after chicanes were measured and the effect of the three criteria was investigated, characterising: the traffic management scheme, road design parameters, landscape elements present in the surroundings of the transition zone and visibility conditions. Based on the confirmation of logical tautology of many pre-selected factors, one aggregate parameter was proposed for the assessment of the practicable level of speed reduction, combining the effect of the selected factors in the above-mentioned criteria. Statistical analysis of the obtained results confirmed a statistically significant relationship between both the speed reduction value and the speed reduction index, and the aggregate parameter proposed by the authors. Factors related to the surrounding landscape and visibility conditions were found to have the greatest direct effect on speed reduction. The chicanes chosen in the final step of the proposed design process should be enhanced by additional solar-powered elements ensuring their improved visibility. These devices should not, however, require any additional energy supply and should not increase the construction or maintenance costs.

Keywords: traffic calming; transition zone; chicane; speed restriction; speed reduction; solar cells

1. Introduction

The increase in traffic volumes observed in recent decades has aggravated transport-related problems in small towns and villages situated on roads carrying through traffic [1–4]. The ever-increasing traffic volumes frequently exceed the capacities of the existing intersections, causing problems with parking spaces in built-up areas [5] and increasing the frequency of road traffic incidents [4,6,7], the level of noise [8–11] and the concentration of exhaust fumes and pollution in the immediate vicinity of roads [4,11,12]. High density housing with narrow fronts lining the through roads, makes the expansion of the existing road system impossible [13]. All these factors have an adverse effect on the acceptance and perception of the road by both the residents and road users [14–18]. Dramatic accumulations of these factors can be encountered, especially on relatively short sections of through roads passing through villages (e.g., according to the data from Danish design guidelines [1], England [2,19,20], Poland [5,21] and an article about design conditions in Iran [22]). The first traffic safety analyses on through roads crossing rural areas were conducted in the 1990s in

Denmark [1], Great Britain [2,19,20,23,24] and later in Canada [17], Germany [3], Sweden [25], Spain [26] and USA [18,27]. To prevent similar cases from occurring on many through roads of lesser importance introduction of traffic calming schemes has been recommended in many countries (e.g., Austria [28], Denmark [1,29], France [30], Germany [3], Great Britain [2,31], Italy [32], Iran [33], Poland [5,21], Sweden [25], Spain [34], Switzerland [35] and USA [27,36–38]), for application both in the transition zone (i.e., a road section where the vehicles are to slow down before entering the village) and over the whole length of the road within the built-up area. The analyses data (according to [4]–40%, [20]–10–15%, [24]–54%, [29]–45%, [39]–40–50%, [13]–8–71%, [40]–40–50%) that thanks to the use of traffic calming measures and various speed management strategies demonstrated that traffic calming measures installed helped to reduce the frequency of road traffic accidents by ca. 40–50%.

Despite the fact that chicanes can help decrease the frequency of road traffic accidents, the design guidelines [3,18,41] recommend that, in order to obtain the desired improvement of the safety of traffic, the design of chicanes in the transition zones should be based on a thorough analysis of the traffic management arrangements, the installed lighting system, the most appropriate location, etc. On the other hand, in [23] attention is additionally drawn to the fact that although horizontal deflections allow to avoid discomfort to the drivers and passengers, which is unavoidable in the case of vertical deflections, when used in rural areas they must both ensure the reduction of the traffic speeds and allow safe passage of oversize heavy agricultural vehicles.

The distances on which speed reduction should be implemented vary between different design guidelines and generally depend on the importance of the two-way through road concerned and on the total length of the road's section within the village limits. For instance, guidelines [40] recommend an analysis of the actual vehicle speeds and next their possible reduction over a 600 m long section of the road. Still, the above-mentioned guidelines recommend that some balance should be maintained between the actual vehicle speeds, the recommended speed reductions and the traffic calming measures to be applied over the whole length of the road section within the village limits. This article deals with the effectiveness of a traffic calming measure installed in the transition zones to villages situated on two-lane single carriageway regional roads with a posted speed limit of 70 km/h.

The concept of traffic calming has three fundamental objectives (Denmark [1], Germany [3], Great Britain [2,19,23,24], Poland [5,21], Sweden [25], USA [26,36]): to reduce vehicle speeds and to improve traffic safety and living conditions of the local residents. Traffic calming schemes involve the use of psychological or physical traffic measures designed to slow down the traffic in order to reduce environmental impact of roads. They directly influence drivers' behaviour, and thus have a beneficial effect on the living conditions of the local residents and on the safety of vulnerable road users. According to [41] traffic calming measures should be selected taking into account the existing road alignment (i.e., the lengths of straight and curved road sections and their parameters), importance of intersections situated within the village boundaries, number of access points to properties, local traffic volumes and vulnerable road users.

Speed of traffic is the main factor having a direct on both the number of road incidents and their severity (Denmark [1,29], Germany [3], Great Britain [2,19,20,23,24,39,41], Poland [5,21,42], Sweden [25], USA [27,36,37]). From the analysis the conclusions from various studies (Denmark [1,29], Germany [3], Great Britain [2,19,20,23,24,39,41], Poland [5,21,42], Sweden [25], USA [27,36,37]) it appears that drivers tend to exceed speed limits in the transition zones to villages. Hence, any traffic calming scheme designed to reduce the speed of through traffic should involve signalling the boundaries of the built-up area and the applicable speed limit to the through-traffic drivers both by means of informative measures, such as traffic signs, and by dynamic measures, that is by the use of traffic calming devices. The specialist literature on traffic calming (Denmark [1], Germany [3,43], Great Britain [2,19,23,31], Poland [5,21], 13, Sweden [25], USA [27,36,37]) usually indicates three elements used in transition zones, namely warning measures (road humps, road markings, rumble devices), village gateways and central islands (i.e., road narrowing measures, including chicanes, pinch-points or overrun areas). All the above-mentioned measures contribute to speed reduction. However, to ensure that the reduced speed is maintained over a longer section of the through road or over the whole length of the passing the village, other traffic calming treatments need to be employed along the whole section. These include physical road narrowing, horizontal deflection of the travel lane, small and mini roundabouts, pedestrian refuges, etc. The classification of road chicanes in terms of the shape and arrangement of individual islands was given as early as in 1998 in [24]. According to the classification given and the terminology used in [24], in this article the authors focused mainly on the speed reducing effect of chicanes, i.e., the central islands affecting one lane, installed in the transition zones to villages.

In projects involving retrofitting of traffic calming measures on existing roads, designers should take into account the expectations of road owner or operators in this respect. In some cases, the road right-of-way cannot be widened as needed to accommodate complicated traffic calming schemes, as it would require acquisition of additional stretch of land along the road, resulting in a major impact on the project cost. Therefore, it is necessary to initially assess the effectiveness of a chicane to be installed from the point of view of its primary objective, i.e., achieving speed reduction. Due to the above, the Danish [1] and UK [2,19] guidelines introduced categorisation of speed reduction, giving the principles for selection of the most appropriate traffic calming treatment, depending on the expected degree of speed reduction. Furthermore, the guidelines [19] give concrete relationships between speed reduction and the chicane parameters, i.e., the length of stagger, the width if the narrowed lane, initial carriageway width, free view width, etc. The publication [20] presents the results of the experiments carried out in 1990s on the test track of the Transport Research Laboratory TRL in Crowthome with chicanes having different parameters. These parameters included, for instance, travel lane width, varying free view width and varying height of visual obstructions placed at the carriageway and island edges. The above-mentioned experiments demonstrated that the factors relevant to the speed reduction included not only free view width but also the length of stagger, travel lane width, height of visual obstructions placed at the island and carriageway edges (i.e., kerb height, traffic signs placed on the island, safety barriers and traffic posts).

The German guidelines [3] are related more to the geometrical parameters of the chosen measure and the dimensions of the chicane itself. According to the design guidelines of [3] a correctly designed chicane, located in the central area of a village should feature adequate proportions between the length of stagger, free view width and the resulting travel lane width at a constant lane deflection angle of 45°. As far as chicanes located in the transition zones to villages are concerned, the main assumption given in the guidelines [3] is based on reducing the traffic speed down to the built-up area speed limit of 50 km/h, with the specified range of travel lane width. Guidelines [3] give tabled geometric parameters of chicanes, depending on the chicane type and the shape of horizontal deflection of the lane. The minimum allowable island widths are also given, which, according to the opinion expressed in the design guidelines [3], play the primary role in speed reduction at a specific length of stagger. In [43] we can find close relationships between the expected speed reduction and the vehicle's speed at the chicane, and the amount of horizontal deflection of the lane. According to these relationships a speed reduction down to 70 km/h can be achieved by installing a central island laterally deflecting one travel lane by 1 m. Deflection of the lane by 3 m can result in a reduction of the traffic speed down to 60 km/h and only with deflections by more than 3 m may speed reductions down to values below 50 km/h be expected. Furthermore, the length of road over which the achieved speed reduction is maintained is also assessed in [43] as being dependent on the amount of deflection, which confirms the previously stated thesis that additional traffic calming treatments must be employed to maintain the achieved speed reduction on the way through the village.

The Swedish guidelines [25], in turn, besides the recommended stagger length, also specify the recommended radii of curves on the approach, at the exit and in the middle part of the central island. In guidelines [25] the central island recommended for installation in village transition zones has the shape of elongated converging lens, deflecting the travel lane by at least two metres. Similar recommendations for semi-circular central islands shifting only the inbound lanes to village are given in the Polish guidelines [5,21], except that they recommend a deflection of the lane by as much as ca. 5 metres.

A recent study has investigated the optimum geometry of different kinds of chicanes provided on dual carriageways and two-lane roads in built-up areas in terms of the expected speed reduction [44]. The conclusions drawn by the authors of the above-mentioned article concern primarily the speed through the chicane. As regards two-lane roads, it was established that the greatest discipline in keeping to the travel lane and the associated speed reduction is achieved by staggered narrowing of the carriageway down to one lane width of 2.7 m, accompanied by trapezoid bulb-outs, imposing a deflection angle of 30°. Although novel and interesting, these result are, however, of little use as regards the subject of this article, i.e., transition zones to villages on two-lane through roads. This is because carriageways of through roads subject to 70 km/h speed limit in transition zones cannot be narrowed to one travel lane as it would considerably impair the road capacity and probably also the safety of traffic. Another reason why the results of research given in [44] cannot be used for the purposes hereof is the presence of a pedestrian crossing between the chicanes. The transition zones analysed in this article included no road infrastructure elements related to pedestrian traffic.

2. Materials and Methods

2.1. Background

The problem of traffic calming basically involves limiting an adverse effect of vehicular traffic on the environment within built-up areas. It first aroused interest in the 1970s and research on the effectiveness of various traffic calming measures has been conducted since that time. The earlier research concerned primarily the use of these elements in built-up areas, with only a few studies investigating the application of traffic calming measures in transition zones. Table 1 shows the data such as parameters, locations and results of the hitherto studies on the application of various traffic calming measures.

The analysis of the abovementioned results demonstrated fundamental differences between individual results and numerous discrepancies in the repeatability of v results in various countries. The comparison of the results of the studies conducted on a test track, within an urban area, in transition zones on the entries to towns and villages by roads of different classes and when using chicanes of various shapes revealed significant discrepancies in the compared research results. Furthermore, limiting the comparison of the results of the above studies presented in Table 1 just to the transition zones to villages on roads with posted speed limit of 70 km/h the authors have found:

- the absence of an analysis of the actual conditions of the traffic management features used (i.e., location of the upright traffic signs associated with the built-up area),
- the absence of considering various road parameters and road infrastructure elements used in built-up areas,
- the absence of an analysis of visibility of the road further ahead and village skyline,
- the absence of an analysis of landscape elements in the surroundings of the road (approach through the open rural area, through the rural area with a tree line or small groves of trees, approach through a forest area).

The above elements have been considered in the research described in this article.

Table 1. Considered parameters, area locations and results of previous one's research.

No.	Considered Parameters	Research Result	References
	Area location of the research site: research track Transport Research Laboratory TRL		
1	lane width, free view width, visual obstruction, length of stagger	speed reduction results Δv are tabulated depending on the size of all the parameters under consideration	[23]
	Area location of the research site: residential road in centre city		
2	length of stagger, narrow lane width, road width, free view width	speed reduction results Δv are tabulated depending on the size of the parameters under consideration for each type of vehicle (passenger cars, HGV, buses)	[1,3,31]
3	12 different geometric chicanery schemes, constant lane width (lane width = 4 m), different widths of the chicane island (2.7, 3, 3.3 and 5 m) and the leading/trailing face angles of these islands (30°, 45° and 60°)	(case study on a driving simulator-on dual carriageways and two-lane street), speed reduction results of speed v and speed reduction Δv graphically summarized depends on 12 schemes, with the road narrowing to one lane	[44]
4	various measures of traffic calming: raised pedestrian crossing, lane narrowing, speed cameras, warning signs, chicanes, middle islands	(case study)—the results Δv are given depending on the option of the tested traffic calming measures on different street classes	[26,40,45–48]
5	horizontal deflection min. 2 m, length of stagger, the radii rounding the curves of the chicane	recommended radii values are listed according to the speed limit on the street	[25]
	Area location of the research site: transition zones		
6	chicanes shape, width of the available land, narrowing lane width, horizontal deflection	speed reduction results Δv are tabulated depending on the road class	[3]
7	length of stagger l_v, horizontal deflection t_v	(case study of 4 different island shapes)-Δv results compiled according to the quotient l_v/t_v	[28]
8	chicanes shape, horizontal deflection	speed reduction results Δv are compiled depending on the v^{before} of a given road	[43]
9	angle horizontal deflection and curb height at central islands gateways	(case study of 12 locations of different gateways)—the analysis concerns: Δv, reducing the number of accidents and reducing driver distraction	[39]
10	different types of gateways: street trees, upgraded pavement treatments, median, lighting, signage and graphics, sculptures or public art	driving simulator case study - research on the influence of various gateways on v^{befor} and v^{after}	[18,49–51]

2.2. Description of the Test Sections

For the purpose of the survey of vehicle speeds on two-lane roads the authors chose from among almost one hundred of the analysed entry zones only nine test sections located in the entry zones to the village, with a speed limit of up to 70 km/h specified on denote B-33 traffic sign. All sections are characterized by the same width of traffic road and the carriageway, the same road class, the same permissible speed, slight similar traffic volume and heave vehicle participation, the same condition of the pavement (capital a renovation on all sections was carried out a year earlier). On all the test sections the travel lanes were 3 m wide.

A view seen by a driver approaching the village under study is shown in Figure 1. Six test sections included a central island horizontally deflecting one lane by a = 2 m (Figure 1a,b,d–g), one section had a central island deflecting both lanes by a = 1 m (Figure 1c) and the two remaining sections were used for comparison only, and thus did not include any chicanes (Figure 1h,i).

The analysis of the main features of the transition zones under study showed that on the basis of the general traffic management scheme trends, as described in [1,2,5,19,43,53], two configurations of positioning vertical signs can be identified: basic, classic (Figure 2) and the one used in the test sections (Figure 3). The first traffic management scheme, presented in Figure 3 relates to the transition zones where the greatest speed reductions were noted, with the chicanes placed between the traffic signs E-17 and D-42 (Figure 3a). Smaller speed reductions were recorded where a central island horizontally deflecting one lane was located before the traffic signs No. E-17 'city' and No. D-42 'built-up area' (Figure 3b).

(a) (b) (c) (d) (e) (f) (g) (h)

Figure 1. *Cont.*

(i)

Figure 1. View ahead seen by the driver approaching the village on the respective test sections in the order of reduction of the 85th percentile speed Δv_{85}: (**a**) No. 1, open rural area, very good visibility of the village skyline and of nearby buildings; (**b**) No. 2, open rural area, view of distant village buildings located along the opposite lane of travel; (**c**) No. 3, open rural area, poor visibility of the village skyline and of nearby buildings; (**d**) No. 4, forest terrain, limited visibility of dispersed buildings; (**e**) No. 5, rural area, along a row of trees, poor visibility of the village skyline; (**f**) No. 6, forest area, village skyline and buildings not visible; (**g**) No. 7, open rural area, good visibility of the village skyline in the distance; (**h**) No. 8, open rural area, village skyline not visible; (**i**) No. 9, open rural area, poorly visible village skyline (source of images in Figure 1b,h,i: Google Earth–Street View [52]).

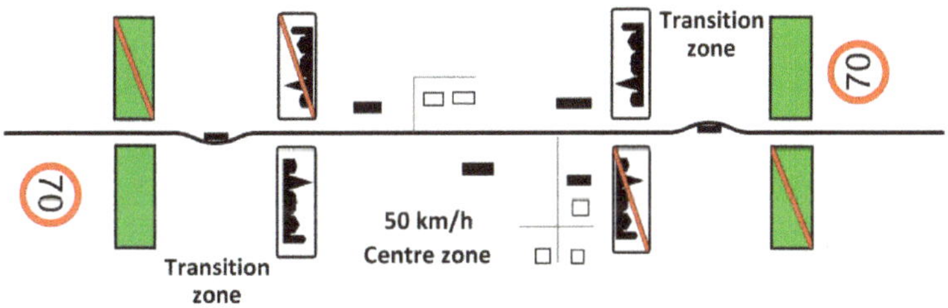

Figure 2. Traffic management scheme with 70 km/h speed limit on test sections with installed on-road chicanes (based on the analysis of the recommendations of [1,2,5,19,42,53]).

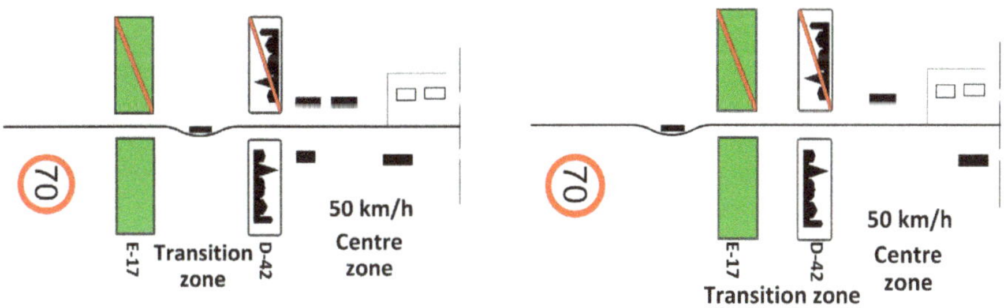

(**a**) (**b**)

Figure 3. Example layout of the E-17 and D-42 traffic signs: (**a**) on sections where a considerable speed reduction was recorded; (**b**) on the remaining test sections.

2.3. Adopted Measurement Methodology

The first step of the assessment of the effectiveness of chicanes in the transition zones to villages involved an analysis of the speeds measured on the test section before and after the chicane. In this part of the research, momentary free-flow and stable-flow speeds were measured with a special device [54] which simultaneously measured the traffic volume and vehicle speeds in both directions of traffic.

This device can identify free-flow conditions in either direction of traffic, which enables selection of values needed for the purpose of the performed analysis. Due to small hourly volumes, comparable on all the test sections (from 150 to 340 veh/h), the differences in the speed distribution parameters between free-flow and stable-flow situations were also small. Taking the above into consideration, further in the article the authors analyse the free-flow values only.

2.4. Choosing Appropriate Statistical Tests

Analysing the effect of various factors on the speed of travel and the amount of its reduction the authors performed a number of statistical tests. In addition to the significance test, goodness-of-fit test and the test for equality of variances, in relation to the measured speeds, i.e., the standard tests (Appendices A and B), also performed were nonparametric tests, i.e., tests of independence and median tests. The chi-square (χ^2) test of independence was carried out first, to verify independence of two tested features (not necessarily measurable). The null hypothesis of the chi-square test of independence was to answer the question whether or not the speed distributions depend on the place of measurement (i.e., before or after the chicane). The inequality $\chi^2 > \chi_\alpha{}^2$ should lead to rejecting the null hypothesis postulating independence of the analysed features, this meaning that the populations of speeds measured before and after the chicane depended on the measurement point. Such results were obtained for each of the analysed chicanes (Appendix C).

The median test was chosen as the second test for identification of determinants, which is suitable for situations where one of the features is possibly non-measurable. It does not require confirmation of the distribution of population, and thus can be used as an alternative to the two-sample means test. In this test, the medians are calculated from all the results of the two compared samples. The result of the median test $\chi^2 < \chi_\alpha{}^2$ is positive when the number of results of the tested feature above the median equals the number of results below the median, i.e., both the compared samples come from the same population and variation of the speed level is not confirmed (i.e., there were no grounds for rejecting the null hypothesis of H_0: $F_1(x) = F_2(x)$). If inequality $\chi^2 > \chi_\alpha{}^2$ was obtained in the median test, the null hypothesis H_0 postulating the independence of the tested features should be rejected, because the speed samples were found to depend on the measurement point and a difference in their value was noted (H_1: $F_1(x) \neq F_2(x)$). Also in this test positive results were obtained, confirming the relevance of the measurement point, meaning change of speed through the chicane (Appendix D).

2.5. Fundamental Assumptions Taken in the Qualitative Analysis Concerning Determination of the Speed Reduction Determinants

All the obtained speed results were estimated in the ranges, as shown in Figure 4. On two sections, section No. 8 and section No. 9, which did not include chicanes, speeds were measured before and after the road sign communicating the boundary of the village and the built-up area. Considering that in the transition zone traffic calming schemes are applied primarily to reduce the speed at which the vehicles enter the village, Figure 4 gives also the value of 85th percentile speed measured after the chicane $v_{85}{}^{after}$. The analysis of the data presented in Figure 4 showed a considerable variation of the speed values despite comparable amounts of lateral shift on all the test sections. The noted speed variations were particularly scattered at chicane No. 6, which is approached through a long straight section (over 2 km long), surrounded by farm fields and rows of trees which largely hide the village skyline (Figure 1f). Straight approach sections of a similar length surrounded by farm fields lead also to the chicanes No. 1 (1.5 km) and No. 4 (2 km) (Figure 1a,d). Also chicanes No. 5 and No. 7 are approached through straight sections of a similar length (Figure 1e,g). In these cases, however, the Approach Sections go through a patch of forest.

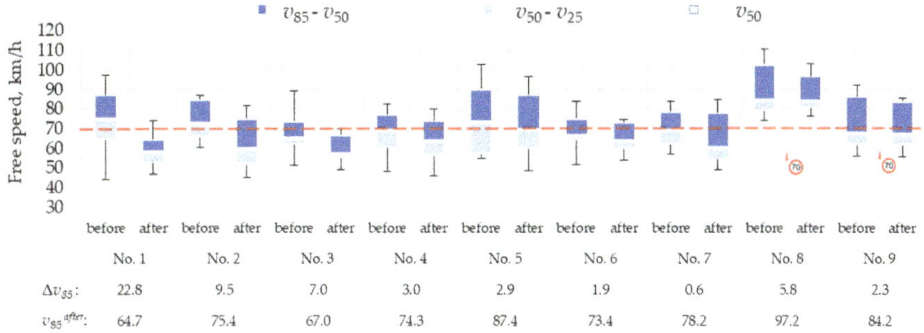

Figure 4. Distribution of free-flow speed ranges on the test section according to the speed reduction value Δv_{85}.

The analysis of the scattering of speed variations on the above-mentioned sections showed that they are much smaller, except for the chicane No. 1, which indicates an existence of some other determinants that affect the drivers' perception and, as consequence, also the village inbound speeds. The two analysed additional cases, i.e., approaches No. 8 and No. 9 (Figure 1h,i) also had opposite inbound speed values, and therefore, in this case we can also assume existence of additional determinants in the transition zone, which influence the drivers' perception, and thus the vehicle speeds.

The data presented in Figure 4 show that the large variation of the 85th percentile speed reductions can be associated with various factors influencing the speed values before v^{before} and after v^{after} the chicane. And these determinants can influence the vehicle speeds and the obtained reductions Δv to a varying degree. Also the results compiled in Figure 4 do not confirm the thesis put forward in [43] (Table 1—position 6) that central islands shifting one lane by up to 3 m, situated on the village inbound lane influence the speed and the amount of its reduction to a similar extent in all the situations. On all the analysed sections, except for the section No. 3, the horizontal deflection of the travel lane was a = 2 m, and yet both the village entry speeds and the amounts of speed reduction varied considerably. Also the results speed after the chicane v^{after} and speed reduction Δv (Figure 4) do not confirm the thesis put forward in study [28] (Table 1—position 7), that $\Delta v = f(l_v/t_v)$, because in all sections tested, except No. 3, l_v/t_v = const, the results speed after the chicane v^{after} and speed reduction Δv are very diverse. Similarly, the results presented in Figure 4 do not confirm theses formulated in the publications [18,39,43,49–51] (Table 1—position 8, 9, 10), because the road class, lane width, chicanes shape, angles, curb heights were the same in all tested sections, except No. 3, and the obtained v^{after} and speed reduction Δv were very diverse (v^{after} = 64.7–87.4 km/h and Δv = 0.6–22.8 km/h). Based on these observations, the authors analysed different factors noted in the transition zones selected in three criteria concerning: Traffic management scheme, road design parameters, surrounding landscape and visibility conditions.

2.6. Genesis of the Aggregate Parameter z and ts Combined Effect on the Traffic Parameters

The analysis of the diversity and variation of the factors noted in the transition zones of the analysed test sections in relation to the traffic control, road design parameters, surrounding landscape and visibility conditions indicated a need for detailed analysis of the effect of the respective factors on the vehicle speeds and on their reduction. The authors in the publication [53] presented preliminary considerations of factors likely to contribute to the speed reduction. However, further consideration of possible determinants allowed for the developed an innovative analytical approach and introduced an independent variable z to consider the criteria adopted in the analyses (Figure 5). The proposed independent variable z is actually an aggregate parameter, introduced by the authors of this article, to quantify the overall impact of the surrounding landscape on the driver's perception, leading to the reduction of the village inbound speed (Figure 6).

$$z_i = \Sigma(z_{o\,i}, z_{d\,i}, z_{zw\,i}) = \Sigma z_{o\,i,j} + \Sigma z_{d\,i,j} + \Sigma z_{zw\,i,j}. \quad (1)$$

The analyses assume that the qualitative indirect factors chosen in a given criterion will be assigned the assumed 'quantitative measures' in the binary system, which are directly related to the confirmation of existence of a given factor, i.e., confirmed logical tautology. For instance, if a given factor is confirmed in field, it is assigned the value 1 (i.e., logical tautology is confirmed). If, on the other hand, a given factor is not observed, then it is assigned the value 0 (i.e., logical tautology is not confirmed). In some cases, especially related to quantitative-qualitative information, such as clearance distance to buildings or visual obstruction, it was necessary to introduce an intermediate quantitative measure and then the assignment of the value 0.5 is proposed. If, for example, the village skyline is visible while passing through the chicane at a distance smaller than 100 m, then logical tautology is confirmed and when it is not visible the logical tautology is rejected. There was a need for an intermediate measure when the village skyline was visible from other distance, for example 300–500 m.

Figure 5. Proposed scheme for determining the aggregate parameter z with a division to the three adopted criteria.

Figure 6. Aggregate parameter z_i on the i-th test section. Designations: $z_{o\,i}$—total value of the score for the traffic management scheme, $z_{d\,i}$—total score for the road design parameters, $z_{zw\,i}$—total score for the surrounding landscape and visibility conditions, $z_{o\,i,j}$—score of a given factor j for the traffic management scheme, $z_{d\,i,j}$ – score of the given factor j for road design parameters, $z_{zw\,i,j}$—score of the given factor j for the surrounding landscape and visibility conditions.

The intermediate factors related to the traffic management scheme criterion, preliminarily chosen for the analyses, included the distance to the nearest residential buildings located close to the road edge measured along the road from the D-42 traffic sign designating the built-up area limit, location of pedestrian pavements or other road infrastructure elements in relation to the D-42 traffic sign and the location of the B-33 speed limit sign in relation to the chicane axis (Figure 7).

In the criterion concerning the road design parameters also hourly traffic volume on both lanes and on the inbound lane alone were considered. The analysis demonstrated that the speeds measured before the chicane did not depend on the hourly traffic volume which, on the analysed roads, did not vary much (correlation factor of only $R = 0.4$). Whereas, the correlation factor between the speeds measured after the chicane and hourly volume of traffic was $R = 0.6$. The above facts indicate the existence of some other determinants influencing reduction of speed after the chicane, and therefore, the effect of the hourly traffic volume on the amount of speed reduction by the chicane was disregarded in the further analyses in the road design criterion. One of the main factors in the criterion relating to the road design parameters, in accordance with the conclusions given in [19,43] was the amount of the lateral deflection of the inbound lane. On the basis of site visits and Google Earth imagery [52] the qualitative factors were expanded with elementary road data related to curved sections, road infrastructure, accesses and junctions. Selected road design factors and the adopted engineering interpretation of their qualification measure are presented in Figure 8.

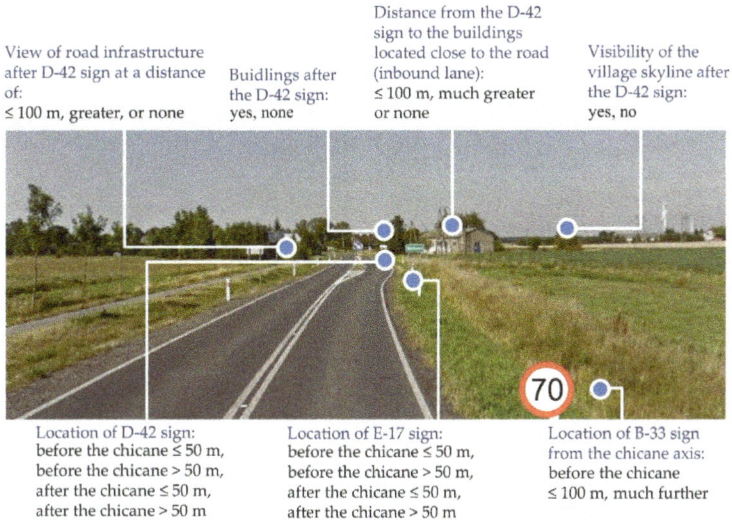

Figure 7. Factors related to the traffic management scheme chosen for the analyses (blue font designates qualitative factors and the black font–the adopted interpretation of the quantitative measure).

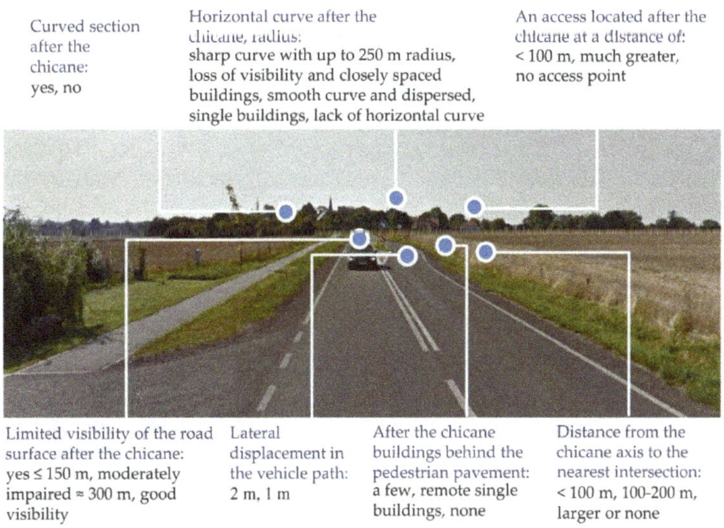

Figure 8. Factors related to the road design parameters chosen for the analyses (blue font designates qualitative factors and the black font designates the adopted interpretation of the quantitative measure).

The last criterion concerning the surrounding landscape and visibility conditions, includes, as independent variables, various qualitative/quantitative factors defining the space around chicanes and elements determining the visibility conditions. In the selection and description of the visibility conditions the authors used the results of fixation points distribution analysis in different driving conditions, as presented in [14] (Figure 9). Referring the fixation points to the more distant view of the road on the test section with 70 km/h speed limit (Figure 9b) the authors found out that the driver tends to focus the eyesight firstly on the road further ahead and on the existing chicane and only secondly on the travel lane and on the upcoming lateral deflection of the lane.

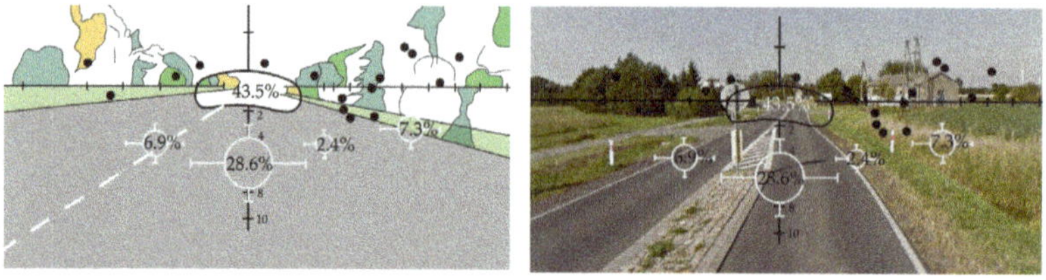

(a) (b)

Figure 9. Distribution of fixation points in different driving conditions: (**a**) using the data from [14]; (**b**) with 70 km/h speed limit indicated by the B-33 sign).

Next, on the basis of the resulting visual attention area of the driver, described in [14], the authors referred it on a likewise basis to the analysed test sections, as displayed in Figure 10. Taking the above into consideration, the authors made a visualisation of the central and peripheral vision areas, as described, for example in [17,55], the results of which are presented in Figure 11. The analysis of the results of this visualisation enabled the authors to determine what are the elements on which car drivers travelling at a speed of 70 km/h or 100 km/h focus their attention, as presented in Figure 11c,d and Figure 12.

Based on the analysis of the central field of vision area and the road infrastructure elements present within its limits, of the view of the road further ahead and the elements of the landscape surrounding the road at the chicane the authors proposed appropriate factors and assigned quantitative measures to them, also with the application of logical tautology, as presented in Figure 12.

Summing up the above considerations, in relation to the above-described three criteria (Figure 5), the authors chose to perform statistical inference associated with the determination of the strength of a correlation between the above-described qualitative variables (z_o, z_d, z_{zw}), the parameters representing the speed (v, Δv, w) and also the general aggregate parameter z. To this end, they decided that the numerical value of the aggregate parameter z should be determined using the scheme presented in Figures 5 and 6. In the statistical analyses regarding parameters characterising speeds the authors decided to analyse all the estimated speed distribution parameters, measured before and after the chicanes and the corresponding speed reductions.

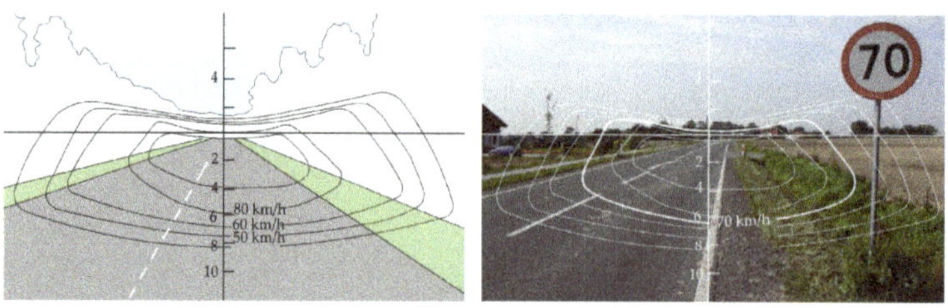

(a) (b)

Figure 10. Driver's visual attention area: (**a**) using the data from [15]; (**b**) with a 70 km/h speed limit indicated on the B-33 sign.

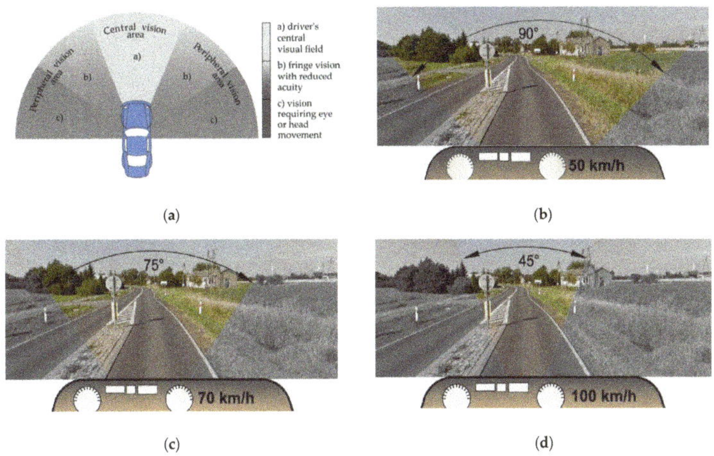

Figure 11. Central and peripheral vision area of a driver: (**a**) on the basis of the data from [55]; (**b**) at speeds up to 50 km/h: (**c**) at the speed of 70 km/h; (**d**) at the speed of 100 km/h.

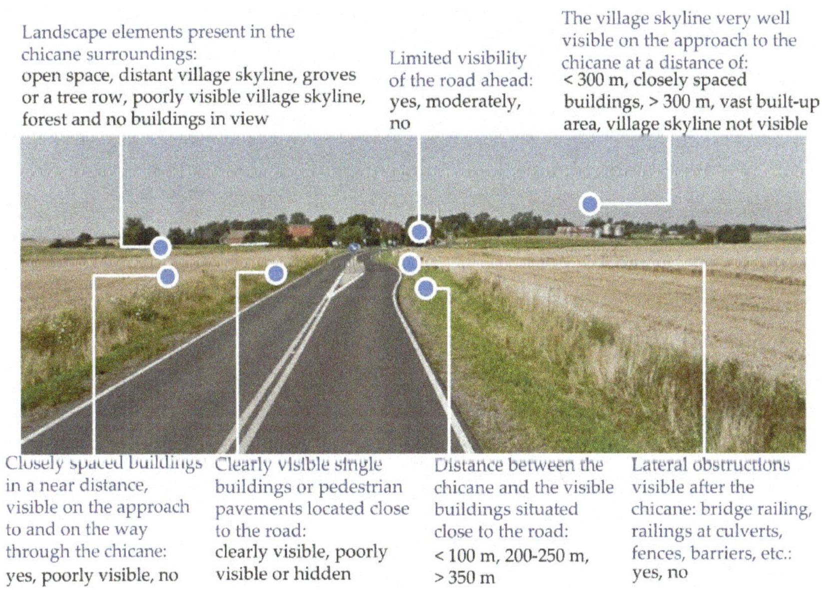

Figure 12. Factors related to the surrounding landscape and visibility chosen for the analyses (blue font designates qualitative factors and the black font designates the adopted interpretation of the quantitative measure).

Furthermore, they decided to analyse the speed reduction index in relation to the 85th percentile speed $w\ (v_{85})$ (Equation (2)) and the mean free-flow speed $w\ (v_{av})$ (Equation (3)), calculated with the following standard equations:

$$w\ (v_{85}) = (v_{85}{}^{before} - v_{85}{}^{after})\ /\ v_{85}{}^{before}, \quad (2)$$

$$w\ (v_{av}) = (v_{av}{}^{before} - v_{av}{}^{after})\ /\ v_{av}{}^{before}. \quad (3)$$

3. Results

3.1. Outcomes of the Assessment of the Factors Associated with the Traffic Management Criterion

For the sake of clarity, the adopted quantification measures, in accordance with the logical tautology, i.e., confirmation or lack of confirmation of a given factor (Figure 7), initially expressed in the binary numeral system were represented by colours (Figure 13), with the dark blue colour designating the measure of location equal to 1, confirming the presence of a given factor or the corresponding small distance, etc. The light blue colour designates the quantification measure equal to 0, i.e., lack of the presence of a given factor, for example a large distance, etc. The intermediate shade designates, for example, location of E-17 and D-42 signs after the chicane at a distance exceeding 100 m. In the case of the additional test sections, No. 8 and No. 9, which did not include chicanes, no weights were assigned to the traffic management related factors conditioned by the presence of a chicane (and hence the lack of colour in the chart).

Figure 13. Outcomes of the assessment of the factors associated with the traffic management criterion. Designations: o_j—a subsequent factor in consideration, Σo_j—the sum of confirmed logical tautologies on the test sections in relation to the considered j-th factor, $z_{o\,i}$—the sum of quantitative measures in all factors on the i-th test section.

The analysis of the values of the quantitative measures represented in Figure 13 demonstrated consistency of a number of selected factors with the speed reduction values. To be specific, such consistency was obtained between speed reduction and the factor representing: location of the B-33 sign in relation to the chicane axis, confirmation of the presence of buildings after the sign location and confirmation of the presence of buildings in close proximity to the road edge. A consistency between the greatest reduction of the 85th percentile speed Δv_{85} with the visibility of the nearby village skyline or road infrastructure after the D-42 sign (Figure 13) was confirmed in only one case.

For factors associated with the location of the E-17 and D-42 signs the authors noted lack of consistency with speed reduction. It is likely that the view of the nearby village buildings seen after the E-17 and D-42 signs affects the driver's perception, resulting, in consequence, in the reduction of driving speed (Figure 1a). If after the E-17 and D-42 signs located at different distances after the chicane the village skyline or adjacent buildings are not visible, especially on the inbound lane side (Figure 1b,d,e), the traffic signs indicating the beginning of the built-up area do not have any mobilising effect on motorists' behaviour who continue driving without slowing down.

Taking the above into account, in relation to the four factors concerning location of the E-17 and D-42 signs the logical tautology was represented by colours, however no numerical values were assigned to them, since the distances d_{E-17} of the E-17 sign location from the chicane ranged from 5 m before the chicane up to 120 m after the chicane, and the distance d_{D-42} of the D-42 sign ranged from 29 m to 290 m after the chicane. Such a high data dispersion prompted the authors to determine the coefficient of correlation between the above-mentioned distances and the speed reductions on the sections under analysis. The values of the correlation coefficient were respectively $R = 0.4$ ($f(d_{E-17}, \Delta v_{85})$) and $R = 0.2$ ($f(d_{D-42}, \Delta v_{85})$), which indicated a lack of statistically significant relationship between the amount of speed reduction and the distance of the sign location from the chicane.

Taking the above into consideration we can conclude that the distance at which the E-17 and D-42 signs are located from the chicane, if it is not related to the visibility of buildings, has no significant effect on the actual vehicle speeds measured after the chicane. However, if the chicane is preceded by the B-33 sign denoting a speed limit of 70 km/h then consideration of logical tautology related to the close vicinity of the village buildings (located along the street) would not be in compliance with the Polish highway code [56] and the guidelines [57] and, moreover, it would pose problems with defining logical tautology related to assignment of a value to a given distance. Therefore, in the outcomes of the assessment of the factors related with the traffic management criterion no quantitative measures are assigned to the above-mentioned four factors (Figure 12).

Summing up the above considerations we can conclude that in the case of chicanes located in the transition zones with the posted speed limit of 70 km/h the determinants effectively reducing the speed include: placement of the B-33 sign before the chicane at a distance of up to 100 m and visible close village buildings situated after the D-42 sign at a distance less than 100 m.

3.2. *Outcomes of the Assessment of the Factors Associated with the Road Design Parameters Criterion*

The second criterion considered in identification of the determinants of the effectiveness of chicanes in the transition zones is associated with the factors defining the road design parameters. In this criterion, various dependency relationships between road design parameters and reduction of the 85th percentile speed Δv_{85} were confirmed. A representative relationship of the analysed variables was confirmed in relation to the amount of lateral deflection of the lane. However, the selected test sections included primarily central islands of rectangular shape horizontally deflecting one lane with the deflection of a = 2 m (Figure 14a). A central island horizontally deflecting both lanes with the deflection of a = 1 m (Figure 14b) was located on only one test section.

Figure 14. Types of road chicanes provided in village entry transition zones and the resulting horizontal deflection of the travel lane from the straight line *a*: (**a**) central island providing horizontal deflection of one lane – $a = 2$ m; (**b**) central island providing horizontal deflection of both lanes – $a = 1$ m.

Another factor with a varying effect on traffic speed reduction was also a visible access. When an access led to a cluster of buildings then a greater speed reduction was noted. Still the scale of the reduction varied considerably, ranging from 3.5 km/h to 22.8 km/h (Figure 15). In the case of a private accesses serving a single family building the speed reductions also varied but over a smaller range (Figure 15–Δv_{85} = 1.9–3 km/h). A similar varied effect was noted for an intersection with a side road, in which case the speed reduction was in the range of Δv_{85} = 1.9–9.5 km/h.

The analysis of the variation of the sum of quantitative measures representing the road design parameters z_d and of the speed reduction amount Δv_{85} revealed a probable consistency between the considered variables (Figure 15). The only exception in the analysed case is the test section No. 7 where a smaller number of significant road factors was noted, as compared to the other sections.

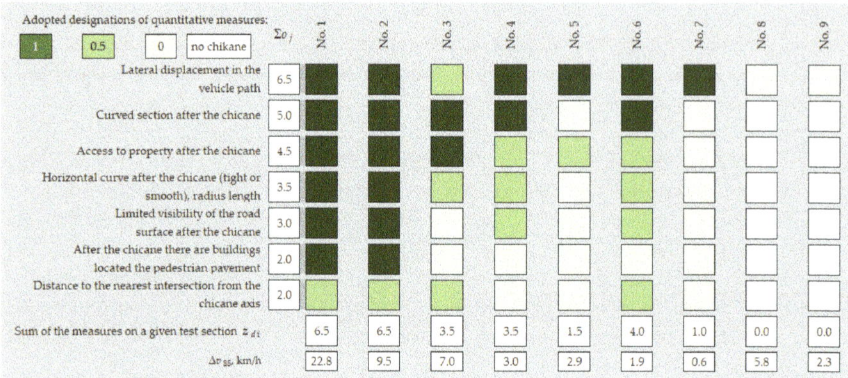

Figure 15. Outcomes of the assessment of the factors associated with the road design parameters criterion. Designations: o_j—a subsequent factor in consideration, Σoj—sum of confirmed logical tautologies on the test sections in relation to the considered j-th factor, $z_{d\,i}$—sum of quantitative measures in all factors on the i-th test section.

3.3. Outcomes of the Assessment of the Factors Associated with the Landscape and Visibility Criterion

The last criterion considered in the identification of the determinants of the effectiveness of chicanes in the transition zones are the factors related to the landscape features in the immediate vicinity of the entry section of the road and to visibility conditions. In this criterion very important dependencies were confirmed between the speed reduction Δv_{85} and the sum of quantitative measures z_{ZW} in particular in relation to the first three test sections (Figure 16). Figure 1a,b show the village skyline seen from a distance of 300 m by a driver approaching the chicane located on each of the first two test sections, i.e., section No. 1 and section No. 2.

Figure 16. Outcomes of the assessment of the factors associated with the landscape and visibility criterion. Designations: o_j—a subsequent factor in consideration, Σo_j—the sum of confirmed logical tautologies on the test sections in relation to the considered j-th factor, $z_{zw\,i}$—the sum of quantitative measures in all factors on the i-th test section.

Figure 17a,b depict the village skyline as seen by a driver passing through the chicane. In the case of section No. 1, the driver sees nearby buildings after the sign D-42 and a horizontal curve limiting the view of the road further ahead (Figure 17a) and this is probably the main cause of the considerable reduction of the free flow speed noted at this section, equal to Δv_{85} = 22.8 km/h. In the second case presented in Figure 17b the driver on the approach to the chicane sees the distant village skyline and remote buildings situated at a distance of ca. 300 m at the other side of the road. In this case the E-17 & D-42 signs are located after the chicane, which results in a much smaller reduction of the free-flow traffic speed (Δv_{85} = 9.5 km/h). Figure 17c,d, show the approach to the chicane on section No. 3 and the view of the road ahead the driver sees when going through the chicane. In this case, the free-flow speed reduction is Δv_{85} = 7 km/h. This can be attributed to the lack of visibility of the village skyline and of nearby buildings and to the fact that the lateral deflection a of the lane is only 1 m. Furthermore, the traffic management infrastructure, namely the sign No. D-42 denoting the beginning of the built-up area is placed as far as 300 m after the chicane.

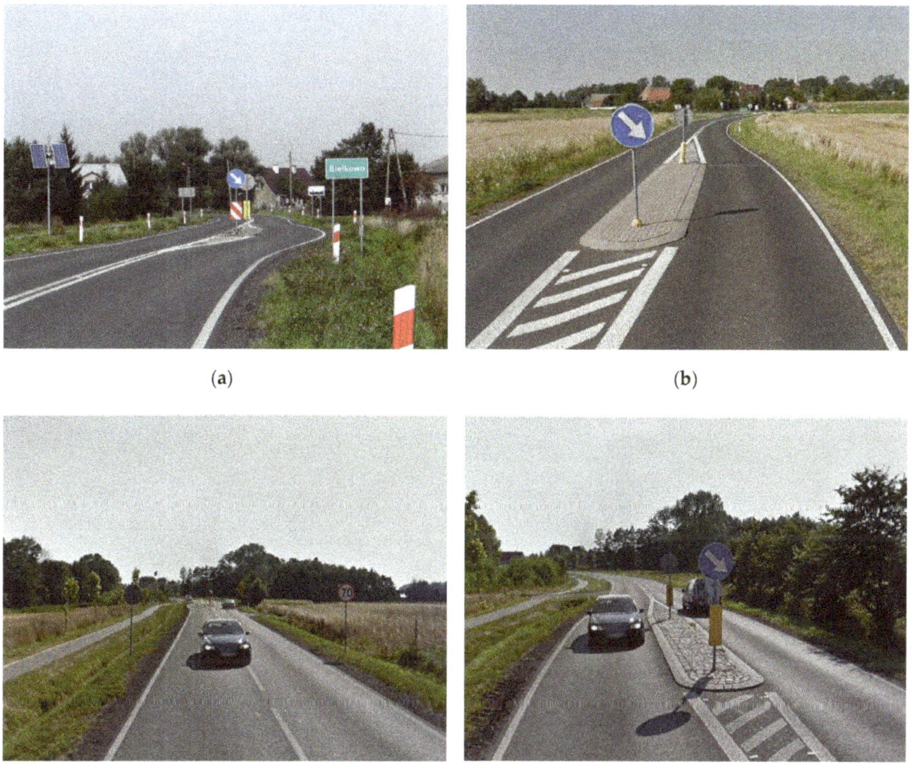

(**a**) (**b**)

(**c**) (**d**)

Figure 17. View of the village skyline of village as seen by a driver approaching the chicane: (**a**) View of nearby buildings at chicane No. 1, sign E-17 located 5 m before the chicane and sign D-42 located 30 m after the chicane; (**b**) Distant view of the village skyline near the chicane No. 2, sign E-17 located 100 m after the chicane and sign D-42 located 150 m after the chicane; (**c**) Lack of the village skyline view on the approach to the chicane No. 3, being a central island deflecting both lanes; (**d**) Lack of the view of village buildings from the chicane No. 3 location, sign E-17 located 150 m after the chicane and sign D-42 located 300 m after the chicane.

On the remaining test sections (No. 4, No. 5, No. 6 and No. 7) the authors noted high variability between the assigned quantitative measures and speed reduction values. The partial sums of the four primary factors, represented in Figure 16 have similar values, yet their distribution in the analysed sections is very diverse, just as diverse are the speed reduction values. A very wide variation in the respective quantitative measures is also observed in the case of the two last factors represented in Figure 16, where visual obstructions or buildings situated close to the chicane were confirmed only in one case, while in the remaining three cases visible buildings were noted however they were situated at a further distance from the chicane. Still, comparing the sums of quantitative measures z_{zw} with the speed reduction values Δv_{85} on the respective test sections, we can detect a certain consistency in the succession of changes, as represented in the radar charts.

3.4. Outcomes of the Assessment of the Aggregate Parameter z and Its Overall Effect on the Traffic Parameters

Summing up the above analyses, Figure 18 below represents the probable consistency between the qualitative variable z, i.e., the aggregate overall effect of the factors related to traffic management (z_o), road design parameters (z_d) and landscape and visibility conditions (z_{zw}), and the calculated free-flow speed reductions Δv_{85} and Δv_{av} on the analysed test sections with the applied horizontal deflection of the lane(s). Figure 18b represents the consistency of the aggregate parameter z with the values of speed reduction indices.

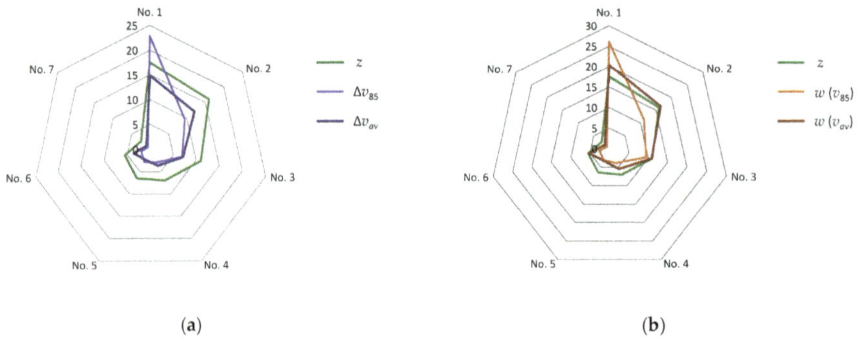

Figure 18. Compilation of changes in the analysed variables: (**a**) consistency of the speed reduction variations Δv_{85} and Δv_{av} and aggregate parameter z ($R = 0.76$ and $R = 0.93$ respectively); (**b**) consistency of variation of the speed reduction index w (v_{85}) and w (v_{av}) and aggregate parameter z ($R = 0.80$ and $R = 0.94$, respectively).

The analysis of variations of the analysed values exhibits the greatest consistency between the free-flow speed v_{av} and the speed reduction index w (v_{av}) and the proposed aggregate parameter z representing a combination of factors, including traffic management, road design parameters and the surrounding landscape elements and visibility conditions.

3.5. Regression Analysis of the Outcomes of Assessment of the Combined Effect of the Selected Factors and Speed Distribution Parameters

In the next step, the authors carried out appropriate regression analyses in relation to the investigated parameters. In the regression analyses, whether confirming or failing to confirm the effectiveness of the applied chicanes, the authors considered both the aggregate values of the respective components of qualitative variables (z_{zw}, z_d and z_o) and the qualitative variable z representing the combined effect of the respective factors in the transition zone. In relation to the speed distribution parameters before and after the chicane the analysis considered both the free-flow 85th percentile speed v_{85}, the average free-flow speed v_{av} and the stable-flow speed $v_{av}{}^{pp}$, and also the corresponding speed reduction values Δv. Along with the speed distribution parameters the authors considered also the speed reduction index w, often taken into account in traffic calming analyses and representing the percentage speed reduction in relation to the initial value ($w = \Delta v / v^{before}$). The correlation coefficient values are given in Tables 2 and 3. In Tables 2 and 3 bold numerals denote cases when the correlation coefficient R was greater than 0.9, which according to the statistical principles given in [58,59] indicates a statistically significant relationship between the analysed variables.

Table 2. Coefficients of correlation R between the quantitative measures and the chosen speed distribution parameters (including additional sections where chicanes were not installed).

	v_{85}^{before}	v_{av}^{before}	$v_{av}^{pp\ before}$	v_{85}^{after}	v_{av}^{after}	$v_{av}^{pp\ after}$	Δv_{85}	Δv_{av}	Δv_{av}^{pp}	$w\ (v_{85})$	$w\ (v_{av})$
z_{zw}	−0.14	−0.21	−0.12	−0.63	−0.68	−0.61	0.76	**0.93**	**0.93**	0.80	**0.94**
z_d	−0.38	−0.36	−0.29	−0.77	−0.78	−0.67	0.68	0.89	0.75	0.71	**0.90**
z_o	−0.18	−0.31	−0.24	−0.67	−0.69	−0.71	0.76	0.81	**0.90**	0.79	0.82
z	−0.24	−0.31	−0.23	−0.73	−0.76	−0.69	0.77	**0.94**	**0.90**	0.81	**0.95**

Table 3. Coefficients of correlation R between the quantitative measures and the chosen speed distribution parameters (excluding additional sections where chicanes were not installed).

	v_{85}^{before}	v_{av}^{before}	$v_{av}^{pp\ before}$	v_{85}^{after}	v_{av}^{after}	$v_{av}^{pp\ after}$	Δv_{85}	Δv_{av}	Δv_{av}^{pp}	$w\ (v_{85})$	$w\ (v_{av})$
z_{zw}	0.39	0.42	0.64	−0.49	−0.74	−0.74	0.81	**0.93**	**0.93**	0.82	**0.94**
z_d	0.18	0.26	0.53	−0.64	−0.83	−0.48	0.77	**0.92**	0.71	0.78	**0.92**
z_o	0.62	0.52	0.73	−0.40	−0.56	−0.61	**0.90**	0.82	**0.96**	**0.90**	0.82
z	0.40	0.42	0.68	−0.57	−0.79	−0.64	0.89	**0.98**	**0.94**	**0.90**	**0.99**

Appropriate regression analyses were carried out considering the obtained positive results of the correlation coefficients in relation to the searched-for relationship between the speed reduction Δv or the speed reduction index w and the magnitude of the effect of the qualitative variable z. The results of the regression analysis in relation to the dependent variables v_{85} and w (v_{85}) are represented in Figure 19.

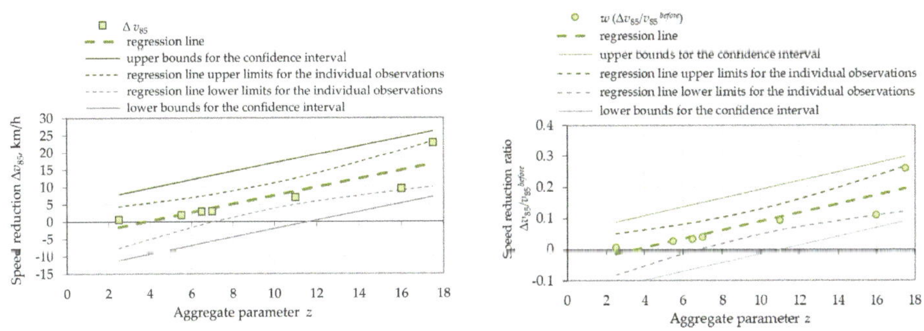

(**a**) (**b**)

Figure 19. The linear regression function defining the relationships between: (**a**) free-flow speed reduction Δv_{85} and the aggregate parameter z, ($R = 0.89$); (**b**) free-flow speed reduction index w (v_{85}) and the aggregate parameter z ($R = 0.90$).

The relationships represented in Figure 19 prove the hypotheses previously put forward by the authors that the reduction of speed through the chicane depends on a number of factors characterizing the spatial features of the surrounding area, which jointly influence the drivers' behaviour and their decision to slow down in the transition zone section. Referring to both foreign guidelines (Denmark [1], Great Britain [2,19,31,39,41], USA [27,36,37]) and Polish publications [5,21,42] which, depending on the country, give various data regarding the percentage reduction of speed of vehicles passing through chicanes, it can be concluded that the theses proposed in these materials have been confirmed by the results obtained by the authors in this research. Still, it must be noted that the amount of speed reduction depends on several external factors which were for the first time identified and estimated in this article.

Figure 20 represents the same regression analyses carried out in relation to the free-flow average speed v_{av}. The analysis of the data shown in Figure 20 indicates that the average free-flow speed v_{av} statistically significantly depends on the combined magnitude of the effect of the landscape elements present within the transition zone, which is confirmed by the high values of the correlation coefficients (Tables 2 and 3). The more of various landscape elements are present within the transition zone to the village in the vicinity of the on-road chicane, the greater speed reduction Δv can be expected. The above observations can indicate to land planners and designers the most suitable locations in which to design and install chicanes effectively reducing traffic speeds, subject to considering the maximum number of relevant factors.

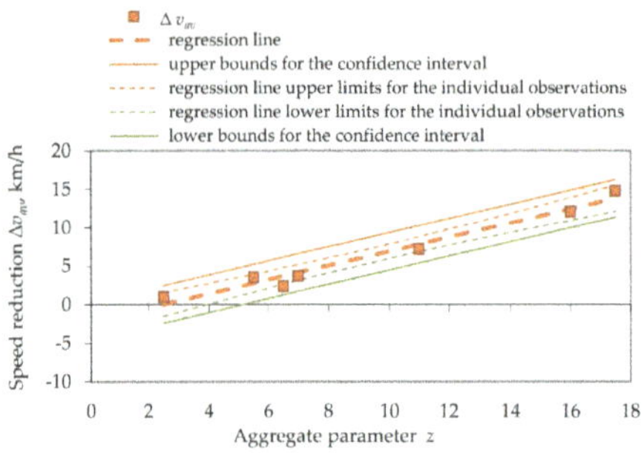

(a)

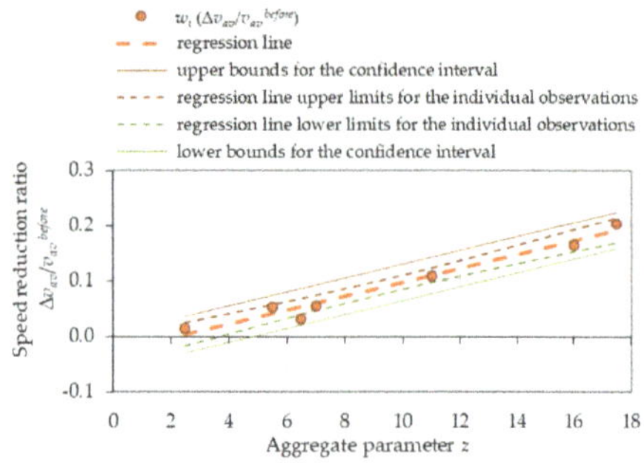

(b)

Figure 20. Linear regression function defining the relationship between: (**a**) the free-flow speed reduction Δv_{av} and the aggregate parameter z, ($R = 0.98$); (**b**) the free-flow speed reduction index w (v_{av}) and the aggregate parameter z ($R = 0.99$).

If there are buildings or elements of highway structures (i.e., visual obstructions) within the transition zone and they are situated close to the road edge, then the chicane should be designed at a location as close as possible to these elements. Furthermore, in order to obtain effective speed reduction, horizontal deflection of the lane by min. 2 m should be considered, located at a distance of up to 100 m from the visible nearby buildings of the village. If the transition zone is approached through a long straight section running through an open area, then for effective speed reduction, the amount of the horizontal deflection of the lane, should be designed taking into account the view of the village skyline and the distance of the chicane from the village buildings in view, measured along the road. The best results in terms of traffic speed reduction can be achieved at locations where the village buildings are situated close to the inbound lane.

4. Discussion

This article analyses seven sections of roads leading through villages with chicanes installed in the transition zones. The analysis of the performed speed survey data represented in Figure 4 demonstrated a very wide and varied dispersion of speeds despite the fact that in almost all of the cases the analysed chicanes provided the same amount of the horizontal deflection of the carriageway lane. The comparison of the results obtained with speed survey data for two other sections, No. 8 and No. 9, without any chicanes installed, demonstrates that speed reductions were noted also on these sections. In view of the above, it is clear there must have been some other determinants resulting in speed reduction in these locations.

In the course of the analysis the authors formulated ten-odd such determinants which could have an effect on the obtained speed reduction. Based on comprehensive analysis of factor assessment values obtained in the individual criteria proposed and of the final values of the aggregate parameters z on the test sections, in recapitulation we can identify the main determinants. For instance, in the traffic management criterion these will be the factors directly related to the view seen by the driver after the D-42 sign (Figure 13) and in the road design criterion, the factors related to horizontal curvature and its parameters (Figure 15). The greatest number of factors which obtained confirmation of logical tautology in the landscape and visibility criterion were related to the surrounding conditions in the vicinity of the road on the approach to the village and in the transition zone (Figure 17). On the test sections with the greatest speed reductions confirmed also other factors with positive logical tautology were confirmed, relating to the visibility conditions and the elements that can be visible to the driver when approaching and passing through the chicane, namely the village skyline, a view of single buildings situated close to the road or a clear view of some road infrastructure elements associated with the built-up area. Summing up the above considerations, in the opinion of the authors the obtained results confirm the postulated hypotheses and proposals regarding the introduction of the aggregate parameter z, which evidently has an effect on the magnitude of speed reduction of vehicles which enter the village (Tables 2 and 3 and Figures 19 and 20).

Limiting the analysis to only the test sections located in open rural areas, we can confirm that visibility conditions have a major effect on drivers' decision to reduce speed. These include the view of the village skyline on the approach to the transition zone, of nearby buildings and road infrastructure elements such as pedestrian pavements, accesses, junctions, visual obstructions e.g., bridge railings, and also the presence of horizontal curvature and the resulting limited view of the road ahead and its direct vicinity. Therefore, in choosing the location for a chicane to be provided as part of a traffic calming scheme, designers and landscape planners should, in the first place, consider the logical tautologies of the above given factors. The precise location of the chicane should also take into account the logical tautologies relating to the view of the village skyline, nearby buildings or visible elements of the road infrastructure in relation to the existing D-42 sign denoting the beginning of built-up area (Figure 21). If a confirmation of logical tautologies was obtained for all the above-mentioned elements, then the chicane should be located as close as practicable to the D-42 sign. With these conditions met, one can expect that the desired speed reduction Δv_{85} = ca. 1020 km/h will be achieved within the transition zone with the 70 km/h speed limit signalled on the B-33 sign (Figure 4, Sections No. 1, No. 2 and No. 3).

The analytical approach proposed by the authors for application by land planners and designers of traffic calming schemes in transition zones to villages is presented in Figure 21. In the first step (Figure 21–Step 1, background) consideration should be given to the type of terrain and landscape elements present in the area surrounding the road in the transition zone to the village. In this study the authors proposed the three following types of road surroundings: open rural area, rural area with groves or a tree row and a forest terrain. The above-mentioned landscape conditions in the surroundings of the road have a determining effect on the vehicle's speed of approach to the transition zone, driver's perception and, as a consequence, on the amount of speed reduction on the way through the chicane. Next, based of the logical tautologies confirmed during the site visits or when reviewing the available Google Earth or Street View imagery [52], the land planners should determine the value of the aggregate parameter z (Figure 21—Step 2—the analytical process regarding the conditions on the approach to the village's transition zone). Further, a similar analysis should be carried out for the landscape elements present in the surroundings of the road in the vicinity of the planned chicane location (Figure 21—Step 3—preliminary location of chicane). This should involve an analysis of the road alignment along a straight section or horizontal curvature, assessment of the visibility of the road ahead and of existing locations of the vertical signs E-17 and D-42, etc. Having initially assessed the landscape conditions in the surroundings of the approach to the transition zone and the planned chicane location, the next step should involve estimation of the individual factors in the three above-mentioned criteria and the value of the aggregate parameter z (Figure 21–Step 4–the assessment of the aggregate parameter z). After determining the aggregate parameter z we can move to the final step in designing a traffic calming scheme in a village transition zone with 70 km/h speed limit, namely to final determination of the chicane location and adoption of its parameters (Figure 21—Step 5—chicane design).

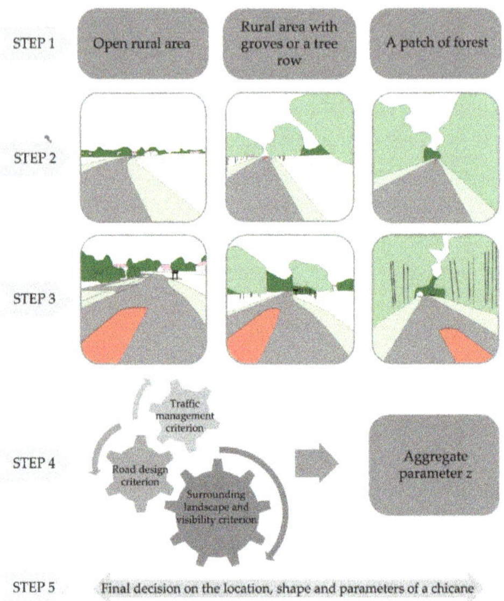

Figure 21. Proposed design process–i.e., selection of the chicane location within the transition zone to the village, with 70 km/h posted speed limit.

If in an open rural area there is no village skyline or nearby buildings in view, then a small speed reduction can also be expected even without a chicane on sections with horizontal curvature hiding the view of the road ahead, as confirmed by the measured vehicle speeds on the section No. 9 (Figures 4 and 1i).

With a greater number of confirmed logical tautologies (i.e., when the value of the aggregate parameter z is higher than 10), according to the authors, chicanes resulting in horizontal deflection of one lane only can be considered for installation in the transition zones if the deflection of the lane from the straight line is at least 2 m. With a smaller value of aggregate parameter z, in order to achieve satisfactory speed reduction, the chicane location should be planned closer to the existing buildings and road infrastructure elements and the applied lane deflection should be considerably larger (Figure 22). In cases when satisfactory speed reduction has not been achieved with chicanes already in place in transition zones, an additional traffic calming measure should be planned for installation in the vicinity of nearby village buildings.

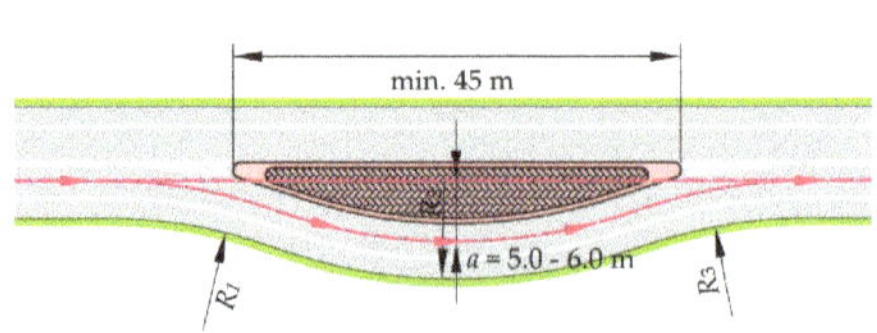

Figure 22. Semi-circular chicane horizontally deflecting one lane proposed for village transition zones approached through a patch of forest (based on the guidelines [21,24,25]).

If installation of a chicane deflecting one lane only is planned in the transition zone to the village situated on road section going through a patch of forest where the driver can see neither the village skyline nor nearby buildings, the authors propose to locate the chicane as close as possible to the D-42 sign and use semi-circular chicanes providing deflection of the lane considerably greater than 2 m, i.e., according to the recommendations given in the guidelines [21,24,25] (Figure 22). Guidelines [25] give also detailed recommendations regarding the size of the radius. For example, for the speed of 70 km/h the entry radius R_1 should not be smaller than 100 m and the subsequent radii (R_2, R_3) should be either equal or greater than the entry radius R_1. The corners of the chicane should be rounded and the accompanying hatched road markings area should end with a pointy tip. The other dimensions of the chicane should be accommodated within the road design parameters, i.e., available right-of-way width (conditioning the size of desired horizontal lane deflection) and national regulations concerning road surface markings.

An example of semi-circular chicane deflecting one lane, installed in a village transition zone in a partly forested area is shown in Figure 23. The effectiveness of semi-circular chicanes installed in transition zones to villages and the resultant further applications of supplementary traffic calming measures provided on the passage of the regional road through the village area was described by the authors in [60].

(a) (b)

Figure 23. Example of semi-circular chicane horizontally deflecting one carriageway lane: (**a**) overview; (**b**) visible details of the semi-circular chicane island.

However, in the opinion of the authors, the process of designing chicanes in village transition zones should not be limited to the selection of the chicane location and its geometrical parameters but it should include proposing application of the latest technologies which are gaining increasingly wide use in the field of traffic engineering. This concerns traffic controls, including both upright signs and pavement markings, where solar technology is being introduced, resulting in considerable energy conservation in built-up areas and in significant improvement in traffic safety. The problem of the effect of LED lighting and its relationship with the speed and improved view of the road ahead has been tackled in a number of publications. For instance in [61] the authors described the results of studies on the use of an active system for delineating the road ahead in order to enhance the ability of drivers to control their vehicle safely on winding roads. A simulator experiment compared night-time driving on a country road under the three following conditions: on an unlit road, on a road illuminated on curves by typical road lighting luminaires, and on a road with an active lane delineation application, where self-luminous road studs are turned on to outline the lane and road edges as the driver approaches and passes the curves. It was concluded that the tested LED application enhanced the ability of drivers to control the virtual car, as compared to an unlit road or road lighting. Considering that transition zones to villages on roads carrying through traffic usually are not provided with road lighting the authors have analysed the results of the research described in the publication [62]. Light conditions are essential factors in traffic safety, but the relationship between light conditions and vehicle speed is not fully understood and has rarely been examined. In the course of the study, the authors of [62] investigated three questions: (a) if vehicle speed between brighter and darker conditions in clear weather will be different; (b) if vehicle speeds are lower during rain and snow than in clear weather conditions and if so; (c) if the speed reduction in rainy and snowy weather conditions is more substantial on roads without road lighting in darkness. The results of the study [62] demonstrated no evidence of consistent patterns of speed differences with respect to lighting conditions (darkness, daylight, twilight or road lighting) under clear weather conditions and no effect of road lighting on speed reduction in rainy and snowy weather. Another study described in publication [63] examined the effect of the light emitting diode (LED) road lighting of a residential road section with speed humps on speed reduction. The study results indicated that the LED road lighting that was placed directly above or in front of the hump achieved the highest luminance. Still, the study did not reveal any significant differences in vehicle speed attributable to light conditions per se.

In the recent years solar components are increasingly being used in solar powered traffic signs (Figure 24). The conditions prevailing on the test sections described in this article indicated that it might turn out to be important to use upright signs equipped with light emitting diodes (LED), which according to the results of the research described in [64] score a better detection ratio than conventional upright signs. Solar LED lights are high capacity light sources powered with batteries self-charging during the day. Depending on the weather conditions during the day, solar energy is converted to electricity by means of photovoltaic cells and stored in ultracapacitors capable of supplying impulse power according to an established program, both overnight and during the day. These devices are wireless, switched on automatically, low maintenance and eliminate the need to extend the lighting system serving the village centre, i.e., they allow to considerably lower the consumption of energy. All the materials used to manufacture the presented solar devices are recyclable, so they do not create any environmental burden and hence require no extra energy input. The operating temperature range of these devices ranges from –40 °C to +70 °C. Their usual warranty period is around 10–12 years, depending on the manufacturer. Also according to the conclusions formulated in guidelines [4] the D-6a sign (Figure 24b) on the approach nose of the central island horizontally deflecting one lane is a better option than using the U-5 bollard in this location (Figure 24a) since the D-6a sign forms a kind of a visual obstruction having an effect on drivers' perception and increasing the reduction in vehicle speeds by additional 1–3 km/h. Still the amount of speed reduction depends mainly on the width of the deflected travel lane and the size of horizontal deflection applied.

Additional solar elements installed in the face of the entrance island, on the outer edge of the carriageway and on the guide posts (Figure 25) are another element related to energy conservation in village environments evidently contributing to the reduction of vehicle speeds in village transition zones. They are recommended by the authors for use in particular in transition zones situated in forest terrain. These are SolarCAP elements, i.e., self-contained LED lighting devices. They are not equipped with additional batteries, which makes them less expensive, and apart from contributing to conservation of energy they also reduce construction and maintenance costs and improve traffic safety owing to their better visibility, and hence a greater impact on drivers' perception. They are produced from environmentally friendly materials and are suitable for recycling. Furthermore, SolarCAP devices are maintenance free for at least 10 years from installation. They are resistant to mechanical damage and frost and water resistant.

Their principle of operation is similar to that described above, i.e., during the day solar energy is converted to electricity by means of photovoltaic cells which is subsequently stored in EnergyCache ultracapacitors capable of providing impulse power according to an established program, both at night and at day time. At dusk, a Solar CAP device switches on automatically and emits directional light produced in LEDs. The charging time to full charge is ca. 1 h in harsh sun or 6–8 h under cloudy weather. A fully charged SolarCAP can operate continuously for over 12 h.

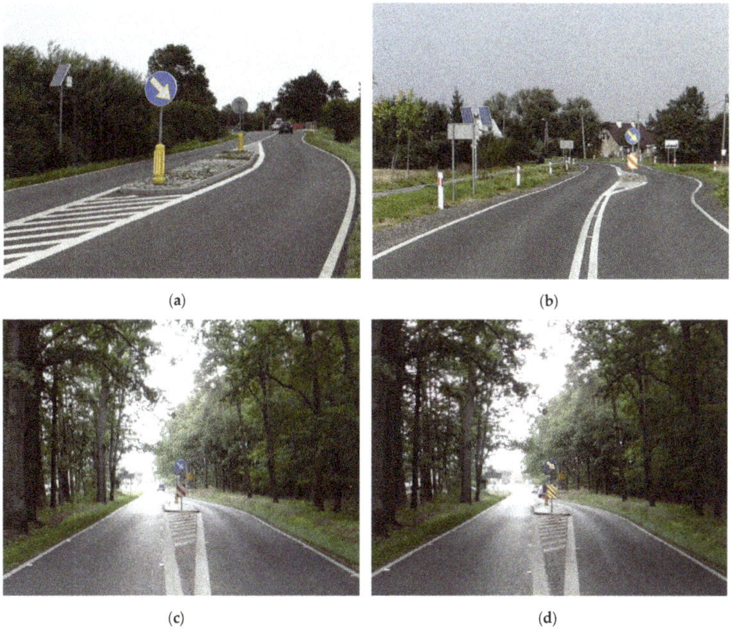

Figure 24. Application of the latest solar technology: (**a**) solar panel powering the LED arrow on bollards U-5 and C-9; (**b**) solar panel illuminating LED lines on the striped bollards D-6a and C-9; (**c**) view of unilluminated signs D-6a and C-9; (**d**) view of signs D-6a and C-9 illuminated with LED lights.

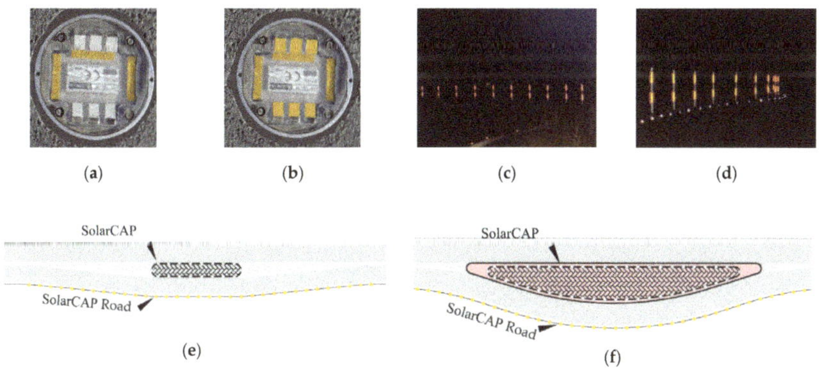

Figure 25. Recommended use of the latest solar technology: (**a**) SolarCAP mounted in the face of the island kerb–white; (**b**) SolarCAP Road installed in the outer edge of the road–yellow; (**c**) red LED lights used in exceptional cases on guide posts placed behind the outer edge of the road; (**d**) recommended yellow LED lights mounted on guide posts placed behind the outer edge of the road in a forest terrain; (**e**) proposed layout of Solar CAP and SolarCAP Road elements placed on a rectangular central island horizontally deflecting one lane; (**f**) proposed layout of Solar CAP and SolarCAP Road elements installed on a semi-circular central island horizontally deflecting one lane.

Based on the conclusions from the tests carried out on the TRL track [24], in locations with poorer visibility conditions, for instance in forest terrains or in rural areas with local groves, the authors can recommend placement of guide posts behind the outer edge of the road, spaced by maximum 0.5 m from the edge of the road (Figure 25c,d). Such arrangements are particularly recommended in the case of varied topography around the chicane (i.e., presence of embankments or drainage ditches). That said, in line with the recommendations of [3], in rural areas, where oversize farm equipment (such as combine-harvesters) traffic was recorded in traffic surveys, the road shoulder should be additionally strengthened over up to 1 m width over the whole length of the lateral deflection of the lane. Such strengthening can be made of (irregular) cobblestones. This measure will prevent deterioration of the dirt shoulder, as shown in Figure 26a. Surfacing the shoulder with asphalt millings is not recommended, as car drivers may mistakenly take such surface for paved shoulder and try and use it for normal travel (Figure 26b).

(a) (b)

Figure 26. Examples of design errors over the length of the deflected lane section: (**a**) deterioration of the dirt shoulder over the length of lateral deflection due to the lack of strengthening; (**b**) asphalt millings placed on the dirt shoulder as an overrun area for oversize agricultural equipment along the horizontal carriageway deflection.

Summing up the discussion conducted in reference to the obtained research results, the authors postulate that there is a need for further analyses and for confirmation of the proposed hypothesis on the identification of determinants of the effectiveness of chicanes on other roads, for example on roads subject to a 50 km/h speed limit, or for use differently designed road chicanes in the transition zones to villages.

5. Conclusions

Recapitulating the analyses carried out under this research, it can be concluded that the authors have obtained a confirmation of the existence of the identified determinants and of their effect on speed reduction in the transition zones to villages on roads with a 70 km/h speed limit indicated by the B-33 traffic signs. The important determinants identified in this research, which have a considerable effect on speed reduction are associated primarily with the view available to the driver on the approach to the village and directly in the transition zone, including: the village skyline, nearby clustered buildings or clearly visible elements of road infrastructure related to the built-up area. The size of provided horizontal deflection of the travel lane is another important factor, particularly significant with regard to rectangular chicanes horizontally deflecting one travel lane, installed on inbound lanes to villages.

Author Contributions: Conceptualization A.S., D.K. Methodology A.S. Formal analysis A.S., D.K. Data curation D.K. Writing—original draft preparation A.S. Writing—review and editing A.S. Visualization A.S., D.K. Supervision A.S. Funding acquisition D.K. All authors have read and agreed to the published version of the manuscript.

Funding: The researches included in the paper were partly financed by grants of WBiA ZUT: DKD Decision No. 517-02-033-6728/17.

Conflicts of Interest: The authors declare no conflict of interest.

Appendix A

Table A1. Results of standard statistical tests on test sections.

Test Sections	Goodness-of-Fit Test H_0: $F(v) = F_0(v)$ H_1: $F(v) \neq F_0(v)$			Significance Test u H_0: $v^{before} = v^{after}$ H_1: $v^{before} \neq v^{after}$		Significance Test t H_0: $v^{before} = v^{after}$ H_1: $v^{before} > v^{after}$		Test for Equality of Variances F H_0: $s_{before}^2 = s_{after}^2$ H_1: $s_{before}^2 > s_{after}^2$	
	$\lambda^{(v\ before)}$	$\lambda^{(v\ after)}$	$\lambda_{\alpha},\ \alpha = 0.05$	u	$u_{\alpha},\ \alpha = 0.05$	t	$t_{\alpha},\ \alpha = 0.05$	F	$F_{\alpha},\ \alpha = 0.05$
No. 1	0.37	0.58	1.36	5.85	1.96	5.85	1.6	3.03	1.61
No. 2	0.45	0.48	1.36	6.33	1.96	6.33	1.6	1.75	1.54
No. 3	0.63	0.70	1.36	3.62	1.96	3.62	1.6	1.72	1.63
No. 4	0.49	0.24	1.36	1.63	1.96	1.63	1.6	1.02	1.62
No. 5	1.24	0.68	1.36	0.73	1.96	0.73	1.6	1.31	1.61
No. 6	0.72	0.15	1.36	1.92	1.96	1.92	1.6	2.38	1.64
No. 7	0.63	0.71	1.36	3.76	1.96	3.76	1.6	1.68	1.54
No. 8	1.17	0.78	1.36	1.22	1.96	1.22	1.6	2.29	1.55
No. 9	0.73	0.58	1.36	0.88	1.96	0.88	1.6	1.39	1.55

Appendix B

Table A2. Results of goodness-of-fit test on test sections (free speed v and vehicle flow speed v^{pp}).

Test Sections	v^{before} and v^{after}		$v^{pp\ before}$ and $v^{pp\ after}$	
	λ	$\lambda_{\alpha},\ \alpha = 0.05$	λ	$\lambda_{\alpha},\ \alpha = 0.05$
No. 1	3.04	1.36	4.25	1.36
No. 2	2.55	1.36	2.61	1.36
No. 3	2.20	1.36	2.23	1.36
No. 4	0.98	1.36	0.96	1.36
No. 5	0.50	1.36	1.25	1.36
No. 6	0.89	1.36	0.37	1.36
No. 7	2.12	1.36	2.46	1.36
No. 8	0.42	1.36	0.37	1.36
No. 9	0.61	1.36	0.55	1.36

Designations colour in the table results of goodness-of-fit test (free speed): H_0: $F(v^{before}) = F(v^{after}) \rightarrow \lambda < \lambda_\alpha$ H_1: $F(v^{before}) \neq F(v^{after}) \rightarrow \lambda \geq \lambda_\alpha$. Designations colour in the table results of goodness-of-fit test (vehicle flow speed): H_0: $F(v^{pp\ before}) = F(v^{pp\ after}) \rightarrow \lambda < \lambda_\alpha$;H_1: $F(v^{pp\ before}) \neq F(v^{pp\ after}) \rightarrow \lambda \geq \lambda_\alpha$.

Table A3. Results of goodness-of-fit test on subsequent test sections (free speed v^{after} and vehicle flow speed $v^{pp\ after}$).

Test Section i and i+1	v^{after}_i and v^{after}_{i+1}		$v^{pp\ after}_i$ and $v^{pp\ after}_{i+1}$	
	λ	$\lambda_{\alpha},\ \alpha = 0.05$	λ	$\lambda_{\alpha},\ \alpha = 0.05$
No.1 and No. 2	1.04	1.36	2.76	1.36
No. 2 and No. 3	0.99	1.36	1.19	1.36
No. 3 and No. 4	1.46	1.36	1.66	1.36
No. 4 and No. 5	1.32	1.36	1.02	1.36
No. 5 and No. 6	1.49	1.36	0.90	1.36
No. 6 and No. 7	1.05	1.36	1.19	1.36
No. 7 and No. 8	4.48	1.36	4.86	1.36
No. 8 and No. 9	3.72	1.36	3.61	1.36

Designations colour in the table results of goodness-of-fit test (free speed): H_0: $F(v^{after}{}_i) = F(v^{after}{}_{i+1}) \rightarrow \lambda < \lambda_\alpha$; H_1 : $F(v^{after}{}_i) \neq F(v^{after}{}_{i+1}) \rightarrow \lambda \geq \lambda_\alpha$. Designations colour in the table results of goodness-of-fit test (vehicle flow speed): H_0: $F(v^{pp\ after}{}_i) = F(v^{pp\ after}{}_{i+1}) \rightarrow \lambda < \lambda_\alpha$; H_1: $F(v^{pp\ after}{}_i) \neq F(v^{pp\ after}{}_{i+1}) \rightarrow \lambda \geq \lambda_\alpha$.

Appendix C

Table A4. Results of test of independence on test sections (free speed v and vehicle flow speed v^{pp}).

Test Section	v^{before} and v^{after}		$v^{pp\ before}$ and $v^{pp\ after}$	
	χ^2	χ_α^2, $\alpha = 0.05$	χ^2	χ_α^2, $\alpha = 0.05$
No. 1	25.7	3.84	28.5	3.84
No. 2	15.6	3.84	16.8	3.84
No. 3	9.7	3.84	14.3	3.84
No. 4	4.3	3.84	4.2	3.84
No. 5	1.00	3.84	7.6	3.84
No. 6	0.98	3.84	0.8	3.84
No. 7	3.80	3.84	6.1	3.84
No. 8	0.04	3.84	1.1	3.84
No. 9	0.03	3.84	0.08	3.84

Designations colour in the table results of test of independence (free speed): H_0: $P\{X = v^{before}{}_i, Y = v^{after}{}_i\} = P\{X = v^{before}{}_i\}, P\{Y = v^{after}{}_i\} \rightarrow \chi^2 < \chi_\alpha{}^2$; H_1: $P\{X = v^{before}{}_i, Y = v^{after}{}_i\} \neq P\{X = v^{before}{}_i\}, P\{Y = v^{after}{}_i\} \rightarrow \chi^2 \geq \chi_\alpha{}^2$. Designations colour in the table results of test of independence (vehicle flow speed): H_0: $P\{X = v^{pp\ before}{}_i, Y = v^{pp\ after}{}_i\} = P\{X = v^{pp\ before}{}_i\}, P\{Y = v^{pp\ after}{}_i\} \rightarrow \chi^2 < \chi_\alpha{}^2$; H_1: $P\{X = v^{pp\ before}{}_i, Y = v^{pp\ after}{}_i\} \neq P\{X = v^{pp\ before}{}_i\}, P\{Y = v^{pp\ after}{}_i\} \rightarrow \chi^2 \geq \chi_\alpha{}^2$.

Table A5. Results of test of independence on subsequent test sections (free speed v^{after} and vehicle flow speed $v^{pp\ after}$).

Test Section i and i+1	$v^{after}{}_i$ and $v^{after}{}_{i+1}$		$v^{pp\ after}{}_i$ and $v^{pp\ after}{}_{i+1}$	
	χ^2	χ_α^2, $\alpha = 0.05$	χ^2	χ_α^2, $\alpha = 0.05$
No.1 and No. 2	4.64	3.84	17.8	3.84
No. 2 and No. 3	7.23	3.84	12.3	3.84
No. 3 and No. 4	7.87	3.84	12.9	3.84
No. 4 and No. 5	2.66	3.84	0.34	3.84
No. 5 and No. 6	3.39	3.84	1.19	3.84
No. 6 and No. 7	0.11	3.84	0.00	3.84
No. 7 and No. 8	76.44	3.84	84.0	3.84
No. 8 and No. 9	52.46	3.84	37.4	3.84

Designations colour in the table results of test of independence (free speed): H_0: $P\{X = v^{after}{}_i, Y = v^{after}{}_{i+1}\} = P\{X = v^{after}{}_i\}, P\{Y = v^{after} y_{i+1}\} \rightarrow \chi^2 < \chi_\alpha{}^2$; H_1: $P\{X = v^{after}{}_i, Y = v^{after}{}_{i+1}\} \neq P\{X = v^{after}{}_i\}, P\{Y = v^{after}{}_{i+1}\} \rightarrow \chi^2 \geq \chi_\alpha{}^2$. Designations colour in the table results of test of independence (vehicle flow speed): H_0: $P\{X = v^{pp\ after}{}_i, Y = v^{pp\ after}{}_{i+1}\} = P\{X = v^{pp\ after}{}_i\}, P\{Y = v^{pp\ after} y_{i+1}\} \rightarrow \chi^2 < \chi_\alpha{}^2$;$H_1$: $P\{X = v^{pp\ after}{}_i, Y = v^{pp\ after}{}_{i+1}\} \neq P\{X = v^{pp\ after}{}_i\}, P\{Y = v^{pp\ after}{}_{i+1}\} \rightarrow \chi^2 \geq \chi_\alpha{}^2$.

Appendix D

Table A6. Results of median tests on test sections (free speed v and vehicle flow speed v^{pp}).

Test Section	v^{before} and v^{after}		$v^{pp\ before}$ and $v^{pp\ after}$	
	χ^2	χ_α^2, $\alpha = 0.05$	χ^2	χ_α^2, $\alpha = 0.05$
No. 1	21.6	3.84	178.2	3.84
No. 2	15.9	3.84	17.7	3.84
No. 3	12.8	3.84	16.3	3.84
No. 4	3.53	3.84	1.4	3.84
No. 5	7.95	3.84	5.1	3.84
No. 6	4.82	3.84	2.0	3.84
No. 7	16.06	3.84	24.0	3.84
No. 8	0.20	3.84	1.2	3.84
No. 9	0.36	3.84	0.3	3.84

Designations colour in the table results of median tests: H_0: $F_1(x) = F_2(x) \rightarrow \chi^2 < \chi_\alpha{}^2$; H_1: $F_1(x) \neq F_2(x) \rightarrow \chi^2 \geq \chi_\alpha{}^2$.

Table A7. Results of median tests on subsequent test sections (free speed v^{after} and vehicle flow speed $v_{pp\ after}$).

Test Section i and i+1	$v^{after}{}_i$ and $v^{after}{}_{i+1}$		$v^{pp\ after}{}_i$ and $v^{pp\ after}{}_{i+1}$	
	χ^2	$\chi_\alpha{}^2$, α = 0.05	χ^2	$\chi_\alpha{}^2$, α = 0.05
No.1 and No. 2	4.95	3.84	31.8	3.84
No. 2 and No. 3	6.31	3.84	3.5	3.84
No. 3 and No. 4	15.98	3.84	9.5	3.84
No. 4 and No. 5	16.58	3.84	0.97	3.84
No. 5 and No. 6	5.63	3.84	4.0	3.84
No. 6 and No. 7	4.90	3.84	13.4	3.84
No. 7 and No. 8	67.33	3.84	90.6	3.84
No. 8 and No. 9	41.10	3.84	37.4	3.84

Designations colour in the table results of median tests: H_0: $F_1(x) = F_2(x) \rightarrow \chi^2 < \chi_\alpha{}^2$; H_1: $F_1(x) \neq F_2(x) \rightarrow \chi^2 \geq \chi_\alpha{}^2$.

References

1. *Urban Traffic Areas—Part 7—Speed Reducers*; Vejdirektoratet-vejregeludvalget: Copenhagen, Denmark, 1991.
2. *Traffic Calming Guidelines*; Devon County Council Engineering & Planning Department: Devon, UK, 1992.
3. *Directives for the Design of Urban Roads. RASt 06*; Road and Transportation Research Association, Working Group Highway Design, FGSV: Köln, Germany, 2006.
4. Harvey, T. *A Review of Current Traffic Calming Techniques*; Universitat of Leeds: Leeds, UK, 2013.
5. Krystek, R. *Principles of Traffic Calming on the Roads of the Pomorskie Region in Poland, Part 1: Street Layouts in Towns and Cities*; GAMBIT Pomorski: Gdańsk, Poland, 2008.
6. World Health Organization (WHO). *Global Status Report on Road Safety 2010–2018*; World Health Organization: Geneva, Switzerland, 2018.
7. Proctor, S. Accident reduction through area-wide traffic schemes. *Traffic Eng. Control.* **1991**, *32*, 566–573.
8. Paige, M. *Speed and Road Traffic Noise*; UK Noise Association: Chatham, UK, 2009. Available online: http://www.ukna.org.uk/uploads/4/1/4/5/41458009/speed_and_road_traffic_noise.pdf (accessed on 1 July 2019).
9. Ellebjerg, L. *Noise Reduction in Urban Areas from Traffic and Driver Management—A Toolkit for City Authorities*; Silence: Brussels, Belgium, 2008.
10. Ellebjerg, L. *Noise Control through Traffic Flow Measures—Effect and Benefits*; Raport 151; Danish Road Institute: Hedehusene, Denmark, 2007.
11. King, R. Noise and Speed—A Guest Blog from UK Noise Association. 2019. Available online: http://www.20splenty.org/noise_and_speed (accessed on 18 May 2020).
12. Ghafghazi, G.; Hatzopoulou, M. Simulating the air quality impacts of traffic calming schemes in a dense urban neighbourhood. *Transp. Res. Part D Transp. Environ.* **2015**, *35*, 11–22. [CrossRef]
13. Lantieri, C.; Lamperti, R.; Simone, A.; Costa, M.; Vignali, V.; Sangiorgi, C.; Dondi, G. Gateway design assessment in the transition from high to low speed areas. *Transp. Res. Part F Traffic Psychol. Behav.* **2015**, *34*, 41–53. [CrossRef]
14. Babkov, V.F. *Road Design Parameters and the Safety of Traffic*; WKŁ: Warszawa, Poland, 1975.
15. Hobbs, F.D.; Richardson, B.D. *Traffic Engineering*; WKŁ: Warszawa, Poland, 1971.
16. Lunenfeld, H. *Evaluation of Traffic Operations, Safety, and Positive Guidance Project*; Federal Highway Administration, Office of Traffic Operations: Michigan, MI, USA, 1980.
17. Crevier, C. *Les Aménagements En Modération De La Circulation, Étude Et Applications*; École De Technologie Supérieure Université Du Québec: Montréal, QC, Canada, 2007.
18. Dixon, K.; Zhu, H.; Ogle, J.; Brooks, J.; Hein, C.; Aklluir, P.; Crisler, M. *Determining Effective Roadway Design Treatments for Transitioning from Rural Areas to Urban Areas on State Highways*; Final Report SPR 631; Federal Highway Administration: Washington, DC, USA, 2008.
19. *Roads Development Guide*; South Ayrshire Council: Strathclyde Roads, Ayrshire, UK, 1995.
20. Mackie, A.M.; Ward, H.A.; Walker, R.T. *Urban Safety Project, Part 3: Overall Evaluation of Area Wide Schemes*; TRRL Report 263; Transport and Road Research Laboratory: Crowthorne, UK, 1990.

21. Krystek, R. *Principles of Traffic Calming on the Roads of the Pomorskie Region in Poland, Part 2: Sections of Major Roads Through Towns and Villages*; GAMBIT Pomorski: Gdańsk, Poland, 2008.
22. Abdi, A.; Rad, H.B.; Azimi, E. Simulation and Analysis of Traffic Flow for Traffic Calming. In *Proceedings of the Institution of Civil Engineers—Municipal Engineer*; Thomas Telford Ltd.: London, UK, 2017; Volume 170, pp. 16–28. [CrossRef]
23. Sayer, I.A.; Parry, D.I. *Speed Control Using Chicanes—A trial at TRL*; TRL Project Report PR 102; Transport Research Laboratory: Crowthorne, UK, 1994.
24. Sayer, I.A.; Parry, D.I.; Barker, J.K. *Traffic Calming—An Assessment of Selected On-Road Chicane Schemes*; TRL Report 313; Transport Research Laboratory: Crowthorne, UK, 1998.
25. *Safe Road Design Manual*; Amendments to the WB Manual, Transport Rehabilitation Project ID PO75207; Consulting Services for Safe Road Design: Loan, Sweden, 2011.
26. Hernández, E.; Abadía, X.; París, A.C. *Criterios de Movilidad ZONAS 30*; Fundación RACC: Barcelona, Spain, 2007.
27. *Guidelines for Traffic Calming*; City of Sparks, Public Works, Traffic Division: Reno, NV, USA, 2007.
28. Berger, W.J.; Linauer, M. *Speed Reduction at City Limits by Using Raised Traffic Islands*; Institut fuer Verkehrswesen (Institute for Transport Studies), Universitaet fuer Bodenkultur A-1190: Vienna, Austria, 1998.
29. Prato, C.G.; Rasmussen, T.K.; Kaplan, S. Risk Factors Associated with Crash Severity on Low-Volume Rural Roads in Denmark. *J. Transp. Saf. Secur.* **2014**, *6*, 1–20. [CrossRef]
30. Vahl, H.G.; Giskes, J. *Traffic Calming through Integrated Urban Planning*; Amarcande: Paris, France, 1990.
31. *Local Transport Note 01/07*; Traffic Calming, Department for Transport, Department for Regional Development (Northern Ireland), Scottish Executive, Welsh Assembly Government: Scottish, UK, 2007.
32. Seneci, F.; Avesani, F.; Bonomi, I. *Piani particolareggiati per mobilita' ciclabile e pedonale e sicurezza stradale*; Comune di Bassano del Grappa: Verona, Italy, 2012.
33. Sadeghi-Bazargani, H.; Saadati, M. Speed Management Strategies; A Systematic Review. *Beat* **2016**, *4*, 126–133. [PubMed]
34. González, D.D. Evaluación de las Zonas 30 en Europa y definición de una Zona 30 revisada. Ph.D. Dissertation, Infraestructura del Transporte y del Territorio (ITT), Universitat Politècnica de Catalunya, Barcelona, Spain, 2012.
35. *Le Temps des Rues*; IREC, Federal Technical University of Lausanne: Lausanne, Switzerland, 1990.
36. Bahar, G.B. *Guidelines for the Design and Application of Speed Humps*; Institute of Transportation Engineers: Washington, DC, USA, 2007.
37. Hallmark, S.L.; Peterson, E.; Fitzsimmons, E.; Hawkins, N.; Resler, J.; Welch, T. *Evaluation of Gateway and Low-Cost Traffic-Calming Treatments for Major Routes in Small Rural Communities*; Institute for Transportation, Iowa State University: Iowa, IA, USA, 2007.
38. *City of Seattle Staff Directory, Streetscape Design Guidelines, Chapter 6*; City of Seattle Staff Directory: Seatle, DC, USA, 2020. Available online: http://www.seattle.gov/rowmanual/manual/6_5.asphttp://www.seattle.gov/rowmanual/manual/pdf/08/chapter6.pdf (accessed on 18 July 2020).
39. *Department of the Environment, Transport and the Regions, Traffic Advisory Leaflet 5/01*; Traffic Calming Bibliography; DETR: London, UK, 2001.
40. Jateikienė, L.; Andriejauskas, T.; Lingytė, I.; Jasiūnienė, V. Impact Assessment of Speed Calming Measures on Road Safety. *Transp. Res. Procedia* **2016**, *14*, 4228–4236. [CrossRef]
41. *Setting Local Speed Limits*; Department for Transport: London, UK, 2013.
42. Zalewski, A. *Traffic Calming as a Transport Engineering Problem*; Publishing House of the Technical University of Łódź: Łódź, Poland, 2011; Scholary Papers, issue No. 1104 habilitation monograph 414.
43. *Wirksamkeit geschwindigkeitsdämpfender Maßnahmen außerorts*; Dezernat Verkehrssicherheit und Verkehrstechnik, Hessisches Landesamt für Straßen—und Verkehrswesen: Hessen, Germany, 1997.
44. Akgol, K.; Gunay, B.; Aydin, M.M. Geometric optimisation of chicanes using driving simulator trajectory data, Transport ICE Publishing. In *Proceedings of the Institution of Civil Engineers—Transport*; Thomas Telford Ltd.: London, UK, 2019.
45. Distefano, N.; Leonardi, S. Effects of speed table, chicane and road narrowing on vehicle speeds in urban areas. In Proceedings of the VI International Symposium New Horizons 2017 of Transport and Communications, Sarajevo, Bosnia and Herzegovina, 17–18 November 2018.
46. Gonzalo-Ordena, H.; Pérez-Acebob, H.; Unamunzagaa, A.L.; Arcea, M.R. Effects of traffic calming measures in different urban areas. *Transp. Res. Procedia* **2018**, *33*, 83–90. [CrossRef]

47. FDM. *202-Speed Management*; FDOT Design Manual: New York, NY, USA, 2019. Available online: https://fdotwww.blob.core.windows.net/sitefinity/docs/default-source/roadway/fdm/2019/2019fdm202speedmgmt.pdf?sfvrsn=129ec9ff_4 (accessed on 8 July 2019).
48. Yassin, H.H. Livable city: An approach to pedestrianization through tactical urbanism. *Alex. Eng. J.* **2019**, *58*, 251–259. [CrossRef]
49. Afghari, A.P.; Haque, M.M.; Washington, S. Applying fractional split model to examine the effects of roadway geometric and traffic characteristics on speeding behawior. *Bus. Med. Eng. Traffic Inj. Prev.* **2018**. [CrossRef] [PubMed]
50. Ariën, C.; Brijs, K.; Brijs, T.; Ceulemans, W.; Vanroelen, G.; Jongen, E.; Daniels, S.; Wets, G. Does the effect of traffic calming measures endure over time? A simulator study on the influence of gates. *Geogr. Psychol. Transp. Res. Part Ftraffic Psychol. Behav.* **2013**. [CrossRef]
51. Theeuwes, J.; Horst, A.V.D.; Kuiken, M. Designing Safe Road Systems: A Human Factors Perspective. *Engineering* **2017**. [CrossRef]
52. Google Earth. Available online: http://www.earth.google.com (accessed on 1 September 2020).
53. Kacprzak, D.; Sołowczuk, A. Effectiveness of road chicanes in access zones to a village at 70 km/h speed limit. In Proceedings of the World Multidisciplinary Civil Engineering—Architecture—Urban Planning Symposium, WMCAUS 2018, Prague, Czech Republic, 18–22 June 2018; Volume 471, p. 062010, IOP Conf. Series: Materials Science and Engineering. [CrossRef]
54. *Technical Documentation of a Device for Measurement of Speed and Vehicle Composition of Traffic*; MART: Szczecin, Poland, 1998; (Funded as Part of the Instrumentation Purchasing Program of the Polish State Committee for Scientific Research (KBN), Decision No. 1829/IA/108/96, Application No. IA/926/96).
55. Qu'est-ce que l'angle mort? *Ornikar*. 2014. Available online: https://www.ornikar.com/permis/conseils-conduite/controle-visuel/angle-mort (accessed on 3 September 2018).
56. *Polish Highway Code*; Polish Official Journal of Laws: Warszawa, Poland, 2020.
57. *Detailed Requirements for the Design and Placement of Road Signs and Signals and Traffic Safety Devices*; Journal of Laws: Warszawa, Poland, 2003.
58. Greń, J. *Mathematical Statistics*; Models and Problems; PWN: Warszawa, Polnad, 1982.
59. Taylor, J.R. *An Introduction to Error Analysis, The Study of Uncertainties in Physical Measurements*, 2nd ed.; University Science Books Sausalito: California, CA, USA, 1997.
60. Kacprzak, D.; Sołowczuk, A. Synergy Effect of Speed Management and Development of Road Vicinity in Wrzosowo. In Proceedings of the World Multidisciplinary Civil Engineering—Architecture—Urban Planning Symposium, WMCAUS 2019, Prague, Czech Republic, 17–21 June 2019. [CrossRef]
61. Shahar, A.; Brémond, R.; Villa, C. Can light emitting diode-based road studs improve vehicle control in curves at night? A driving simulator study. *Sage J.* **2016**, *50*, 266–281. [CrossRef]
62. Jägerbrand, A.K.; Sjöbergh, J. Speed responses of trucks to light and weather conditions. *J. Cogent Eng.* **2019**, *6*. [CrossRef]
63. Jägerbrand, A.K.; Johansson, M.; Laike, T. Speed Responses to Speed Humps as Affected by Time of Day and Light Conditions on a Residential Road with Light-Emitting Diode (LED) Road Lighting. *Safety* **2018**, *4*, 10. [CrossRef]
64. Jun, W.; Ha, J.; Jeon, B.; Lee, J.; Jeong, H. LED traffic sign detection with the fast radial symmetric transform and symmetric shape detection. *IEEE Xplore* **2015**. [CrossRef]

energies

MDPI

Article

The Economic Effects of Electromobility in Sustainable Urban Public Transport

Oliwia Pietrzak * and Krystian Pietrzak *

Faculty of Engineering and Economics of Transport, Maritime University of Szczecin, 1-2 Wały Chrobrego St., 70-500 Szczecin, Poland
* Correspondence: o.pietrzak@am.szczecin.pl (O.P.); k.pietrzak@am.szczecin.pl (K.P.)

Abstract: This paper focuses on effects of implementing zero-emission buses in public transport fleets in urban areas in the context of electromobility assumptions. It fills the literature gap in the area of research on the impact of the energy mix of a given country on the issues raised in this article. The main purpose of this paper is to identify and analyse economic effects of implementing zero-emission buses in public transport in cities. The research area was the city of Szczecin, Poland. The research study was completed using the following research methods: literature review, document analysis (legal acts and internal documents), case study, ratio analysis, and comparative analysis of selected variants (investment variant and base variant). The conducted research study has shown that economic benefits resulting from implementing zero-emission buses in an urban transport fleet are limited by the current energy mix structure of the given country. An unfavourable energy mix may lead to increased emissions of SO_2 and CO_2 resulting from operation of this kind of vehicle. Therefore, achieving full effects in the field of electromobility in the given country depends on taking concurrent actions in order to diversify the power generation sources, and in particular on increasing the share of Renewable Energy Sources (RES).

Keywords: electromobility; emission costs; electric vehicles; zero-emission buses; air pollution; public transport; sustainable transport; transport management; renewable energy sources

Citation: Pietrzak, O.; Pietrzak, K. The Economic Effects of Electromobility in Sustainable Urban Public Transport. *Energies* **2021**, *14*, 878. https://doi.org/10.3390/en14040878

Academic Editor: Elżbieta Macioszek
Received: 5 January 2021
Accepted: 4 February 2021
Published: 8 February 2021

1. Introduction

Electromobility is a vital area addressed in the economic and environmental context by individual countries worldwide. In recent years, it has gradually been gaining importance also in the area of transport, in particular road transport. The issues addressed in relation to road transport pertain to climate change mitigation, local air pollution reduction, and decreasing the dependence on foreign oil [1]. This results from the fact that the transport system is one of the biggest sources of global greenhouse gas (GHG) emissions [2,3] as well as ensuing climate changes and illnesses [4]. Transport stands for nearly 25% of Europe's GHG emissions and is the main source of air pollution in urban areas [5]. What is important is that, out of all modes of transport, road carriage is the biggest emitter of pollutants, accounting for over 70% of all GHG emissions generated by transportation systems in Europe [6].

In connection with the negative impacts of air pollution on human life and health as well as the whole ecosystem, emissions of air pollutants from transport need to be drastically reduced without delay [5]. In answer to the challenge, Europe has proposed a shift to low-emission mobility. Pursuant to "A European Strategy for Low-Emission Mobility" (ESLEM) of 2016, three areas of actions have been distinguished in relation to transport: optimising and improving efficiency of the transportation system, increasing the use of low-emission energy, and the share of zero-emission vehicles in the total number of vehicles [5]. It should be noted that important elements of those measures include adoption of zero-emission technologies in relation to city buses. The document also underlines that public procurement undertaken by local authorities constitutes an important tool to create markets for innovative solutions, and it should be applied to create the demand for zero-emission vehicles [5].

Cities are still the places of the increasing concentration of the global population [7], rising numbers of road vehicles, and consequently increased air pollution and ensuing negative impacts on inhabitants' health [8]. In view of such risks, urban public transport—in particular urban public bus transport—constitutes an important area for implementing and developing electromobility in transport. According to the data provided in the literature, an electric bus reduces petroleum consumption by 85–87% compared to a diesel bus and achieves a 32–46% reduction in fossil fuel use as well as a 19–35% reduction in CO_2 emissions, from a life-cycle perspective. A cleaner power grid and an increase in system charging efficiency would enhance the future benefits resulting from implementation of electric buses [9] in urban fleets. It is also important that urban buses are identified as a prioritised group of vehicles for electrification, as they have regular routes, so they can utilise smaller batteries and it is easier to plan charging infrastructures for them [10].

In order to promote a wider use of electric vehicles, not only by private, but to a large extent also by public users, various kinds of programmes and legal regulations are adopted on both international and national level. In Poland, the key legal act in that respect is the Electromobility Development Program [11] and the resulting Act of 11 January 2018 on Electromobility and Alternative Fuels (AEAF) [12]. The main goal of the Act is to set up a legal framework for development of electromobility and making use of other alternative fuels in national transport systems. Moreover, and particularly important for the purpose of this paper, AEAF identifies the detailed obligations of local governments and conditions (type, time horizon, quantity) for implementing zero-emission buses to any urban public transport fleet.

In view of the above, the topic addressed in this paper is both current and important. The main purpose assumed for this paper was to identify and analyse economic effects of implementing zero-emission buses in urban public transport, based on the example of the city of Szczecin, Poland. The study assumed the number and type of zero-emission vehicles in accordance with the provisions specified in AEAF [12]. The area of research was the city of Szczecin, Poland. The research was conducted based on the source data as of the end of 2018, projected up to 2035.

In view of the purpose of the study, three research hypotheses were adopted:

Hypothesis 1 (H1). *Implementation of zero-emission buses is an important tool to reduce external costs generated by urban public transport fleets.*

Hypothesis 2 (H2). *Economic benefits resulting from implementing zero-emission buses in urban public transport in Szczecin are limited by the current energy mix structure in Poland.*

Hypothesis 3 (H3). *Achieving full effects of electromobility in Poland as a result of implementing zero-emission buses in urban transport fleets depends on taking concurrent actions aimed at diversifying the sources of power generation in Poland (changing the energy mix), including in particular a wider use of Renewable Energy Sources (RES).*

The rest of this article is divided in a specific order. Section 2 focuses on the literature review, addressing issues of sustainable urban transport, pollution, electromobility in transport, and transport solutions aimed at shifting to low-emission mobility. Section 3 presents the individual stages of the research process; it also specifies the applied methods and data sources. Section 4 contains characteristics of the research area—the city of Szczecin, with a particular focus on its public transport system. Section 5 presents the research results regarding the financial and economic effects of implementing zero-emission vehicles in the public bus fleet in Szczecin. The calculations were made taking into account the legal requirements imposed on Polish cities by AEAF, assumptions presented in "Blue Book. Public transport sector in cities, agglomerations and regions" of 2015 (BB2015) [13], the current energy mix in Poland, the data regarding the public bus fleet structure in Szczecin, and the results of research studies completed by Pietrzak and Pietrzak [14] regarding volumes of pollutant emissions generated by the public transport in Szczecin. Section 6 presents a discussion of the findings obtained in the research process. The last section contains conclusions that summarise the findings.

The research results presented in this paper are part of a wider scope of research studies carried out by the authors with regard to effects of implementing electromobility in urban public transport in Polish cities. The calculations of the projected emissions of NHMC/NMVOC, NO_x, PM, SO_2, and CO_2 in Szczecin in the years 2021–2035, being relevant information for the research process carried out for the purposes of this paper, were made by Pietrzak and Pietrzak [14].

2. Literature Review

Urbanisation is one of the most significant global change processes [15]. The number of people living in urban areas has been rising, and the urban population is forecast to continue growing. It is estimated that in 2018 over 55% of the world population lived in cities, and the percentage is expected to rise to 60% by 2030 [16]. The phenomenon results in increased traffic flows in urban areas, generated by both passenger and freight transport. The increase in the number of vehicles in urban areas leads to negatives impacts for both people and the environment.

According to Tang et al. [17], traffic is one of the most important air pollution sources in urban areas. Traffic congestion, along with its harmful impacts on people's health, the economy, and the environment, constitutes a major problem faced by metropolitan areas [18]. Provision of transport services is connected with considerable emissions of harmful substances which include NO_x, CO, CO_2, PM_{10}, $PM_{2.5}$, SO_2 [19–23]. Numerous studies point to a relationship between traffic air pollution and various diseases, such as asthma, COPD/chronic bronchitis, cardiovascular disease, and acute changes in blood pressure [24–27].

Moreover, the fast growth of demand for transport services, in terms of both passenger and cargo flows, has a direct effect on the number of road accidents and their outcomes [28–30], increased noise and vibrations [31,32], uncontrolled land consumption [33,34], and residents' life quality. A major problem in that respect is that internal combustion engine vehicles (ICEVs) still dominate the transport services market, and transport remains very dependent on oil—oil-derived fuels account for 95% of energy consumption in transport [35].

Therefore, policy makers must develop feasible strategies to reduce GHG emissions [2], and it is not surprising that development of electromobility is a priority in transport policies of many European countries [36]. Electromobility, as a viable alternative for conventional vehicles, has been gaining importance throughout the EU [37]. According to Sarigiannis et al. [38] promoting "green transport" in cities via, inter alia, making use of electric vehicles, may provide considerable monetary savings resulting from decreased exposure to pollutants such as PM_{10}, $PM_{2.5}$, NO_2. In its broad sense, electromobility becomes an important element in building sustainable transport systems in urban areas. It may be considered one of the vital tools to be used in measures taken in order to reduce the negative impacts of transport on the environment [39]. The significance of electromobility is shown by the EU documents. These include: ESLEM, the White Paper 2011 "Roadmap to a Single European Transport Area—Towards a competitive and resource efficient transport system" (WP2011), and the Green Paper—Towards a new culture for urban mobility of 2007 (GP2007).

The document that specifies the principles of implementing electromobility in Poland is AEAF. AEAF is a national legislative instrument regulating the area of changes proposed by the EU, aimed at reducing the negative environmental impacts of road transport. It is aimed at stimulating the development of electromobility and wider use of alternative fuels in the Polish transport sector. AEAF specifies, inter alia, the principles of development and functioning of the infrastructure for making use of alternative fuels in transport, indicates the duties of public entities with regard to developing alternative fuel infrastructures or clean transport area functioning.

AEAF specifically addresses services connected with public transport. The document stipulates that Local Government Units (LGUs) are obliged to provide public transport services or outsource them to an entity with a fleet where the share of zero-emission buses operated within the given LGU is at least 30%. As the process of purchasing zero-emission vehicles may be costly and time-consuming, the provisions of AEAF stipulate that increasing the share of zero-emission vehicles in the fleet will be phased. The percentage of such vehicles in the fleet of individual LGUs should be [12]:

- 5% from the beginning of the year 2021,
- 10% from the beginning of the year 2023,
- 20% from the beginning of the year 2025,
- 30% from the beginning of the year 2028.

In the context of the studies completed as part of the research project, it is important that AEAF explicitly defines the concept of "zero-emission bus". Pursuant to its provisions, a zero-emission bus is a vehicle powered exclusively by electric power produced in hydrogen fuel cells installed in the vehicle or exclusively by an engine whose operating cycle does not lead to emissions of any greenhouse gases or any other substances covered by the GHG emissions management system. Pursuant to AEAF, the definition of "zero-emission bus" excludes CNG/LNG and hybrid vehicles.

According to Melkonyan et al. [40], urbanisation trends and the increasing demands of urban mobility create new challenges for urban planners. Therefore, it is necessary to develop integrated and sustainable urban mobility policies. Transportation management strategies aimed at reduction of air pollution, according to Pinto et al. [41], may contribute to building sustainable cities in the future.

Environmental sustainability is a requirement for modern urban freight transport systems [42], as well as modern urban public transport systems. In view of the above, in urban areas it is possible to notice that numerous, innovative solutions are introduced successively, which are aimed at providing mobility while accounting for the economy, environment, and human life. The solutions may also pertain to vehicles, infrastructure, or changes in habits of transport system users.

In the case of Sustainable Urban Freight Transport (SUFT), the main idea is to reduce freight traffic in the city centre. A popular solution in that respect is organising local Urban Consolidation Centres (UCCs) [43–45], located outside city centres, from which deliveries may be made via, among other things: Electric Freight Vehicles (EFVs), Light Electric Freight Vehicles (LEFVs) [46–50], Small Sized Electric Vehicles (SEVs) [51], or e-cargo bikes [52]. The tasks completed by these vehicles may also be supported by a vehicle category called Non-Motorised Transport (NMT), including cargo bikes/cargo cycles [53,54], cycle rickshaws, or handcarts [55]. Even though in the literature there are studies analysing the possibility of using rail transport to handle cargo flows [56–58], in practice its use is still significantly limited. This results mainly from the specific features of this mode of transport. The use of trams in serving urban cargo flows is also limited [59,60].

As for Sustainable Urban Public Transport (SUPT), it is also necessary to mention all the organisational and infrastructural measures aimed at limiting individual mobility in favour of public transport. Such solutions include, e.g., Park&Ride car parks [61–65] or Bike&Ride facilities [66–68]. The Bike&Ride solution is very often supported by municipalities that set up a public system of bike rentals as part of a Bike-Sharing System (BSS) [69–72]; it is more and more often also supplemented by an Electric Scooter Sharing system [73,74]. Municipalities also actively engage in promoting public transport, e.g., via providing Free Fare Public Transport (FFPT) [75,76] and taking organisational measures to favour public transport, such as e.g., dedicated lanes for buses, trams, or light rail [77,78] or designating Low Emission Zones (LEZs) [79,80] and Zero Emission Zones (ZEZs) in city centres [81]. Research studies show that significant positive changes regarding environmental protection in urbanised areas may also be provided by ecodriving [82–85] as well as appropriate organisation and multimodal integration of passenger transport, including the special role to be played by rail transport [86,87].

The European transport policy assumes that transport systems that are organised should respect the sustainability principles. Shifting to a more sustainable mobility system is perceived as the main challenge for the decades to come, if we want to avoid or at least mitigate the harm done by transport [88]. According to Dyr et al. [89], applying alternative fuels on a wider scale may be the fundamental instrument of that policy.

In the case of passenger transport, an important measure aimed at mitigating negative effects of transport and reducing the consumption of fossil fuels is replacing the currently used public transport diesel buses with alternative vehicles. These include hybrid buses, Compressed Natural Gas (CNG) and Liquefied Natural Gas (LNG) buses, Fuel Cell Vehicles (FCV), and electric buses.

Hybrid vehicles may be classified as ICEVs which in order to reduce pollutant emissions are additionally equipped with a supporting electric engine. As shown in the studies carried out by Tzeng et al. [90], vehicles of this type may constitute a significant alternative to conventional buses, particularly for the time of improving the electric vehicle technologies. It is worth noting some research studies regarding the potential possibilities of reducing harmful substances emissions via providing fleets with this type of vehicles. The results of such studies are presented, e.g., in Xu et al. [91], Lijewski et al. [92], Hallmark et al. [93], and Pawełczyk & Szumska [94].

Other types of vehicles that form an alternative to diesel buses are CNG and LNG vehicles. The benefits and limitations of using CNG buses are shown in the studies completed by, inter alia, Milojevic et al. [95], Ivković et al. [96], Merkisz et al. [97], Yue et al. [98], and Tica et al. [99]. The study done by Dyr et al. [100] shows interesting results of the analysis of costs and benefits of using CNG buses in public transport, whereas Jurkovič et al. [101] analysed the possibility of decreasing the chosen environmental indicators by introducing LNG buses into bus transport.

Promising results with regard to reducing negative effects of transport may also be provided by using FCVs. In this case, it is worth consulting the studies performed by Zhang et al. [102], Hua et al. [103], Lee et al. [104], or Langford & Cherry [105]. Relevant data for studying this issue may also be found in the results of technical reports published by Eudy and Post [106–108].

From the point of view of the main subject of this paper, its goals and the variants being the object of detailed analysis, it is electric buses that may be an exceptionally important solution and a viable alternative to diesel buses. The literature provides a number of studies related to application of electric buses to handle passenger transport in cities. The issues related to electric vehicle operation were addressed by, inter alia, Mahmoud et al. [109], Quarles et al. [110], Kivekas et al. [111], or Logan et al. [112].

Based on the studies completed in several Swedish cities, Borén [113] estimated the social costs savings resulting from operating electric buses in the cities. In view of the limited driving range of electric buses, Kunith et al. [114] analysed various ways of battery charging, indicating advantages and disadvantages of each of the described solutions. In relation to the problem, Csiszár et al. [115] pointed out that convenient locations for EV charging were P&R areas, whereas Chao and Xiaohong [116] presented a model to optimise the electric bus scheduling. Problems related to optimal charging of bus batteries were also studied by Czerepicki et al. [117]. With regard to electric buses operation, it is also important to analyse battery longevity and battery use optimisation. The ones worth mentioning include papers published by Astaneh et al. [118] or Logan et al. [119].

In the context of analysing how zero-emission buses operate, it is also worth considering trolleybuses which have long been functioning in urban areas. Studies focusing on this type of buses were completed by, inter alia, Zavada et al. [120], Wołek et al. [121,122], or Bartłomiejczyk & Połom [123]. Interesting research results pertaining to operation of this type of vehicle were also obtained by Dziubinski et al. [124], Grygar et al. [125], and Jakubas et al. [126].

Other valuable studies include comparative analyses of effects of operating various types of buses. These include studies carried out by Correa et al. [127,128], Stempien and Chan [129], Misanovic et al. [130], Imam et al. [131], and Tzeng et al. [90]. The last of the studies is also a very valuable, concise description of each type of city bus available on the market.

Even though there are numerous studies regarding operation of buses being an alternative to diesel buses, they often ignore the impact of the current energy mix of a given country on the assumed financial and economic results of implementing the alternative solution. The studies completed as part of the research project described in this paper have shown that the energy mix of a given country may significantly limit the economic benefits resulting from purchasing and operation of electric buses. Moreover, full achievement of assumed electromobility goals depends on taking concurrent actions to diversify the sources of power generation in Poland (changing the current energy mix), including in particular increasing the share of Renewable Energy Sources (RES). Therefore, this article fills the existing literature gap.

3. Methods and Study Stages

The research process was carried out in several subsequent stages, where each of them took into account the results of the previous ones. The graphical presentation of the research process completed for the purposes of this paper is presented in Figure 1.

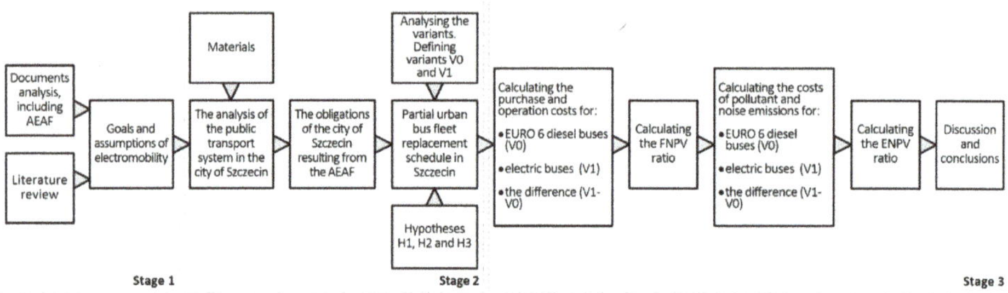

Figure 1. Stages of the research study.

Stage 1 consisted in analysing the literature and legal acts specifying the research background in the electromobility area and the assumptions of the electromobility policy implemented on the international and national level. On completion of this stage, we specified the goals and assumptions of the electromobility policy in relation to the public transport in Poland.

Stage 2 included the analysis of the public transport system in the city of Szczecin, and the analysis of alternative solutions available on the market, which could be used instead of diesel buses: hybrid buses, CNG/LPG buses, hydrogen buses, electric buses. The outcome of this stage was defining the fleet renewing variant (V0) and investment variant (V1), as well as a partial urban bus fleet replacement schedule in Szczecin.

Taking into account the outcomes of stages 1 and 2 as well as the study results obtained by Pietrzak and Pietrzak [14] with regard to the forecast pollution and noise levels in Szczecin for the years 2020–2035, Stage 3 focused on the financial and economic study of the effects of replacing the urban bus fleet with electric vehicles in the research area—the city of Szczecin. The analysis was completed in two steps. First, based on the costs of purchase and operation of urban buses in Szczecin calculated for variant V0 and variant V1 respectively as well as the difference (V1 − V0), we computed the financial flows and then the Financial Net Present Value (FNPV) ratio. Next, based on the costs of pollution and noise emissions generated by urban buses in Szczecin calculated for variant V0 and variant V1 respectively as well as the difference (V1 − V0), we calculated the economic flows and then the Economic Net Present Value (ENPV) ratio. The last part of the research carried out as part of Stage 3 covered the discussion of the results and the formulation of the conclusions.

The case study, for the purposes of the research carried out to prepare this paper, covered the city of Szczecin, Poland. The materials obtained from the following sources constituted the source data for the study:

- Szczecin City Hall (SCH),
- Roads and Public Transport Authority in Szczecin (RPTAS),
- Bus transport operators providing public transport in Szczecin: Szczecińskie Przedsiębiorstwo Autobusowe "Klonowica" Sp. z o.o. (SPAK), Szczecińskie Przedsiębiorstwo Autobusowe "Dąbie" Sp. z o.o. (SPAD), Szczecińsko-Polickie Przedsiębiorstwo Komunikacyjne Sp. z o.o. (SPPK) and Przedsiębiorstwo Komunikacji Samochodowej w Szczecinie Sp. z o.o. (PKS)
- Tram transport operator providing public transport in Szczecin Tramwaje Szczecińskie Sp. z o.o. (TS).

The study was conducted based on the source data as of the end of 2018, projected up to the year 2035.

Moreover, conducting the research for the purposes of this paper required some knowledge about the EU electromobility policy and legal regulations on implementing electromobility in Poland. The legal regulations pertain in particular to: the kind of zero-emission vehicles, the share of zero-emission vehicles in the public transport fleet, and deadlines for meeting the individual objectives of electromobility. The sources of data in this respect were: AEAF, Electromobility Development Plan for Poland: "Energy for the Future" of 2017, BB2015, ESLEM, WP2011, and GP2007.

In view of the provisions of AEAF, in Section 5 of this paper we specify the number (prescribed by the law) of zero-emission buses that must be purchased by a LGU (i.e., the city of Szczecin), the time limits for purchasing the specific number of zero-emission vehicles by the city of Szczecin, and we select the purchase options.

Moreover, it should be noted that AEAF also stipulates how any LGU should justify the need for replacing a part of the public transport fleet with zero-emission vehicles. AEAF indicates the need for LGUs to carry out a financial analysis, estimation of environmental effects in connection with emissions harmful for the natural environment and human health, and an economic analysis that includes estimation of costs related to harmful substances emissions [12]. The computations made by the Authors for the research area in Section 5 of this paper comply with the AEAF requirements in that respect. Additionally, the volumes of pollutants emitted by internal combustion and zero-emission buses were calculated using the tool "Calculator of pollutant emissions and climate costs for public means of transport—Excel spreadsheet", unified for transport projects, developed and made available by the Centre for EU Transport Projects (CEUTP). The detailed results were published by Pietrzak and Pietrzak [14]. The results also constituted the source of data for the purposes of the financial and economic study of the effects of replacing the urban bus fleet with electric vehicles in the city of Szczecin, performed as part of this research project.

4. Characteristics of the Research Area

The area of research for the purposes of this paper was the city of Szczecin, Poland. Szczecin is located in the north-west of Poland. It is the biggest city in the West Pomerania region both in terms of the surface area and the population size. Szczecin is also the capital city of the West Pomerania Voivodeship and the centre of the Szczecin Metropolitan Area (SMA). According to the statistical data as at 31 December 2019, Szczecin took up a surface area of 301 km^2, and its population amounted to 401,907 [132].

In the context of the issues addressed in this paper, the object of the analysis was the urban public transport system in Szczecin, with a particular focus on bus transport. Public transport service within the city area (which falls within the municipality's scope of duties) is provided on the basis of the provisions of the Act of 16 December 2010 on Public Collective Transport (APCT) [133] and of the Act of 8 March 1990 on Local Self-Government

(ALSG) [134]. Pursuant to APCT, a municipality is responsible for organising its public collective transport and to this end it performs the three basic tasks: planning the transport development, organising the public collective transport, and managing the public collective transport [133]. On behalf of the Municipality of Szczecin, RPTAS (as an entity established for this purpose) acts in the capacity of the public transport organiser. Within the area of the city of Szczecin, public transport uses two kinds of vehicles: trams and buses. Public transport services within the city of Szczecin are provided on the basis of agreements concluded with RPTAS by the following five operators: TS, SPAK, SPAD, SPPK and PKS.

Moreover, rail connections are planned to be commissioned in 2022 as part of the Szczecin Metropolitan Railway (SMR) system [135]. Due to the well-developed rail network infrastructure within the city of Szczecin, in addition to connections between Szczecin and the adjacent localities being part of the SMA, SMR will also be used as an urban rail service that makes it possible for the city residents to move between individual districts of Szczecin. Thus, the public transport services in the SMA and the city of Szczecin itself will be extended to offer passengers the possibility of using rail to move about the city.

In view of the purpose of this paper, the public bus fleet operated in Szczecin was analysed. According to the 2018 data, the total number of vehicles operated by bus operators within the city of Szczecin was 296, of which 104 buses were used by SPAD, 102—by SPAK, 58—by SPPK and 32—by PKS. All the vehicles were internal combustion, compression-ignition engine vehicles using diesel fuel. An important aspect in the context of obtaining the research purpose was the analysis of the bus fleet structure in terms of the EURO pollution standards which define the permissible exhaust emissions limits for new vehicles. The detailed data in this respect are presented in Figure 2.

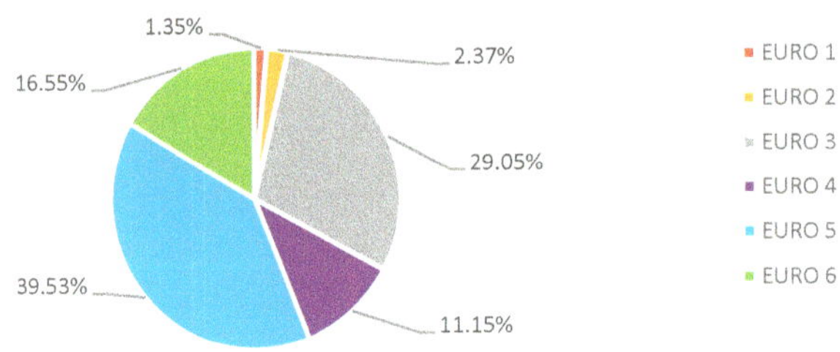

Figure 2. Breakdown of the public bus fleet in Szczecin in 2018 by European exhaust emission standard [%].

As shown by the data presented in Figure 2, the breakdown of public transport bus fleet operated in Szczecin is unfavourable in view of the EURO pollution standard. The analysed fleet consists of vehicles of all emission standard groups: from EURO 1 to EURO 6. Whereas the vehicles with the highest standard (i.e., EURO 6) accounted for merely ca. 16.5% of the fleet, the ones with the lowest emission standards (EURO 1, 2 and 3) constituted nearly 33% of all the bus fleet.

In the context of the purpose of this paper, it is also important to analyse the age structure of the bus fleet used in the public transport in Szczecin. The detailed data in this respect are presented in Figure 3. Based on the obtained data, the most numerous group (more than 32%) was the one where the buses were manufactured in the years 2009–2011, so they were aged 7–9 years in 2018. The buses from the two youngest groups—manufactured in the years 2012–2018, and thus aged 0–6 in 2018—constituted 24% of the fleet, the buses produced in 2003–2008 (aged 10–15 in 2018) accounted for slightly over 30% of the fleet, the buses dating back to 2000–2002 (16–18 years old)—ca. 10.5%, whereas the oldest vehicles which in 2018 were 19 and more years old accounted for 2.70% of the total fleet. The mean age of the buses used in public transport in Szczecin in 2018 was 10.11 years.

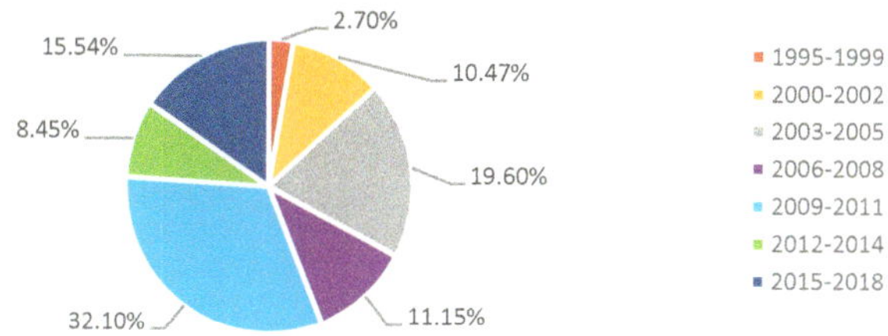

Figure 3. Breakdown of the public bus fleet in Szczecin in 2018 by year of manufacture [%].

Taking into account the assumptions of the BB2015 [13] that the expected operation period of diesel buses is 10 years, the Szczecin bus fleet age structure indicates that it will be necessary to gradually replace some of the vehicles used in the city's public transport system.

The structure of the bus fleet in question has a direct impact on the volume of pollutant emissions generated by the vehicles. Based on the data regarding the structure of the analysed bus fleet (type, EURO standard, age, size) and also the actual kilometrage data of each of the buses, the amount of pollutant emissions generated by the buses used in the public transport in Szczecin in 2018 was computed. The calculations were made using the tool "Calculator of pollutant emissions and climate costs for public means of transport—Excel spreadsheet", unified for transport projects, developed, and made available by the CEUTP. The results are presented in Table 1.

Table 1. Amounts of pollutant emissions generated by public bus fleet in Szczecin in 2018.

Number of Vehicles (pcs.)	Transport Activity in 2018 (Vehicle Kilometres Travelled)	Amounts of Pollutant Emissions in 2018 (Tonnes)				
		NMHC/NMVOC	NO_x	PM	SO_2	CO_2
296	17,420,335	37.76	215.19	3.269	0	21,214.37

5. Financial and Economic Study of the Effects of Replacing the Bus Fleet with Electric Vehicles in the City of Szczecin—Research Results

In view of the characteristics of the bus fleet operated in Szczecin (age and EURO standard), described in Section 4 of this paper, implementation of the investment process must be completed in stages so as to replace some of the vehicles with new ones. The need for partial replacement of the bus fleet in Szczecin is therefore compliant with the assumptions of AEAF which imposes an obligation to phase in zero-emission buses in public transport fleets.

In view of the purpose of this paper, the comparative analysis method was applied in order to compare effects of implementing two variants. The base (fleet renewing) variant of fleet replacement consisted in purchasing new diesel buses meeting the EURO 6 standard. For the purposes of the comparative analysis, the second variant was the investment variant which assumed making investments pursuant to the AEAF requirements. The application of the difference-based method made it possible to compare the financial and economic effects of both (fleet renewing and investment) variants. According to the Authors, comparing the two variants makes it possible to properly evaluate the economic effects resulting from implementing zero-emission buses in urban public transport in the research area.

When choosing the investment variant, four alternative solutions to diesel buses were considered: hybrid buses, CNG/LPG buses, hydrogen buses, electric buses. The choice was made based on the features of the particular solution (Table 2):

- technology—market experience (presence and prevalence of the technology on the national market),

- meeting the AEAF requirements (a particularly important feature as the legal act defines which vehicles are considered zero-emission vehicles in Poland).

Table 2. Selection of the investment variant.

	Prevalence of the Technology	Meeting AEAF Requirements	Recommendations—Variant
Hydrogen buses	in development	+	rejected
CNG/LPG buses	present on the market	–	rejected
Hybrid buses	present on the market	–	rejected
Electric buses	present on the market	+	investment (V1)

When choosing the investment variant (V1), the following options were rejected:

- Hydrogen buses—as this technology is still not very wide-spread, this is connected with a risk that the fleet replacement will not be completed by the time limits specified in AEAF,
- CNG/LPG buses—as AEAF does not list this type of vehicles in the zero-emission vehicle group,
- Hybrid buses—as AEAF does not list this type of vehicles in the zero-emission vehicle group.
- In view of the above, the following two variants were the objects of further comparative studies:
- fleet renewing variant (V0)—it assumed that some of the diesel buses operated in the city would be replaced with new diesel buses meeting the strictest standards for ICEVs, namely EURO 6,
- investment variant (V1)—it assumed that some of the diesel buses operated in the city would be replaced with new zero-emission buses.

Pursuant to the provisions of AEAF, the authors computed the required share of zero-emission buses in the public transport fleet in Szczecin. Based on the quantity of vehicles (296 as at the end of 2018) it was assumed that the target share of the zero-emission vehicles (in 2028) should be no less than 30% of that quantity, i.e., minimum 89 vehicles. As already described in Section 3 of this paper, pursuant to the schedule specified in AEAF, the zero-emission vehicles may be phased in. For the purposes of the study, it was therefore assumed that the 89 vehicles would be purchased in tranches as described below, so that in each of the years specified by AEAF the city of Szczecin could put into operation a specific number of zero-emission vehicles:

- in 2021: 15 zero-emission buses will be commissioned,
- in 2023: 15 zero-emission buses will be commissioned,
- in 2025: 30 zero-emission buses will be commissioned,
- in 2028: 29 zero-emission buses will be commissioned.

Moreover, for the purposes of the research study it was assumed that the purchase of the buses as well as any indispensable supporting infrastructure (plug-in chargers and pantograph chargers) would each time be completed in the years directly preceding the years of commissioning the respective tranches of buses. Thus, the purchase items: "purchase of vehicles" and "purchase of supporting infrastructure" were planned for 2020, 2022, 2024 and 2027.

The other assumptions made for the purposes of the research study were:

- The vehicles to be replaced will come from only two bus operators: SPAK and SPAD. This is mainly connected with the routes of the specific bus lines. The point was to enable zero-emission buses to serve the routes that run through the city centre and districts characterised by considerable population density and building coverage ratio.
- The two categories of vehicles will be subject to replacement: MAXI buses—which are 10.5–13 m long and MEGA buses—exceeding 13 m in length.

- Based on the actual mean annual kilometrage of buses operated by the two bus operators: SPAK and SPAD (as at 2018), the estimated annual kilometrage for each new bus was: 59,994 km for SPAK and 69,491 km for SPAD.
- Based on the declarations made by manufacturers of the vehicles and infrastructure, data and experience of the operators, governmental, scientific and advisory bodies, as well as based on the observations of tender proceedings results in Poland, the estimated costs of purchasing new vehicles and infrastructure were as follows:
    - diesel bus MAXI (12 m)—PLN 1100 k,
    - diesel bus MEGA (18 m)—PLN 1400 k,
    - electric bus MAXI (12 m)—PLN 2200 k,
    - electric bus MEGA (18 m)—PLN 3000 k,
    - plug-in charger—PLN 100 k,
    - pantograph charger—PLN 500 k,
    - battery—PLN 500 k.
- Based on the declarations made by the vehicle manufacturers, the operators' data and experience, also with regard to Polish cities already operating zero-emission buses (Warsaw, Krakow, Jaworzyna, Tarnów), the demand for energy was estimated as follows:
    - diesel bus MAXI (12 m)—37.5 L/100 km,
    - diesel bus MEGA (18 m)—47.2 L/100 km,
    - electric bus MAXI (12 m)—125 kWh/100 km,
    - electric bus MEGA (18 m)—150 kWh/100 km.
- The costs of purchase were estimated as follows: 1 litre of diesel fuel: PLN 4.24 net, 1 kWh of electric power: PLN 0.40 net.

As the research study was carried out for a selected Polish city, the calculations were made in the local currency, i.e., the Polish zloty (PLN). In the context of applying the research results to other European countries, it can be assumed that 1 PLN $\approx$ 0.22 euro.

Additionally, for the purposes of performing the financial and economic analysis, the following assumptions were made to account for the provisions of the BB2015 [13]:

- the analysis covers the 2020–2035 period,
- the analysis was based on fixed prices, without taking into account any inflation,
- the analysis was made in net prices, without taking into account any VAT,
- the analysis was carried out using the difference-based method, whereby the difference between the variant required pursuant to AEAF (V1—investment variant) and the base variant (V0—fleet renewing variant) was calculated.

As for the variable items of the operation costs, the differentiating costs were the costs of diesel fuel (for variant V0) and of electric power (for variant V1), whereas any other costs including tyre exchange, insurance or after-sale service were considered as ceteris paribus.

5.1. Financial Analysis

The first step taken as part of the financial study of effects of replacing the bus fleet with electric vehicles in the city of Szczecin was a comparative analysis of vehicle and supporting infrastructure purchase costs for variants V0 and V1 (Table 3). Based on the long-standing experience of bus operators as well as the assumptions resulting from the BB2015, it was assumed that the mean operation period of diesel buses was 10 years. Therefore, the analysis made for the years 2020–2035 for variant V0 also took into account the subsequent replacement purchases, namely in the years: 2030, 2032 and 2034. Due to lack of analogous data regarding the operation period of electric buses, the study was based on the data provided by the electric buses manufacturers and assumptions provided in the BB2015. It was assumed that the operation period of electric buses would be longer than in the case of diesel buses, and it may be similar to the assumed operation period for trolleybuses, i.e., up to 20 years. Therefore, in the analysis made for the years 2020–2035, in the case of variant V1 it was not necessary to take into account any subsequent replacement purchases of electric vehicles. However, other costs had to be taken into account; namely, it was necessary to consider the cost of plug-in and pantograph chargers. In the course of the operation period of the electric buses (after ca. 10 years) it is also necessary to purchase new batteries, which was included in the computations. This assumption was made on the basis of the first tender proceedings held in Poland with regard to a delivery of electric buses, where manufacturers offered a maximum 10-year guarantee for this piece of equipment.

Table 3. Comparing the costs of purchasing the vehicles and supporting infrastructure for variants V0 and V1.

	V0			V1						
	BUSES		Total Purchase Costs (k PLN)	BUSES		INFRASTRUCTURE			Total Purchase Costs (k PLN)	V1 − V0 (k PLN)
	MAXI (pc)	MEGA (pc)		MAXI (pc)	MEGA (pc)	Batteries (pc)	Plug-in Chargers (pc)	Pantograph Chargers (pc)		
2020	15	0	16,500	15	0		8	3	35,300	18,800
2021			0						0	0
2022	5	10	19,500	5	10		7	3	43,200	23,700
2023			0						0	0
2024	21	9	35,700	21	9		15	5	77,200	41,500
2025			0						0	0
2026			0						0	0
2027	19	10	34,900	19	10		15	4	75,300	40,400
2028			0						0	0
2029			0						0	0
2030	15	0	16,500			15			7500	−9000
2031			0						0	0
2032	5	10	19,500			15			7500	−12,000
2033			0						0	0
2034	21	9	35,700			30			15,000	−20,700
2035			0						0	0

While interpreting the results presented in the last column in Table 3, marked "V1 − V0", it must be noted that a negative value in a given year means that the investment costs incurred for variant V1 were lower than the costs incurred for variant V0. A positive value, in turn, means that the investment costs incurred for variant V1 exceeded the analogous costs incurred for variant V0.

The second step taken as part of the financial study of effects of replacing the bus fleet with electric vehicles in the city of Szczecin was a comparative analysis of the new vehicles operation costs for variants V0 and V1 (Table 4).

Table 4. Comparing the vehicle operation costs for variants V0 and V1.

	q-ty of Vehicle Kilometres (vkm)		V0	V1	V1 − V0 (k PLN)
	MAXI Buses (vkm)	MEGA Buses (vkm)	Operation Costs—Purchase of Fuel (k PLN)	Operation Costs—Purchase of Electric Power (k PLN)	
2020	0	0	0.00	0.00	0.00
2021	899,910	0	1430.86	449.96	−980.90
2022	899,910	0	1430.86	449.96	−980.90
2023	1,199,880	599,940	3108.46	959.90	−2148.55
2024	1,199,880	599,940	3108.46	959.90	−2148.55
2025	2,573,718	1,139,886	6373.44	1970.79	−4402.65
2026	2,573,718	1,139,886	6373.44	1970.79	−4402.65
2027	2,573,718	1,139,886	6373.44	1970.79	−4402.65
2028	3,837,065	1,834,796	9772.87	3019.41	−6753.46
2029	3,837,065	1,834,796	9772.87	3019.41	−6753.46
2030	3,837,065	1,834,796	9772.87	3019.41	−6753.46
2031	3,837,065	1,834,796	9772.87	3019.41	−6753.46
2032	3,837,065	1,834,796	9772.87	3019.41	−6753.46
2033	3,837,065	1,834,796	9772.87	3019.41	−6753.46
2034	3,837,065	1,834,796	9772.87	3019.41	−6753.46
2035	3,837,065	1,834,796	9772.87	3019.41	−6753.46

As per the adopted time schedule resulting from the AEAF provisions, operation of the vehicles would start in 2021 (following the purchase of the first tranche in 2020). The planned phased commissioning of subsequent vehicles in the following years was accounted for in the accruing amount of vehicle kilometres covered by the new vehicles. The computations also accounted for the different amounts of vehicle-kilometres for the bus operators: SPAK and SPAD.

While interpreting the results presented in the last column in Table 4, marked "V1 − V0", it must be noted that a negative value in a given year means that the vehicles operation costs incurred for variant V1 were lower than the costs incurred for variant V0. There was no instance of a positive value.

The comparative analysis of the purchase costs of the vehicles and the supporting infrastructure (Table 3) has shown that variant V1 requires more expenditure than the V0 (fleet renewing) variant. Should the city of Szczecin decide to implement variant V1, apart from purchasing the electric buses (whose market prices are much higher than those of diesel buses) it would also be required to purchase the infrastructure which is necessary for their operation. A different situation takes place in the case of the vehicle operation costs (Table 4); where variant V1 may bring specific financial benefits.

Comparing the rationale for purchasing the vehicles by the city of Szczecin as per the individual variants (V0 and V1), financial flows were calculated, which took into account the indicated investment costs (Table 3) and operation costs (Table 4) related to the vehicles. The annual financial flows for the years 2020–2035 are presented in Table 5.

Table 5. Annual financial flows for the years 2020–2035.

	Investment Costs V1 − V0 (k PLN)	Operation Costs V1 − V0 (k PLN)	Financial Flows (k PLN)
2020	18,800	0.00	−18,800.00
2021	0	−980.90	980.90
2022	23,700	−980.90	−22,719.10
2023	0	−2148.55	2148.55
2024	41,500	−2148.55	−39,351.45
2025	0	−4402.65	4402.65
2026	0	−4402.65	4402.65
2027	40,400	−4402.65	−35,997.35
2028	0	−6753.46	6753.46
2029	0	−6753.46	6753.46
2030	−9000	−6753.46	15,753.46
2031	0	−6753.46	6753.46
2032	−12,000	−6753.46	18,753.46
2033	0	−6753.46	6753.46
2034	−20,700	−6753.46	27,453.46
2035	0	−6753.46	6753.46

While interpreting the results shown in the last column of Table 5, marked "financial flows", it should be noted that the result for each year was computed on the basis of the cost difference (V1 − V0). A negative value in a given year means that the costs (investment + operation) incurred for variant V1 exceeded the analogous costs for variant V0. A positive value, in turn, means that the costs (investment + operation) incurred for variant V1 were lower than the costs incurred for variant V0.

The results presented in Table 5 show that in the analysed period the annual financial flows in most years are more favourable for the V1 investment variant. This is due to the considerably lower operation costs of electric vehicles compared to diesel buses. However, from the perspective of the full analysed period (2020–2035), it is possible to notice that variant V1 is less cost-effective compared to variant V0. This is due to the need to incur considerable investment costs connected with the higher unit prices of electric buses and the need to purchase the supporting infrastructure.

Based on the financial flows shown in Table 5, the Financial Net Present Value (FNPV) was calculated. Pursuant to the assumptions of the BB2015 [13], the discount rate of 4% was adopted in the calculations. The calculations were made using the following formula:

$$\text{FNPV} = \sum_{t=1}^{N} \frac{C_t}{(1+r)^t} - C_0 \tag{1}$$

where:

- C_t—cash flow in period t;
- r—discount rate;
- C_0—initial outlay.

$$\text{FNPV} = -30,764.58\ [\text{k PLN}].$$

It should be noted that a negative value of the FNPV, computed with the difference-based method using the financial flows, means that variant V1 is less cost-effective in comparison with variant V0. In accordance with the adopted methodology, a positive value of the FNPV means, in turn, that variant V1 is more cost-effective compared to variant V0. Thus, when interpreting the obtained FNPV value (−30,764.58 k PLN), it is possible to conclude that implementation of the investment according to variant V1, covering purchase and operation of electric buses in the city of Szczecin, will be less favourable in financial terms compared to variant V0 which covers purchase and operation of diesel buses meeting the EURO 6 standard.

It is possible to notice that as a result of electromobility implementation, in the future the difference in price between an electric bus and a diesel bus may decrease, which may be reflected in FNPV values. Such a situation may arise as a result of reduced production costs of electric buses as well as of increased diesel bus prices, due to e.g., the need to pay environmental fees by manufacturers of diesel buses. In the future it may also be possible that production of diesel buses will be limited or totally abandoned. Therefore, it would be necessary to conduct further studies in that respect.

Below, this article continues to describe an economic analysis which was performed to additionally take into account the social cost and benefit resulting from implementation of the analysed variants (V1 and V0).

5.2. Economic Analysis

Preparing an economic analysis as well as a financial analysis is required under AEAF. Economic effects of an investment in practice are not connected with any cash flows for the investor; they pertain to expected effects of the investment, which have not been accounted for in the financial analysis but may have an impact on the society, economy, and natural environment. The monetised nonmarket effects included in the analysis took into account emissions of: NHMC/NMVOC, NO_x, PM, SO_2, and CO_2 and also noise.

The calculation of the pollutant emissions costs (PEC) generated by the individual substances (NHMC/NMVOC, NO_x, PM, SO_2 and CO_2) for each of the analysed years was carried out as per the formula:

$$\text{PEC} = \text{UC} * \text{E} \tag{2}$$

where:

- UC—unit cost (PLN/tonne);
- E—emissions volume (tonne).

The following data were used in the computations:

- unit costs of individual substances (NHMC/NMVOC, NO_x, PM, SO_2, and CO_2) for each examined year were calculated according to CEUTP [136];
- volumes of emissions of the individual substances (NHMC/NMVOC, NO_x, PM, SO_2, and CO_2) for each examined year were based on Pietrzak and Pietrzak [14], where

the volumes of the individual substances' emissions were estimated for the analysed variants V0 and V1.

The noise emissions costs (NEC) for each of the examined years were computed as per the following formula:

$$NEC = UC * QVK \tag{3}$$

where:

- UC—unit cost (PLN/vkm);
- QVK—quantity of vehicle kilometres (vkm).

The following data were used in the computations:

- The noise-generated unit costs for each of the examined years were calculated according to CEUTP [136]
- The quantity of vehicle kilometres for each of the examined years was adopted on the basis of the results presented in Table 4.

The calculation results regarding the projected emissions costs of NHMC/NMVOC, NO_x, PM, SO_2, CO_2, and noise for the two variants (V1 and V0) are presented in Table 6 and Figures 4–9. The calculations have accounted for the commissioning of subsequent buses as required by AEAF, and thus for the rising quantity of vehicle kilometres in the subsequent years.

While interpreting the results presented in Table 6 in columns marked "V1 − V0", it must be noted that a negative value in a given year means that the costs of emissions of individual substances incurred for variant V1 were lower than the costs incurred in the case of variant V0. A positive value, in turn, means that the costs of emissions of individual substances incurred for variant V1 exceeded the analogous costs incurred for variant V0.

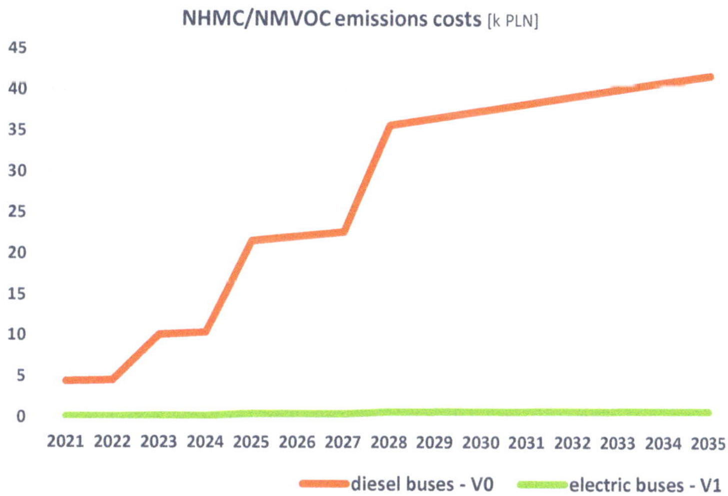

Figure 4. Projected amounts of NHMC/NMVOC emissions costs of the public bus fleet in Szczecin in the 2021–2035 period for variants V0 and V1.

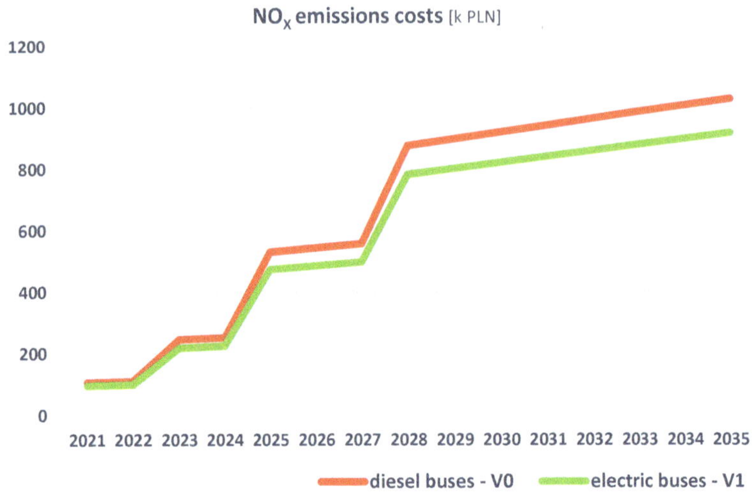

Figure 5. Projected amounts of NO_x emissions costs of the public bus fleet in Szczecin in the 2021–2035 period for variants V0 and V1.

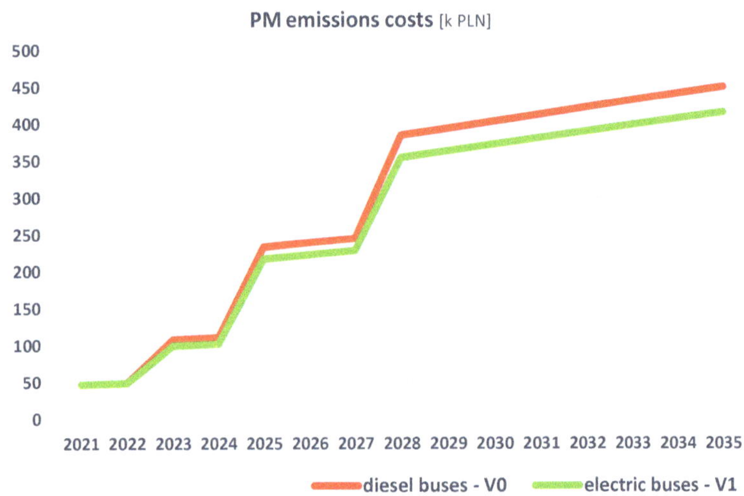

Figure 6. Projected amounts of PM emissions costs of the public bus fleet in Szczecin in the 2021–2035 period for variants V0 and V1.

Figure 7. Projected amounts of SO_2 emissions costs of the public bus fleet in Szczecin in the 2021–2035 period for variants V0 and V1.

Figure 8. Projected amounts of CO_2 emissions costs of the public bus fleet in Szczecin in the 2021–2035 period for variants V0 and V1.

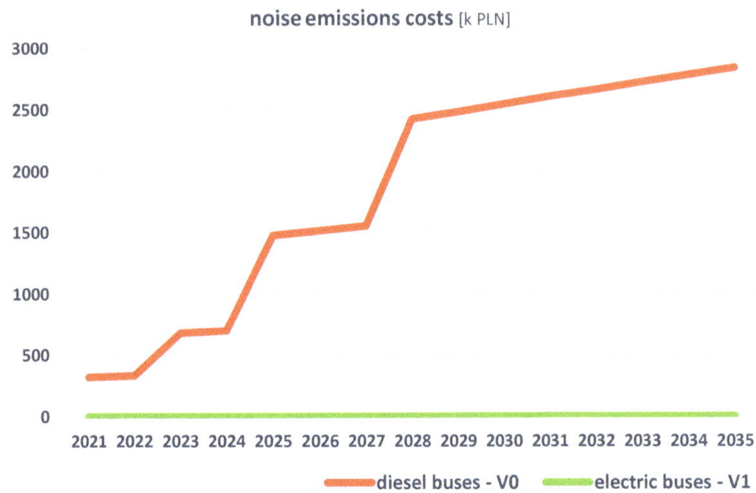

Figure 9. Projected amounts of noise emissions costs of the public bus fleet in Szczecin in the 2021–2035 period for variants V0 and V1.

In view of the obtained results, it should be stressed that choosing the V1 investment variant may bring measurable benefits—namely reduction of NMHC/NMVOC, NO_x, PM, and noise emissions—and thus decrease the costs of emissions generated by the said substances, which confirms Hypothesis H1. Importantly, choosing variant V1 for the city of Szczecin may concurrently lead to increased emissions of other harmful substances, i.e., CO_2 and SO_2, and consequently to increased costs of emissions of the said substances. According to the results of the research carried out for the city of Szczecin by Pietrzak and Pietrzak [14], purchase and operation of electric buses (variant V1) may be connected with:

- Increased emissions of CO_2 in variant V1 by more than 200 tonnes a year compared to variant V0 (starting from 2028, when—pursuant to the AEAF assumptions—the last tranche of zero-emission buses is to be phased in),
- Appearance of SO_2 emissions in variant V1, estimated to amount to nearly 20 tonnes a year (as from 2028).
- Increased emissions of CO_2 and appearance of SO_2 as a new harmful substance considerably decrease the assumed economic effects expected for investments in the area of electromobility. The reason for this phenomenon, which is unexpected in the context of a shift to zero-emission vehicles, may be the current balance of the energy mix in Poland, where electric power generation is dominated by coal-fired power stations. The structure of electric power generation in Poland in the years 2018 and 2019 is presented in Figure 10.

As shown in Figure 10, the share of coal in electric power generation in Poland was respectively: 78.2% in 2018 and 73.6% in 2019. In the same period, the share of RES in electric power production was as small as 12.70% and 15.40% respectively. These data unambiguously show that the energy mix in Poland is unfavourable.

In order to compute the ENPV for the analysed case study, the economic flows were calculated—the results are presented in Table 7. Similarly as in the case of the previous computations, the difference-based method was applied. The economic analysis was based on the data resulting from the financial analysis and the economic costs generated by NMHC/ NMVOC, NO_x, PM, SO_2, CO_2 and noise emissions. As for the financial analysis, market prices constitute an appropriate frame of reference for the purposes of evaluating the project financial results for both private and public investors. However, they are not appropriate when the goal is evaluation of the project's contribution to the benefit of the society. To that end, all market prices should be converted into shadow prices, which better reflect the benefit for the society. Market prices are converted into shadow prices by using Conversion Factors (CF). Pursuant to the BB2015 recommendations, adjustments were made and the following CFs were applied [13]:

- 0.83 for investment costs connected with purchasing the infrastructure,
- 0.87 for investment costs connected with purchasing the vehicles,
- 0.78 for investment costs connected with the vehicles' operation.

While interpreting the results shown in the last column of Table 7, marked "economic flows", it should be noted that the result for each year was calculated on the basis of the costs computed with the difference-based method (V1 − V0). A negative value in a given year means that the economic costs incurred for variant V1 exceeded the analogous costs incurred for variant V0. A positive value, in turn, means that the economic costs incurred for variant V1 were lower than the costs incurred for variant V0.

Table 6. Costs of NHMC/NMVOC, NOx, PM, SO_2, CO_2 and noise emissions for: V0, V1 and V1 − V0.

	NMHC/NMVOC			NO_x			PM		
	Costs of Emissions (k PLN)			Costs of Emissions (k PLN)			Costs of Emissions (k PLN)		
	V0	V1	V1 − V0	V0	V1	V1 − V0	V0	V1	V1 − V0
2020	0.00	0.00	0.00	0.00	0.00	0.00	0.00	0.00	0.00
2021	4.36	0.09	−4.28	107.22	96.90	−10.32	47.13	47.13	0.00
2022	4.49	0.09	−4.40	110.39	99.76	−10.63	48.53	48.53	0.00
2023	10.07	0.19	−9.88	245.95	219.08	−26.86	107.93	99.63	−8.30
2024	10.33	0.19	−10.13	252.31	224.75	−27.56	110.72	102.20	−8.52
2025	21.52	0.41	−21.11	530.99	474.44	−56.54	233.03	217.01	−16.02
2026	22.06	0.42	−21.65	544.45	486.47	−57.98	238.94	222.51	−16.43
2027	22.61	0.43	−22.18	557.90	498.49	−59.41	244.84	228.01	−16.83
2028	35.65	0.67	−34.98	877.14	782.96	−94.18	384.22	354.43	−29.80
2029	36.51	0.68	−35.83	898.36	801.90	−96.46	393.52	363.00	−30.52
2030	37.40	0.70	−36.70	920.22	821.41	−98.81	403.10	371.83	−31.26
2031	38.29	0.71	−37.57	942.02	840.87	−101.15	412.64	380.64	−32.00
2032	39.17	0.73	−38.44	963.72	860.24	−103.48	422.15	389.41	−32.74
2033	40.04	0.75	−39.30	985.28	879.48	−105.79	431.59	398.12	−33.47
2034	40.91	0.76	−40.15	1006.65	898.56	−108.09	440.96	406.76	−34.20
2035	41.77	0.78	−40.99	1027.79	917.43	−110.36	450.21	415.30	−34.91

	SO_2			CO_2			Noise		
	Costs of Emissions (k PLN)			Costs of Emissions (k PLN)			Costs of Emissions (k PLN)		
	V0	V1	V1 − V0	V0	V1	V1 − V0	V0	V1	V1 − V0
2020	0.00	0.00	0.00	0.00	0.00	0.00	0.00	0.00	0.00
2021	0.00	246.63	246.63	163.18	170.72	7.54	321.27	0.00	−321.27
2022	0.00	253.93	253.93	167.72	175.47	7.75	330.27	0.00	−330.27
2023	0.00	568.23	568.23	372.94	384.69	11.75	678.53	0.00	−678.53
2024	0.00	582.92	582.92	382.74	394.79	12.06	696.53	0.00	−696.53
2025	0.00	1227.49	1227.49	804.86	831.39	26.53	1474.30	0.00	−1474.30
2026	0.00	1258.61	1258.61	825.00	852.19	27.19	1511.44	0.00	−1511.44
2027	0.00	1289.70	1289.70	845.13	872.99	27.86	1548.57	0.00	−1548.57
2028	0.00	2025.04	2025.04	1326.68	1369.14	42.46	2421.88	0.00	−2421.88
2029	0.00	2074.03	2074.03	1357.55	1401.00	43.45	2478.60	0.00	−2478.60
2030	0.00	2124.50	2124.50	1388.42	1432.86	44.43	2540.99	0.00	−2540.99
2031	0.00	2174.83	2174.83	1419.24	1464.66	45.42	2603.38	0.00	−2603.38
2032	0.00	2224.93	2224.93	1450.11	1496.52	46.41	2660.10	0.00	−2660.10
2033	0.00	2274.70	2274.70	1480.98	1528.38	47.40	2722.49	0.00	−2722.49
2034	0.00	2324.04	2324.04	1511.79	1560.17	48.38	2779.21	0.00	−2779.21
2035	0.00	2372.84	2372.84	1542.67	1592.04	49.37	2835.93	0.00	−2835.93

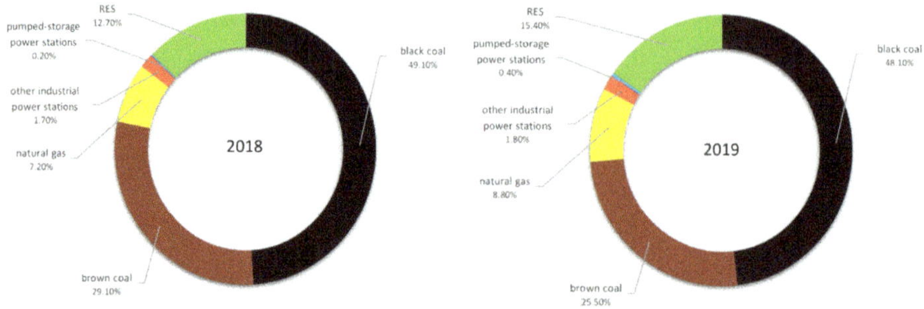

Figure 10. Breakdown of electricity production in Poland in the years 2018–2019 [%] (data from references [137,138]).

Table 7. Annual economic flows for the years 2020–2035.

	Adjusted Financial Costs (V1 − V0) (k PLN)	Costs of Emissions (V1 − V0) (k PLN)						Economic Flows (k PLN)
		NHMC	NO_x	PM	SO_2	CO_2	Noise	
2020	16,264.00	0.00	0.00	0.00	0.00	0.00	0.00	−16,264.00
2021	−765.10	−4.28	−10.32	0.00	246.63	7.81	−321.27	846.53
2022	19,765.90	−4.40	−10.63	0.00	253.93	8.86	−330.27	−19,683.38
2023	−1675.87	−9.88	−26.86	−8.30	568.23	11.75	−678.53	1819.47
2024	34,269.13	−10.13	−27.56	−8.52	582.92	12.06	−696.53	−34,121.37
2025	−3434.07	−21.11	−56.54	−16.02	1227.49	26.53	−1474.30	3748.03
2026	−3434.07	−21.65	−57.98	−16.43	1258.61	27.19	−1511.44	3755.75
2027	31,573.93	−22.18	−59.41	−16.83	1289.70	27.86	−1548.57	−31,244.49
2028	−5267.70	−34.98	−94.18	−29.80	2025.04	42.46	−2421.88	5781.05
2029	−5267.70	−35.83	−96.46	−30.52	2074.03	43.45	−2478.60	5791.63
2030	−13,397.70	−36.70	−98.81	−31.26	2124.50	44.43	−2540.99	13,936.53
2031	−5267.70	−37.57	−101.15	−32.00	2174.83	45.42	−2603.38	5821.56
2032	−16,007.70	−38.44	−103.48	−32.74	2224.93	46.41	−2660.10	16,571.12
2033	−5267.70	−39.30	−105.79	−33.47	2274.70	47.40	−2722.49	5846.66
2034	−23,876.70	−40.15	−108.09	−34.20	2324.04	48.38	−2779.21	24,465.92
2035	−5267.70	−40.99	−110.36	−34.91	2372.84	49.37	−2835.93	5867.69

Based on the economic flows shown in Table 7, the Economic Net Present Value (ENPV) was calculated. Pursuant to the assumptions of the BB2015 [13], the social discount rate (SDR) of 4.5% was adopted in the calculations. The calculations were made using the following formula:

$$\mathrm{ENPV} = \sum\nolimits_{t=1}^{N} \frac{C_t}{(1+r)^t} - C_0 \tag{4}$$

where:

- C_t—economic cash flow in period t;
- r—social discount rate;
- C_0—initial outlay.

$$\mathrm{ENPV} = -27{,}727.67\ [\mathrm{k\ PLN}]$$

It should be noted that a negative value of the ENPV, computed with the difference-based method using the economic flows, means that variant V1 is economically less cost-effective in comparison with variant V0. In accordance with the adopted methodology, a positive value of the ENPV means, in turn, that variant V1 in economic terms is more cost-effective compared to variant V0. Thus, when interpreting the obtained ENPV value (−27,727.67 k PLN), it is possible to conclude that implementation of the investment according to variant V1, covering purchase and operation of electric buses in the city of

Szczecin, will be less favourable in economic terms compared to variant V0 which covers purchase and operation of diesel buses meeting the EURO 6 standard.

However, it should be noted that ENPV > FNPV, which means that the economic costs included in the calculations and resulting from NHMC/NMVOC, NO_x, PM, SO_2, CO_2 as well as noise emissions have improved the cost-effectiveness of variant V1 in comparison with variant V0.

Putting electric buses into operation in a given city may reduce NMHC, NO_x, PM, and noise emissions, so it may have a positive impact on the society, economy, and natural environment. However, due to the unfavourable energy mix in Poland (Figure 10) and the prevailing share of coal, the assumed full effects of variant V1 implementation are in fact difficult to achieve. The computed higher economic costs resulting directly from increased emissions of SO_2 and CO_2 may decrease the effectiveness of implementing electromobility in the transport sector. This in turn may have a negative impact on the decarbonisation process defined as reduction in (total and transport) carbon intensity of the whole economy [139]. The results have confirmed Hypothesis H2.

The results of the studies completed for the purposes of this research paper have shown that purchasing electric buses in specific conditions (the current energy mix in Poland) may be insufficient for full and effective achievement of electromobility objectives. The authors' observations in this respect are compliant with the results of the studies completed by Dillman et al. [140], where it was shown that the probability that EVs will lead to lower life-cycle GHG emissions compared to a diesel vehicle is much lower in Poland than in most other European countries. Therefore, it must be concluded that it is necessary for the process of replacing the bus fleet with electric buses to be combined with other supplementary measures, in particular with diversification of energy sources in Poland and increasing the share of RES, which confirms Hypothesis H3. Without such supplementary measures, the "zero-emission" feature of electric buses commissioned and operated in public transport systems in Polish cities may be deemed a substantial simplification. Even though operation of such vehicles may bring measurable benefits on a local scale (local zero emission mobility in a given city), in terms of the whole country this may merely be called "geographical shift of emission", which is not the main goal of electromobility.

6. Discussion

The main purpose of this paper was to identify and analyse economic effects derived from implementing zero-emission buses in urban public transport, based on the example of the city of Szczecin, Poland. In view of the purpose of this paper, the authors carried out a comparative analysis of effects of implementing the two variants:

- Fleet renewing variant (V0) which assumed that some of the diesel buses operated in Szczecin would be replaced with new diesel buses meeting the strictest standards for ICEVs, namely EURO 6,
- Investment variant (V1) which assumed that some of the diesel buses operated in Szczecin would be replaced with new, zero-emission buses.

In addition, FNPV and ENPV ratios were calculated. In order to compare the financial and economic effects of both variants, the difference-based method was applied. The computed ratios made it possible to conclude that investment implementation according to variant V1, covering purchase and operation of electric buses in the city of Szczecin, would be less cost-effective than variant V0 covering purchase and operation of diesel buses meeting the EURO 6 standard. The obtained result is substantially due to the two phenomena attributable to the current energy mix in Poland (Figure 10): the increased emissions of CO_2 in relation to electric buses (Figure 8) and appearance of SO_2 emission (Figure 7).

The research study focusing on estimation of economic effects of electromobility in sustainable urban public transport in Polish cities made it possible to demonstrate that:

- Implementation of zero-emission buses is an important tool to reduce external costs generated by urban public transport fleets.
- Economic benefits resulting from implementing zero-emission buses in urban public transport in Szczecin are limited by the current energy mix structure in Poland.
- Achieving full effects of electromobility in Poland as a result of implementing zero-emission buses in urban transport fleets depends on taking concurrent actions aimed at diversifying the sources of power generation in Poland (changing the energy mix), including in particular a wider use of Renewable Energy Sources (RES).

However, it is worth discussing the results in the aspect of the impact of the energy mix in a given country (in this case in Poland) on economic effects resulting from implementation of zero-emission buses in urban public transport (in this case in the city of Szczecin). As already mentioned, when calculating the emissions costs for each of the environmental pollutants (NHMC/NMVOC, NO_x, PM, SO_2 and CO_2) and noise, unit costs were applied, which were calculated according to CEUTP. In order to demonstrate the role of the energy mix in effective achievement of electromobility goals, further, alternative computations of ENPV were made for the analysed case study. Similarly as in the case of the previous computations, the difference-based method was applied for this purpose. Taking into account the fact that it is CO_2 and SO_2 that are responsible for the result that is unfavourable for electric buses, the subsequent calculations took into account reductions in the costs of the two substances (CO_2 and SO_2) emissions. ENPV was calculated based on the assumption that the current unit costs of CO_2 and SO_2 emissions would be reduced by, respectively: 25%, 50%, 75% up to a hypothetical situation when the costs are totally reduced. The results were as follows:

- a decrease in unit costs of CO_2 and SO_2 emissions by 25%:

$$\text{ENPV} = -21,462.73\ [\text{k PLN}]$$

- a decrease in unit costs of CO_2 and SO_2 emissions by 50%:

$$\text{ENPV} = -15,199.07\ [\text{k PLN}]$$

- a decrease in unit costs of CO_2 and SO_2 emissions by 75%:

$$\text{ENPV} = -8935.41\ [\text{k PLN}]$$

- a decrease in unit costs of CO_2 and SO_2 emissions by 100%:

$$\text{ENPV} = -2671.75\ [\text{k PLN}]$$

As shown by the ENPV calculation results for the analysed situations, the reduction of unit costs of CO_2 and SO_2 emissions may effectively level out the difference between variant V1 and variant V0. Thus, it may have a significant impact on the increased cost-effectiveness of variant V1, and thus the possibility of achieving the full assumed effects of electromobility in transport.

7. Conclusions

In the context of the research problem handled in this paper, it is important to point out that the condition for success of electromobility development is not only providing the basis for the electromobility ecosystem but also coordination of activities in the area of electromobility industry development and stimulation of demand for electric vehicles. The analyses completed in the course of the research project made it possible to formulate the following conclusions:

- An important role in the process of electromobility development in urban transport is played by the government and the local authorities. The former is responsible for an appropriate energy policy (striving for a large share of RES in the energy mix) and fiscal policy (a system of subsidies and incentives for purchase and operation of zero-emission vehicles). The latter may create appropriate local infrastructural and organisational conditions to support and privilege zero-emission vehicle users, e.g., via: organisation of an EV charging system, providing free-of-charge car parks in the city centre, or designating LEZs and ZEZs. The task for local authorities is also propagation of using zero-emission buses as a vital element of Sustainable Urban Public Transport.

- Achieving the assumed electromobility effects is predicated not only on purchasing appropriate vehicles or organising a charging system for them but also on a big share of RES in the country's energy mix. A small share of RES may result in limited benefits of zero-emission vehicle operation.
- Purchase and operation of electric buses in urban public transport may contribute to improving life quality in cities via reduction of local emissions of pollutants to the environment. Thus, as opposed to diesel buses, vehicles of this type contribute to meeting the electromobility goals in local terms.
- Purchase and operation of electric buses in urban public transport may also contribute to meeting the electromobility goals and transport decarbonisation in supralocal terms; however, this depends on an appropriate share of RES in the country's energy mix. Failure to provide electric power derived from renewable energy sources may only result in a geographical shift of emissions.

Analysing the world trends in the area of production and operation of zero-emission vehicles, we can conclude that over the next years the demand for electric power will be rising intensively. This trend may have an adverse effect on economies characterised by a big share of fossil fuels in the energy mix (this may also be the case in Poland). If the power engineering industry in any given country is unable to quickly adapt to the growing demand (via diversification of energy generation sources), this may lead to insufficiency of the given country's system or the need to meet the demand via increased electric power production in the existing coal-fired power plants. This situation may paradoxically lead to further growth of the share of fossil fuels in the country's energy mix, and thus further growth of unit costs of emissions of substances that are harmful to people and the environment. The research study described in this paper is of practical value, as it indicates the need to adapt the energy market to the changes resulting from electromobility.

The changes observed on the transport market, which result from the electromobility policy assumptions, require further studies, in particular with regard to external effects of zero-emission vehicles operation.

Author Contributions: Conceptualization, O.P. and K.P.; data curation, O.P. and K.P.; formal analysis, O.P. and K.P.; investigation, O.P. and K.P.; methodology, O.P. and K.P.; visualization, K.P.; writing—original draft, O.P. and K.P.; writing—review and editing, O.P. All authors have read and agreed to the published version of the manuscript.

Funding: The research presented in this article was carried out in the Maritime University of Szczecin under the Grant 1/S/KGMiST/2021 "Transport 4.0".

Institutional Review Board Statement: Not applicable.

Informed Consent Statement: Not applicable.

Conflicts of Interest: The authors declare no conflict of interest.

References

1. Nordelöf, A.; Romare, M.; Tivander, J. Life Cycle Assessment of City Buses Powered by Electricity, Hydrogenated Vegetable Oil or Diesel. *Transp. Res. D Transp. Environ.* **2019**, *75*, 211–222. [CrossRef]
2. Chang, C.-C.; Liao, Y.-T.; Chang, Y.-W. Life Cycle Assessment of Alternative Energy Types—Including Hydrogen—For Public City Buses in Taiwan. *Int. J. Hydrogen Energy* **2019**, *44*, 18472–18482. [CrossRef]
3. Ivković, I.; Čokorilo, O.; Kaplanović, S. The Estimation of GHG Emission Costs in Road and Air Transport Sector: Case Study of Serbia. *Transport* **2016**, *33*, 260–267. [CrossRef]
4. Liimatainen, H. Measures for Energy Efficient and Low Emission Private Mobility. In *Encyclopedia of the UN Sustainable Development Goals*; Springer International Publishing: Cham, Switzerland, 2020; pp. 1–12. [CrossRef]

5. European Commission. *A European Strategy for Low-Emission Mobility*; European Commission: Brussels, Belgium, 2016.
6. European Commission. Available online: https://ec.europa.eu/clima/policies/transport_en (accessed on 30 November 2020).
7. Foltyński, M. Sustainable Urban Logistics Plan—Current Situation of the City of Poznań. *Transp. Res. Procedia* **2019**, *39*, 42–53. [CrossRef]
8. Kumar, P.; Morawska, L.; Martani, C.; Biskos, G.; Neophytou, M.; Di Sabatino, S.; Bell, M.; Norford, L.; Britter, R. The Rise of Low-Cost Sensing for Managing Air Pollution in Cities. *Environ. Int.* **2015**, *75*, 199–205. [CrossRef] [PubMed]
9. Zhou, B.; Wu, Y.; Zhou, B.; Wang, R.; Ke, W.; Zhang, S.; Hao, J. Real-World Performance of Battery Electric Buses and Their Life-Cycle Benefits with Respect to Energy Consumption and Carbon Dioxide Emissions. *Energy* **2016**, *96*, 603–613. [CrossRef]
10. Mutter, A. Obduracy and Change in Urban Transport—Understanding Competition Between Sustainable Fuels in Swedish Municipalities. *Sustainability* **2019**, *11*, 6092. [CrossRef]
11. Poland's State Assets Ministry. Available online: https://www.gov.pl/web/aktywa-panstwowe/elektromobilnosc-w-polsce (accessed on 30 November 2020).
12. Act of 11 January 2018 on Electromobility and Alternative Fuels (Ustawa z Dnia 11 Stycznia 2018 r. o Elektromobilności i Paliwach Alternatywnych, Dz. U. 2018, poz. 317). Available online: https://isap.sejm.gov.pl (accessed on 30 November 2020).
13. Blue Book. Public Transport Sector in Cities, Agglomerations and Regions, Jaspers. 2015. Available online: https://www.cupt.gov.pl/ (accessed on 30 November 2020).
14. Pietrzak, K.; Pietrzak, O. Environmental Effects of Electromobility in a Sustainable Urban Public Transport. *Sustainability* **2020**, *12*, 1052. [CrossRef]
15. Haase, D.; Güneralp, B.; Dahiya, B.; Bai, X.; Elmqvist, T. Global Urbanization. In *Urban Planet: Knowledge towards Sustainable Cities*; Cambridge University Press: Cambridge, UK, 2018; pp. 19–44.
16. United Nations Department of Economic and Social Affairs. *The World's Cities in 2018—Data Booklet*; (ST/ESA/ SER.A/417); United Nations: New York, NY, USA, 2018.
17. Tang, J.; McNabola, A.; Misstear, B. The Potential Impacts of Different Traffic Management Strategies on Air Pollution and Public Health for a More Sustainable City: A Modelling Case Study from Dublin, Ireland. *Sustain. Cities Soc.* **2020**, *60*, 102229. [CrossRef]
18. Abbasi, M.; Hadji Hosseinlou, M.; JafarzadehFadaki, S. An Investigation of Bus Rapid Transit System (BRT) Based on Economic and Air Pollution Analysis (Tehran, Iran). *Case Stud. Transp. Policy* **2020**, *8*, 553–563. [CrossRef]
19. Talla, A.; Ngohe-Ekam, P.S.; Nkeumaleu, A.T. Evaluation of the Impact of Motorcycles in Urban Transport on Air Pollution: A Case of Douala City in Cameroon. *J. Sci. Res. Rep.* **2018**, *20*, 1–11. [CrossRef] [PubMed]
20. Titos, G.; Lyamani, H.; Drinovec, L.; Olmo, F.J.; Močnik, G.; Alados-Arboledas, L. Evaluation of the Impact of Transportation Changes on Air Quality. *Atmos. Environ.* **2015**, *114*, 19–31. [CrossRef]
21. Wang, Y.; Yang, D. Impacts of Freight Transport on PM2.5 Concentrations in China: A Spatial Dynamic Panel Analysis. *Sustainability* **2018**, *10*, 2865. [CrossRef]
22. Gao, J.; Chen, H.; Dave, K.; Chen, J.; Jia, D. Fuel Economy and Exhaust Emissions of a Diesel Vehicle under Real Traffic Conditions. *Energy Sci. Eng.* **2020**, *8*, 1781–1792. [CrossRef]
23. Delgado, J.; Moura, P.; de Almeida, A.T. Electric Mobility: Key Technology to Decarbonize the Economy and Improve Air Quality. In *Encyclopedia of the UN Sustainable Development Goals*; Springer International Publishing: Cham, Switzerland, 2020; pp. 1–18. [CrossRef]
24. Migliaretti, G.; Cadum, E.; Migliore, E.; Cavallo, F. Traffic Air Pollution and Hospital Admission for Asthma: A Case-Control Approach in a Turin (Italy) Population. *Int. Arch. Occup. Environ. Health* **2005**, *78*, 164–169. [CrossRef]
25. Lindgren, A.; Stroh, E.; Montnémery, P.; Nihlén, U.; Jakobsson, K.; Axmon, A. Traffic-Related Air Pollution Associated with Prevalence of Asthma and COPD/Chronic Bronchitis. A Cross-Sectional Study in Southern Sweden. *Int. J. Health Geogr.* **2009**, *8*, 2. [CrossRef] [PubMed]
26. Raaschou-Nielsen, O.; Andersen, Z.J.; Jensen, S.S.; Ketzel, M.; Sørensen, M.; Hansen, J.; Loft, S.; Tjønneland, A.; Overvad, K. Traffic Air Pollution and Mortality from Cardiovascular Disease and All Causes: A Danish Cohort Study. *Environ. Health* **2012**, *11*, 60. [CrossRef]
27. Weichenthal, S.; Hatzopoulou, M.; Goldberg, M.S. Exposure to Traffic-Related Air Pollution during Physical Activity and Acute Changes in Blood Pressure, Autonomic and Micro-Vascular Function in Women: A Cross-over Study. *Part. Fibre Toxicol.* **2014**, *11*. [CrossRef]
28. Soehodho, S. Motorization in Indonesia and Its Impact to Traffic Accidents. *IATSS Res.* **2007**, *31*, 27–33. [CrossRef]
29. Sun, L.-L.; Liu, D.; Chen, T.; He, M.-T. Road Traffic Safety: An Analysis of the Cross-Effects of Economic, Road and Population Factors. *Chin. J. Traumatol.* **2019**, *22*, 290–295. [CrossRef]
30. Retallack, A.E.; Ostendorf, B. Current Understanding of the Effects of Congestion on Traffic Accidents. *Int. J. Environ. Res. Public Health* **2019**, *16*, 3400. [CrossRef]
31. Jacyna, M.; Wasiak, M.; Lewczuk, K.; Karoń, G. Noise and Environmental Pollution from Transport: Decisive Problems in Developing Ecologically Efficient Transport Systems. *J. Vibroengineering* **2017**, *19*, 5639–5655. [CrossRef]
32. Jeon, J.; Hong, J.; Kim, S.; Kim, K.-H. Noise Indicators for Size Distributions of Airborne Particles and Traffic Activities in Urban Areas. *Sustainability* **2018**, *10*, 4599. [CrossRef]
33. McCahill, C.; Garrick, N. Automobile Use and Land Consumption: Empirical Evidence from 12 Cities. *URBAN Des. Int.* **2012**, *17*, 221–227. [CrossRef]

34. Shin, Y.E.; Vuchic, V.R.; Bruun, E.C. Land Consumption Impacts of a Transportation System on a City: An Analysis. *Transp. Res. Rec.* **2009**, *2110*, 69–77. [CrossRef]
35. Available online: https://www.eea.europa.eu/themes/transport/term/increasing-oil-consumption-and-ghg (accessed on 30 November 2020).
36. Skrúcaný, T.; Kendra, M.; Stopka, O.; Milojević, S.; Figlus, T.; Csiszár, C. Impact of the Electric Mobility Implementation on the Greenhouse Gases Production in Central European Countries. *Sustainability* **2019**, *11*, 4948. [CrossRef]
37. Milojević, S.; Skrucany, T.; Milošević, H.; Stanojević, D.; Pantić, M.; Stojanović, B. Alternative Drive Systems and Environmentaly Friendly Public Passengers Transport. *Appl. Eng. Lett. J. Eng. Appl. Sci.* **2018**, *3*, 105–113. [CrossRef]
38. Sarigiannis, D.A.; Kontoroupis, P.; Nikolaki, S.; Gotti, A.; Chapizanis, D.; Karakitsios, S. Benefits on Public Health from Transport-Related Greenhouse Gas Mitigation Policies in Southeastern European Cities. *Sci. Total Environ.* **2017**, *579*, 1427–1438. [CrossRef] [PubMed]
39. Auvinen, H.; Järvi, T.; Kloetzke, M.; Kugler, U.; Bühne, J.-A.; Heinl, F.; Kurte, J.; Esser, K. Electromobility Scenarios: Research Findings to Inform Policy. *Transp. Res. Procedia* **2016**, *14*, 2564–2573. [CrossRef]
40. Melkonyan, A.; Koch, J.; Lohmar, F.; Kamath, V.; Munteanu, V.; Alexander Schmidt, J.; Bleischwitz, R. Integrated Urban Mobility Policies in Metropolitan Areas: A System Dynamics Approach for the Rhine-Ruhr Metropolitan Region in Germany. *Sustain. Cities Soc.* **2020**, 102358. [CrossRef]
41. Pinto, J.A.; Kumar, P.; Alonso, M.F.; Andreão, W.L.; Pedruzzi, R.; Espinosa, S.I.; de Almeida Albuquerque, T.T. Kriging Method Application and Traffic Behavior Profiles from Local Radar Network Database: A Proposal to Support Traffic Solutions and Air Pollution Control Strategies. *Sustain. Cities Soc.* **2020**, *56*, 102062. [CrossRef]
42. Muñoz-Villamizar, A.; Santos, J.; Montoya-Torres, J.R.; Velázquez-Martínez, J.C. Measuring Environmental Performance of Urban Freight Transport Systems: A Case Study. *Sustain. Cities Soc.* **2020**, *52*, 101844. [CrossRef]
43. Chwesiuk, K.; Kijewska, K.; Iwan, S. Urban Consolidation Centres for Medium-Size Touristic Cities in the Westpomeranian Region of Poland. *Procedia Soc. Behav. Sci.* **2010**, *2*, 6264–6273. [CrossRef]
44. Van Rooijen, T.; Quak, H. Local Impacts of a New Urban Consolidation Centre—The Case of Binnenstadservice.Nl. *Procedia Soc. Behav. Sci.* **2010**, *2*, 5967–5979. [CrossRef]
45. Quak, H.; van Duin, R.; Hendriks, B. Running an Urban Consolidation Centre: Binnenstadservice 10 Years Back and Forth. *Transp. Res. Procedia* **2020**, *46*, 45–52. [CrossRef]
46. Taefi, T.T.; Kreutzfeldt, J.; Held, T.; Konings, R.; Kotter, R.; Lilley, S.; Baster, H.; Green, N.; Laugesen, M.S.; Jacobsson, S.; et al. Comparative Analysis of European Examples of Freight Electric Vehicles Schemes—A Systematic Case Study Approach with Examples from Denmark, Germany, The Netherlands, Sweden and the UK. In *Dynamics in Logistics*; Springer International Publishing: Cham, Switzerland, 2016; pp. 495–504. [CrossRef]
47. Teoh, T.; Kunze, O.; Teo, C.-C.; Wong, Y. Decarbonisation of Urban Freight Transport Using Electric Vehicles and Opportunity Charging. *Sustainability* **2018**, *10*, 3258. [CrossRef]
48. Fiori, C.; Marzano, V. Modelling Energy Consumption of Electric Freight Vehicles in Urban Pickup/Delivery Operations: Analysis and Estimation on a Real-World Dataset. *Transp. Res. D Transp. Environ.* **2018**, *65*, 658–673. [CrossRef]
49. Lambas, M.E.L.; Ricci, S. The Environmental and Economic Effects of Innovative Measures in Urban Parcels Delivery. In *Urban Transport XX*; WIT Press: Southampton, UK, 2014. [CrossRef]
50. Moolenburgh, E.A.; van Duin, J.H.R.; Balm, S.; van Altenburg, M.; van Amstel, W.P. Logistics Concepts for Light Electric Freight Vehicles: A Multiple Case Study from the Netherlands. *Transp. Res. Procedia* **2020**, *46*, 301–308. [CrossRef]
51. Melo, S.; Baptista, P.; Costa, Á. Comparing the Use of Small Sized Electric Vehicles with Diesel Vans on City Logistics. *Procedia Soc. Behav. Sci.* **2014**, *111*, 350–359. [CrossRef]
52. Gruber, J.; Kihm, A. Reject or Embrace? Messengers and Electric Cargo Bikes. *Transp. Res. Procedia* **2016**, *12*, 900–910. [CrossRef]
53. Niels, T.; Hof, M.T.; Bogenberger, K. Design and Operation of an Urban Electric Courier Cargo Bike System. In Proceedings of the 21st International Conference on Intelligent Transportation Systems (ITSC), Maui, HI, USA, 4–7 November 2018. [CrossRef]
54. Zhang, L.; Matteis, T.; Thaller, C.; Liedtke, G. Simulation-Based Assessment of Cargo Bicycle and Pick-up Point in Urban Parcel Delivery. *Procedia Comput. Sci.* **2018**, *130*, 18–25. [CrossRef]
55. Gupta, S. Role of Non -Motorized Transport in Distribution of Goods in the Metropolitan City of Delhi. *Transp. Res. Procedia* **2017**, *25*, 978–984. [CrossRef]
56. Kelly, J.; Marinov, M. Innovative Interior Designs for Urban Freight Distribution Using Light Rail Systems. *Urban Rail Transit* **2017**, *3*, 238–254. [CrossRef]
57. Gonzalez-Feliu, J. A Joint Freight Catchment and Cost Benefit Analysis to Assess Rail Urban Logistics Scenarios. In *Lecture Notes in Business Information Processing*; Springer International Publishing: Cham, Switzerland, 2018; pp. 14–27. [CrossRef]
58. Pietrzak, O.; Pietrzak, K. The Role of Railway in Handling Transport Services of Cities and Agglomerations. *Transp. Res. Procedia* **2019**, *39*, 405–416. [CrossRef]
59. Zych, M. Identification of Potential Implementation of the Cargo Tram in Warsaw: A First Overview. *Procedia Soc. Behav. Sci.* **2014**, *151*, 360–369. [CrossRef]
60. Orczyk, M.; Tomaszewski, F. Freight Tram Concept for the City of Poznań. *Transp. Econ. Logist.* **2018**, *80*, 169–178. [CrossRef]
61. Ortega, J.; Tóth, J.; Péter, T. Mapping the Catchment Area of Park and Ride Facilities within Urban Environments. *ISPRS Int. J. Geoinf.* **2020**, *9*, 501. [CrossRef]

62. Macioszek, E.; Kurek, A. The Use of a Park and Ride System—A Case Study Based on the City of Cracow (Poland). *Energies* **2020**, *13*, 3473. [CrossRef]
63. Kimpton, A.; Pojani, D.; Sipe, N.; Corcoran, J. Parking Behavior: Park 'n' Ride (PnR) to Encourage Multimodalism in Brisbane. *Land Use Policy* **2020**, *91*, 104304. [CrossRef]
64. Buchari, E. Transportation Demand Management: A Park and Ride System to Reduce Congestion in Palembang City Indonesia. *Procedia Eng.* **2015**, *125*, 512–518. [CrossRef]
65. Fierek, S.; Bieńczak, M.; Zmuda-Trzebiatowski, P. Multiple Criteria Evaluation of P&R Lots Location. *Transp. Res. Procedia* **2020**, *47*, 489–496. [CrossRef]
66. Hamidi, Z.; Camporeale, R.; Caggiani, L. Inequalities in Access to Bike-and-Ride Opportunities: Findings for the City of Malmö. *Transp. Res. Part A Policy Pract.* **2019**, *130*, 673–688. [CrossRef]
67. Pritchard, J.P.; Stępniak, M.; Geurs, K.T. Equity Analysis of Dynamic Bike-and-Ride Accessibility in The Netherlands. In *Measuring Transport Equity*; Elsevier: Amsterdam, The Netherlands, 2019; pp. 73–83. [CrossRef]
68. Martens, K. Promoting Bike-and-Ride: The Dutch Experience. *Transp. Res. Part A Policy Pract.* **2007**, *41*, 326–338. [CrossRef]
69. Bieliński, T.; Kwapisz, A.; Ważna, A. Bike-Sharing Systems in Poland. *Sustainability* **2019**, *11*, 2458. [CrossRef]
70. Macioszek, E.; Świerk, P.; Kurek, A. The Bike-Sharing System as an Element of Enhancing Sustainable Mobility—A Case Study Based on a City in Poland. *Sustainability* **2020**, *12*, 3285. [CrossRef]
71. Ricci, M. Bike Sharing: A Review of Evidence on Impacts and Processes of Implementation and Operation. *Res. Transp. Bus. Manag.* **2015**, *15*, 28–38. [CrossRef]
72. Chang, S.; Song, R.; He, S.; Qiu, G. Innovative Bike-Sharing in China: Solving Faulty Bike-Sharing Recycling Problem. *J. Adv. Transp.* **2018**, *2018*, 1–10. [CrossRef]
73. Bieliński, T.; Ważna, A. Electric Scooter Sharing and Bike Sharing User Behaviour and Characteristics. *Sustainability* **2020**, *12*, 9640. [CrossRef]
74. Ilkevich, S.V. The Sources of Competitive Advantages of Electric Scooter Sharing Services. Стратегические Решения и Риск-Менеджмент **2019**, *10*, 238–251. [CrossRef]
75. Tomanek, R. Free-Fare Public Transport in the Concept of Sustainable Urban Mobility. *Transp. Probl.* **2018**, *12*, 95–105. [CrossRef]
76. Štraub, D.; Jaroš, V. Free Fare Policy as a Tool for Sustainable Development of Public Transport Services. *Hum. Geogr.* **2019**, *13*, 45–59. [CrossRef]
77. Tsao, H.-S.J.; Wei, W.; Pratama, A. Operational Feasibility of One-Dedicated-Lane Bus Rapid Transit/Light Rail Systems. *Transp. Plan. Technol.* **2009**, *32*, 239–260. [CrossRef]
78. Li, J.-Q.; Song, M.K.; Li, M.; Zhang, W.-B. Planning for Bus Rapid Transit in Single Dedicated Bus Lane. *Transp. Res. Rec.* **2009**, *2111*, 76–82. [CrossRef]
79. Ku, D.; Bencekri, M.; Kim, J.; Leec, S.; Leed, S. Review of European Low Emission Zone Policy. *Chem. Eng.* **2020**, 78. [CrossRef]
80. Da Silva, F.N.; Custódio, R.A.L.; Martins, H. Low Emission Zone: Lisbon's Experience. *J. Traffic Logist. Eng.* **2014**, *2*, 133–139. [CrossRef]
81. Biswas, D.; Ghosh, S.; Sengupta, S.; Mukhopadhyay, S. A Predictive Supervisory Controller for an HEV Operating in a Zero Emission Zone. In Proceedings of the IEEE Transportation Electrification Conference and Expo, Detroit, MI, USA, 19–21 June 2019. [CrossRef]
82. Savković, T.; Gladović, P.; Miličić, M.; Pitka, P.; Ilić, S. Effects of Eco-Driving Training: A Pilot Program in Belgrade Public Transport. *Tehnicki Vjesnik* **2019**, *26*, 1031–1037. [CrossRef]
83. Strömberg, H.K.; Karlsson, I.C.M. Comparative Effects of Eco-Driving Initiatives Aimed at Urban Bus Drivers—Results from a Field Trial. *Transp. Res. D Transp. Environ.* **2013**, *22*, 28–33. [CrossRef]
84. Barla, P.; Gilbert-Gonthier, M.; Lopez Castro, M.A.; Miranda-Moreno, L. Eco-Driving Training and Fuel Consumption: Impact, Heterogeneity and Sustainability. *Energy Econ.* **2017**, *62*, 187–194. [CrossRef]
85. Ho, S.-H.; Wong, Y.-D.; Chang, V.W.-C. What Can Eco-Driving Do for Sustainable Road Transport? Perspectives from a City (Singapore) Eco-Driving Programme. *Sustain. Cities Soc.* **2015**, *14*, 82–88. [CrossRef]
86. Mankowski, C.; Weiland, D.; Abramović, B. Impact of Railway Investment on Regional Development—Case Study of Pomeranian Metropolitan Railway. *PROMET Traffic Transp.* **2019**, *31*, 669–679. [CrossRef]
87. Abramović, B.; Šipuš, D.; Jurešić, D. Organisation of Integrated Passenger Transport on the Zagreb—Velika Gorica Route. *Transp. Res. Procedia* **2020**, *44*, 342–347. [CrossRef]
88. Reichenbach, M. The Multimodal Transport User—A Challenge for Public Transport? *Transp. Res. Procedia* **2019**, *41*, 357–359. [CrossRef]
89. Dyr, T.; Ziółkowska, K.; Misiurski, P.; Kozłowska, M. Effectiveness of Application Alternative Drive Vehicles in Public Transport. *MATEC Web Conf.* **2018**, *180*, 01002. [CrossRef]
90. Tzeng, G.-H.; Lin, C.-W.; Opricovic, S. Multi-Criteria Analysis of Alternative-Fuel Buses for Public Transportation. *Energy Policy* **2005**, *33*, 1373–1383. [CrossRef]
91. Xu, G.; Li, M.; Zhao, Y.; Chen, Q. Study on Emission Characteristics of Hybrid Buses under Driving Cycles in a Typical Chinese City. *Adv. Mech. Eng.* **2017**, *9*, 168781401772823. [CrossRef]

92. Lijewski, P.; Ziolkowski, A.; Daszkiewicz, P.; Andrzejewski, M.; Gallas, D. Comparison of CO2 Emissions and Fuel Consumption of a Hybrid Vehicle and a Vehicle with a Direct Gasoline Injection Engine. *IOP Conf. Ser. Mater. Sci. Eng.* **2018**, *421*, 042046. [CrossRef]
93. Hallmark, S.L.; Wang, B.; Sperry, R. Comparison of On-Road Emissions for Hybrid and Regular Transit Buses. *J. Air Waste Manag. Assoc.* **2013**, *63*, 1212–1220. [CrossRef]
94. Pawełczyk, M.; Szumska, E. Evaluation of the Efficiency of Hybrid Drive Applications in Urban Transport System on the Example of a Medium Size City. *MATEC Web Conf.* **2018**, *180*, 03004. [CrossRef]
95. Milojevic, S.; Grocic, D.; Dragojlovic, D. CNG Propulsion System for Reducing Noise of Existing City Buses. *J. Appl. Eng. Sci.* **2016**, *14*, 377–382. [CrossRef]
96. Ivković, I.; Kaplanović, S.; Sekulić, D. Analysis of External Costs of Co2 Emissions for Cng Buses in Intercity Bus Service. *Transport* **2019**, *34*, 529–538. [CrossRef]
97. Merkisz, J.; Fuć, P.; Lijewski, P.; Pielecha, J. Actual Emissions from Urban Buses Powered with Diesel and Gas Engines. *Transp. Res. Procedia* **2016**, *14*, 3070–3078. [CrossRef]
98. Yue, T.; Chai, F.; Hu, J.; Jia, M.; Bao, X.; Li, Z.; He, L.; Zu, L. Gaseous Emissions from Compressed Natural Gas Buses in Urban Road and Highway Tests in China. *J. Environ. Sci.* **2016**, *48*, 193–199. [CrossRef] [PubMed]
99. Tica, S.; Živanović, P.; Bajčetić, S.; Milovanović, B.; Nađ, A. Study of the Fuel Efficiency and Ecological Aspects of CNG Buses in Urban Public Transport in Belgrade. *J. Appl. Eng. Sci.* **2019**, *17*, 65–73. [CrossRef]
100. Dyr, T.; Misiurski, P.; Ziółkowska, K. Costs and Benefits of Using Buses Fuelled by Natural Gas in Public Transport. *J. Clean. Prod.* **2019**, *225*, 1134–1146. [CrossRef]
101. Jurkovič, M.; Kalina, T.; Skrúcaný, T.; Gorzelanczyk, P.; L'upták, V. Environmental Impacts of Introducing LNG as Alternative Fuel for Urban Buses—Case Study in Slovakia. *PROMET Traffic Transp.* **2020**, *32*, 837–847. [CrossRef]
102. Zhang, G.; Zhang, J.; Xie, T. A Solution to Renewable Hydrogen Economy for Fuel Cell Buses—A Case Study for Zhangjiakou in North China. *Int. J. Hydrogen Energy* **2020**, *45*, 14603–14613. [CrossRef]
103. Hua, T.; Ahluwalia, R.; Eudy, L.; Singer, G.; Jermer, B.; Asselin-Miller, N.; Wessel, S.; Patterson, T.; Marcinkoski, J. Status of Hydrogen Fuel Cell Electric Buses Worldwide. *J. Power Source* **2014**, *269*, 975–993. [CrossRef]
104. Lee, D.-Y.; Elgowainy, A.; Vijayagopal, R. Well-to-Wheel Environmental Implications of Fuel Economy Targets for Hydrogen Fuel Cell Electric Buses in the United States. *Energy Policy* **2019**, *128*, 565–583. [CrossRef]
105. Langford, B.C.; Cherry, C. Transitioning a Bus Transit Fleet to Hydrogen Fuel: A Case Study of Knoxville Area Transit. *Int. J. Hydrogen Energy* **2012**, *37*, 2635–2643. [CrossRef]
106. Eudy, L.; Post, M. *BC Transit Fuel Cell Bus Project: Evaluation Results Report*; Office of Scientific and Technical Information: Oak Ridge, TN, USA, 2014. [CrossRef]
107. Eudy, L.; Post, M. *BC Transit Fuel Cell Bus Project Evaluation Results: Second Report*; Office of Scientific and Technical Information: Oak Ridge, TN, USA, 2014. [CrossRef]
108. Eudy, L.; Post, M.; Jeffers, M. *American Fuel Cell Bus Project Evaluation: Third Report*; Office of Scientific and Technical Information: Oak Ridge, TN, USA, 2017. [CrossRef]
109. Mahmoud, M.; Garnett, R.; Ferguson, M.; Kanaroglou, P. Electric Buses: A Review of Alternative Powertrains. *Renew. Sustain. Energy Rev.* **2016**, *62*, 673–684. [CrossRef]
110. Quarles, N.; Kockelman, K.M.; Mohamed, M. Costs and Benefits of Electrifying and Automating Bus Transit Fleets. *Sustainability* **2020**, *12*, 3977. [CrossRef]
111. Kivekas, K.; Vepsalainen, J.; Tammi, K. Stochastic Driving Cycle Synthesis for Analyzing the Energy Consumption of a Battery Electric Bus. *IEEE Access* **2018**, *6*, 55586–55598. [CrossRef]
112. Logan, K.G.; Nelson, J.D.; Hastings, A. Electric and Hydrogen Buses: Shifting from Conventionally Fuelled Cars in the UK. *Transp. Res. D Transp. Environ.* **2020**, *85*, 102350. [CrossRef]
113. Borén, S. Electric Buses' Sustainability Effects, Noise, Energy Use, and Costs. *Int. J. Sustain. Transp.* **2019**, 1–16. [CrossRef]
114. Kunith, A.; Mendelevitch, R.; Goehlich, D. Electrification of a City Bus Network—An Optimization Model for Cost-Effective Placing of Charging Infrastructure and Battery Sizing of Fast-Charging Electric Bus Systems. *Int. J. Sustain. Transp.* **2017**, *11*, 707–720. [CrossRef]
115. Csiszár, C.; Csonka, B.; Földes, D.; Wirth, E.; Lovas, T. Urban Public Charging Station Locating Method for Electric Vehicles Based on Land Use Approach. *J. Transp. Geogr.* **2019**, *74*, 173–180. [CrossRef]
116. Chao, Z.; Xiaohong, C. Optimizing Battery Electric Bus Transit Vehicle Scheduling with Battery Exchanging: Model and Case Study. *Procedia Soc. Behav. Sci.* **2013**, *96*, 2725–2736. [CrossRef]
117. Czerepicki, A.; Choromański, W.; Kozłowski, M.; Kazinski, A. Analysis of the Problem of Electric Buses Charging in Urban Transport. *Sci. Technol.* **2020**, *19*, 349–355. [CrossRef]
118. Astaneh, M.; Andric, J.; Löfdahl, L.; Maggiolo, D.; Stopp, P.; Moghaddam, M.; Chapuis, M.; Ström, H. Calibration Optimization Methodology for Lithium-Ion Battery Pack Model for Electric Vehicles in Mining Applications. *Energies* **2020**, *13*, 3532. [CrossRef]
119. Ahmeid, M.; Muhammad, M.; Milojevic, Z.; Lambert, S.; Attidekou, P. The Energy Loss Due to Interconnections in Paralleled Cell Configurations of Lithium-Ion Batteries in Electric Vehicles. In Proceedings of the IEEE 4th International Future Energy Electronics Conference, Singapore, 25–28 November 2019. [CrossRef]

120. Zavada, J.; Blašković Zavada, J.; Miloš, K. Conditions for Implementing Trolleybuses in Public Urban Transport. *PROMET Traffic Transp.* **2012**, *22*, 467–474. [CrossRef]
121. Wołek, M.; Wolański, M.; Bartłomiejczyk, M.; Wyszomirski, O.; Grzelec, K.; Hebel, K. Ensuring Sustainable Development of Urban Public Transport: A Case Study of the Trolleybus System in Gdynia and Sopot (Poland). *J. Clean. Prod.* **2021**, *279*, 123807. [CrossRef]
122. Wołek, M.; Szmelter-Jarosz, A.; Koniak, M.; Golejewska, A. Transformation of Trolleybus Transport in Poland. Does in-Motion Charging (Technology) Matter? *Sustainability* **2020**, *12*, 9744. [CrossRef]
123. Bartłomiejczyk, M.; Połom, M. The Road to the Development of Electromobility in the Czech's Prague: From Electric Buses to... Trolleybuses? *AUTOBUSY Tech. Eksploat. Syst. Transp.* **2019**, *24*, 22–28. [CrossRef]
124. Dziubinski, M.; Siemionek, E.; Adamiec, M.; Drozd, A.; Kolodziej, S. Energy Consumption of the Trolleybuses. In Proceedings of the International Conference on Electromagnetic Devices and Processes in Environment Protection with Seminar Applications of Superconductors, Naleczow, Poland, 3–6 December 2017. [CrossRef]
125. Grygar, D.; Koháni, M.; Štefún, R.; Drgoňa, P. Analysis of Limiting Factors of Battery Assisted Trolleybuses. *Transp. Res. Procedia* **2019**, *40*, 229–235. [CrossRef]
126. Jakubas, A.; Chwastek, K.; Cywiński, A.; Gnatowski, A.; Suchecki, Ł. An Analysis of the Performance of Trolleybus Brushes Developed from Recycled Materials. *Appl. Sci.* **2020**, *10*, 7929. [CrossRef]
127. Correa, G.; Muñoz, P.M.; Rodriguez, C.R. A Comparative Energy and Environmental Analysis of a Diesel, Hybrid, Hydrogen and Electric Urban Bus. *Energy* **2019**, *187*, 115906. [CrossRef]
128. Correa, G.; Muñoz, P.; Falaguerra, T.; Rodriguez, C.R. Performance Comparison of Conventional, Hybrid, Hydrogen and Electric Urban Buses Using Well to Wheel Analysis. *Energy* **2017**, *141*, 537–549. [CrossRef]
129. Stempien, J.P.; Chan, S.H. Comparative Study of Fuel Cell, Battery and Hybrid Buses for Renewable Energy Constrained Areas. *J. Power Source* **2017**, *340*, 347–355. [CrossRef]
130. Misanovic, S.M.; Zivanovic, Z.M.; Tica, S.M. Energy Efficiency of Different Bus Subsystems in Belgrade Public Transport. *Therm. Sci.* **2015**, *19*, 2233–2244. [CrossRef]
131. Imam, R.; Kang, S.-C.; Quezada, D. Exploring Low-Carbon Bus Options for Urban BRT Systems: The Case of Amman. *J. Public Trans.* **2020**, *22*. [CrossRef]
132. Local Data Bank. Available online: https://bdl.stat.gov.pl/BDL/start (accessed on 30 November 2020).
133. Act of 16 December 2010 on Public Collective Transport (Journal of Laws of 2011, No. 5, item 13). Available online: https://isap.sejm.gov.pl/isap.nsf/DocDetails.xsp?id=WDU20110050013 (accessed on 30 November 2020).
134. Act of 8 March 1990 r. on Local Self-Government (Journal of Laws of 1990, No. 16, item 95). Available online: https://isap.sejm.gov.pl/isap.nsf/DocDetails.xsp?id=WDU19900160095 (accessed on 30 November 2020).
135. Szczecin Metropolitan Railway. Available online: http://skm.szczecin.pl/index.php (accessed on 30 November 2020).
136. Center for EU Transport Projects. Tablice Kosztów Jednostkowych do Wykorzystania w Analizach Kosztów i Korzyści. Available online: https://www.cupt.gov.pl/wdrazanie-projektow/analiza-kosztow-i-korzysci/narzedzia/tablice-kosztow-jednostkowych-do-wykorzystania-w-analizach-kosztow-i-korzysci (accessed on 30 November 2020).
137. Energy Forum. Energy Transition in Poland. Edition 2019. Available online: https://forum-energii.eu/en/analizy/transformacja-2019 (accessed on 30 November 2020).
138. Energy Forum. Energy Transition in Poland. Edition 2020. Available online: https://forum-energii.eu/en/analizy/transformacja-2020 (accessed on 30 November 2020).
139. Tapio, P.; Banister, D.; Luukkanen, J.; Vehmas, J.; Willamo, R. Energy and Transport in Comparison: Immaterialisation, Dematerialisation and Decarbonisation in the EU15 between 1970 and 2000. *Energy Policy* **2007**, *35*, 433–451. [CrossRef]
140. Dillman, K.J.; Árnadóttir, Á.; Heinonen, J.; Czepkiewicz, M.; Davíðsdóttir, B. Review and Meta-Analysis of EVs: Embodied Emissions and Environmental Breakeven. *Sustainability* **2020**, *12*, 9390. [CrossRef]

energies

Article

A Back-of-Queue Model of a Signal-Controlled Intersection Approach Developed Based on Analysis of Vehicle Driver Behavior

Elżbieta Macioszek [1,*] and Damian Iwanowicz [2]

[1] Department of Transport Systems, Traffic Engineering and Logistics, Faculty of Transport and Aviation Engineering, Silesian University of Technology, Krasińskiego 8 Street, 40-019 Katowice, Poland
[2] Department of Road and Transport Engineering, Faculty of Civil and Environmental Engineering and Architecture, UTP University of Science and Technology in Bydgoszcz, Prof. S. Kaliskiego 7 Street, 85-796 Bydgoszcz, Poland; damian.iwanowicz@utp.edu.pl
* Correspondence: elzbieta.macioszek@polsl.pl; Tel.: +48-32-603-41-50

Abstract: In smart cities, it is expected that transport, communication as well as the movement of people and goods will take place in the shortest possible time while maintaining a high level of safety. In recent years, due to the significant increase in the number of passengers and vehicles on the road and the capacity limitations of transport networks, it has become necessary to use new technologies for intelligent control and traffic management. Intelligent transport systems use advanced technologies in the field of data gathering, information processing, and traffic control to meet current transport needs. To be able to effectively control and manage road traffic, it is necessary to have reliable mathematical models that allow for a faithful representation of the real traffic conditions. Models of this type are usually the basis of complex algorithms used in practice in road traffic control. The application of appropriate models reflecting the behavior of road users contributes to the reduction of congestion, the vehicles travel time on the transport network, fuel consumption and the emissions, which in turn support broadly understood energy savings. The article proposes a model that allows for the estimation of the maximum queue size at the signal-controlled intersection approach (so-called: maximum back-of-queue). This model takes into account the most important traffic characteristics of the vehicles forming this queue. The verification allowed for the conclusion that the proposed model is characterized by high compliance with the actual traffic and road conditions at the intersections with signal controllers located in built-up areas in Poland. The obtained compliance confirms the possibility of using the model for practical applications in calculating the maximum back-of-queue at signal-controlled intersections located in built-up areas in Poland.

Keywords: signal-controlled intersections; back-of-queue model; traffic monitoring; video analysis; drivers behaviours; traffic engineering; road transport

Citation: Macioszek, E.; Iwanowicz, D. A Back-of-Queue Model of a Signal-Controlled Intersection Approach Developed Based on Analysis of Vehicle Driver Behavior. *Energies* **2021**, *14*, 1204. https://doi.org/10.3390/en14041204

Academic Editor: Stefania Santini

Received: 22 January 2021
Accepted: 18 February 2021
Published: 23 February 2021

1. Introduction

The main measures of assessing traffic conditions at signal-controlled intersections include delays incurred by vehicle drivers and the queue lengths at the approach. In the oldest models, the delays and queue lengths were determined using the theory of mass service (also known as the queue theory). Queue theory models belong to the group of traffic micromodels and, like meso- and macromodels, they can be divided into deterministic and stochastic (with or without elements of randomness). The models of the mass service theory use the Kendall's designation scheme, written as $A/B/m$, where A is the distribution of a random variable in the access process, B is the distribution of a random variable according to which the service process takes place, and m is the total number of positions for handling in the process [1–4]. The queuing process is stochastic, both for the time headways to the system and the times of service by the system [3,5,6].

The idea of the modeling process of delays and the queues at signal-controlled intersections is presented in Figure 1. As the process of servicing vehicles begins, the queue length begins to decrease. It can be concluded that the queue length is completely dependent on the number of arrivals at the approach. It should also be noted that in the cycle presented in Figure 1, there is a period of free inflow of vehicles to the intersection approach while transmitting the effective green signal. In the existing models, the impact of this type of vehicles on the queue formed at the intersection approach is usually not taken into account.

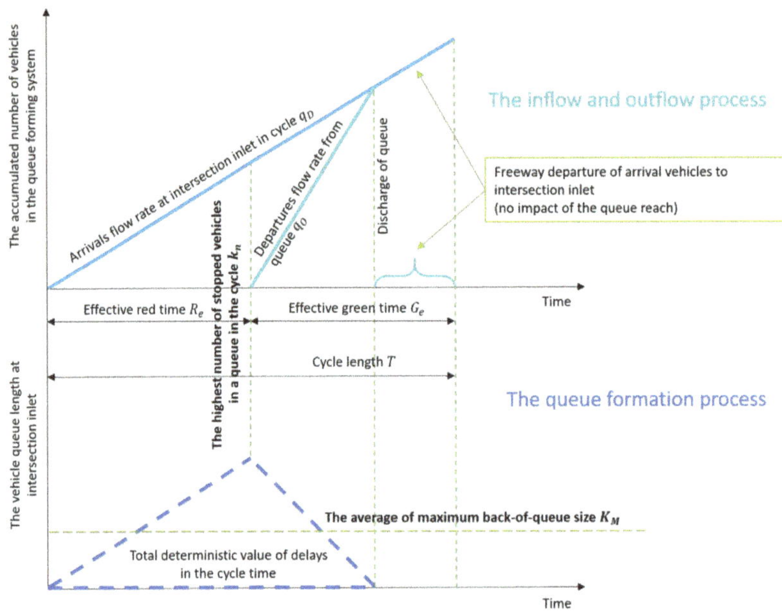

Figure 1. The idea of the modeling of delays and the back-of-queues at signal-controlled intersection approach in a typical period of vehicle service (in the period of no traffic overload). Source: Own research based on [3,4].

The use of queue theory models to describe the road traffic process usually requires the introduction of certain simplifications resulting from an ambiguous interpretation of some phenomena that occur in road traffic along a road section or at an intersection. As an example, you can provide the process of arrivals for servicing in the system, e.g., vehicle access to the intersection. As the model does not represent the road section leading to the intersection approach, service is assumed to be at the stop line. This has its consequences in the access process, where the moments of access by vehicles are replaced by moments of theoretical arrivals at the approach [1–3,7,8].

A certain drawback of mass service models, due to the very high randomness of the road traffic process on the street network, is their reference only to the so-called steady-state (queue in a stochastic equilibrium state), Both the IN and OUT process are constant over time, and therefore the parameters describing the distribution of random variables of these processes are constant. Hence, mass service models cannot be used to describe the road traffic process in situations of the high variability of traffic volumes, which often occur during peak periods in large- and medium-sized cities. Nevertheless, the most popular model that allows estimating the value of delays incurred by vehicle drivers and the average queue lengths at the intersections approach with signal controllers is the Webster model [9]. This model and its subsequent updates [10], until now constitute the canon of the methodology for estimating the capacity and measures for assessing traffic conditions at signal-controlled intersections. The undoubted disadvantage of these models was their use only when the intersection was loaded with traffic volumes not exceeding its capacity, which, however, did not have a significant impact on those years.

The development of models based on the queue theory was dealt with, inter alia, in [11–13]. The mathematical functions used in these works were indirectly based on the method of estimating delays with the use of waiting queues. The subsequent evolution of mathematical models allowed these models to be classified into the group of time-dependent models. In this case, the technique of coordinate transformation was used concerning the delay curve derived for the steady-state [4,11,14]. This modeling approach assumes that for small values of load levels for a given lane, in unsaturated traffic conditions, the queue length is approximately equal to the product of the constant intensity of reports and the time of arrival. On the other hand, in the conditions of high traffic load, i.e., in the conditions of saturated and oversaturated demand flow rate, the maximum back-of-queue can be approximated with the use of a linear model, which assumes proportionality to the analysis time and the amount of overload. The use of a coordinate transformation "shifts" the asymptotic of the estimated delay curve (queue length), from the equilibrium point between the inflow and outflow intensity (volume-to-capacity ratio $X = 1$) for the steady-state towards the deterministic overload loss line. The algorithm of this solution was included, inter alia, in the works [11,15].

Further effects of work on the improvement time-dependent models of delays and queues are presented, inter alia, in the works [16–28]. The result of this work was the penultimate significant update of the HCM guidelines [29], considered one of the most reliable studies on road infrastructure design guidelines in the world. The mathematical model developed as part of this update for estimating the average delays and the average back-of-queues at signal-controlled intersections is still used in practice by most countries around the world. Only some countries have developed their calculation methods based on the guidelines presented in HCM.

The models for estimating the maximum queue length usually consist of two separate estimators:

- the length of the queue formed during the red signal (k_Z),
- the length of the queue remaining after the end of green signal (k_P).

It is worth emphasizing that from the aforementioned design guidelines, only the American HCM model from 2010 [30] includes an estimator taking into account the condition of the queue from the period preceding the analysis period. This is the so-called initial queue (k_0). Other models do not have such an estimator in their formula or take this queue into account indirectly (e.g., in the HCM 2000 guidelines [30] in the estimator of the remaining queue length).

Based on the analysis of the form of mathematical models concerning the queue length at signal-controlled intersections from Australia [31,32], Canada [33], Germany [34,35], Poland [36], Denmark, France, Ireland and Sweden, Great Britain [10,15] and USA [29,30,37] the following conclusions were made:

- calculations of the average back-of-queues for a given computational group of traffic lanes consist in determining the queue length that may form in the analyzed period of analysis in an average signaling cycle,
- for the given initial conditions, the results of calculating the maximum queue lengths using various methods applied in the world do not differ significantly from each other (except for the Canadian [33] and American [29,30,37] methods),
- most of the calculation models of the average queue length are based mainly on the simplified description of vehicle operation in the "system", which takes place in the analyzed period of the analysis, without taking into account the earlier periods and significant fluctuations in the value of the inflow intensity in short periods (equal to or less than 15 min); the only exception are US models [29,30,37]; however, the specificity of the process of forming the queue of vehicles by such maneuvers as starting, stopping and moving in a column of vehicles during the signaling cycle is not taken into account at all, which may affect the length of the maximum queue in

a completely different way than in the case of data only about demand and supply (traffic volume/arrival flow rate/of vehicles moving to the intersection approach and the capacity/saturation flow rate/of the computational group of lanes),

- the existing calculation methods do not take into account with sufficient accuracy the process of shifting vehicle queues between individual signaling cycles and their accumulation, which is of significant importance, in particular, for cases of saturation or oversaturation traffic for a long time; hence, in this respect—to the best of the authors' knowledge—none of the analyzed guidelines for calculating traffic conditions at traffic light intersections do not include a sufficiently precise method for calculating the initial lengths of vehicle queues [38–40],
- the calculation models do not take into account the phenomenon of interaction between vehicle streams of different computational groups of traffic lanes (operated at the same or different times) before intersection approach, which has a significant impact on the actual degree of traffic load for individual lanes (so-called: volume-to-capacity ratio).

The article presents a model that allows for the estimation of the maximum back-of-queues at signal-controlled intersections in a single cycle. This model takes into account the most important traffic characteristics of vehicles forming the queue. For this purpose, the process of starting vehicles in the maximum queue in the states of saturation and oversaturation with the traffic of the signal-controlled intersection approach was subjected to a detailed analysis. The conducted analyzes allowed for the conclusion that the intensity of starting has a direct impact on the final result of the process of forming the maximum queue range, taking into account the remaining queue lengths and the initial queue lengths.

The article consists of seven sections. After the introduction, the second section presents a literature review on the subject in the field of modeling the queue length at the signal-controlled intersections. The third section defines in detail the subject of research, the testing ground, and the measurement method used, which by definition had to enable the simultaneous measurement of all phases of the vehicle queue formation, synchronized with all traffic light signals in individual signaling cycles. The next, fourth section presents the analysis of the research results. Section five presents the developed model of the maximum back-of-queue at the signal-controlled intersection approach. In turn, the verification and calibration of the proposed model is presented in section six. The paper ends with the discussion and conclusions section.

2. Literature Review

The problem of modeling the back-of-queue size at signal-controlled intersections is mainly related to the improvement of calculation methods and their adaptation to the needs of adaptive traffic control at city intersections. Models used to estimate the queue lengths are most often developed based on the following mathematical formulas:

- Markov chain—used, among others in Alfa [41], Chen et al. [42], Viti et al. [43],
- time series methods—used, among others in Gasz [44],
- fuzzy logic—used, among others in Shou et al. [45], Kafash et al. [46], Musci et al. [47],
- probability distribution—used, among others in von Zuylen et al. [48], Viti et al. [49],
- shockwave analysis—used, among others in Cao et al. [50], Ng et al. [51],
- cell transmission model—used, among others in Srivastawa et al. [52].

The mathematical models are often theoretical and suitable for practical implementation as part of the traffic control program algorithm using signal heads. So far, a significant group of works has been devoted to researching the process of forming queues of vehicles at the approach of insulated signal-controlled intersections, e.g., Hao et al. [53], Anokye et al. [54], as well as elements of main streets in the city, e.g., Bie et al. [55] and Fu et al. [56]. A frequently discussed research issue is also the search for a more precise than before method of traffic control using adaptive light signaling, e.g., Kafash et al. [46], Viti et al. [57], Yu et al. [58] or network analyses, dealing with the problem of selection of the optimal vehicle speed trajectory on a signalized arterial with consideration of vehicle queue, e.g., He et al. [59], Zhang et al. [60], Cao et al. [61], Zhao et al. [62] as well as works on microscopic simulation for signal-controlled intersections, e.g., Naghawi et al. [63], Ma et al. [64], or Li et al. [65].

Verifications of models used to estimate the queue length at signal-controlled intersections together with proposals for updates are presented, among others in the works of Anantharam [66], Chaudhry et al. [67], Szczuraszek et al. [39], and Viti [68]. Often, models of the queue lengths at intersections with signal controllers are a component of complex virtual reality models of the simulated road traffic process, e.g., Liu et al. [69], Xu et al. [70] and Vigos et al. [71].

Nowadays, the emerging advanced intelligent transportation systems, such as connected vehicles and cooperative vehicle-to-infrastructure systems, attract much research attentions. Scientific descriptions on the improvement of automated driving systems and the dynamic development of autonomous vehicle technology have resulted in the need to research the impact of autonomous vehicles on the process of queuing at signal-controlled intersections. In this group of works, we can distinguish works in the field of microscopic analyzes taking into account key operational parameters like vehicles speed, acceleration, inter-vehicle headway, and key performance metrics like degrees of load and delay, e.g., Le Vine et al. [72], Ramezani et al. [73], Yang et al. [74]. One can also distinguish works on macroscopic analyses aimed at optimizing the speed of connected autonomous vehicles and coordinating traffic light programs in transport networks by minimizing the queue lengths at intersections with signal controls, e.g., Wang et al. [75], Kong et al. [76], Iwasaki [77] and Liu et al. [78].

Moreover, in order to improving traffic control at intersections, it is necessary to have a reliable information collection system and apply modern effective methods of processing the collected information. Remote sensing tools are an effective way to detect vehicles and traffic conditions on the transport network. Digital camera images it is well suited to derive various traffic parameters such as average vehicle speed, vehicle density, beginning and end of congestion, length of congestion or for other traffic monitoring applications e.g., Anusha et al. [79,80], or Cai et al. [81]. This method is based on the vehicle detection on the road segment by change detection between two images with a short time lag, the usage of a priori information such as road data base, vehicle sizes and road parameters and a simple linear traffic model based on the spacing between vehicles. The studies are also used standalone video processing algorithms and based on this information is stored in databases, for example, trajectories of traffic streams of vehicles, lengths of queues of vehicles, etc. Data from passive remote sensing methods were used to describe the process of forming queues of vehicles at signal-controlled intersections, e.g., in the works of Cetin et al. [82], Chang et al. [83], Hao et al. [84], Mukhopadhyay et al. [85], Iwanowicz et al. [86], while the data from active remote sensing methods were used e.g., in the works of Liu et al. [87], Milla et al. [88], Khan et al. [89], Sun et al. [90], and Xie et al. [91]. Most often, the result of these works were proposals for the methodology of testing road traffic characteristics or proposals for improving the existing methods of estimating traffic conditions at signal-controlled intersections or improving traffic control programs.

In recent years, light unmanned aerial vehicles have also been willingly used to collect data on vehicle queues at traffic light intersections, e.g., Zhou et al. [92], Wan et al. [93] or Salvo et al. [94].

3. Materials and Methods

3.1. *Subject of Study and Testing Ground*

The subject of the research was the process of forming queues of vehicles on a single traffic lane of the intersection approach with signal controllers. The examination of this process must take place both at the approach and before the approach of a given intersection, along the longest formed queue reach, so that it is possible to fully observe the individual phases of the queue forming and discharging process in time. This means that the tests should be conducted to select signal-controlled intersections, on which the section of the road leading to the approach has a relatively large accumulation zone of vehicles, undisturbed by access points, i.e., additional traffic inflows and outlets (intersections without traffic controllers), parkings or bus stops and of course crossing for pedestrians or cyclists.

The traffic lane at the approach of the signal-controlled intersection was adopted as the model section for the tests, which is devoid of additional side on and off streets (switches), on which vehicles of the same route through the intersection are accumulated, having the same sequence of transmitted traffic light signals. The service of vehicles for all lanes at the approach is collision-free. Only under such conditions is it possible to obtain an undisturbed system of both inflow and outflow of the vehicle stream and it will be possible to observe the actual changes formed in particular phases of vehicle queues in time. An example of a measuring section adopted as a reference section for conducting research is shown in Figure 2.

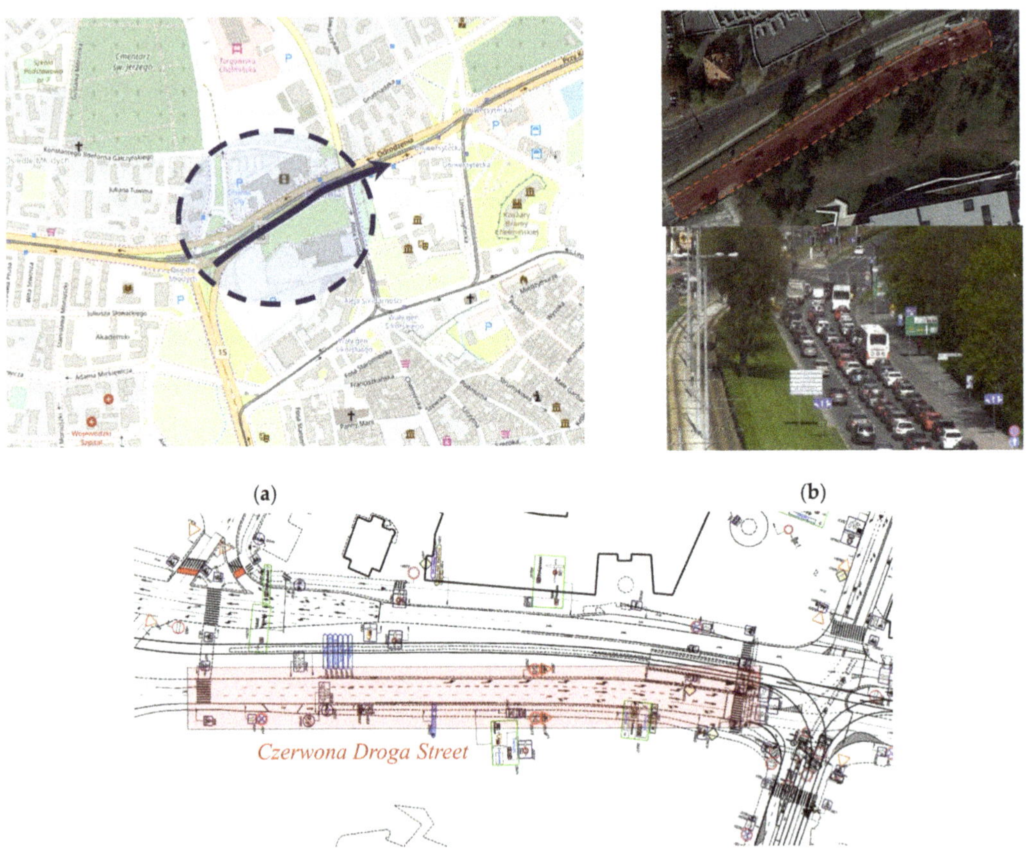

Figure 2. An example of a measuring section adopted as a model for research (Poland, Torun city, intersection approach Czerwona Droga Street/west/): (**a**). measuring section location; (**b**). view of the queue of the vehicle on a measuring section; (**c**). situational plan. Source: Own work based on Open Street map, Google Earth and WZDR System [95].

The testing ground consisted of a total of 20 signal-controlled intersections located in cities in Poland with a large (>0.50 million inhabitants) or average population (0.20 ÷ 0.50 million inhabitants). The test grounds were located in cities such as Bydgoszcz (TGa1 ÷ TGa8, TGb1

÷ TGb8), Torun (TGb9, TGb10) and Warsaw (TGb11, TGb12). The collected data was divided into two groups. Based on the data from 12 test sites, a model was developed to calculate the numerical values of the maximum range of back-of-queue (TGb1 ÷ TGb12). However, this model was verified on the basis of data from 8 research polygons (TGa1 ÷ TGa8). On these proving grounds, only the throught movements were analyzed. The traffic control parameters did not change during the measurements, i.e., the traffic at the intersections was controlled by means of fixed-time signaling. Measurements were carried out in 2014–2019. The research covered the following features of the process of forming queues of vehicles at intersections with signal controllers in relation to a single signaling cycle as well as the features of the traffic streams flow:

- arrival (demand) flow rate,
- departure flow rate,
- starting-up flow rate of vehicles from the maximum queue,,
- stopping flow rate of vehicles from the remaining (initial) queue,

and:

- the fact of the formed range of the maximum back-of-queue in a given signaling cycle,
- generic groups of vehicles of a given traffic stream (arrival, departue, starting, stopping).

The collected research material also made it possible to determine the basic characteristics of the traffic signal program (start and change times of individual signals of the analyzed period). During the processing of the collected research material, the starting maneuvers that took place during the transmission of the green signal were also clearly distinguished from those that were performed after the commencement of transmitting the red signal in the next cycle.

Measurements were carried out in weather conditions favorable to traffic (no precipitation, good visibility). The sample sizes for the analyzes were selected from the Lapunov formula assuming a significance level of $\alpha = 0.05$. Table 1 presents the characteristics of the most important features of the studied intersection approaches and the collected data. In further analyzes, data on only the group of passenger cars was used, without taking into account the characteristics of other generic groups of vehicles (e.g., starting-up or slowing down in the queues of heavy vehicles and buses), i.e., type structure of vehicles [96,97]. The number of vehicles in the queues takes into account all vehicle types.

Table 1. The summary of the characteristics of the examined intersections and the collected data.

Parameters Type	Parameter Name	The Scope of Data
General	Function of a road in spatial development	Collective and distributing
	Entry movements through the intersection	Through, through and left, through and right
Geometric	Length of the accumulation zone [m]	250–620
	Lane width [m]	3.0–3.5
Traffic control signals	Type of control	Fixed-time
	Signaling cycle time [s]	96.0–140.0
	Green signal time [s]	18.0–34.0
	Traffic movements while servicing	There is no conflict movements
Traffic volumes	Arrival flow rate [Veh./s]	0.00–0.25
	Departure flow rate [Veh./s]	0.32–0.61
	Starting-up flow rate [Veh./s]	0.37.1.35
	Stopping flow rate [Veh./s]	0.18–0.93
	Remaining queue range [Veh./cycle]	0–49
	Maximum back-of-queue range [Veh./cycle]	5–61

3.2. Measurement Method

The basic condition for selecting the measurement method was to ensure that all phases of the vehicle queue formation process were measured simultaneously, synchronized precisely with all traffic light signals in individual signaling cycles. Moreover, the observations of the road traffic process had to take place practically simultaneously along the entire length of the interstitial section between the established cross-section of the inflow and outflow.

The characteristics of traffic signal programs and traffic characteristics were tested with the use of digital cameras, which are classified as passive remote sensing tools. This test method is characterized by very high measurement accuracy. Moreover, this method is often used to study road traffic characteristics at various intersections, including signal-controlled intersections [98]. The measurements were performed with first-generation YI Action digital cameras, which enable recording images in Full HD 1920 × 1080 px resolution with a recording speed of at least 24 fps. This resolution is sufficient to scale the video image and add or arrange additional graphics on it. These cameras are connected via a Wi-Fi network, enabling observe the image recorded by them, e.g., on the screen of a smartphone or tablet. For 24-h continuous measurement, it was necessary to equip the camera with an external battery with a capacity of 20,000 mAh with lithium-ion cells (so-called power-bank).

At all test sites, digital cameras were installed on streetlights at a height of 5.5 ÷ 7.5 m above the level of traffic lanes so that the image recorded with their help captures as wide a fragment of the road section as possible (Figure 3). Such a location of digital cameras ensured good visibility of each vehicle and the spaces between them (the vehicle was not obstructed by the bodies of larger vehicles). For such a height, it was established that the distance between the cameras should be within 40 ÷ 60 m from each other, due to the need to overlay the images of frames from subsequent cameras. In most cases, this distance is the distance between two or three consecutive street lamps. The first of the cameras was installed as close as possible to the stop line cross-section. That was possible to correctly read the sequence changes of the traffic signal times.

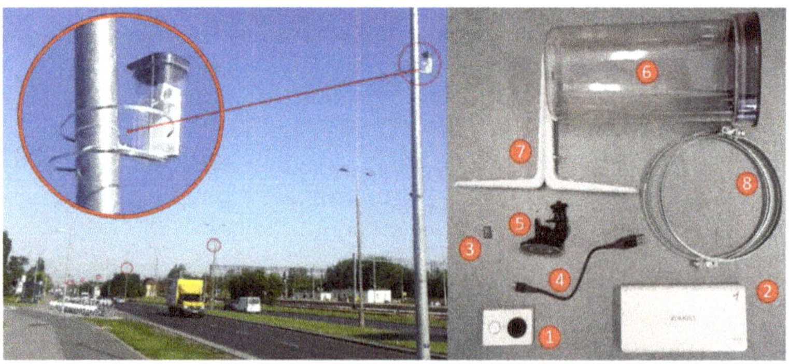

Figure 3. Arrangement of measurement kits along Kamienna Street in Bydgoszcz, Poland (where: 1—camera, 2—power bank, 3—memory card, 4—power supply cable, 5—camera holder, 6—weatherproof enclosure, 7—support sections, 8—clamps).

After the measurements, the collected video material was compiled from film frames from individual cameras into one main image. Thus, it was possible to observe the flow of the vehicle streams along the entire research section, recorded on the frames from individual cameras using one monitor. The collected measurement material was processed using the DVideo Player. Details of the measurement method were developed in [86].

4. Model of the Maximum Back-of-Queue

The following definitions were adopted to characterize the model of the process of forming queues at intersection approach with signal controllers:

- The queue at the signal-controlled intersection approach on a single lane is a stream of vehicles made up of vehicles arriving at the intersection, constituting a compact column of vehicles in the impact space of this intersection, waiting to be able to pass through the intersection (service). In the queue, vehicles may be:
    - stopped—this applies to vehicles which, due to conditioned by red signal duration on a signal heads, have stopped and their drivers are waiting for the possibility of passing in green signal duration,
    - partially in motion, (crawl speed)—this applies to vehicles moving at a very low speed ($\leq$8 km/h [37]) because they attach a standing column of vehicles in front of the intersection approach or they are vehicles whose drivers alternately start and stop maneuvers due to the oversaturation of the queue (only a change of position in the queue without departure from the intersection after starting-up maneuver),
    - in motion—this applies to vehicles whose drivers have starting-up maneuvered to leave the intersection at the green signal duration, but are still in the intersection approach area (not serviced yet); this also applies to those vehicles which have failed to leave the intersection approach during the green signal interval and which are moving towards the approach after the red signal has started in order to come to a closer position in front of intersection in the queue than before starting-up maneuver.
- A vehicle in the queue is understood as the one that reaches the intersection and leaves it with a delays resulting from the process of servicing vehicles controlled by signal heads. Thus, it is a vehicle that forms a stopped queue or is only under the influence of the queue formed at the intersection approach (losing speed). A vehicle that crosses the intersection on the green signal duration without being influenced by the queue of vehicles (undisturbed flow) is not a queue forming vehicle,
- Back-of-queue length—k—means the actual number of vehicles in the queue at the intersection approach on a single lane [Veh.], [m],
- Back-of-queue range (size)—K—means the hypothetical length of the queue, measured by the hypothetical number of vehicles between the stop line at the intersection approach and the last stopped vehicle in the queue on a single lane [Veh.], [m],
- Maximum back-of-queue range ("maximum queue")—K_M—means the highest theoretical number of stopped vehicles that would line up in the lane from the stop line to the last stopped vehicle in the queue on a given signaling cycle and on a single lane. The maximum queue most often occurs after the start of the green signal duration. The number of vehicles in the maximum queue is a theoretical value due to the dispersion (inertia) of vehicles from the beginning of the queue that performs the starting-up maneuver after a green signal duration, while the vehicles from the end of the queue have not yet performed this maneuver (and can even only join the queue as a arrivals in back-of-queue interaction). Hence, the range of the queue takes up a space composed of hypothetically stopped vehicles [Veh.], [m].

The geometric dependencies of the physical process of forming the maximum queue at the signal-controlled intersection approach are shown schematically in Figure 4.

The meaning of the notations used in the description of the process of the volatility of the back-queue size in time and the formation of the maximum back-queue size in a given signaling cycle i and also used on the Figure 4 have been presented in Table 2.

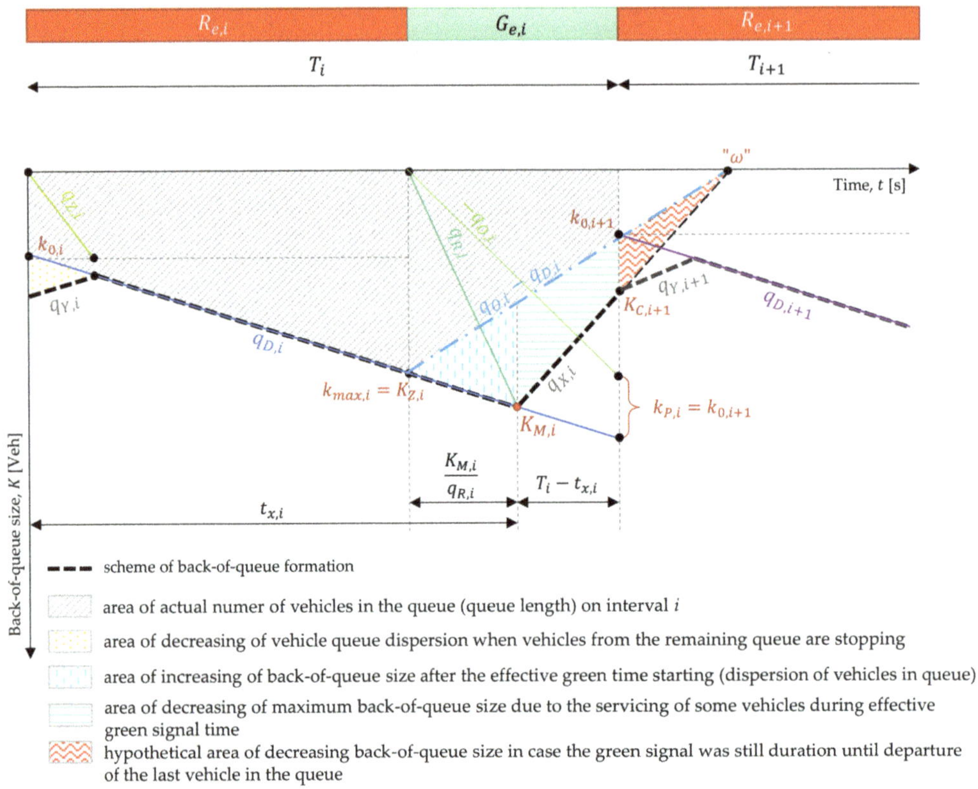

Figure 4. Diagram of the process of the volatility of the back-of-queue size in time and the formation of the maximum back-of-queue size in a given signaling cycle *i*.

Table 2. The summary of the characteristics of the examined intersections and the collected data.

Parameter	Parameter Description	Unit
i	Number of signaling cycle (interval of analysis)	[−]
T_i	Cycle length of interval i	[s]
$R_{e,i}$	Effective red time of interval i	[s]
$G_{e,i}$	Effective green time of interval i	[s]
$q_{D,i}$	Arrival flow rate of interval i	[Veh./s]
$q_{O,i}$	Departure flow rate on interval i, (i.e., saturation flow rate)	[Veh./s]
$q_{R,i}$	Starting-up flow rate of vehicles in the maximum queue of interval i	[Veh./s]
$q_{Z,i}$	Stopping flow rate of vehicles in the initial queue of interval i	[Veh./s]
$q_{X,i}$	Decreasing flow of maximum queue range during effective green time $G_{e,i}$ due to service time at intersection approach of interval i	[Veh./s]

Table 2. *Cont.*

Parameter	Parameter Description	Unit
$q_{Y,i}$	Decreasing flow of maximum queue range after start effective red time $R_{e,i}$ including vehicles arriving to queue $q_{D,i}$ on interval i	[Veh./s]
$K_{Z,i}$	Back-of-queue size at the moment of starting effective green time duration $G_{e,i}$ on interval i; generally the number of vehicles in $K_{Z,i}$ is equal to maximum queue length $k_{max,i}$,	[Veh.]
$K_{C,i+1}$	Back-of-queue size at the moment of starting effective red time duration $R_{e,i}$ on next interval i+1; the number of vehicles in $K_{C,i+1}$ is equal to remaining queue length ($k_{P,i}$)	[Veh.]
$K_{M,i}$	Maximum back-of-queue size on interval i	[Veh.]
$t_{x,i}$	Formation time of the maximum queue size on interval i	[s]
$k_{0,i}$	Initial queue lenght on interval i	[Veh.]
$k_{mac,i}$	Maximum queue length (maximum number of stopped vehicle) on interval i	[Veh.]
$k_{P,i}$	Remaining queue length on interval i, equivalent with initial queue length on interval i + 1	[Veh.]
ω	A hypothetical point in time defining the end of the service process of vehicles from the maximum queue, if the green signal was still duration at the intersection approach	[s]
c_i	Lane capacity of interval i	[Veh./s]
λ_i	Green effective period ratio	[−]
ρ_i	Arrivals-to-Starts ratio	[−]
η_i	Capacity to Starts ratio	[−]
X_i	Volume-to-Capacity ratio	[−]

The maximum queue can be also formed when the duration of the next interval (red or green effective signal time in i + 1 cycle), and even in subsequent cycles. In this most likely scenario, the maximum queue size will be formed at the crossing of the two functions below:

- starting-up of the stopped vehicles in the queue at the intersection approach after the effective green time began,
- arriving of the inflow vehicles taking into account the initial queue length at the intersection approach.

The assumption is that the course of these functions is approximately linear i.e., there are constant time headways between vehicles in both of these processes. Hence, to obtain the crossing of these two functions, one should solve the following system of equations:

$$\begin{cases} K(t) = q_{R,i} \cdot (t_{x,i} - R_{e,i})[\text{Veh.}] \\ K(t) = k_{0,i} + q_{D,i} \cdot t_{x,i}[\text{Veh.}] \end{cases} \quad (1)$$

Taking the above into account, we obtain a dependence that allows determining the maximum queue formation time $t_{x,i}$:

$$\begin{gathered}
q_{R,i} \cdot (t_{x,i} - R_{e,i}) = k_{0,i} + q_{D,i} \cdot t_{x,i} \\
q_{R,i} \cdot t_{x,i} - q_{D,i} \cdot t_{x,i} = k_{0,i} + q_{R,i} \cdot R_{e,i} \\
t_{x,i} = \frac{k_{0,i} + q_{R,i} \cdot R_{e,i}}{q_{R,i} - q_{D,i}} \\
t_{x,i} = \frac{k_{0,i}}{q_{R,i} - q_{D,i}} + \frac{q_{R,i} \cdot R_{e,i}}{q_{R,i} - q_{D,i}} [\mathrm{s}]
\end{gathered} \tag{2}$$

In Equation (2), two components were obtained. The first represents the extension of the maximum queue formation time due to the presence of the initial queue k_0. The second component of Equation (2) represents the maximum queue formation time in a given signaling cycle i. In this second component, the queue consists only of vehicles arriving at the intersection approach during the analyzed signaling cycle.

Assuming the following variable designations:

$$\rho_i = \frac{q_{D,i}}{q_{R,i}} [-] \tag{3}$$

$$R_{e,i} = T_i - G_{e,i} [\mathrm{s}] \tag{4}$$

$$\lambda_i = \frac{G_{e,i}}{T_i} [-] \tag{5}$$

Equation (2) can be written as follows:

$$\begin{gathered}
t_{x,i} = \frac{k_{0,i}}{q_{R,i} \cdot \left(1 - \frac{q_{D,i}}{q_{R,i}}\right)} + \frac{q_{R,i} \cdot T_i \cdot (1 - \lambda_i)}{q_{R,i} \cdot \left(1 - \frac{q_{D,i}}{q_{R,i}}\right)} \\
t_{x,i} = \frac{k_{0,i}}{q_{R,i} \cdot (1 - \rho_i)} + \frac{T_i \cdot (1 - \lambda_i)}{1 - \rho_i} \\
t_{x,i} = \frac{1}{1 - \rho_i} \cdot \left[\frac{k_{0,i}}{q_{R,i}} + T_i \cdot (1 - \lambda_i)\right] [\mathrm{s}]
\end{gathered} \tag{6}$$

Therefore, the maximum back-of-queue size on the cycle i will be equal:

$$\begin{gathered}
K_{M,i} = q_{R,i} \cdot (t_{x,i} - R_{e,i}) \\
K_{M,i} = q_{R,i} \cdot \left[\frac{k_{0,i} + q_{R,i} \cdot R_{e,i}}{q_{R,i} - q_{D,i}} - R_{e,i}\right] \\
K_{M,i} = q_{R,i} \cdot \left[\frac{k_{0,i} + q_{R,i} \cdot R_{e,i} - q_{R,i} \cdot R_{e,i} + q_{D,i} \cdot R_{e,i}}{q_{R,i} - q_{D,i}}\right] \\
K_{M,i} = q_{R,i} \cdot \left[\frac{k_{0,i} + q_{D,i} \cdot R_{e,i}}{q_{R,i} - q_{D,i}}\right] \\
K_{M,i} = \frac{1}{1 - \rho_i} \cdot [k_{0,i} + q_{D,i} \cdot T_i \cdot (1 - \lambda_i)] [\mathrm{Veh.}]
\end{gathered} \tag{7}$$

However, assuming the following variable designations:

$$\eta_i = \frac{c_i}{q_{R,i}} [-] \tag{8}$$

$$X_i = \frac{q_{D,i}}{c_i} [-] \tag{9}$$

where:

c_i—lane capacity on the cycle i, [Veh./s], hence the final form of the model takes the form:

$$K_{M,i} = \frac{1}{(1 - X_i \cdot \eta_i)} \cdot [k_{0,i} + q_{D,i} \cdot T_i \cdot (1 - \lambda_i)] [\mathrm{Veh.}] \tag{10}$$

Unlike the existing models, the proposed model of the maximum queue at the signal-controlled intersection approach in a given lane in the i-th signaling cycle (10), additionally takes into account:

- an initial queue length, which is usually omitted in most of the existing models (the exception is the model [30,37], which takes into account k_0 in the calculations—but in a different way),
- an intensity of starting-up of stopped vehicles in the maximum queue q_R on the cycle i—after starting an effective green time using the relationship of a saturation flow rate.

To define a model that allows for the determination of numerical values of the initial queue length for the i-th cycle, which is the remaining queue length for the $i-1$ cycle, the geometric dependencies of this process were used, presented in Figure 5 (where all variables are as previously defined). Taking into account the data presented on Figure 5, can be written:

$$k_{0,i} = k_{P,i-1}[\text{Veh.}] \tag{11}$$

$$k_{P,i} = \max\{0; [q_{D,i} \cdot T_i - q_{O,i} \cdot G_{e,i} + k_{0,i}]\}[\text{Veh.}] \tag{12}$$

and because of:

$$q_{O,i} = \frac{c_i}{\lambda_i}[\text{Veh./s}] \tag{13}$$

the form of the model can be determined:

$$\begin{gathered} k_{P,i} = \max\left\{0; \left[q_{D,i} \cdot T_i - \frac{c_i}{\lambda_i} \cdot \lambda_i \cdot T_i + k_{0,i}\right]\right\} \\ k_{P,i} = \max\{0; [k_{0,i} + q_{D,i} \cdot T_i - c_i \cdot T_i]\} \\ k_{P,i} = \max\{0; [k_{0,i} + c_i \cdot T_i \cdot (X_i - 1)]\} = k_{0,i+1}[\text{Veh.}] \end{gathered} \tag{14}$$

Although it contains some common features and identical variables, the proposed model (14) allows for the determination of numerical values of the initial queue length for the i-th cycle, is a completely different estimate of the remaining queue length compared to the models that can be found in various existing design guidelines (models based on of work [11,15]). A similar mathematical formula to the model (14) can be found in [29]. It is used to determine the initial queue length for a given analysis period (0.25 h or 1.00 h), indirectly used in the calculation of the remaining queue length and the volume of lane group inflow rate.

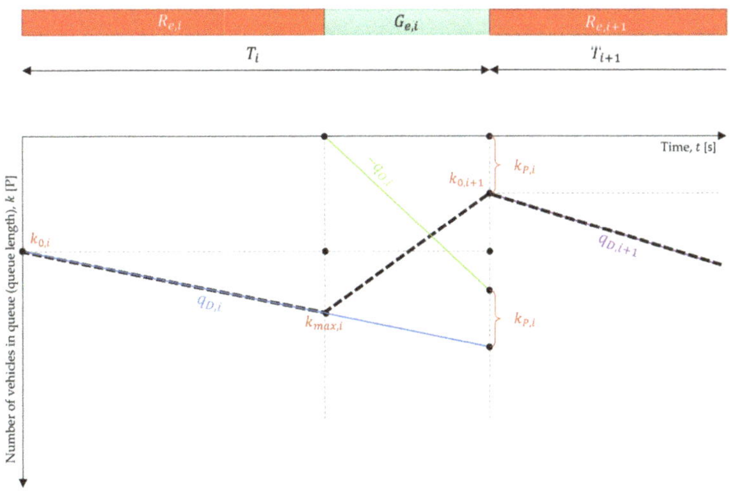

Figure 5. Diagram of the process of forming the remaining queue length in a given signaling cycle i (or a process of forming the initial queue length for the next signaling cycle $i+1$).

5. Field Survey Data Analysis

The collected data made it possible to carry out detailed statistical analyzes of the processes taking place at signal-controlled intersections, including in particular the process of starting vehicles in the maximum queue, which hypothetically determines the final range of the queue in a given signaling cycle. The analyzes used the values of the time headways on the lanes at the signal-controlled intersections in the appropriate signaling cycles. In the analyzes, using the 95% confidence interval, statistically uncertain results i.e., gross errors. Additionally, to confirm the validity of rejection of gross errors, the D. Dixon test was performed, assuming the significance level $\alpha = 0.05$, which also allowed to identify of outliers in the sample. In the end, 2.5% of the shortest and 2.5% of the longest intervals between vehicles in the approach queue were rejected. In total, 1791 signaling cycles were used to build the model, in which 18,377 vehicles reported for service at the intersection approach. These vehicles performed a total of 24,641 starting maneuvers in the given signaling cycles. From Figures 6–8, selected examples of the raw results of the research on the process of forming queues of vehicles at the signal-controlled intersections approach are presented.

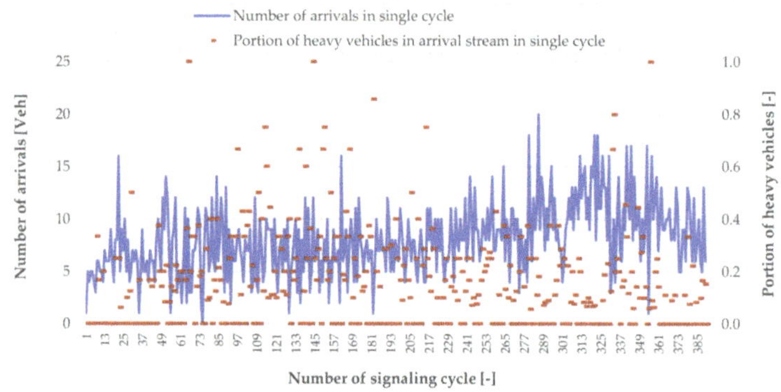

Figure 6. Distribution of the number of vehicles from the arrival stream in signaling cycles in a given measurement period taking into account the portion of heavy vehicles.

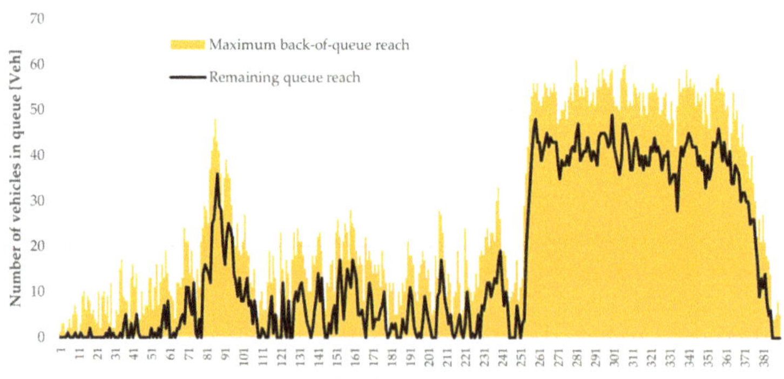

Figure 7. Distribution of the number of vehicles in the maximum back-of-queue for a given lane at the approach in individual signaling cycles in a given measurement period (including vehicles Figure 8. Starting-up times for the vehicles forming the maximum back-of-queues for a given lane controlled by traffic signal heads with the same signal control parameters, i.e., green signal and cycle length.

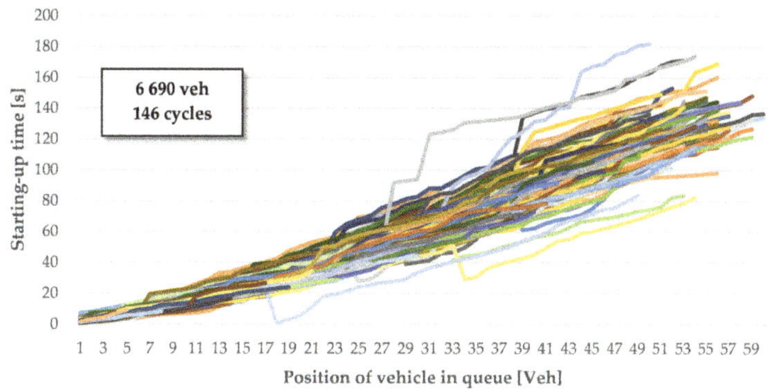

Figure 8. Starting-up times for the vehicles forming the maximum back-of-queues for a given lane controlled by traffic signal heads with the same signal control parameters, i.e., green signal and cycle length.

The average values of cumulative time intervals of the stream of vehicles starting from the maximum queue after the start of transmitting a green signal were analyzed (Figure 9). During the analysis of the research results, it was found that the intensity of starting may depend on the level of service at the green signal duration. In connection with the above, it was checked whether there is a correlation between the average values of the starting-up flow rate during green signal and the average values of the maximum departure flow rate (and so saturation flow rate) in the given traffic lanes (Figure 10).

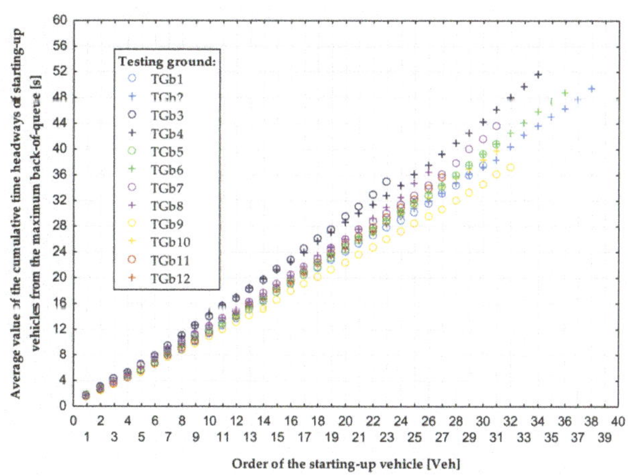

Figure 9. Results of the analysis of average values of cumulative time headways in the stream of starting-up vehicles from the maximum back-of-queue after turning on the green signal in a given cycle for subsequent vehicles.

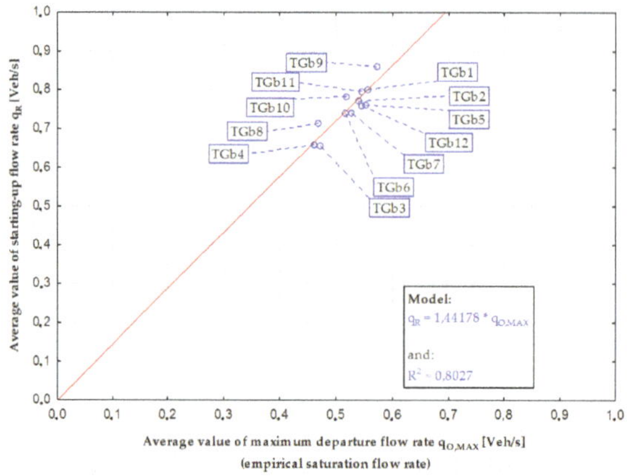

Figure 10. Results of the correlation analysis of the average values of the intensity of starting vehicles from the maximum queue as a function of the average values of the maximum intensity of the service.

The analysis of the dependence of the average values of the starting-up flow rate from the maximum back-of-queue in green signal duration as a function of the average values of the maximum departure flow rate in the given lanes was performed for the state of saturation and oversaturation traffic conditions using the least-squares method. Due to the small share of heavy vehicles in a modal split in cities, the analysis did not take into account the impact of these types of vehicles. The time headways analysis of the starting process concerned only passenger cars.

The form of the linear regression function was established without the intercept, which corresponds to the situation where no vehicles are departing from the queue, both in real traffic conditions and in the conducted analyzes. As a result of the analysis, the following model describing the average value of the intensity of starting-up vehicles from the queue was obtained:

$$\overline{q}_R = 1.45 \cdot \overline{q}_{O,MAX}\ [\text{Veh./s}],\ R^2 = 0.8027 \tag{15}$$

where: $\overline{q}_R$—average value of starting-up flow rate [Veh./s] and $\overline{q}_{O,MAX}$—average value of maximum departure flow rate during green signal in saturated and oversaturated traffic condition at intersection approach [Veh./s].

The results of the regression and correlation analysis confirm the existence of a relationship between the average value of the intensity of starting vehicles from the maximum queue and the average maximum value of the outflow intensity during green signal interval. This relationship is reflected in reality and represents the behavior of vehicle drivers. Each progressive vehicle from the queue takes part in the stream of maximum outflow, i.e., in the conditions of the so-called saturation flow rate. Even if the driver is unable to perceive the displayed signal on the signalization e.g., due to its distant position in the queue or the possible obstruction of the field of view by a vehicle of larger dimensions, when performing a starting-up maneuver, it does so under the influence of the vehicle in front of it. It is related to the behavior of drivers in line with the model of the so-called "Driving behind the leader". In the case under consideration, the analysis of vehicles starting-up from a stopped queue, the vehicle in front of it which has already performed such a maneuver. From Figure 9 it can be seen that the approximation of the starting-up flow rate function can be applied by a simple linear model. However, a slight decrease in this intensity is noticeable along with the increase in the number of vehicles in a queue (after about 20–25 position).

6. Model Verification

The proposed model of the maximum queue size at signal-controlled intersection approach on a given lane has been verified. The verification of the model consisted of a comparative analysis of the empirical values of the maximum back-of-queue size ($K_{M,i(emp)}$) with the theoretical values ($K_{M,i}$) determined based on the model (10). The verification was carried out on data obtained from remote sensing of the process of forming queues of vehicles at eight signal-controlled intersections. Comparative analyzes were performed for two states of traffic conditions at the intersection approach, i.e., in unsaturated inflow (ie when the volume-to-capacity ratio X did not exceed one) and in saturated or oversaturated inflow, when remaining queues formed at the approach (ie when the volume-to-capacity ratio X was equal to or greater than one) and jointly for both states. In total, 1406 signaling cycles were verified, including 655 in unsaturated traffic conditions and 751 in saturated or oversaturated traffic conditions occurring at intersection approach. The values of errors in estimating the maximum back-of-queues are presented in Table 3.

Table 3. Values of errors in estimating the maximum back-of-queue size at the signal-controlled intersection approach [Veh.].

Testing Ground	Traffic Conditions at the Intersection Approach:		
	Unsaturated Inflow [Veh.]	**Saturated or Oversaturated Inflow [Veh.]**	**All Together [Veh.]**
TGa1	2.50	3.28	2.95
TGa2	1.24	8.55	6.03
TGa3	1.74	3.09	2.75
TGa4	1.03	8,25	5.78
TGa5	1.97	4.67	4,19
TGa6	1.77	4.97	3.34
TGa7	1.22	5.74	4.51
TGa8	1.75	2.97	2.08
In all together	1.76	5.23	4.01

The results of the regression analysis for the maximum back-of-queue, determined basis of the proposed model, and empirical data are in Figure 11. The mean value of the starting-up flow rate determined from model (15) was used for each i-th signaling cycle. To fully reflect the process of back-of-queue formation at the signal-controlled intersection approach, the linear regression equation was determined without the expression of an intercept function, which corresponds to the situation that there is no maximum queue in reality and the theoretical maximum queue in the calculations.

At a later stage of the analysis, an error distribution graph was made (Figure 12). Ideally, the relative error values should be distributed around the mean value of the Gaussian curve. Thus, the errors should compensate each other.

Using the χ^2 Pearson test (assuming the significance level $\alpha = 0.05$), the proportion of the error distribution with the normal distribution was tested. Statistical compliance tests showed a positive result in all analyzed cases. It was also found that the values of the maximum queue, calculated using the proposed model, are in most cases slightly lower than the empirical values. Based on the analysis of the causes, it was found that the lower values of the maximum queue size, calculated with the variables of the model, are not affected by the traffic signals control parameters, the initial queue length, or the value of the starting-up intensity.

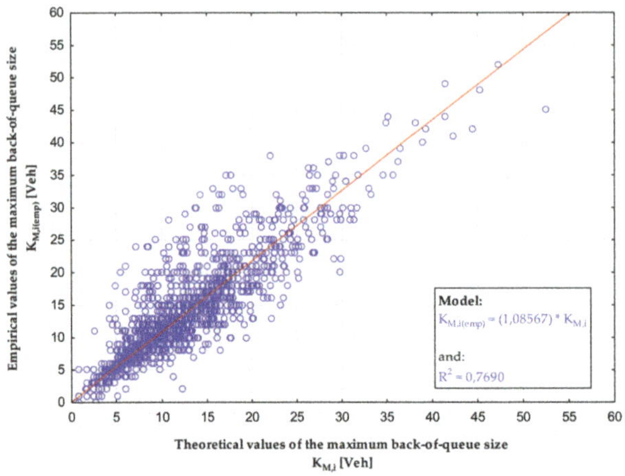

Figure 11. Results of regression and correlation analysis between empirical values of maximum queue size and theoretical values determined by the proposed model.

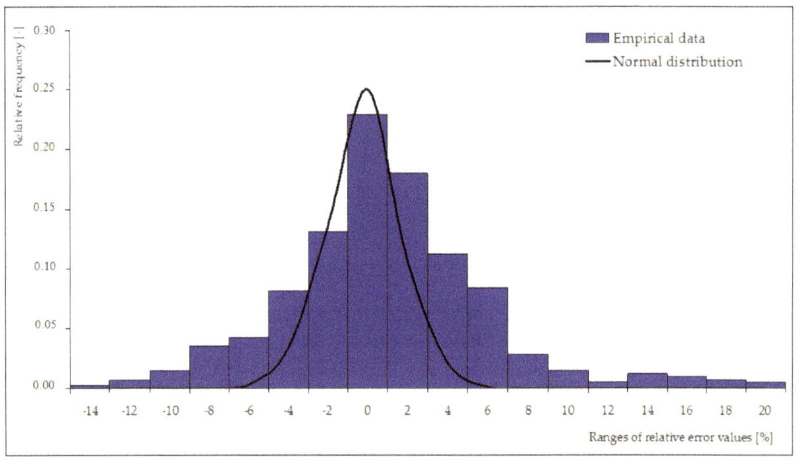

Figure 12. Distribution of errors in estimating the maximum queue size at signal-controlled intersection approach obtained by the proposed model.

The values of the maximum back-of-queue size at signal-controlled intersection approach determined by the model were corrected by applying the appropriate coefficient expressing the increase in the empirical value of the maximum queues compared to the value determined based on the model. The primary value for determining the correction coefficient was the maximum back-of-queue size at the signal-controlled intersection approach in the ideal road-traffic conditions. Hence, the correction factor f_k was determined as the quotient of the empirical maximum queue size to the value calculated using the model:

$$f_k = \frac{K_{M,i(emp)}}{K_{M,i}} [-] \tag{16}$$

where: f_k—correction factor after model verification $[-]$ and $K_{M,i(emp)}$—empirical value of the maximum back-of-queue size [Veh.].

The value of correction factor f_k was calculated as f_k = 1.08 [−]. Hence, the value of the maximum back-of-queue at signal-controlled intersection approach on a given lane was finally determined from the relationship:

$$K_{M,i(emp)} = 1.08 \cdot K_{M,i}[\text{Veh.}/\text{s}],\ R^2 = 0.769 \tag{17}$$

The accuracy of the model was analyzed again for that corrected model. In Table 4, the values of errors in estimating the maximum queues after model calibration are presented. In Figure 13, the results of linear regression analysis for the maximum back-of-queue size at signal-controlled intersection approach are presented, determined based on the corrected model and empirical data. As a result of regression and correlation analysis, high values of the correlation coefficient (R = 0.877) and determination (R^2 = 0.769) were obtained. The obtained value of the correlation coefficient shows a strong relationship between the values estimated based on the model presented in the article and those observed in reality. This is also confirmed by a regression coefficient equal to one.

Table 4. Values of errors in estimating the maximum back-of-queue size at the signal-controlled intersection approach [Veh.] after calibration.

Testing Ground	Traffic Conditions at the Intersection Approach:		
	Unsaturated Inflow [Veh.]	Saturated or Oversaturated Inflow [Veh.]	Unsaturated Inflow [Veh.]
TGa1	2.98	4.24	3.72
TGa2	1.28	7.65	5.42
TGa3	1.95	3.12	2.82
TGa4	1.10	7.49	5.27
TGa5	2.35	4.33	3.95
TGa6	1.73	4.11	2.86
TGa7	1.34	5.03	3.98
TGa8	1.89	2.64	2.08
In all together	1.95	4.91	3.83

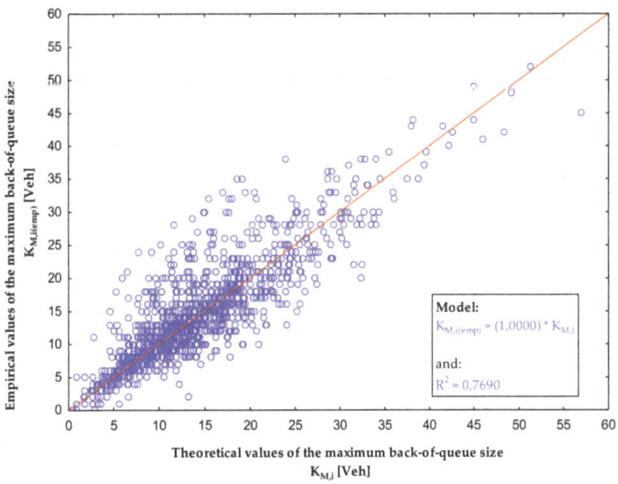

Figure 13. Results of regression and correlation analysis between empirical values of maximum queue size and calibrated theoretical values determined by the proposed model.

The distribution of errors in estimating the maximum back-of-queue size at signal-controlled intersection approach using a calibrated model is presented on the Figure 14.

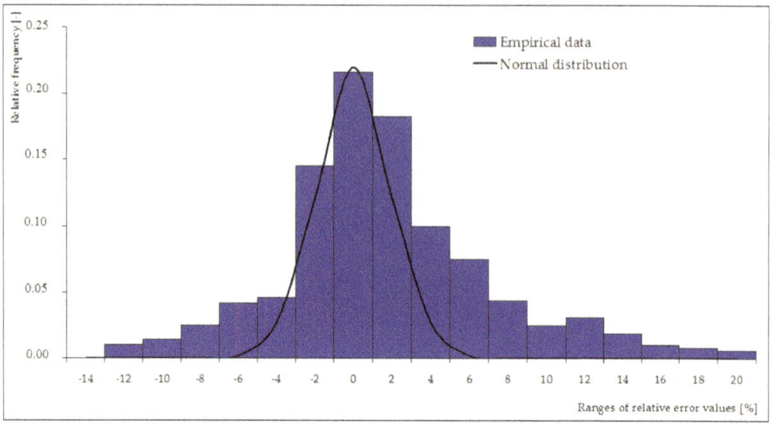

Figure 14. Distribution of errors in estimating the maximum queue size at signal-controlled intersection approach obtained by the proposed model after calibration.

Based on the obtained results, it can be concluded that in most cases the relative error is less than 10%. The low values of these errors prove that the maximum queue size at signal-controlled intersection approach, estimated by the calibrated model, is very consistent with the empirical results. An additional part that ensured the most accurate mapping of traffic at signal-controlled intersections was performed with passive remote sensing tools, such as digital cameras, which significantly increased the accuracy of recording and understanding the measured values. The compliance of the error distribution with the normal distribution was re-examined using the χ^2 Pearson test (assuming the significance level $\alpha = 0.05$). As before, statistical compliance tests showed a positive result in all analyzed cases. After the model calibration, the values of the χ^2 test statistic had lower values than the values previously. That proves the improvement of the quality of the model fit to empirical data. Also, the descriptive parameters of the error distribution, such as the mean and standard deviation, assume lower values in each case.

The results obtained from the verification prove the high compliance of the constructed model with the real road and traffic conditions at the signal-controlled intersections located in urban areas. It confirms a satisfactory level of model accuracy. The obtained compliance proves the possibility of using the model for practical applications in the calculation of the maximum back-of-queue size at signal-controlled intersection approach.

7. Discussion and Conclusions

The accuracy of estimating the maximum queue sizes at signal-controlled intersection approach with the proposed model in the article proves its good adjustment to the empirical values. It is confirmed by both the high value of the coefficient of determination ~0.8 and the relatively small errors in calculations, amounting to ~4 vehicles per signaling cycle. The use of digital cameras in measurements probably contributed to the construction of a model that allows estimating the maximum back-of-queue size at signal-controlled intersections with high accuracy, which ensured high accuracy of the research, including in particular measurements parameters of the process of the starting-up flow rate of vehicles stopped in the queue. The arrangement of cameras on the sections of the intersection approach made it possible to study the analyzed process along the entire section of the road, which was influenced by increasing and decreasing the queue size. The analysis presented in the article concerned the so far unrecognized from the mathematical point of view of the process of starting progressive vehicles stopped in the queue at the intersection approach in a given lane.

At a further stage of work on calculations of the maximum queue sizes at signal-controlled intersection approach, the authors plan to take into account the influence of subsequent factors on the course of the maximum queue formation process, as well as to characterize the course of this process during a single cycle and a given analysis period, e.g., 15 min. Moreover, further work will also concern the study of the configuration and form of functions describing the relationships between the variables, i.e., the headways between vehicles in streams of arriving, departure, starting-up of the maximum queue, and stopping of the remaining (initial) queue. The analysis performed so far is based on averaged values of both the intensity of starting-up and saturation flow rate after starting the effective green time.

It is also recognized that this model may be useful for adaptive control at intersections approach, which uses a green signal length adjustment in its traffic control algorithms. It may be based on the prediction of road traffic characteristics in short time horizons (e.g., 5–15 min), using the maximum back-of-queue size forecast for this purpose. Despite a much smaller share of heavy vehicles in a modal split in urban traffic than commuter traffic, conduct further detailed work in this area and taking into account the influence of heavy vehicles and buses on the course of the analyzed queue formation process is also planned.

The results of the estimation are considered promising, despite the above-mentioned simplifications applied at the stage of model construction. In particular, that is in the case of the worst traffic conditions at urban streets, i.e., conditions of oversaturated vehicle flow rates. Thence, the presented model can be the basis for further analysis. The purpose of these analyzes will be a develop more precise relationships characterizing the process of formation the back-of-queue size at signal-controlled intersections.

Author Contributions: Conceptualization, D.I. and E.M.; methodology, D.I. and E.M.; software, D.I.; validation, D.I. and E.M.; formal analysis, E.M.; investigation, D.I.; resources, D.I. and E.M.; data curation, D.I.; writing—original draft preparation, D.I. and E.M; writing—review and editing, E.M.; visualization, D.I.; supervision, E.M.; project administration, E.M.; funding acquisition, E.M. All authors have read and agreed to the published version of the manuscript.

Funding: This research received no external funding.

Institutional Review Board Statement: Not applicable.

Informed Consent Statement: Not applicable.

Data Availability Statement: Not applicable.

Acknowledgments: Authors would like to thank you very much the late Professor Tomasz Szczuraszek, founder and long term Head of the Department of Road Engineering and Transport at the University of Technology and Life Sciences in Bydgoszcz, for valuable substantive advice and thorough scientific care, which allowed to explore the topic of the process of forming queues of vehicles at the signal-controlled intersections. Honor His memory.

Conflicts of Interest: The authors declare no conflict of interest.

References

1. Drew, D.R. *Traffic Flow Theory & Control*; Series in Transportation; McGRAW-HILL: New York, NY, USA, 1968.
2. Homburger, W.S.; Hall, J.W.; Loutzenheiser, R.C.; Reilly, W.R. *Fundamentals of Traffic Engineering*, 14th ed.; University of California: Berkeley, CA, USA, 1996.
3. Newell, G.F. *Application of Queuing Theory*, 2nd ed.; Chapman and Hall: London, UK, 1982.
4. Rouphail, N.; Tarko, A.; Li, J. Traffic flow at signalized intersections (Chapter 9). In *Revised Monograph on Traffic Flow Theory*; original publication: Special Report 165: "Traffic Flow Theory", published in 1975; Federal Highway Administration Research and Technology: Washington, DC, USA, 1992.
5. Mathew, T.V. *Transportation Systems Engineering. Lecture Notes (Online Reading)*; Indian Institute of Technology Bombay: Maharashtra, India, 8 March 2017; Available online: https://www.civil.iitb.ac.in/tvm/1100_LnTse/ceTseLn/ceTseLn.html (accessed on 30 September 2018).

6. Szczuraszek, T.; Macioszek, E. Analysis of time intervals distribution between vehicles on the roadway around central island of small roundabouts. *Drog. Mosty* **2010**, *9*, 87–99.
7. Macioszek, E. Changes in Values of Traffic Volume—Case Study Based on General Traffic Measurements in Opolskie Voivodeship (Poland). In *Directions of Development of Transport Networks and Traffic Engineering*; Macioszek, E., Sierpiński, G., Eds.; Springer International Publishing: Cham, Switzerland, 2019; Volume 51, pp. 66–76.
8. Macioszek, E. Analysis of driver behaviour at roundabouts in Tokyo and the Tokyo surroundings. In *Modern Traffic Engineering in the System Approach to the Development of Traffic Networks*; Macioszek, E., Sierpiński, G., Eds.; Springer International Publishing: Cham, Switzerland, 2020; Volume 1083, pp. 216–227.
9. Webster, F.V. *Traffic Signal. Settings*; Road Research Technical paper No 39; Department of Scientific and Industrial Research: New Delhi, India, 1958. Available online: https://www.sinaldetransito.com.br/artigos/traffic_signals_webster.pdf (accessed on 21 November 2017).
10. Webster, F.V.; Cobbe, B.M. *Traffic Signals*; Road Research Technical paper No 56; Road Research Laboratory: London, UK; H.M.S.O.: London, UK, 1966.
11. Kimber, R.M.; Hollis, E.M. *Traffic Queues and Delays at Road Jucttions*; TRRL Report. LR 909; Transport and Road Research Laboratory (TRRL): Wokingham, UK, 1979.
12. Miller, A.J. Settings for fixed-cycle traffic signals. *Oper. Res. Q.* **1963**, *14*. [CrossRef]
13. Newell, G.F. Approximation methods for queues with application to the fixed-cycle traffic light. *SIAM Rev.* **1965**, *7*, 223–240. [CrossRef]
14. Van Zuylen, H.J.; Taale, H. Dynamic and Stochastic Aspects of Queues at Signal Controlled Intersections. Available online: https://citeseerx.ist.psu.edu/viewdoc/download?doi=10.1.1.541.7437&rep=rep1&type=pdf (accessed on 20 November 2018).
15. Kimber, R.M.; Daly, P.; Barton, J.; Giokas, C. Predicting time-dependent distributions of queues and delays for road traffic at roundabouts and priority junctions. *J. Oper. Res. Soc.* **1986**, *37*. [CrossRef]
16. Akçelik, R.; Besley, M. *Microsimulation and Analytical Methods for Modelling Urban Traffic*; Conference papers from Advance Modeling Techniques and Quality of Service in Highway Capacity Analysis; Sidra Solutions: Truckee, CA, USA, 2001.
17. Akçelik, R.; Besley, M. Queue discharge flow and speed models for signalized intersections. In Proceedings of the 15th International Symposium on Transportation and Traffic Theory, Adelaide, Australia, 16–18 July 2002.
18. Akçelik, R.; Rouphail, N.M. Overflow queues and delays with random and platooned arrivals at signalized intersections. *J. Adv. Transp.* **1994**, *28*, 227–251. [CrossRef]
19. Akçelik, R. Progression factor for queue length and other queue-related statistics. *Transp. Res.* **1996**. [CrossRef]
20. Akçelik, R. *Progression Factors in the HCM 2000 Queue and Delay Models for Traffic Signals*; Technical note; Akcelik and Associates Pty Ltd.: Greythorn, Australia, 2001.
21. Akçelik, R. Stops at traffic signals. In Proceedings of the 10th ARRB Conference, Sydney, Australia, 25–29 August 1980; Volume 10.
22. Akçelik, R. The Highway Capacity Manual delay formula for signalised intersections. *ITE J.* **1988**, *56*, 23–27.
23. Akçelik, R. *Time-Depent Expressions for Delay, Stops Rate and Queue Length at Traffic Signals*; Internal Report AIR 367-1; Australian Road Research Board: Port Melbourne, Australia, 1980.
24. Olszewski, P. Modelling of queue probability distribution at traffic signals. In Proceedings of the 11th International Symposium on Transportation and Traffic Theory, Yokohama, Japan, 18–20 July 1990.
25. Olszewski, P. Modelling probability distribution of delay at signalized intersections. *J. Adv. Transp.* **1994**, *28*. [CrossRef]
26. Olszewski, P. Overall delay, stopped delay, and stops at signalized intersections. *J. Transp. Eng.* **1993**, *119*. [CrossRef]
27. Wu, N. Estimation of queue lengths and their percentiles at signalized intersections. In *Proceedings of the Third International Symposium on Algorithmic Number Theory*; Springer: Berlin/Heidelberg, Germany, 1998.
28. Wu, N. Wartezeit und Leistungsfähigkeit von Lichtsignalanlagen unter Berücksichtigung von Instationarität und Teilgebundenheit des Verkehrs. Ph.D. Thesis, Schriftenreihe des Lehrstuhls. für Verkehrswesen der Ruhr-Universität Bochum, Bochum, Germany, 1990.
29. *Highway Capacity Manual 2000 (HCM 2000)*; Transportation Research Board: Washington, DC, USA, 2000.
30. *Highway Capacity Manual 2010 (HCM 2010)*; Transportation Research Board: Washington, DC, USA, 2010.
31. Akçelik, R. *Traffic Signals: Capacity and Timing Analysis*; 7th reprint; Australian Road Research Board: Port Melbourne, Australia, 1998.
32. *Guide to Traffic Management Part 3: Traffic Studies and Analysis*; Austroads: Sydney, Australia, 2020.
33. *Canadian Capacity Guide for Signalized Intersections*, 3rd ed.; The Institute of Transportation Engineers: Washington, DC, USA, 2008.
34. *Handbuch für die Bemessung von Straßenverkehrsanlagen (HBS)*; Forschungsgesellschaft für Straßen- und Verkehrswesen e. V.: Berlin, Germany, 2001.
35. *Handbuch für die Bemessung von Straßenverkehrsanlagen (HBS 2015)*; Forschungsgesellschaft für Straßen- und Verkehrswesen e. V.: Berlin, Germany, 2015.
36. Tracz, M.; Chodur, J.; Gaca, S.; Gondek, S.; Kieć, M.; Ostrowski, K. *Metoda Obliczania PrzepustowośCI SkrzyżOwań Z Sygnalizacją śWietlną*; Generalna Dyrekcja Dróg Krajowych i Autostrad: Krakow, Poland, 2004.
37. *Highway Capacity Manual 2016, Sixth Edition: A Guide for Multimodal Mobility Analysis (HCM 2016)*; Transportation Research Board: Washington, DC, USA, 2016.

38. Iwanowicz, D.; Szczuraszek, T. Dylematy przy ustalaniu stanu przesycenia ruchem wlotów skrzyżowań z sygnalizacją świetlną. *Czas. Inż. Lądowej Śr. Architekt.* **2016**, *63*. [CrossRef]
39. Szczuraszek, T.; Iwanowicz, D. The impact of the analysis period on the estimation accurary of queue lengths at intersection inlets with traffic lights. In *Recent Advances in Traffic Engineering for Transport Networks and Systems*; Lecture Notes in Networks and Systems; Macioszek, E., Sierpiński, G., Eds.; Springer International Publishing: Berlin/Heidelberg, Germany, 2018; Volume 21.
40. Iwanowicz, D. Assessment of Selected Methods of Estimating the Maximum Queue Length in Intersection Inlets with Traffic Lights. In 66. Scientific Conference Krynica-Zdrój. Poland. 2020; (publication under review).
41. Alfa, A.S. Modelling traffic queues at a signalized intersection with vehicle-actuated control and marcovian arrival processes. In *Computers & Mathematics with Applications*; Elsevier Science Ltd.: Amsterdam, The Netherlands, 1995; Volume 30, pp. 105–119.
42. Chen, P.; Liu, H.; Qi, H.; Wang, F. Analysis of delay variability at isolated signalized intersections. *J. Zhejiang Univ.* **2013**, *14*, 691–704. [CrossRef]
43. Viti, F.; van Zuylen, H.J. Markov mesoscopic simulation model of overflow queues at multilane signalized intersections. *Adv. OR AI Methods Transp.* **2005**, 475–481. Available online: http://www.iasi.cnr.it/ewgt/16conference/ID87.pdf (accessed on 18 November 2019).
44. Gasz, K. Zastosowanie szeregów czasowych do opisu długości kolejek na wlotach skrzyżowań. In Proceedings of the 50. Jubileuszowa Konferencja Naukowa Komitetu Inżynierii Lądowej i Wodnej PAN i Komitetu Nauki PZITB "Krynica 2004", Krynica, Poland, 12–17 September 2004.
45. Shou, Y.; Xu, J. Multi-objective optimization of oversaturated signalized intersection based on fuzzy logic. In Proceedings of the 8th World Congress on Intelligent Control and Automation, Jinan, China, 6–9 July 2010.
46. Kafash, M.; Sharif, M.J.; Menhaj, M.B.; Maleki, A. Designing fuzzy controller for traffic lights to reduce the length of queues in according to minimize extension of green light time and reduce waiting time. In Proceedings of the 13th Iranian Conference on Fuzzy Systems (IFSC), Qazvin, Iran, 27–29 August 2013; pp. 1–6.
47. Mucsi, K.; Khan, A.M.; Ahmadi, M. An adaptive neuro-fuzzy inference system for estimating the number of vehicles for queue management at signalized intersections. *Trans. Res. Part C* **2011**, *19*, 1033–1047. [CrossRef]
48. Van Zuylen, H.J.; Viti, F. Uncertainty and the dynamics of queues at controlled intersections. *IFAC Proc.* **2003**, *36*, 43–48. [CrossRef]
49. Viti, F.; van Zuylen, H.J. The dynamics and the uncertainty of queues at fixed and actuated controls: A probabilistic approach. *J. Intell. Transp. Syst. Technol. Plan. Oper.* **2009**, *13*, 39–51. [CrossRef]
50. Cao, J.; Hu, D.; Hadiuzzaman, M.; Wang, X.; Qiu, T.Z. Comparison of queue estimation accuracy by shockwave-based and input-output-based models. In Proceedings of the 17th International IEEE Conference on Intelligent Transportation Systems (ITSC), Qingdao, China, 8–11 October 2014; pp. 2687–2692.
51. Ng, K.M.; Reaz, M.I. A Comparative Study of the LWR-IM Traffic Model and Shockwave Analysis. In Proceedings of the 2015 IEEE Conference on Systems, Process and Control (ICSPC 2015), Bandar Sunway, Malaysia, 18–20 December 2015; pp. 39–43.
52. Srivastawa, A.; Jin, W.L.; Lebacque, J.P. A modified Cell Transmission Model with realistic queue discharge features at signalized intersections. *Transp. Res. Part B Methodol.* **2015**, *81*, 302–315. [CrossRef]
53. Hao, J.; Hou, Z.; Bu, X. The iterative learning approach for vehicle queuing length balanced-control of the signalized isolated intersection. In Proceedings of the 30th Chinese Control Conference, Yantai, China, 22–24 July 2011; pp. 5556–5561.
54. Anokye, M.; Abdul-Aziz, A.R.; Annin, K.; Oduro, F.T. Application of queuing theory to vehicular traffic at signalized intersection in Kumasi-Ashanti Region, Ghana. *Am. Int. J. Contemp. Res.* **2013**, *3*, 23–29.
55. Bie, Y.; Mao, C.; Yang, M. Development of vehicle delay and queue length models for adaptive traffic control at signalized roundabout. *Procedia Eng.* **2016**, *137*, 141–150. [CrossRef]
56. Fu, L.; Hellinga, B.; Zhu, Y. An adaptive model for real-time estimation of overflow queues on congested arterials. In Proceedings of the 2001 IEEE Intelligent Transportation Systems Conference Proceedings, Oakland, CA, USA, 25–29 August 2001; pp. 219–226.
57. Viti, F.; van Zuylen, H.J. Probabilistic models for queues at fixed control signals. *Transp. Res. Part B* **2010**, *44*, 120–135. [CrossRef]
58. Yu, X.; Sulijoadikusumo, G.; Prevedouros, P. Analysis of downstream queues on upstream capacity expansion of urban signalized intersection. *J. Transp. Syst. Eng. Inf. Technol.* **2012**, *12*, 98–108. [CrossRef]
59. He, X.; Liu, H.X.; Liu, X. Optimal vehicle speed trajectory on a signalized arterial with consideration of queue. *Transp. Res. Part C Emerg. Technol.* **2015**, *61*, 106–120. [CrossRef]
60. Zhang, H.; Liu, H.X.; Chen, P.; Yu, G.; Wang, Y. Cycle-Based End of Queue Estimation at Signalized Intersections Using Low-Penetration-Rate Vehicle Trajectories. *IEEE Trans. Intell. Transp. Syst.* **2019**, *21*, 3257–3272. [CrossRef]
61. Cao, P.; Fan, Q.; Liu, X. Real-time detection of end-of-queue shockwaves on freeways using probe vehicles with spacing equipment. *IET Intell. Transp. Syst.* **2018**, *12*, 1227–1235. [CrossRef]
62. Zhao, Y.; Zheng, J.; Wong, W.; Wang, X.; Meng, Y.; Liu, H.X. Various methods for queue length and traffic volume estimation using probe vehicle trajectories. *Transp. Res. Part C Emerg. Technol.* **2019**, *107*, 70–91. [CrossRef]
63. Naghawi, H.; Idewu, W. Analyzing delay and queue length using microscopic simulation for the unconventional intersection design superstreet. *J. S. Afr. Inst. Civ. Eng.* **2014**, *56*, 100–107.
64. Ma, W.; Wan, L.; Yu, C.; Zou, L.; Zheng, J. Multi-objective optimization of traffic signals based on vehicle trajectory data at isolated intersections. *Transp. Res. Part C Emerg. Technol.* **2020**, *120*, 102821. [CrossRef]

65. Li, Y.; Ma, S.; Li, W.; Wang, H. Microscopic urban traffic simulation with multi-agent system. In Proceedings of the Fourth International Conference on Information, Communications and Signal Processing, 2003 and the Fourth Pacific Rim Conference on Multimedia, Kuching, Malaysia, 1–4 December 2013; pp. 1835–1839.
66. Anantharam, V. Queuing analysis with traffic models based on deterministic dynamical systems. *Comput. Sci. J. Mold.* **2004**, *12*, 154–170.
67. Chaudhry, M.S.; Ranjitkar, P. Delay Estimation at Signalized Intersections with Variable Queue Discharge Rate. *Proc. East. Asia Soc. Transp. Stud.* **2013**, *9*, 1764–1775.
68. Viti, F. Overflow queues at urban corridors and with time-dependent control. In *Trail conference proceedings 2001, A World of Transport, Infrastructure and Logistics*; Bovy, P.H.L., Ed.; Dup Science: Delft, The Netherlands, 2004; pp. 399–421.
69. Liu, H.; Liang, W.; Rai, L.; Teng, K.; Wang, S. A Real-Time Queue Length Estimation Method Based on Probe Vehicles in CV Environment. *IEEE Access* **2019**, *7*, 20825–20839. [CrossRef]
70. Xu, H.; Ding, J.; Zang, Y.; Hu, J. Queue length estimation at isolated intersections based on intelligent vehicle infrastructure cooperation systems. In Proceedings of the 2017 IEEE Intelligent Vehicles Symposium (IV), Los Angeles, CA, USA, 11–17 June 2017; pp. 655–660.
71. Vigos, G.; Papageorgiou, M. A Simplified Estimation Scheme for the Number of Vehicles in Signalized Links. *IEEE Trans. Intell. Transp. Syst.* **2010**, *11*, 312–321. [CrossRef]
72. Le Vine, S.; Liu, X.; Zheng, F.; Polak, J. Automated cars: Queue discharge at signalized intersections with 'Assured-Clear-Distance-Ahead' driving strategies. *Transp. Res. Part C* **2016**, *62*, 35–54. [CrossRef]
73. Ramezani, M.; Geroliminis, N. Exploiting probe data to estimate the queue profile in urban networks. In Proceedings of the 16th International IEEE Annual Conference on Intelligent Transportation Systems, The Hague, The Netherlands, 6–9 October 2013; pp. 1817–1822.
74. Yang, H.; Rakha, H.; Ala, M.V. Eco-cooperative adaptive cruise control at signalized intersections consedering queue effects. *IEEE Trans. Intell. Transp. Syst.* **2017**, *18*, 1575–1585.
75. Wang, P.; Jiang, Y.; Xiao, L.; Zhao, Y.; Li, Y. A joint control model for connected vehicle platoon and arterial signal coordination. *J. Intell. Transp. Syst. Technol. Plan. Oper.* **2020**, *24*, 81–92. [CrossRef]
76. Kong, J.; Hou, Z.; Ren, Y. Findings on Queue Length Based Macroscopic Fundamental Diagrams with Enhanced Floating Car Estimation Method. In Proceedings of the 37th Chinese Control Conference (CCC), Wuhan, China, 25–27 July 2018; pp. 7874–7879.
77. Iwasaki, Y. An image processing system to measure vehicular queues and an adaptive traffic signal control by using the information of the queues. Proceedings of Conference on Intelligent Transportation Systems, Boston, MA, USA, 9–12 November 1997; pp. 195–200.
78. Liu, X.; Dai, S.; Lu, S. Macroscopic model for interrupted traffic flow of signal controlled intersection. In Proceedings of the 2009 ISECS International Colloquium on Computing, Communication, Control, and Management, Sanya, China, 8–9 August 2009; pp. 78–81.
79. Anusha, S.P.; Vanajakshi, L.D.; Sharma, A. A simple method for estimation of queue length. *Civ. Eng. Fac. Publ.* **2013**, *12*, 1–5.
80. Anusha, S.P.; Vanajakshi, L.; Subramanian, A.C.; Sharma, A. Performance comparison of two model based schemes for estimation of queue and delay at signalized intersections. In Proceedings of the IEEE Intelligent Vehicles Symposium (IV), Seoul, Korea, 29 June–1 July 2015; pp. 988–993.
81. Cai, Q.; Wang, Z.; Guo, X.; Wu, B. New calculating method for HCM 2000 queue length estimation procedures with the application of floating car data. *Procedia Soc. Behav. Sci.* **2013**, *96*, 2201–2210. [CrossRef]
82. Cetin, M.; Comert, G. Estimating queues at signalized intersections: Value of location and time data from instrumented vehicles. In Proceedings of the Intelligent Vehicles Symposium, Istanbul, Turkey, 13–15 June 2007; pp. 1138–1143.
83. Chang, J.; Talas, M.; Muthuswamy, S. A simple methodology to estimate queue lengths at signalized intersections using detector data. *Transp. Res. Rec. J. Transp. Res. Board* **2014**, 1–14. [CrossRef]
84. Hao, P.; Ban, X. Long queue estimation for signalized intersections using mobile data. *Transp. Res. Part B Methodol.* **2015**, *82*, 54–73. [CrossRef]
85. Mukhopadhyay, S.; Pramod, M.J.; Kumar, A. An approach for analysis of mean delay at a signalized intersection with indisciplined traffic. In Proceedings of the Intelligent Transportation System Workshop, Bangalore, India, 6–10 January 2015; pp. 1–6.
86. Iwanowicz, D.; Chmielewski, J. Analysis of the Methods of Traffic Evaluation at the Approaches of Urban Signalised Intersections. In *Nodes in Transport Networks—Research, Data Analysis and Modelling*; Macioszek, E., Kang, N., Sierpiński, G., Eds.; Springer International Publishing: Berlin/Heidelberg, Germany, 2020; pp. 180–198.
87. Liu, H.X.; Wu, X.; Ma, W.; Hu, H. Real-time queue length estimation for congested signalized intersections. *Transp. Res. Part C* **2009**, *17*, 412–427. [CrossRef]
88. Milla, J.M.; Toral, S.L.; Vargas, M.; Barrero, F.J. Dual-rate backgraund subtraction approach for estimating traffic queue parameters in urban scenes. *IET Intell. Transp. Syst.* **2013**, *7*, 122–130. [CrossRef]
89. Khan, M.A.; Ectors, W.; Bellemans, T.; Janssens, D.; Wets, G. Unmanned Aerial Vehicle-Based Traffic Analysis: A Case Study for Shockwave Identification and Flow Parameters Estimation at Signalized Intersections. *Remote Sens.* **2018**, *10*, 458. [CrossRef]
90. Sun, Z.; Ban, X. Vehicle trajectory reconstruction for signalized intersections using mobile traffic sensors. *Transp. Res. Part C Emerg. Technol.* **2013**, *36*, 268–283. [CrossRef]

91. Xie, X.; van Lint, H.; Verbraeck, A. A generic data assimilation framework for vehicle trajectory reconstruction on signalized urban arterials using particle filters. *Transp. Res. Part C Emerg. Technol.* **2018**, *92*, 364–391. [CrossRef]
92. Zhou, J.; Jin, L.; Wang, X.; Sun, D. Resilient UAV Traffic Congestion Control Using Fluid Queuing Models. *IEEE Trans. Intell. Transp. Syst.* **2020**, 1–12. [CrossRef]
93. Wang, L.; Chen, F.; Yin, H. Detecting and tracking vehicles in traffic by unmanned aerial vehicles. *Autom. Constr.* **2016**, *72*, 294–308. [CrossRef]
94. Salvo, G.; Caruso, L.; Scordo, A. Urban Traffic Analysis through an UAV. *Procedia Soc. Behav. Sci.* **2014**, *111*, 1083–1091. [CrossRef]
95. Chmielewski, J. Zastosowanie systemu WZDR do zarządzania bezpieczeństwem ruchu drogowego. *Drog. Bud. Infrastrukt.* **2011**, *2*, 52–61.
96. Macioszek, E. The Passenger Car Equivalent Factors for Heavy Vehicles on Turbo Roundabouts. *Front. Built Environ. Sect. Transp. Transit. Syst.* **2019**, *5*, 68. [CrossRef]
97. Granà, A.; Giuffrè, T.; Macioszek, E.; Acuto, F. Estimation of Passenger Car Equivalents for two-lane and turbo roundabouts using AIMSUN. *Front. Built Environ.* **2020**, *6*, 86. [CrossRef]
98. Macioszek, E. Roundabout entry capacity calculation—A case study based on roundabouts in Tokyo, Japan, and Tokyo surroundings. *Sustainability* **2020**, *12*, 1533. [CrossRef]

energies

MDPI

Article

Investigations of the Dynamic Travel Time Information Impact on Drivers' Route Choice in an Urban Area—A Case Study Based on the City of Bialystok

Robert Ziółkowski [1,*] and Zbigniew Dziejma [2]

1 Faculty of Civil Engineering and Environmental Sciences, Bialystok University of Technology, Street Wiejska 45E, 15-351 Bialystok, Poland
2 Municipal Road Authority, Street Składowa 11, 15-399 Bialystok, Poland; zdziejma@um.bialystok.pl
* Correspondence: robert.ziolkowski@pb.edu.pl

Abstract: Increasing traffic volumes in cities leads to common traffic congestions building up, especially during peak hours. To protect city dwellers from excessive fuel exhaust and traffic noise and to prevent drivers from time loss due to overloaded routes, it is important to inform them about real-time traffic conditions and possible delays in advance. Effectively influencing drivers' decisions to divert from an original route choice in case of traffic hinderance is essential, and application of dynamic travel information in the form of variable message signs (VMSs) is believed to be effective in these terms. The paper examines drivers' willingness to divert from an initial route choice due to the information provided on VMS boards. Their behavior was analyzed in terms of their response to everyday and artificially elongated travel times displayed on the VMSs. Maximum simulated elongation reached 200% and 300% of the initial state, depending on the characteristics of the pre-peak conditions. To assess the effectiveness of VMSs, the changes in traffic intensities were statistically analyzed. In general, apart from few significant differences, the results revealed drivers' ignorance of the travel time information provided on the VMS, regardless of the extension of the original times.

Keywords: variable message signs; driver's behavior; traffic diverting

Citation: Ziółkowski, R.; Dziejma, Z. Investigations of the Dynamic Travel Time Information Impact on Drivers' Route Choice in an Urban Area—A Case Study Based on the City of Bialystok. *Energies* **2021**, *14*, 1645. https://doi.org/10.3390/en14061645

Academic Editor: Mario Marchesoni

Received: 18 February 2021
Accepted: 13 March 2021
Published: 16 March 2021

1. Introduction

The main concepts of a smart city assume the broadly understood care for the residents' quality of life in various areas of life. One of the basic ones is the improvement of living surroundings and reliability of travelling. Actions taken in the field of road users' safety are mostly related to the implementation of effective speed management solutions from a group of traffic-calming measures [1,2]. Improving the comfort and reliability of travel is associated with the implementation of modern intelligent transportation systems (ITS) technologies supporting the management of traffic flows. That is particularly important due to the continuous dynamic development of motorization that contributes to the growing communication problems in cities. To overcome increasing delays and total travel times that drivers spend in a road network, city authorities invest lots of funds in advanced traffic management systems. Modern systems are extensive and consist of several subsystems. One of them is the dynamic travel information subsystem with the provision of actual travel time information for drivers. Two types of dynamic systems are currently widely used: variable message signs (VMSs) and graphical route information panel (GRIP), and the main concern of this article is given to VMS signs.

Variable message signs are part of intelligent transportation systems. The directive of the European Union 2010/40/EU defined ITS as systems in which information and communication technologies are applied in the field of road transport, including infrastructure, vehicles and users, and in traffic management and mobility management, as well as for interfaces with other modes of transport [3]. ITS contributes to the improvement of transport efficiency in a number of situations, i.e., road transport, traffic management, mobility, etc. [4,5]. The goal of the development and application of ITS solutions is to support, control, and manage transport processes.

To date, the effectiveness of ITS has been investigated and confirmed in many respects. The studies of safety considerations [6,7] have shown high potential in accidents and injuries reduction. Improvement of transportation systems performance has been examined [8–10], and essential improvements in average operating speed have been proved. Several researchers investigated possible applications of ITS solutions in vehicles to improve traffic conditions and traffic safety and to reduce transport impact on the environment through better vehicle-to-vehicle communication and interaction [11–14] or by incorporation of speed cameras and section speed systems [15–17].

Weather condition information provided on VMS systems was found to be effective in speed reduction in Sweden [18,19], but research results in terms of accidents reduction were disputable [20,21], and it was argued that the net safety effects of such message systems were not conclusive. They concluded that while messages for adverse weather conditions significantly influenced speed reduction, drivers tended to compensate for the speed reduction by increasing their speeds downstream where such adverse conditions did not exist. Research conducted in Leeds, UK [7,22], employed vehicle simulators, but the results were also inconclusive even though they showed the improvement of drivers alertness to the context of the VMS, and consequently their response to a displayed message became more timely under specific conditions.

Variable message signs, being an integral component of ITS, have an important impact on a motorist's travel behavior, and thus they are a key component in a comprehensive traffic management system. Mostly they are placed at roadside or overhead on roadways and are designed to deliver traffic information covering congestion and accident reports, weather information, lane control, limit speed, etc. They have been widely used to provide information and advice to drivers to increase traffic fluency and reduce time losses. Traffic congestion level and fluency of driving highly influence drivers' behavior, which in turn might affect their route-choice decisions. Typically, as traffic builds, average speed decreases, and travel time and queue length increase, which in turn causes deterioration of driving. To avoid such conditions, VMS are implemented to redirect traffic flows toward less crowded routes. In these terms, scare literature can be found. Considering travel time information Wang et al. [10] analyzed traffic flow on a given link in an urban road network at a signalized intersection based on traffic stream directions. They showed that inclusion of traffic streams, depending on their directional distribution into analyses, can produce more accurate models of dynamic traffic assignment. Other works devoted to travel time estimation models and techniques evaluated their usage in relation to characteristics such as traffic lights parameters, queue length forming, types of vehicle detectors, or type of measurement data [23–25]. Application of numerical simulations have been used to analyze the impact of VMS on the changes of route flow and travel time under recurrent and non-recurrent traffic conditions [26]. It was found that for recurrent congestion VMS can positively influence travelers' decisions, depending on the route length. In the case of non-recurrent congestion, the influence is additionally related to increased traffic demand. Chatterjee et al. [27] developed logistic regression models relating the probability of route diversion to driver, journey, and message characteristics based on questionnaire data. They found that the incident and the message content are important factors influencing the probability of diversion. However, a survey of drivers' actual responses to a message activation showed that only one-third of drivers saw the information presented to them, and few of these drivers diverted. They also noticed that the opposite results of the survey data were obtained for another UK city with a newly installed VMS system which showed that the number of drivers diverting due to VMS information was very similar to that expected from the results of the stated intention questionnaire. Quite similar results were obtained by Erke et al. [26]. During their investigations at two sites on motorways, they found that every fifth driver responded to the VMS information and changed their route choice.

Even though they noticed a higher number of drivers reducing their speed, they explained that as a chain reaction where one vehicle braked and forced the following vehicles to brake or change lanes in order to avoid collisions. More recent articles investigate the exploitation of vehicular ad-hoc network (VANET) architecture to overcome the problem of congested metropolitan areas [28–30]. Conducted studies showed that such systems can effectively reduce traffic level inside cities, exploiting inter-communication among vehicles and supporting infrastructures. Based on vehicle position and speed, drivers using dedicated smartphone applications achieve information on crowded lanes in a short time in order to redirect traffic adequately to avoid overloaded routes.

Previous studies, mostly carried out in developed countries where modern technologies had been developed for many years, have focused on a wide range of ITS solutions, proving their usefulness. In this scope, computer simulations are being increasingly used in research nowadays, and together with questionnaire studies they bring significant contribution to explaining transport problems. However, they cannot replace field research, and there is a lack of investigations conducted in developing countries where ITS solutions have been introduced on a smaller scale.

In this context, keeping in mind that several previous investigations have been conducted into drivers attitudes toward and responses to expressway-based VMSs, we examined the potential and influence of VMSs at inducing motorists to divert during congestion in urban area. For this purpose, research was conducted in Bialystok, a medium-size city in Poland where traffic is managed by a traffic management system. VMSs displaying travel time information are one of the base subsystems designed to divert traffic flows heading for the city center in case of deteriorated traffic conditions. To evaluate effectiveness of this system, we designed investigations in real and simulated on-road conditions.

The manuscript is organized as follows: Section 2 presents the research area where data were collected. In Section 3, we present the research scenarios, while in Section 4 results are presented and discussed. Finally, Section 5 summarizes the obtained conclusions.

2. Research Area and Data Acquisition

The tested area included intersections of city arterial streets being managed by the traffic management system (TMS) and consisted of 5 intersections equipped with VMSs. All intersections were located in a crucial part of the city of Bialystok, Poland. Currently, Bialystok is the only city in Poland where operating TMS covers all intersections with traffic lights, providing good driving conditions along alternative routes. VMSs are part of a traffic management system that controls all signalized intersections in the city [31]. On-trip information for travelers is displayed in the form of on-board announcements (Figure 1) and provides travel time information (TTI) to the center area depending on the chosen direction.

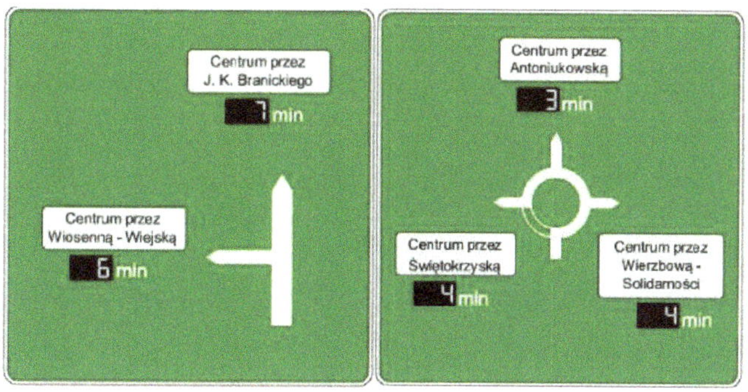

Figure 1. Structure and travel time information (TTI) displayed on the investigated variable message signs (VMS).

The location of all operating VMSs in the city with the ones chosen to detail analyses, marked in circles, is given in Figure 2.

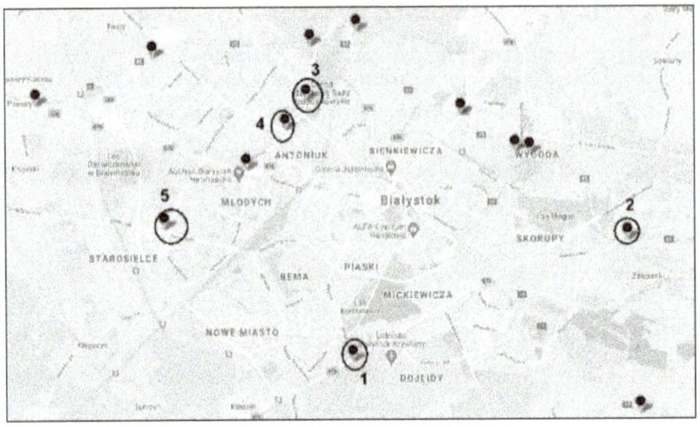

Figure 2. Location of VMSs in Bialystok.

Each of the signs was located at the inlet of an intersection diverting main traffic flows heading to the city center from highly populated residential districts and outskirts regions. The choice of the specific VMSs was discussed with city traffic engineers. Provided travel time information was actuated on the basis of real travel time measured between the beginning and end point of the specific route with the use of ANPR (automatic number plate recognition) cameras. Any disturbances on a primary route caused by excessive traffic or an accident resulted in an extension of travel time displayed on the VMS. Information on traffic volumes was collected during morning peak hours and further analyzed in relation to the traffic directional distribution. The detailed rush hours range selected for analysis was based on information obtained from the city traffic engineers. This allowed specification of an hour range of two hours between 7:00 a.m. and 9:00 a.m. In general, the posted speed limit within the city borders was 50 km/h, but some segments of main dual carriageway arterials had an elevated speed limit of 70 km/h. The analyzed routes are reflected in Table 1. The length of segments with elevated speed limits constituted around 25–30% of their total lengths, and specific segments with elevated speed limits stretched between two consecutive signalized intersections.

Table 1. Description of investigated routes managed by VMSs.

Intersection	Travel Time Displayed for Main (M) and Alternative Routes (A)		Speed Limit (km/h)	Route Length (km)
1	Wiosenna (M)	7–9	50	3.5
	Ciołkowskiego (A)	7–7	50/70	4.1
2	Baranowicka (M)	7–12 (20, 10 min) *	50	5.2
	Sulika (A)	8–17	50/70	6.3
3	Sopoćki (M)	4–7 (9, 7 min) *	50	3.1
	Antoniukowska (A)	4–9 (11, 3 min) *	50	2.4
4	Świętokrzyska (A)	4–9 (13, 1 min) *	50	3.1
	Antoniukowska (M)	3–4	50	3.1
	Wierzbowa (A)	4–7 (9, 4 min)	50	4.3
5	Kopernika (M)	8–10	50/70	4.7
	Sikorskiego (A)	7–2	50/70	4.6

*—the values in the bracketa show the longest TTI displayed on the VMS and its duration time accordingly. (M)—main route. (A)—alternative route.

3. Methodology

The impact of VMSs on drivers' behavior change was analyzed on the basis of changes in traffic flow intensity caused by travel time information displayed on the VMS. Assuming effective impact of the information displayed on the VMS and a positive reaction of drivers as a result, a significant part of them should divert the original direction (marked as a main route) into an alternative one, less crowded with shorter travel time (marked as an alternative route). The logical consequence of that would be differences in traffic intensities. Current TTI values during daily work change automatically upon the real travel time measured between beginning–end points. Minimum and maximum displayed TTI values recorded for everyday conditions are presented in Table 1, and they were gathered with the use of existing ANPR cameras. Maximum elongations recorded during peak hours on three consecutive days (Mon–Tue–Wed) varied distinctly depending on the localization, but in each case noted values were very temporary and were not displayed longer than for 10 min (8% of studied period). For the purpose of the research, TTI values were artificially imposed by the traffic operator at 7:00 a.m. and remained unchanged for the next 2 h. Traffic intensity was measured in 15 min intervals. The evaluation procedure was carried out in three steps:

- Everyday conditions. Getting information about traffic flow in three working days (Mon–Tue–Wed) and analyses of the traffic characteristics (hourly volume traffic intensity, quarterly volume traffic, and peak hour factor (PHF)). Verification of significant differences in traffic volumes between those days in the period 7:00–8:00 (Mon–Tue–Wed) and the period 8:00–9:00 (Mon–Tue–Wed).
- Everyday conditions. Evaluation of hourly traffic volume fluctuations within a day and determining the significance of differences between hourly volumes in two consecutive peak hours (between 7:00–8:00 and 8:00–9:00) to establish the nature and magnitude of traffic fluctuations.
- Artificial conditions of TTI. Determining the significance of changes in traffic volumes after the implementation of elongated travel times to investigate the effect of the VMS operation. Setting the length of travel time information displayed on the VMS board at 7:00 was based on reviewed everyday conditions and was further assumed as a phase "0". Artificially elongated times were assumed to be distinctly longer from the initial times but still should present reliable and reasonable values. In case of too extended values, drivers could treat the information as a hardware/software error instead of accepting the information on actual conditions. The maximum elongation reached 200% (intersections 3 and 4) and 300% (intersections 1, 2, 5) and depended on the initial time in phase "0". The planned TTI values are presented in Figure 3. Traffic volumes were measured in all set conditions accordingly. Traffic measures were conducted between March and June during the same weather conditions. Weeks with public holidays were excluded from the research to keep similar traffic conditions as much as possible.

The research question on the basis of the null hypothesis claimed that the mean traffic flow volumes were the same. The alternative hypothesis stated that the mean traffic flow volumes were different, which entails that the changed conditions (TTI) have an impact on the traffic intensity in a specific direction. To determine the statistical significance of differences between means at each stage, analysis of variance was utilized and a 5% significance level was applied. The significance was verified by Fischer–Snedecor test, and Levene's test was used to verify that assumption of the homogeneity of variances.

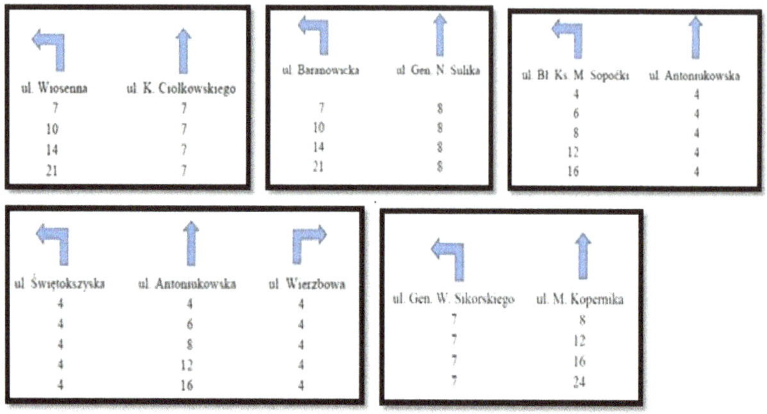

Figure 3. Simulated travel times displayed on the VMSs.

4. Research Results and Discussion

4.1. Characteristic of Traffic Flow Volumes Measured in Weekdays in Everyday Conditions

For the purpose of the research, 2 h traffic volume measurements were conducted during working days peak hours (Mon–Tue–Wed). The values reflecting one hour traffic intensity in consecutive rush hours are presented in Figure 4. As can be seen traffic volumes differ distinctly depending on the localization of the VMS, but do not show distinct differences between consecutive days. The highest difference recorded in Kopernika and Sopoćki streets at 8:00–9:00 reached 12%, but as the statistical analyses show (Table 2), the differences were not statistically significant. In other cases the difference did not exceed 6% and were also insignificant.

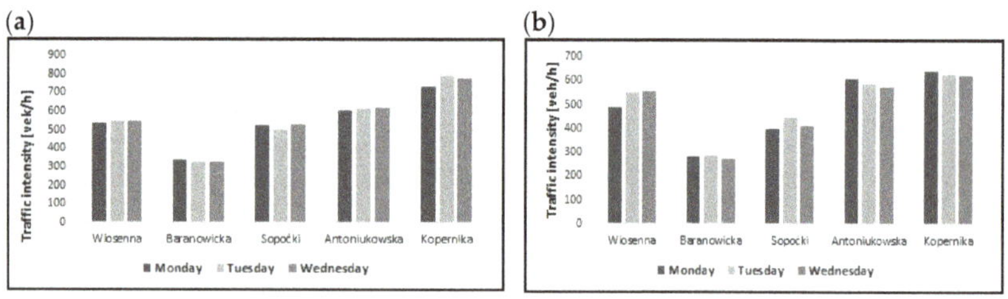

Figure 4. Hourly traffic intensity on main routes (**a**) 7:00–8:00 (**b**) 8:00–9:00.

Table 2. ANOVA results of daily traffic intensities changes.

Time	Wiosenna	Baranowicka	Sopoćki	Antoniukowska	Kopernika	
			$p =$			
7:00–8:00	0.9438	0.8414	0.6926	0.9560	0.6538	
			$p =$			
8:00–9:00	0.5487	0.7068	0.3518	0.7493	0.8911	
Time	**Ciołkowskiego**	**Sulika**	**Antoniukowska**	**Świętokrzyska**	**Wierzbowa**	**Sikorskiego**
				$p =$		
7:00–8:00	0.9896	0.7858	0.9619	0.6001	0.7108	0.7183
				$p =$		
8:00–9:00	0.9325	0.2265	0.5402	0.9273	0.8104	0.7092

The lack of significant differences shows that traffic intensity stabilized on a similar and stable level regardless the day of a week. The explanation for a lack of distinct differences even when they reached 12% may be high fluctuations in traffic flow that can be observed in shorter time intervals. This also indicates that in further analyses measurements could be carried out on any of those days.

4.2. Hourly Traffic Volume Fluctuations Within Peak Hours in Everyday Conditions

It is believed that effective VMSs should redirect traffic flows from crowded directions to less crowded alternatives. Before the artificial values of TTI were set one-hour traffic flow variations in the period between 7:00–8:00 and 8:00–9:00 were checked to avoid biased conclusions arising from natural traffic fluctuations. For this purpose, traffic flow was analyzed in 15-min intervals (Figure 5) on the basis of peak hour factors (PHF) k15 (Figure 6) and average traffic volumes (Qavg). The peak hour factor compares the traffic volume during the busiest 15 min of the peak hour with the total volume during the peak hour. It indicates how consistent traffic volume is during the peak hour.

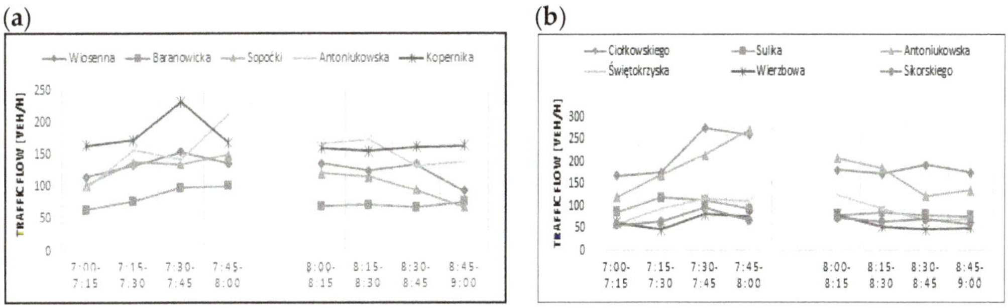

Figure 5. 15-min traffic intensity variations (**a**) main routes (**b**) alternative routes.

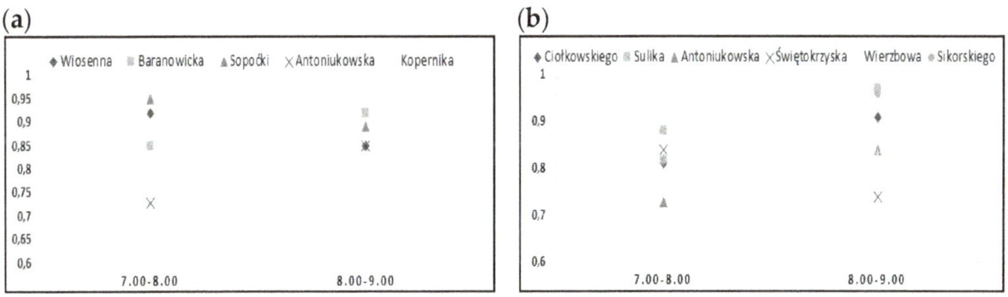

Figure 6. Peak hour factor (**a**) main routes (**b**) alternative routes.

Visible irregularity in main (Kopernika and Antoniukowska streets) and alternative routes (Ciołkowskiego and Antoniukowska streets) correlated with higher upstream traffic and is validated by k15 values. A large discrepancy of PHF values (Figure 6) along with increasing and decreasing trend in traffic volumes in peak hours indicate the randomness of the irregularity phenomena rather than its regularity caused by TTI information displayed on the VMS. This also confirms that everyday traffic volumes are subjected to some natural fluctuations which in most cases are statistically insignificant though (Table 3).

Table 3. ANOVA results for main and alternative routes.

	Main Routes									
	Wiosenna		Baranowicka		Sopoćki		Antoniukowska		Kopernika	
Time	7:00–8:00	8:00–9:00	7:00–8:00	8:00–9:00	7:00–8:00	8:00–9:00	7:00–8:00	8:00–9:00	7:00–8:00	8:00–9:00
Q avg (veh/15 min)	137	133	84	71	129	106	153	150	193	158
SD	14.3	22.8	15.7	5.1	14.4	18.2	45.6	24.7	21.7	11.7
$F =$	0.2025		4.4008		7.6645		0.0245		24.3532	
$p =$	0.6570		0.0545		0.0151		0.8776		0.0001	
	Alternative Routes											

	Ciołkowskiego		Sulika		Antoniukowska		Świętokrzyska		Wierzbowa		Sikorskiego	
Time	7.00–8.00	8.00–9.00	7.00–8.00	8.00–9.00	7.00–8.00	8.00–9.00	7.00–8.00	8.00–9.00	7.00–8.00	8.00–9.00	7.00–8.00	8.00–9.00
Q avg (veh/15 min)	223	182	103	82	205	156	98	88	65	57	74	64
SD	46.6	18.8	13.9	3.3	65.8	29.7	21.2	20.7	14.5	10.5	12.9	7.4
$F =$	7.7931		18.0642		3.8143		0.9874		1.4523		5.2518	
$p =$	0.0106		0.0008		0.0711		0.3388		0.2481		0.0318	

4.3. Determining the Significance of Traffic Intensities Changes Caused by Variations of TTI Displayed on the VMS

Information on traffic flow intensity variations caused by artificially elongated travel time displayed on the VMS boards present similar trends to those found during everyday operation of the VMSs. In the case of main routes (Figure 7), in the first hour traffic intensity volumes show smooth upward trends which in the second hour change into downward ones. In the case of alternative routes (Figure 8), the pattern of traffic flow variations are in line with those achieved in everyday operations.

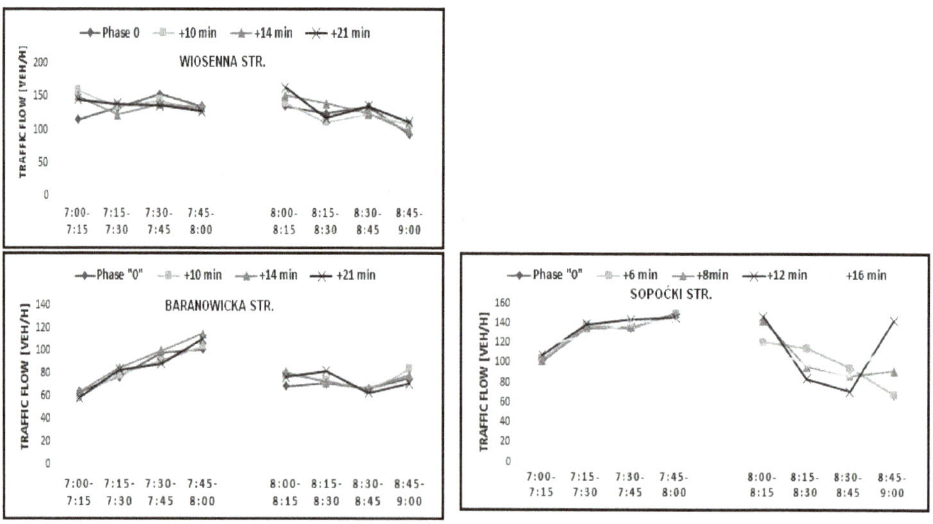

Figure 7. *Cont.*

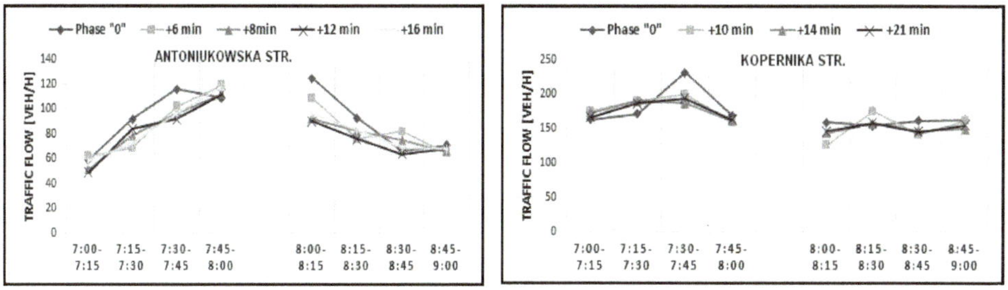

Figure 7. Traffic flow changes in 15-min intervals on main directions.

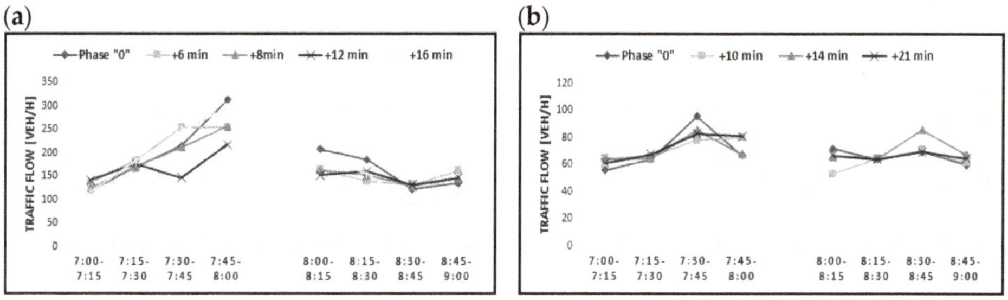

Figure 8. Samples of traffic flow changes in 15 min intervals on alternative directions (**a**) Antoniukowska Str. (**b**) Sikorskiego Str.

To determine the significance of changes of traffic volumes as a result of artificial time extension, the ANOVA analyses was applied. The first step of the analysis contained a comparison of traffic intensities in the first hour of morning rush with different TTI displayed. Then the same procedure was applied for the second hour of morning rush with adequate travel times. Finally, the comparison included traffic flow differences recorded in two consecutive rush hours. The results (Appendix A) show no significant differences between traffic flows either in the first or the second rush hour, regardless of the TTI value. In the group of main routes, the only significant difference was gained in Kopernika route—TTI extended up to 200% and 300% of the initial state brought significant difference between traffic flow intensity in the first and second hour. Box plots represent those changes graphically (Figure 9). Even though the traffic intensities decreased, the reason for that was not diversion of traffic flow into the alternative Sikorskiego route—in that case the changes remain insignificant. It results from Figures 7 and 8 that along with the increase in the traffic intensity in the first hour on Kopernika str, the intensity increases similarly on Sikorskiego str. After the intensity reaches its peak in the third quarter of the first hour, it decreases in both cases. Within alternative routes the differences in traffic flows showed a significant level with the TTI values elongated to 10 min and 14 min (Ciołkowskiego), yet further extension of TTI did not cause such influence. In the case of Sulika, route traffic intensities differed significantly when the TTI was elongated up to 300%. In the case of Ciolkowskiego and Sulika alternative routes, higher local speed limit constituted around 30% of the route length. Hence, it could be expected that this facet should be additional motivation to reroute for drivers facing deteriorated conditions on the main routes. However, the results did not confirm that, and no significant differences in traffic flows changes were stated on those routes. Admittedly, on Ciolkowskiego route we stated a significant difference in the changed traffic intensity, but it is not supported by relevant changes to the main direction (Wiosenna str) so it cannot be the result of the information given of the VMS.

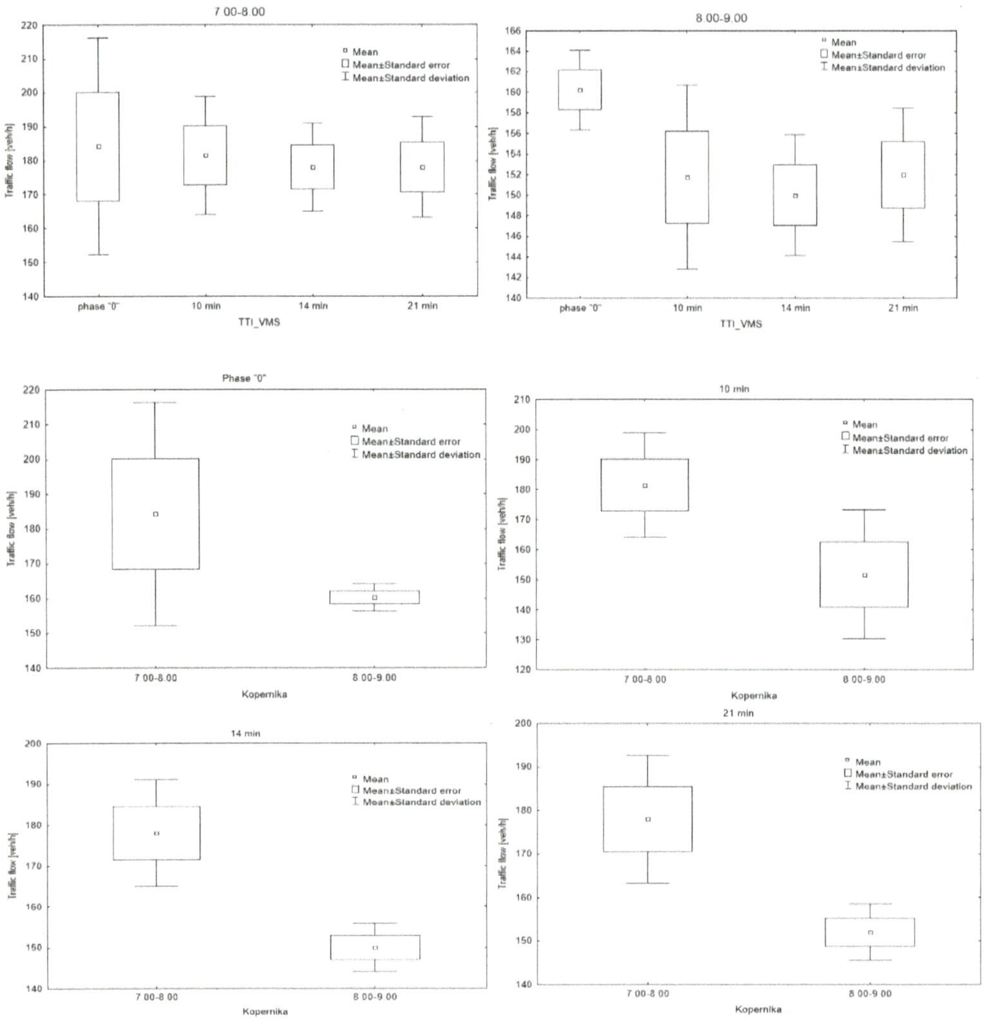

Figure 9. Box plots for Kopernika route with different TTI values.

Considering achieved ANOVA analyses results, two observations are symptomatic and essential in terms of VMS impact on drivers travel route choice. The first one shows that significant difference for the Sulika route was also stated in phase "0". This result seems to be random because the significant difference between the first and second hour at phase "0" becomes insignificant for TTI elongated up to 100%, becoming significant again for TTI elongated to 300%. The second revelation is that significant differences are related with the quite intensive traffic flow decreases in the second rush hour, and the decreasing traffic is a natural process evolving from daily traffic flow changes. The results of this study contradict Benson's [32] and Zhong-Rens et al. [33] findings. Benson found that drivers respond well to simple and reliable information on VMS, while Zhong-Ren stated that 66% of drivers changed their route and their behaviour was correlated with the frequency that a driver encounters an arterial VMS.

The authors' observations suggest that recorded traffic fluctuations are the result of natural traffic intensity fluctuations in traffic flow rather than changes caused by the impact of VMS information. This conclusion is also supported by the data presented in Figure 7. If elongated TTI would effectively influence drivers' route choice decisions, then we should observe the decrease of traffic intensity in the first hour along with increase intensity on the alternative route (i.e., Kopernika–Sikorskiego intersection). However, as it can be seen, this does not happen. Traffic intensity increases smoothly to reach its maximum in the third quarter (7:30–7:45) after which it decreases. This trend is equally present in Kopernika (Figure 7) and Sikorskiego routes (Figure 8).

The lack of drivers' response to the displayed TTI values in everyday conditions cannot be explained by a short display time (it is the case for the longest exposure of TTI in everyday conditions) because in this study each elongated TTI value was displayed for 2 h. A possible explanation for that behavior can be sought in the fact that drivers have no knowledge about the route of the alternative direction (the alternative route is not displayed on the VMS board). Another justification could be that drivers have limited confidence in the information provided, but that should be additionally confirmed in an extended questionnaire survey.

5. Conclusions

Several investigations testing drivers attitudes toward VMS placed on motorways have been conducted so far, but many of them were based on simulator surveys. This paper is based on in situ surveys and focused on the effect of travel time information provided on VMSs on drivers' route choice in the urban area of a middle-size city. The achieved outcomes are in contradiction to earlier research findings and showed a lack of drivers' response to the information displayed on the VMSs, regardless of the length of expected travel time. Drivers did not divert, even with the time elongated up to 200% or 300% of the initial state.

The significant changes observed in traffic flow intensities between consecutive morning rush hours resulted from natural fluctuations in daily traffic, not from expected delays on a given direction. Hence, the directing role of VMS investigated in this paper is very questionable, and the use of VMS providing only simple information about expected travel time is insufficient—drivers seem to ignore such a message. The reason for that may also lie in limited drivers' knowledge about the route of alternative directions. Better route change could be achieved from boards displaying travel time information together with a graphical display of the alternative route. However, recognition of drivers' behavior in this scope could be additionally confirmed through an extended questionnaire study. Similarly, possible use of mobile applications and drivers' preferences and attitudes toward traffic information conveyed through VMSs and their propensity to divert under certain situations could be verified.

Author Contributions: Conceptualization, R.Z. and Z.D.; methodology, R.Z.; formal analysis, R.Z.; investigation, R.Z.; resources, Z.D.; writing—original draft preparation, R.Z.; writing—review and editing, R.Z.; visualization, R.Z. All authors have read and agreed to the published version of the manuscript.

Funding: This research was supported by Project No WZ//WB-IIL/1/2020, and it was financially supported by the Ministry of Science and Higher Education, Poland.

Institutional Review Board Statement: Not applicable.

Informed Consent Statement: Not applicable.

Data Availability Statement: The data presented in this study are available on request from the corresponding author. The data are not publicly available due to legal issues—they are confidential data.

Conflicts of Interest: The authors declare no conflict of interest.

Appendix A

Table A1. ANOVA results for hourly traffic fluctuations (elongated TTI).

	Main Routes									
	Wiosenna		Baranowicka		Kopernika		Sopoćki		Antoniukowska	
7:00–8:00	$F =$	$p =$	$F =$	$p =$	$F =$	$p =$	$F =$	$p =$	$F =$	$p =$
	0.2730	0.8436	0.4078	0.7502	0.0857	0.9665	0.0216	0.9989	0.0017	0.9999
8:00–9:00	0.2769	0.8409	0.3715	0.7750	1.9592	0.1740	0.1305	0.9688	0.6113	0.6608
	$F =$	$p =$	$F =$	$p =$	$F =$	$p =$	$F =$	$p =$	$F =$	$p =$
Phase "0"	0.9095	0.3770	2.4998	0.1649	2.2171	0.1870	2.6608	0.1596	0.0023	9629
10 min	4.566	0.0764	0.5571	0.4836	4.3022	0.0755				
14 min	0.3780	0.5612	2.0486	0.2023	15.4229	0.0077				
21 min	0.1667	0.6972	1.1357	0.3275	10.4000	0.0180				
6 min							4.2852	0.0838	0.3921	0.5542
8 min							2.5366	0.1623	0.0008	0.9772
12 min							1.1348	0.3277	0.3032	0.6017
16 min							5.7907	0.0528	0.1376	0.7233

	Alternative Routes											
	Ciołkowskiego		Sulika		Sikorskiego		Antoniukowska		Świętokrzyska		Wierzbowa	
7:00–8:00	$F =$	$p =$	$F =$	$p =$	$F =$	$p =$	$F =$	$p =$	$F =$	$p =$	$F =$	$p =$
	0.0161	0.9970	0.1021	0.9572	0.0411	0.9883	0.2608	0.8984	0.0944	0.9827	0.0771	0.9881
8:00–9:00	0.3290	0.8044	1.043	0.4085	1.0112	0.4215	0.4264	0.7872	0.3752	0.8227	0.6594	0.6295
	$F =$	$p =$	$F =$	$p =$	$F =$	$p =$	$F =$	$p =$	$F =$	$p =$	$F =$	$p =$
Phase "0"	1.8797	0.2194	7.0617	0.0376	0.2155	0.6588	0.8198	0.4001	0.0806	0.7860	0.6457	0.4522
10 min	8.8113	0.0250	5.4777	0.0578	3.0636	0.1306						
14 min	6.3200	0.0456	5.2086	0.0626	0.0048	0.9465						
21 min	1.5796	0.2552	7.7327	0.0319	1.5455	0.2601						
6 min							3.7376	0.1013	0.0924	0.7713	1.7444	0.2347
8 min							2.2369	0.1853	0.1609	0.7021	5.2224	0.0623
12 min							1.5087	0.2653	0.4622	0.5219	5.1200	0.0643
16 min							2.7145	0.1505	0.2329	0.6464	4.1136	0.0888

References

1. Gonzalo-Orden, H.; Pérez-Acebo, H.; Linares-Unamunzaga, A.; Rojo, M. Effects of traffic calming measures in different urban areas. *Transp. Res. Proc.* **2018**, *33*, 83–90. [CrossRef]
2. Lee, G.; Joob, S.; Oh, C.; Choi, K. An evaluation framework for traffic calming measures in residential areas. *Transp. Res. Part D Transp. Environ.* **2013**, *25*, 68–76. [CrossRef]
3. Directive 2010/40/EU of the European Parliament and of the Council of 7 July 2010. Available online: https://eur-lex.europa.eu (accessed on 15 April 2020).
4. Blythe, P.; Rackliff, T.; Holland, R.; Mageean, J. ITS applications in public transport: Improving the service to the transport system. *J. Adv. Transp.* **2000**, *34*, 325–345. [CrossRef]
5. Stelzer, A.; Englert, F.; Hörold, S.; Mayas, C. Improving service quality in public transportation systems using automated customer feedback. *Transp. Res. Part E Logist. Transp. Rev.* **2015**, *89*, 259–271. [CrossRef]
6. Lee, S.H.; Cho, H.J. A Study on Safety Impacts for VMS Traffic Information. *J. Korea Inst. Intell. Transp. Syst.* **2015**, *14*, 22–30. [CrossRef]
7. Hamish, A.J.; Merat, N. The effectiveness of safety campaign vms messages—A driving simulator investigation. In Proceedings of the Fourth International Driving Symposium on Human Factors in Driver Assessment, Training and Vehicle Design, Washington, DC, USA, 9–12 July 2007; Available online: https://www.researchgate.net/publication/228598600_the_effectiveness_of_safety_campaign_vms_messages_a_driving_simulator_investigation (accessed on 25 January 2021).
8. Kwon, E.; Brannan, D.; Shouman, K.; Isackson, C.; Arseneau, B. Development and Field Evaluation of Variable Advisory Speed Limit System for Work Zones. *Transp. Res. Rec. J. Transp. Res. Board* **2015**, 12–18. [CrossRef]
9. Gao, F.; Li, S.E.; Zheng, Y.; Kum, D. Robust control of heterogeneous vehicular platoon with uncertain dynamics and communication delay. *IET Intell. Transp. Syst.* **2016**, *10*, 503–513. [CrossRef]
10. Wang, D.; Fengjie, F.; Xiaoqin, L.; Sheng, J.; Dongfang, M. Travel time estimation method for urban road based on traffic stream directions. *Transp. A Transp. Sci.* **2016**, *12*, 479–503. [CrossRef]
11. Wang, Z.; Wu, G.; Barth, M.J. Cooperative Eco-Driving at Signalized Intersections in a Partially Connected and Automated Vehicle Environment. *IEEE Trans. Intell. Transp. Syst.* **2020**, *21*, 2029–2038. [CrossRef]
12. Tian, D.; Li, W.; Wu, G.; Barth, M.J. Examining the Safety, Mobility and Environmental Sustainability Co-Benefits and Tradeoffs of Intelligent Transportation Systems. UC Davis: National Center for Sustainable Transportation. 2017. Available online: https://escholarship.org/uc/item/0m49j95r (accessed on 24 April 2020).
13. Jurgen, R. *V2V/V2I Communications for Improved Road Safety and Efficiency*; SAE: Warrendale, PA, USA, 2012; pp. i–viii.
14. Pribyl, O.; Blokpoel, R.; Matowicki, M. Addressing EU climate targets: Reducing CO_2 emissions using cooperative and automated vehicles. *Transp. Res. Part D Transp. Environ.* **2020**, *86*. [CrossRef]
15. Hoye, A. Speed Cameras, Section Control, and Kangaroo Jumps—A Meta-Analysis. *Accid. Anal. Prev.* **2014**, *73*, 200–208. [CrossRef] [PubMed]
16. La Torre, F.; Meocci, M.; Nocentini, A. Safety effects of automated section speed control on the Italian motorway network. *J. Saf. Res.* **2019**, *69*, 115–123. [CrossRef]
17. Ziolkowski, R. Effectiveness of Automatic Section Speed Control System Operating on National Roads in Poland. *Promet-Traffic Transp.* **2019**, *31*, 435–442. [CrossRef]
18. Karlberg, N.O. Road, weather, action. In *Traffic Technology International*; AutoIntermediates Limited: Surrey, UK, 2003.
19. Rama, P.; Kulmala, R. Effects of variable message signs for slippery road conditions on driving speed and headways. *Transp. Res. Part F* **2000**, *3*, 85–94. [CrossRef]
20. Boyle, L.; Mannering, F. Impact of traveller advisory systems on driving speed: Some new evidence. *Trans. Res. Part C* **2004**, *12*, 57–72. [CrossRef]
21. Zhenzhen, Y.; Hongcheng, G.; Shengrong, Z. Simulation analysis for guidance benefit of travel time displayed by variable message signs. *IEEE* **2011**, *5*, 123–127.
22. Cheu, R.L.; Liu, Q.; Lee, D. Arterial Travel Time Estimation Using SCATS Detectors. *Appl. Adv. Technol. Transp.* **2002**, *4*, 32–39. [CrossRef]
23. Elhenawy, M.H.; Chen, H.; Rakha, H.A. Dynamic Travel Time Prediction Using Data Clustering and Genetic Programming. *Transp. Res. Part C Emerg. Technol.* **2014**, *42*, 82–98. [CrossRef]
24. Jamshidnejad, A.; Schutter, B.D. Estimation of the Generalised Average Traffic Speed based on Microscopic Measurements. *Transp. A Transp. Sci.* **2015**, *11*, 525–546. [CrossRef]
25. Shang, H.; Huang, H.; Gao, Z. Impacts of variable message signs on traffic congestion. *Sci. China Ser. E Technol. Sci.* **2009**, *52*, 477. [CrossRef]
26. Erke, A.; Sagberg, F.; Hagman, R. Effects of route guidance variable message signs (VMS) on driver behaviour. *Transp. Res. Part F Traffic Psychol. Behav.* **2007**, *10*, 447–457. [CrossRef]
27. Chatterjee, K.; Hounsella, N.B.; Firminb, P.E.; Bonsall, P.W. Driver response to variable message sign information in London. *Transp. Res. Part C Emerg. Technol.* **2002**, *10*, 149–169. [CrossRef]
28. Zhang, E.; Zhang, X. Road Traffic Congestion Detecting by VANETs. In *Proceedings of the 2nd International Conference on Electrical and Electronic Engineering (EEE 2019)*; Atlantis Press: Paris, France, 2019; Volume 185.

29. Menelaou, C.; Kolios, P.; Timotheou, S.; Panayiotou, C.G.; Polycarpou, M.P. Controlling road congestion via a low-complexity route reservation approach. *Transp. Res. Part C Emerg. Technol.* **2017**, *81*, 118–136. [CrossRef]
30. Jayapal, C.; Roy, S.S. Road traffic congestion management using VANET. In Proceedings of the International Conference on Advances in Human Machine Interaction (HMI), Kodigehalli, India, 3–5 March 2016; pp. 1–7. [CrossRef]
31. Ziółkowski, R.; Anszczak-Pycz, M.; Dziejma, Z. Analysis of red light camera violations—A case study in Bialystok. In Proceedings of the 2018 Smart City Symposium Prague (SCSP), Prague, Czech Republic, 24–25 May 2018; pp. 1–6. [CrossRef]
32. Benson, B.G. Motorist Attitudes about Content of Variable-Message Signs. *Transp. Res. Rec. J. Transp. Res. Board* **1996**, *1550*, 48–57. [CrossRef]
33. Zhong-Ren, P.; Guequierre, N.; Blakeman, J. Motorist Response to Arterial Variable Message Signs. *Transp. Res. Rec. J. Transp. Res. Board* **2004**, *1899*, 55–63. [CrossRef]

energies

MDPI

Article

Modelling the Effects of Traffic-Calming Introduction to Volume–Delay Functions and Traffic Assignment

Jan Paszkowski [1,2,*], Marcus Herrmann [2], Matthias Richter [2] and Andrzej Szarata [1]

1 Politechnika Krakowska, ul. Warszawska 24, 31-155 Kraków, Poland; aszarata@pk.edu.pl
2 Westsächsische Hochschule Zwickau, Kornmarkt 1, 08056 Zwickau, Germany; marcus.herrmann.e54@fh-zwickau.de (M.H.); m.richter@fh-zwickau.de (M.R.)
* Correspondence: jan.paszkowski.krakow@gmail.com

Abstract: Traffic calming is introduced to minimise the negative results of motor vehicle use, for example, low safety level or quality of life, high noise and pollution. It can be implemented through the introduction of road infrastructure reducing the velocity and the traffic volume. In this paper, we studied how traffic-calming influences the traffic assignment. For the research, a traffic-calming measure of speed cushions on the Stachiewicza street in Krakow was taken. A method of extracting trajectories from aerial footage was shown, and it was used to build a model. For a given example, through driving characteristics research and microscopic modelling, volume–delay BPR functions were estimated—for a street with and without traffic calming. Later, a toy network of two roads of the same length, connecting the same origin and destination, was simulated using an equilibrium traffic assignment method. Simulations were conducted both with the use of PTV Vissim and Visum software and through individual calculations. According to the results of this paper, there was a difference in traffic volume according to the equilibrium traffic assignment in the aforementioned toy network as a function of total network traffic volume.

Keywords: traffic calming; traffic modelling; volume–delay functions; traffic assignment

Citation: Paszkowski, J.; Herrmann, M.; Richter, M.; Szarata, A. Modelling the Effects of Traffic-Calming Introduction to Volume–Delay Functions and Traffic Assignment. *Energies* **2021**, *14*, 3726. https://doi.org/10.3390/en14133726

Academic Editor: Elżbieta Macioszek

Received: 29 April 2021
Accepted: 15 June 2021
Published: 22 June 2021

Publisher's Note: MDPI stays neutral with regard to jurisdictional claims in published maps and institutional affiliations.

1. Introduction. Traffic Calming

Road planning, together with intelligent transport systems, helps to minimise the negative impact of motorised traffic. Moreover, data acquired from traffic control, such as traffic flow information or video footage of the road network, give a background for traffic modelling. In this paper, we used the video detection of traffic-calmed roads to produce the trajectories needed to build a microsimulation model. Later, we investigated the influence of the traffic-calming measures on traffic assignment.

According to the definition by the Institute of Transportation Engineers [1], traffic calming is the combination of mainly physical measures that reduce the negative effects of motor vehicle use, alter driver behavior and improve conditions for nonmotorized street users. The reduction in the negative effects comes with the reduction in the main factor causing it—high traffic volume. This allows the reduction in the noise or air pollution and improves the road safety and quality of life. The introduction of traffic calming historically started in the 1960s and 1970s in the Netherlands, Great Britain, Germany and Denmark, and it is connected with the idea to prioritise walking by closing the streets for motorised traffic or to slow car speeds down to walking pace—pedestrianisation [2].

Traffic calming can be categorised by the scales of influences into three levels [3]:

Level I—actions restricted to local, residential areas with low traffic volumes and capacities to restrain traffic velocities and to reduce negative traffic impacts;

Level II—actions dedicated to a corridor, which reduces traffic on that corridor, but not on the network;

Level III—actions introduced in a greater area and having an influence at a macro-level.

Recent progress in extracting the trajectories from video footage and the low cost of aerial videography thanks to the growing availability of unmanned aircraft has enabled easy trajectory recording for a relatively large area. This has made it possible to undertake the following research of the influence of the volume–delay function of a traffic-calmed street in order to create a modelling background to answer the question: Is it possible to model a situation where a traffic-calmed street attracts less traffic?

In this paper, the level III influence was discussed. Research on how traffic calming influences the traffic assignment was undertaken. For a given example, through driving characteristics research and microscopic modelling, BPR functions (volume–delay functions in the shape researched by the Bureau of Public Roads [4]) were estimated—for a street with and without traffic calming. Later, a toy network of two roads of the same length, connecting the same origin and destination, was simulated using an equilibrium traffic assignment method. Simulations were conducted both with the use of PTV Visum software and through individual calculations. The result of this paper was the difference in traffic volume according to the equilibrium traffic assignment in the aforementioned "toy network" as a function of total network traffic volume.

2. Literature Review

2.1. Measurement Methods

The following research requires the fulfilment of two challenges to properly reproduce driver behaviour. The first is the accurate reproduction of vehicle movement, needed for proper calibration of the kinetic parameters of the traffic. The second is connected with the interactions among other vehicles. Identifying leader–follower pairs would allow the estimation of parameters of the car-following model. In the Wiedemann car-following model, these parameters are, for example: distances between successive vehicles, decceleration forces when noticing the approach to the car in front, and the velocity and distance fluctuations when following.

Measurement of the vehicle movement can be divided into two categories: inside and outside the vehicle. First, measurements inside the vehicle (known also as a floating car) are undertaken with the help of the measurement vehicle, which is equipped with a movement tracking device, recording distance and velocity, and possibly also measuring the press of the acceleration and braking pedal. Alternatively, the floating car can be equipped with the GPS device, which also makes it possible to record the complete trajectory (the positions of the vehicle in the time interval). When using this method, the accuracy of the GPS (2–5 m) should be noted [5], which can influence the results negatively. Measurements outside the vehicle are able to record the trajectories of more than one vehicle. These measurements include the velocity measurement on the street, such as radar control or pneumatic tubes for speed measurement [2]; moreover, they can be: speed control, photographing or video recording. Unfortunately, most of these methods collect the data in one specific point and not continuously on the whole measured road link.

To ensure the continuity and recording of every vehicle in the traffic flow, a video recording can be used. With the use of machine learning and computer vision, vehicles, bicycles and pedestrians can be detected automatically without the need to go through the time-consuming process of manual trajectory extraction.

Video recording requires a proper camera positioning to maintain the required angle and an unobstructed view on the traffic. This can be solved with the positioning of a camera vertically pointing down for simple trajectory calculation and to minimize the amount of unwanted objects in the picture. This requires the installation of the camera high above the street or to use aerial recording. In this research, an unmanned aircraft was used.

2.2. Microsimulation Model

The microsimulation model analyses every object (vehicle, pedestrian, etc.) separately, representing the driver behaviour (speed, acceleration, etc.) in the current traffic situation in every time stamp of the simulation [6]. The simulation model should be calibrated correctly, meaning it should reproduce the traffic as close to the measured trajectories as possible, including unimpeded traffic speed, and interactions between vehicles. Based on these parameters, the vehicle (traffic) moves along the simulated road network along the set path. The microsimulation model usually does not produce a traffic assignment (choice of the path in the network), because it has been previously inputted, for example, from the traffic measurements or macrosimulation model. The microsimulation model delivers exact travel times. Every vehicle can be described by the following basic parameters: position, velocity, acceleration in time. Microsimulation models are dynamic models, meaning their product is the traffic information as a function of time. Usually, the simulation time is over a dozen minutes and the simulation time resolution is high (ex. 1 s). Due to the aforementioned accuracy, microscopic models require a precise and complete network reproduction, which comes with a lot of computing power, which makes it difficult or impossible to apply for larger areas, such as cities or regions. Each simulation run represents one example of the stochastic situation. The result of the simulation is both graphical and analytical, such as animation, trajectories (spatial and movement data), average travel times or velocities, unimpeded traffic speed or delay as an effect of growing density.

To simulate the interactions between vehicles in the microsimulation model, a car-following model is used, which defines the leader–follower behaviour. The movement characteristics can be based on the mathematical calculations dependent on the leader [7–10]. However, there have been models developed, which are based on the aforementioned models but also adds the behaviour additional to the reaction to the leader's activity. These relate to the Wiedemann model [11], optimum velocity model [12], and Intelligent Driver model [13].

In the article [14], the authors collected the trajectories, simulated various models, and assessed which gave the best reproduction of the real behaviour. Based on this article, in this research, Vissim software and the Wiedemann model were chosen.

Vissim's traffic flow model is a stochastic, time-step-based, microscopic model that treats driver-vehicle units as basic entities. The car-following model is based on the Wiedemann model. Driving behaviour in this model is divided into four states: free driving, approaching, following, and braking [6,15].

Figure 1 shows the driving behaviour reproduced in the model. The green colour is the situation when no other vehicle is in the distance needed for the reaction (that distance depends on the speed difference between two vehicles), the car is moving in the free flow with the velocity of its choice. When approaching the vehicle in front, after reaching the perception threshold (the moment when the driver notices the car in front), and the distance is small enough according to the difference in velocity, the following car starts to decelerate (light and dark orange colour). After reaching a similar speed and a proper distance, the following car speed oscillates randomly near the vehicle velocity in front of them, trying to adjust to a similar velocity and same distance (white colour) [15].

Detailed driving behaviour is described with Wiedemann model parameters: Wiedemann 74, such as safety distances, lack of attention duration or lack of attention probability [11]. Additionally, in Wiedemann 99: time distribution of the speed-dependent part of the desired safety distance, time of deceleration before reaching safe distance, influence of distance on speed oscillation, desired acceleration when starting from standstill [15].

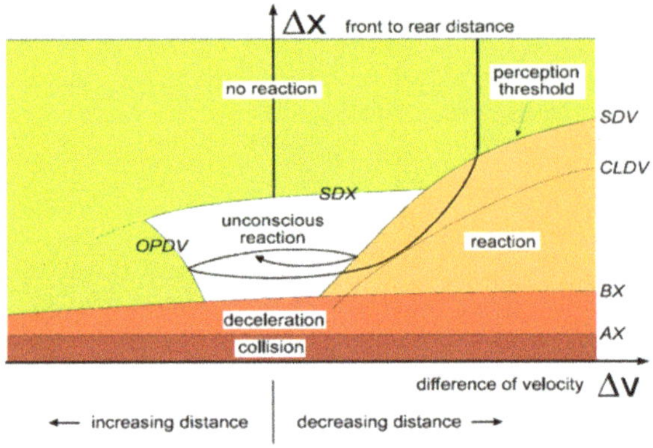

Figure 1. States of PTV Vissim model, source: PTV Vissim 9 manual [15].

2.3. Relationship between Traffic Flow and Velocity

The relationship between traffic flow and velocity has been a research subject since the beginning of mass motorisation. First, a meaningful study of that topic was presented in the 13th Annual Meeting of the Highway Research Board in December 1933 by Greenshields [16]. With the camera and chronometer in sight, the first measurements of density and speed were conducted. One of the most important results of the research was a fundamental diagram showing the relation between traffic volume (named in the research as "density") and speed. That classic diagram is shown on Figure 2.

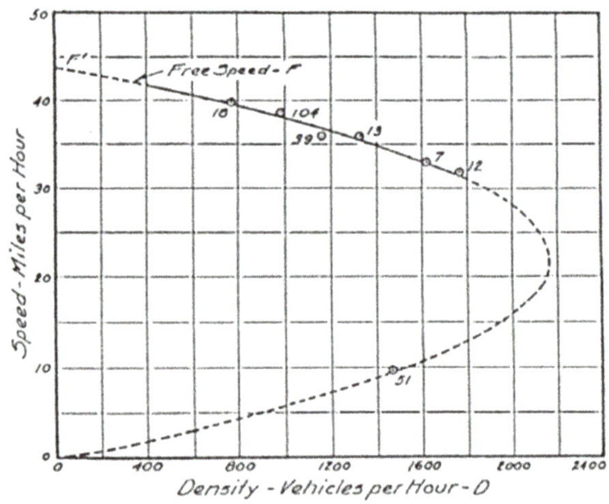

Figure 2. Greenshield's fundamental diagram of traffic flow, source: [16].

On the diagram, there are two states of traffic visible. On the top is the state without congestion. With the growth in the traffic volume (D), there is a moderate decrease in the velocity, which continues to the point of reaching capacity. The other state, as seen on the bottom of the chart, is a low velocity and low traffic volume caused by congestion.

Primarily, fundamental diagrams of traffic flow were researched not for the urban streets but for highway traffic, establishing the relationship between traffic volume, velocity and capacity, level of service, etc. More advanced diagrams to use in urban areas were based on traffic density or a relationship with kinematic wave theory [17,18].

An alternative form of describing the relation between traffic flow and velocity is a volume–delay function. The volume–delay function (VDF) is an approximation of fundamental traffic phenomena such as gridlocks and spillbacks occurring in dynamic traffic models, for macroscopic, static models [6,19]. The VDF, unlike the fundamental diagram, is a function, containing a hypocritical part (where the traffic volume is below the capacity) and a hypercritical one.

One of the historically first and most popular formulations of VDFs are BPR functions, published in 1964 by the Bureau of Public Roads in USA [4]. The other examples of VDFs can be Conical [20] or Akcelik [21] functions. Those function formulas are commonly used in macrosimulation models.

In the macroscopic model, the network is seen as a graph, where edges (links) (optionally nodes) are resistive. The resistance of the edges is described as a relationship between saturation grade (traffic volume to capacity ratio) and travel time (multiplied travel time of empty network), known as the volume–delay function. The volume–delay function is a nondescending, continuous, differentiable (with exception to the point sat = 1, where traffic volume is equal to capacity) function, usually calibrated based on assumptions or observations of the observed travel time and the corresponding volume data. [22]. For unimpeded traffic flow, the result of the volume–delay function equals 1, meaning that vehicles travel with a free flow travel time. With the growth in traffic volume (vehicles per hour), the function values grow. When the traffic volume reaches capacity, the saturation grade equals 1. After crossing the capacity, the function starts to increase significantly. The example of the volume-delay function is shown in Figure 3.

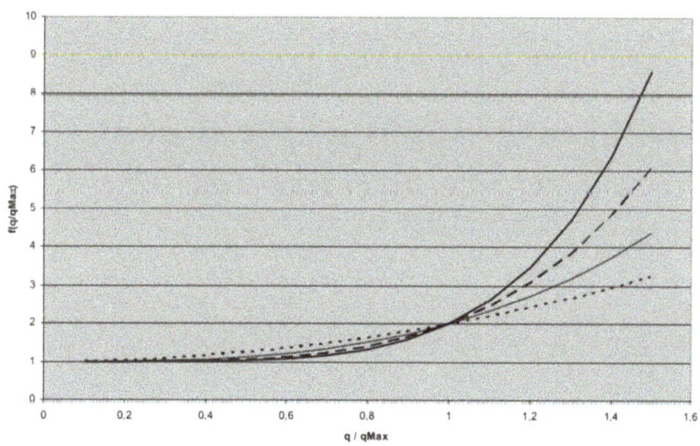

Figure 3. Example of volume–delay functions, source: [23].

2.4. Macrosimulation Model and Traffic Assignment

Among the types of macrosimulation models according to the methods on which they were built, we considered a four-stage model. It consists of four steps: Trip generation, trip distribution, mode choice and traffic assignment [24,25].

Trip generation is the first stage in which spatial analysis is undertaken to determine the amount of trips going in and out of each homogenous area during the peak hour.

Trip distribution creates a matrix of trips between the origin and destination areas based on the gravity function, which determines the probability of a trip existing based on the distance or travel time.

Mode choice determines the amount of trips taken by every available means of transport, such as car, public transport, and bike. It is determined by functions defining the probability of choice of the specific means of transport according to distance, availability, and motivation (home-work, home-school etc.)

Traffic assignment sets the path for every trip generated, distributed and assigned to the transport mode in the previous stage. For the traffic assignment, Wardrop's principle of user equilibrium is usually used [26]. Its assumptions are that every user chooses the path with the lowest cost (cost is represented by travel time), and that the transport system aims to achieve equilibrium—the state when no vehicle can change the path to the other one, with a lower cost. However, for large networks, the simple user-equilibrium algorithm is relatively slow. Therefore, more time-efficient algorithms are used, based on the user-equilibrium rule, for example, the LUCE (Linear User Cost Equilibrium) algorithm, used in PTV Visum, which aims to be 10–100 times faster than other methods [27]. The example of the traffic assignment is shown on Figure 4.

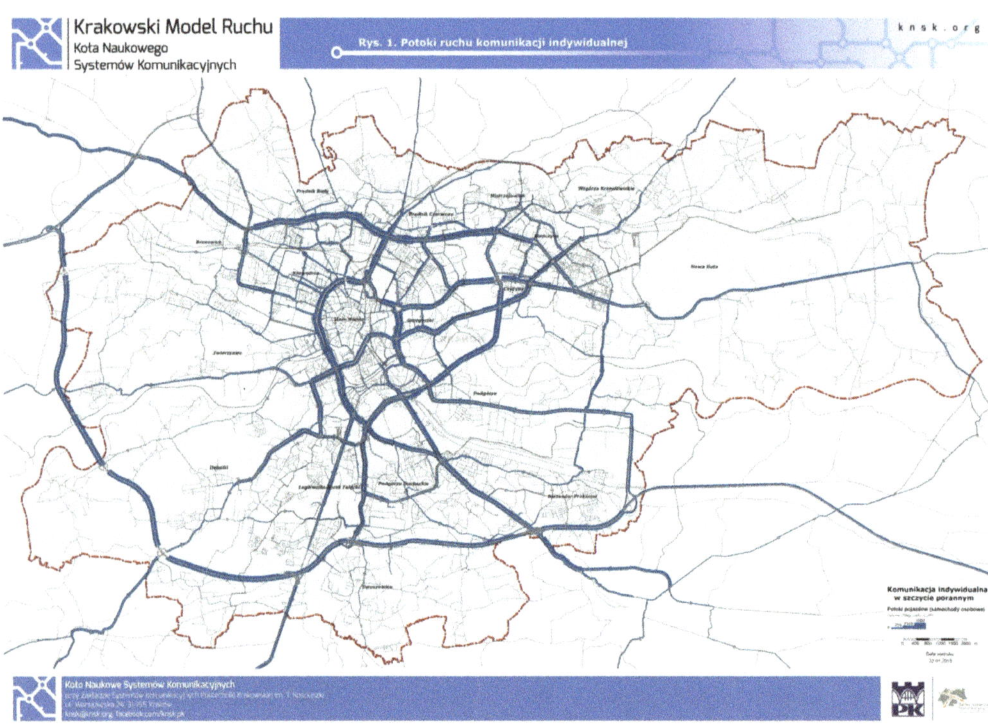

Figure 4. Example of traffic assignment for the city of Kraków. Source: [28].

Equilibrium methods require the iteration of traffic assignments, because the travel time on the link, the main factor to assign the trip to the path, depends on traffic flow, and thus on the amount of trips through the link in the form of volume–delay functions. Iterations are undertaken until the model reaches the equilibrium state.

In this research, only the last step of the four-step model is taken into the consideration, precisely, the traffic assignment reaction on the volume–delay functions existing on traffic-calmed roads.

3. Measurements and Building a Microsimulation Model

3.1. Methodology Overview

To produce the macrosimulation model, the vehicle movement was collected by unmanned aircraft with a video camera. Later, using video detection, the trajectories were produced with the help of machine learning. This allowed the extraction of the speed distributions of the vehicles, acceleration and braking—necessary parameters for the Vissim model, and speed profiles necessary for evaluation of the model. The overview is presented in Figure 5. A more detailed methodology is presented in Sections 3.2–3.5.

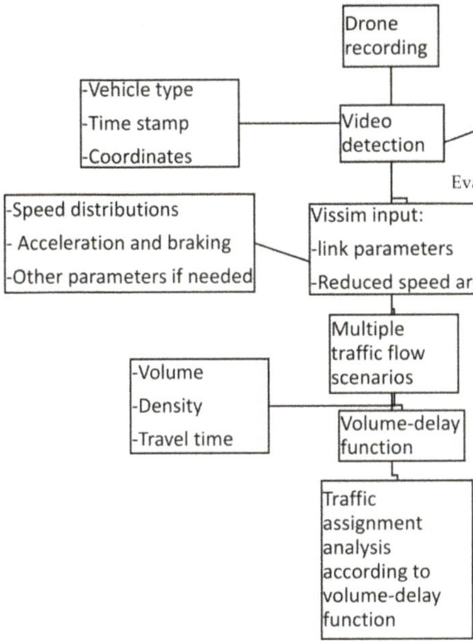

Figure 5. Methodology diagram. Source: own.

3.2. Location

The research area was Stachiewicza street in Kraków, Poland. There are speed cushions installed, two in each direction, allowing vehicles with a larger wheel gauge (such as buses) to pass them without slowdown. The street serves the role of one of the main connectors between the north-western and other districts, therefore carrying numerous and diversified traffic. Further, the direction "1" in this article means northbound, and "2" means southbound.

3.3. Video Data Collection

To provide coverage of the whole measured area, where a high point to place a camera was not present, an unmanned quadcopter (DJI Phantom IV) was used. Its coverage was 110 m of the street with an altitude of 80 m above ground level. The duration of the flight was approximately 25 min, which, after substracting the ascend and descend, left 15–20 min of traffic recording. The video parameters were: 1920 × 1080 pixels resolution and 40 frames per second.

The video footage was processed by the company DatafromSky [29] to obtain the trajectories. The results of the processing delivered a graphical visualisation to watch in the DatafromSky viewer (see Figure 6) and the csv file containing all trajectory data, such as: timestamp, coordinates, object ID, and object type (car, heavy vehicle, bus, bicycle, pedestrian).

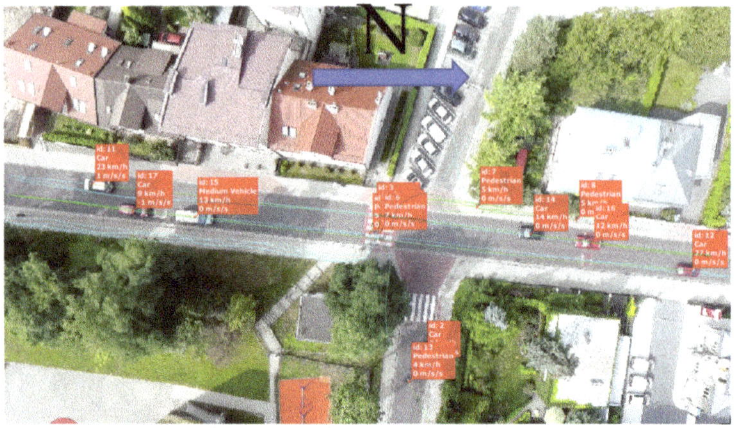

Figure 6. DatafromSky viewer software screenshot with car detection and trajectory data displayed (source: own).

3.4. *Trajectory Analysis*

The processing of the data acquired from the recordings was made as follows. First, the street visible in Figure 6 was divided into measurement sections of 10 cm in length. Instantaneous velocities were identified in each section for every vehicle, with the division to vehicle classes, such as cars (standard and van), busses and heavy vehicles. Then, the sections were put together to create the speed charts showing the speed profiles of the analysed road, considering the velocity change on the speed cushions. For this measurement, 20 min of the recording was used, containing 192 vehicles: 165 cars, 18 vans, 3 heavy vehicles, and 6 busses. There were also 90 pedestrians detected, but these data were not used in this research. Charts of speed profiles are presented below in Figures 7 and 8.

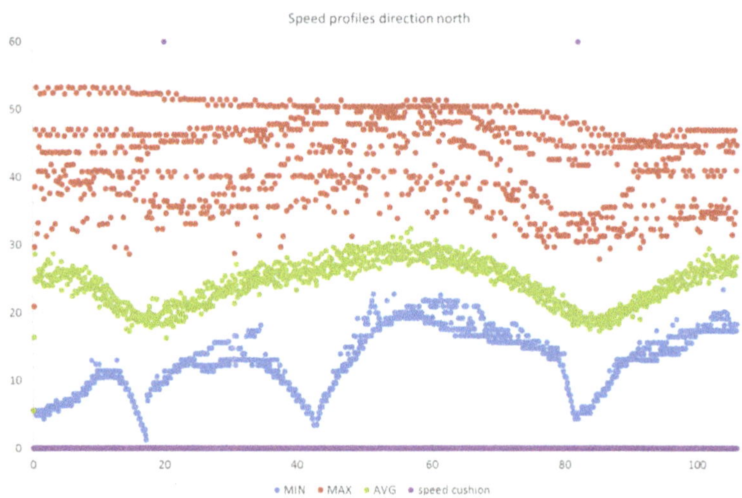

Figure 7. Speed profile for Stachiewicza street, direction northbound (source: own).

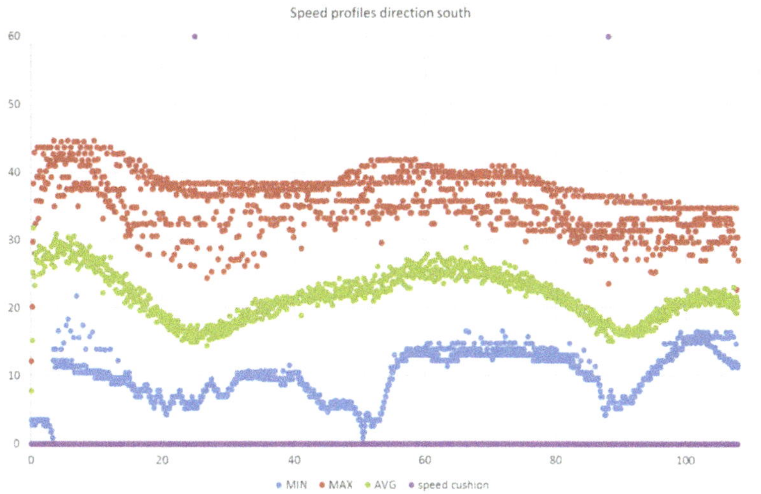

Figure 8. Speed profile for Stachiewicza street, direction southbound (source: own).

On the charts, speed profiles for the studied 107 m length of Stachiewicza Street can be seen. Speed cushions are indicated by violet dots on the top of the chart at 20 and 82 m along this section northbound, whereas southbound—on the 26th and 88th meter of the measured area. The minimum speed in each section is indicated with blue points. The decrease in the minimal velocity between the speed cushions (between 40th and 60th metre) shows the car stopping on the pedestrian crossing. The average velocity is represented with the green points—its decrease in the area of speed cushions is clearly visible. The red points marking the maximal speed show that there were vehicles which did not slow down on speed cushions, especially on the first one in the northbound direction.

3.5. Microsimulation Model

Building the microsimulation model started from preparing the speed distributions. Those were created according to the measured, cumulated frequencies for each vehicle type. There are two kinds of the speed distributions, corresponding to the two situations: near or away from the speed cushions. On the transfer between the speed distributions, the accelerations and braking were identified. In this model, there are two traffic directions, each on the separate link. There are two reduced speed areas in each link, representing the speed cushions. For every direction and for every reduced speed area, a separate speed distribution was inputted. Examples of the speed distributions are presented in Figure 9.

In the vissim microsimulation model, the speed distributions, acceleration and deceleration data were inputted. Later, the simulation was undertaken for the traffic volume equal to that during the measurement, to check the calibration. To compare the simulated trajectory data with the measured ones, average speed profiles for all vehicles from the simulation were exported. The calibration was undertaken using the speed distributions when the traffic volume allows unimpeded traffic to correspond with the simulated velocities. Velocities in the simulation were adjusted to comply with the measurements. For the calibration of the speed profile shape, vehicles' acceleration and braking were estimated. For the first direction of the Stachiewicza street (northbound), the correlation coefficient between the model and the measurement was equal to 0.79. The compared speed profiles are presented in Figure 10.

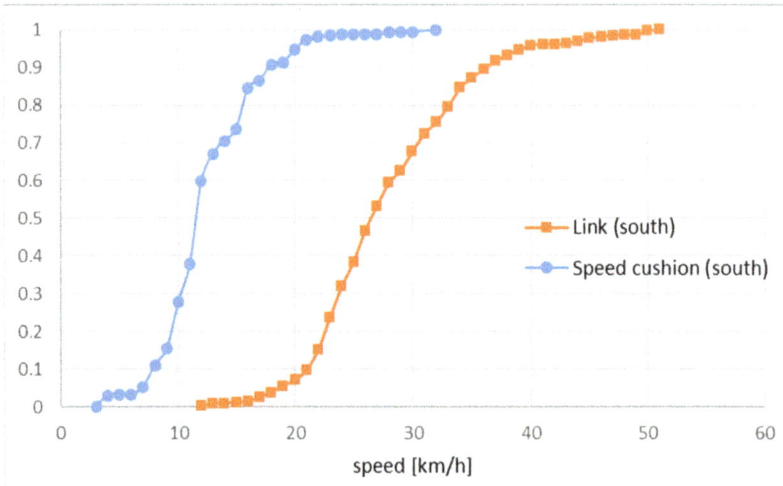

Figure 9. Car speed distributions for the link, direction southbound (orange) and its second speed cushion (blue), source: own.

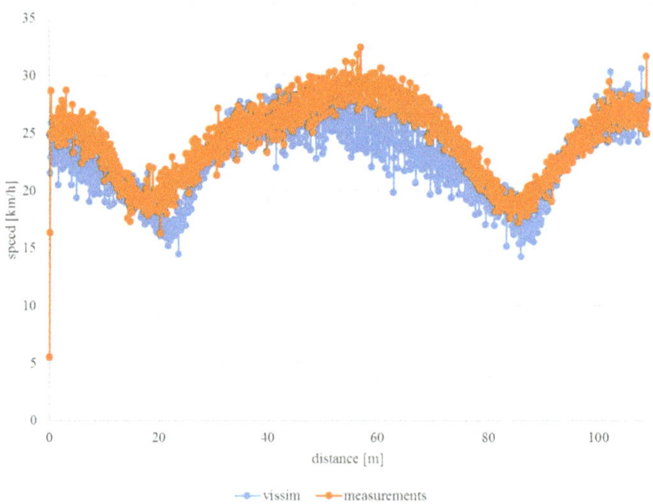

Figure 10. Comparison of speed profiles on the measured street section, heading northbound, source: own.

For the second direction (southbound), the correlation coefficient was 0.71. Figure 11 shows this comparison.

A correlation between measured and simulated speed profiles was clearly visible and acceptable. In case of needing correlation improvement and further calibration, the car-following model (Wiedemann) parameters can be adjusted or the link can be divided into more parts, with their own speed distributions, acceleration and breaking parameters.

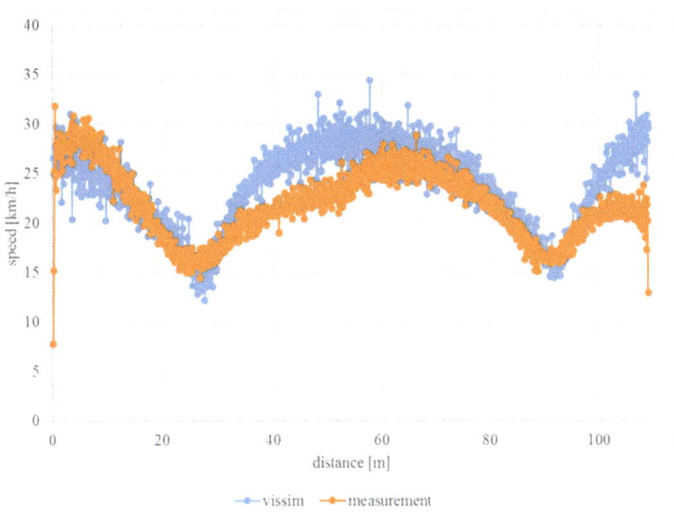

Figure 11. Comparison of speed profiles on the measured street section, heading southbound, source: own.

For the volume–delay function estimation, a variety of traffic volumes were simulated for both traffic-calmed and not-traffic-calmed links. To research the travel time in the traffic volumes exceeding the capacity, for the simulated link, another was added to accumulate the queues. Travel time (later used for the velocity calculation) is the travel time of the link with extensions (plus queue travel time if applicable) minus the travel time on extensions on the free flow. The initial result of that experiment was a relation between traffic volume [veh/h] and velocity [km/h]. This results in the chart shown in Figure 12.

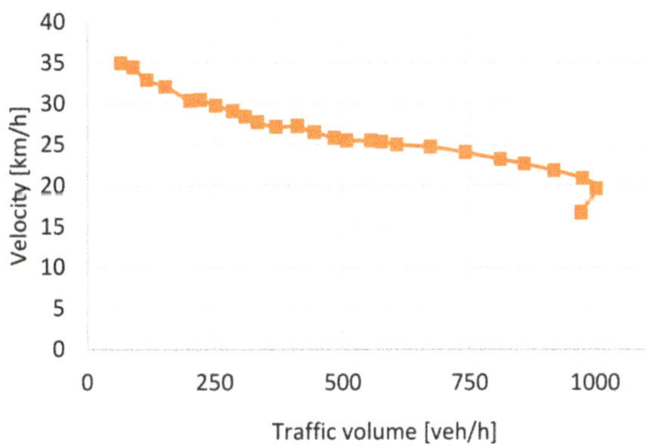

Figure 12. Traffic volume–speed relation, source: own.

On Figure 12, a descent in the speed with the traffic volume can be seen. At the point of reaching capacity, the velocity decreased with the maximal traffic volume, which created a relation complying with a fundamental diagram of traffic flow, where there are two results for velocity with the same traffic volume, depending on whether it is above or below capacity. To create a relation as a function, the traffic volume [veh/h] was replaced with the traffic density [veh/km] [30]. In that case, after reaching the capacity, the density still increased with the decrease in the velocity. This is presented in Figure 13.

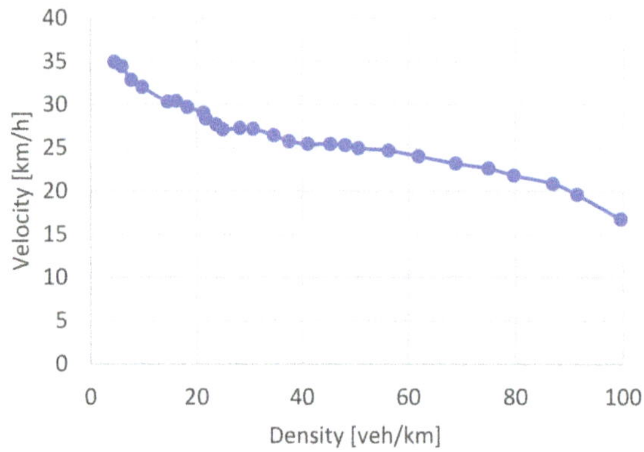

Figure 13. Traffic density–speed relation, source: own.

The form of the travel times was calculated with the form corresponding to the volume–delay function presentation available in PTV Visum, meaning that the relationship between the saturation grade (density divided by the maximum traffic volume density) and the travel time is the t_0 (free flow travel time) multiplied by the rest of the function. In the macroscopic transport model of the city of Kraków, volume–delay functions were estimated using the BPR2 volume–delay functions shape (see Formula (1)):

Formula (1). BPR volume–delay function, source: [4]

$$t_{cur}\ (sat) = \begin{cases} t_0 \cdot \left(1 + a \cdot sat^{b}\right), 0 < sat < 1 \\ t_0 \cdot \left(1 + a \cdot sat^{b'}\right),\ \ sat \geq 1 \end{cases} \tag{1}$$

where:

- t_0—free flow travel time (for the following speeds: with traffic calming: 28 km/h, without traffic calming: 30 km/h);
- a, b, b'—parameters of the function;
- sat—saturation grade (q/q_{max}).

Estimating the function parameters for the best fit was made using the squared difference minimisation. Parameters' values are shown in Table 1.

Table 1. Function parameter values.

Number of Function	Direction 1 with Traffic Calming	Direction 2 with Traffic Calming	Direction 1 without Traffic Calming	Direction 2 without Traffic Calming
a=	0.641359	0.758637	0.527618	0.611864
b=	0.738809	0.643984	0.905478	0.646525
b'=	6.760153	5.292947	7.080322	2.591875
q_{max}	1068	1044	1308	1158

The fitting of the function had a high correlation coefficient, in every case above 0.9. The comparison of the empirical relation with the function fitting can be seen in Figure 14.

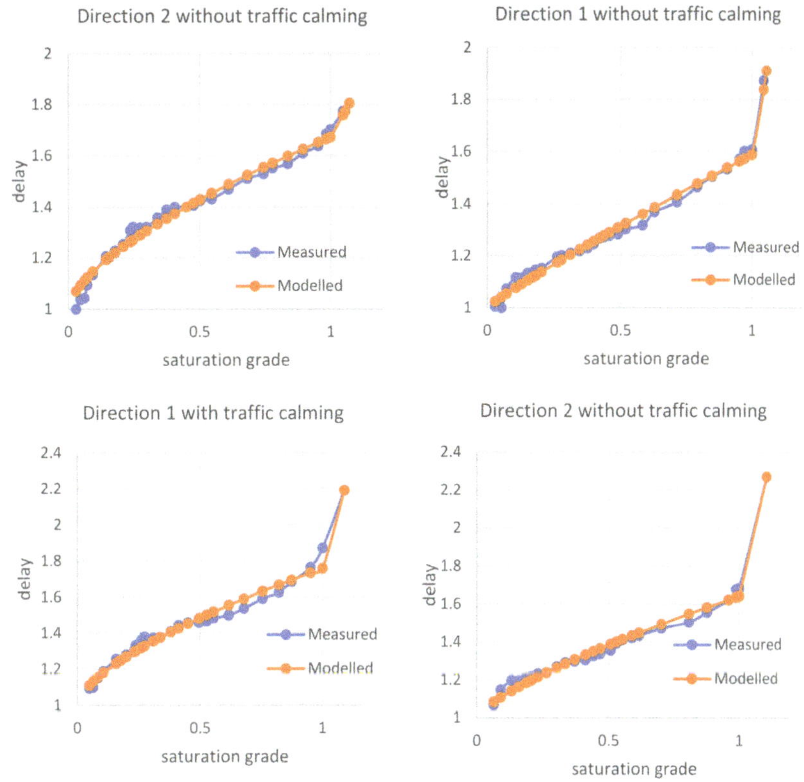

Figure 14. Fitting of the volume–delay functions, source: own.

4. Experiment

In this experiment, a "toy network" of two links of the same length was constructed, connecting a common origin and a destination. The links differed in the volume–delay function defined for them—one with, the other without traffic calming; thus, the travel time is as seen in Formulas (2) and (3):

Formula (2). Travel time with traffic calming. Source: own, based on: [4]

$$t_{wtc} = t_{0(wtc)} \cdot \left(1 + a_{wtc} \cdot \left(\frac{q}{q_{max,wtc}}\right)^{b_{wtc}}\right) \tag{2}$$

Formula (3). Travel time without traffic calming

$$t_{wotc} = t_{0(wotc)} \cdot \left(1 + a_{wotc} \cdot \left(\frac{q_{tot} - q}{q_{max,wotc}}\right)^{b_{wotc}}\right) \tag{3}$$

In the experiment, we researched a situation when our toy network is in the state of equilibrium. That means, for every user, the travel time is the same $t_{wtc} = t_{wotc}$. In that case, the traffic volume would be different for each link, which is written as q_{wtc}: traffic volume for a link with the traffic calming and the remaining traffic (on the link without traffic calming): $q_{tot} - q_{wtc}$, where the total traffic flow is described by q_{tot}. The traffic assignment on this simple network was calculated by solving the equation given in Formula (4): $t_{wtc} = t_{wotc}$.

Formula (4): The equation of the model

$$t_{wtc} = t_{wotc}$$
$$t_{0(wtc)} \cdot \left(1 + a_{wtc} \cdot \left(\frac{q}{q_{max,wtc}}\right)^{b_{wtc}}\right) = t_{0(wotc)} \cdot \left(1 + a_{wotc} \cdot \left(\frac{q_{tot}-q}{q_{max,wotc}}\right)^{b_{wotc}}\right) \quad (4)$$

Calculations of the traffic assignment were made with the help of Microsoft Excel's Solver together with the use of macros to automate the process of simulating various traffic volumes, and these were then confirmed with PTV Visum software (using the equilibrium traffic assignment for the aforementioned model). A graphical description of the model can be seen in Figure 15. There are two pictures with each of them containing different total traffic (q_{tot}): on the left: 500 veh/h, on the right—3000 veh/h. For this model, the BPR function values of Direction 2 stated in Table 1 are used—the top with traffic calming (wtc) and the bottom without traffic calming ($wotc$).

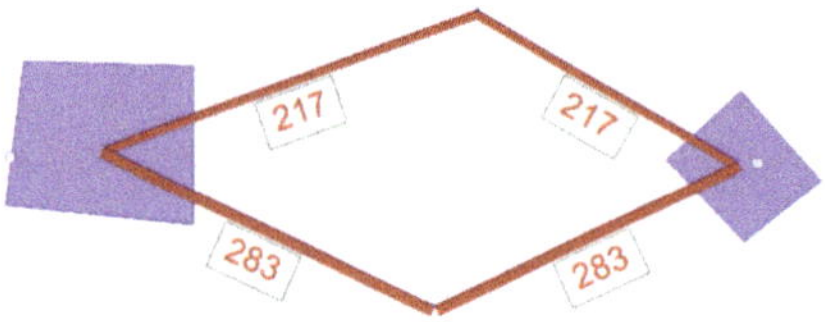

Figure 15. Example results of the traffic assignment, source: own.

5. Results

As a result of the experiment, a chart showing the relation between the total traffic volume (q_{tot}) and the traffic volume on each link was created. The results are shown in Figure 16.

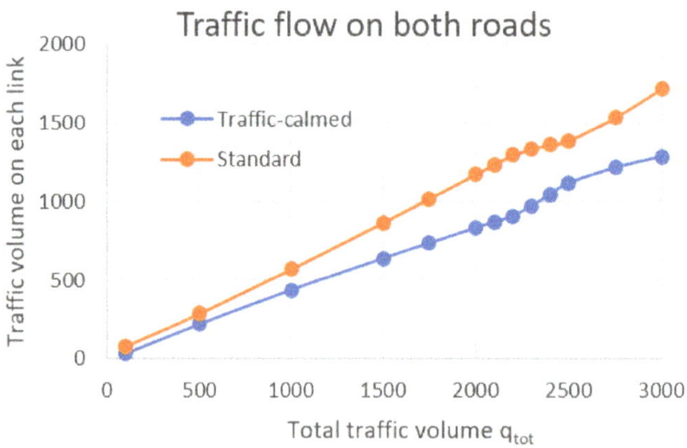

Figure 16. Traffic flow on both roads according to the total traffic volume, source: own.

For a better interpretation of the above chart, it was shown (on Figure 17) as a difference in traffic volume between each link.

The phenomena presented in the chart can be divided into three phases, depending on the state of the network:

1. Both links are below capacity. On the empty street, the speed is lower on the traffic-calmed road (because of the slowdown caused by speed cushions) than on the road without traffic calming. No or a little slowdown is caused by a high traffic volume. More vehicles choose the standard road. $q_{tot} \leq 2200$ veh/h.

2. Because more vehicles choose the standard road, it reaches capacity earlier (despite the capacity being higher than the other link) and causes congestion; thus, drivers choose an alternative road through a traffic-calmed one, which is still under capacity. $2200 < q_{tot} \leq 2500$ veh/h.
3. Traffic-calmed road exceeds capacity, as well as standard road; more vehicles choose the standard road, with higher speeds during congestion. $q_{tot} > 2500$ veh/h.

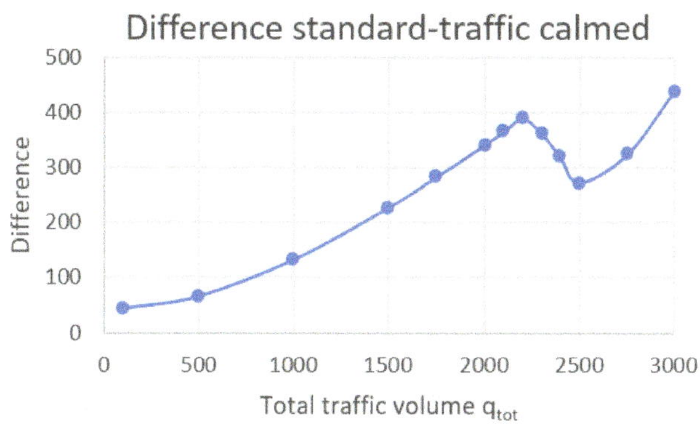

Figure 17. Difference in traffic flow on each link, source: own.

The difference between the traffic volumes on links with the aforementioned volume–delay functions in relation to total traffic volume was modelled as a polynomial function shown in Formula (4): $diff = c \cdot q_{tot}^{a} + b$. Function parameters are presented in Table 2.

Formula (5). Estimation function formula

$$diff\ (q_{tot}) = c \cdot q_{tot}^{a} + b \tag{5}$$

Table 2. Function parameters.

	a	b	c
1. $q_{tot} \leq 2200$ veh/h	1.75075	44.77371	0.00049
2. $2200 < q_{tot} \leq 2500$ veh/h	1.06087	1203.81532	−0.23048
3. $q_{tot} > 2500$ veh/h	2.93892	0.141×10^{-12}	0.259×10^{-8}

The results of the function fitting have been presented in Figure 18.

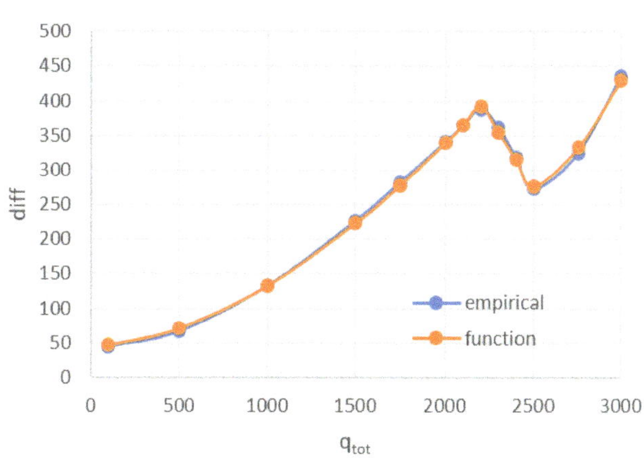

Figure 18. Fitting of the function, source: own.

Comparison with the Results of the Current Solutions

Currently, in existing macroscopic urban models, in most of the cases, there is no modelling of the volume–delay functions for traffic-calmed roads. Either for roads with small traffic (where traffic-calmed ones often belong) are no volume–delay functions defined (meaning that the travel time is not dependent on the traffic volume) or a volume–delay function is standard for the whole link type. It is considered that the traffic on the toy network of two links would be simulated as the same for both types of links unless there is no differentiation in the free flow velocity. However, when that v_0 velocity is different, in the case of a constant volume–delay function, there will be no traffic on the traffic-calmed road, because it is always slower than the other one. In the case of using the same volume–delay function, together with a different v_0 velocity, which is a proper solution in existing models, the result will be as in the Figure 19.

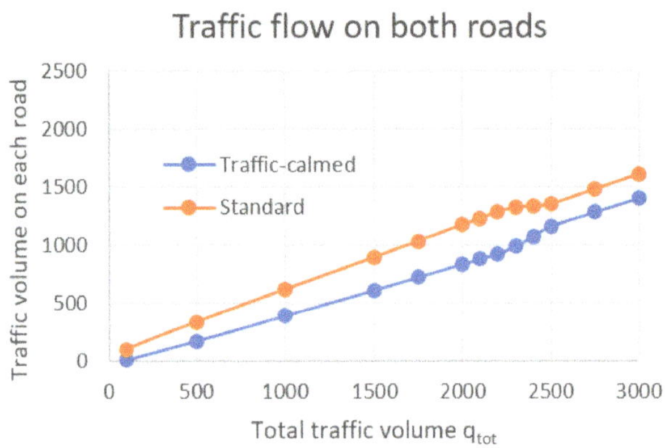

Figure 19. Traffic flow on both roads with the same volume–delay function and different v_0 velocities, source: own.

The difference between traffic volumes is shown on Figure 20.

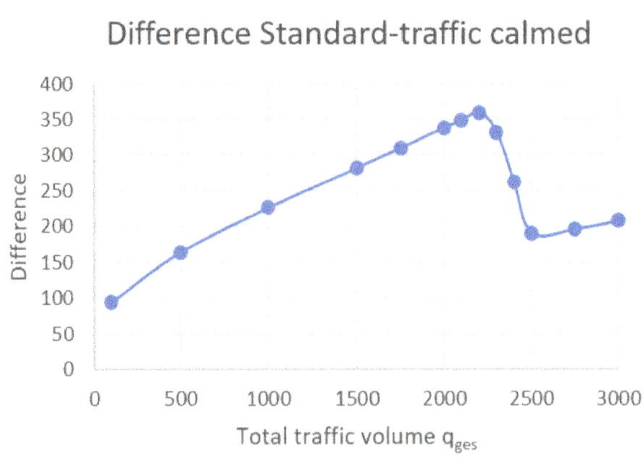

Figure 20. Difference in traffic volume between links with the same volume–delay function and different v_0 velocities, source: own.

Comparing the last two charts with Figures 19 and 20, we can see similarities such as the aforementioned three stages of traffic and similar values below the capacity. This is a result of a similar shape of the volume–delay functions for a standard and traffic-calmed road below capacity. However, the difference in the shape of the volume–delay functions above the capacity corresponds with a different traffic assignment behaviour. When we have different VDFs, the difference grows exponentially; otherwise, it grows linearly.

The described behaviour should occur in this specific case, but it is also the universal reaction of the macroscopic model for the change in the volume–delay functions. The experiment and comparison with the other possible methods show that applying volume–delay functions dedicated for the traffic-calmed areas gives other, possibly more precise traffic assignment results.

6. Summary

The aim of the article was to show the method of modelling a volume–delay function through video-detecting trajectories and microscopic modelling, and to use those functions to research traffic assignment. For the researched street, using this method, volume–delay functions' parameters were estimated. That led to the analysis of the traffic assignment on the "toy network" using the aforementioned functions. It was shown that a traffic-calmed road attracts less traffic compared with that without. Lastly, the article showed that using different volume–delay functions for describing traffic calming for macroscopic modelling makes a difference in a traffic assignment results. For the comparison and discussion with the other research in this field, similar research using methodology used in this was not found. The results of this paper can help in understanding a traffic assignment of traffic-calmed roads. For further research, other traffic-calming measures than speed cushions can be used. This would allow a comparison of various traffic-calming measures regarding the influence of the traffic calming.

Author Contributions: Conceptualization: J.P., M.R. and A.S.; Methodology: J.P. and M.H.; Software: J.P. and M.H.; Validation: J.P.; Formal analysis: J.P. and M.H.; Investigation: J.P., M.R., A.S. and M.H.; Data curation: J.P. and M.H.; Writing—original draft preparation: J.P.; Writing—review and editing: M.R. and A.S.; Visualisation: J.P.; Supervision: M.R. and A.S. All authors have read and agreed to the published version of the manuscript.

Funding: This research received no external funding.

Institutional Review Board Statement: This study did not involve humans or animals.

Informed Consent Statement: Not applicable.

Data Availability Statement: Not applicable.

Acknowledgments: The authors would like to thank the Saxon State Ministry of Science and Art, represented by the State of Saxony, for their partial funding of the research project "Applied research in the future field of digital communication" (diKo19/diKo20), under which this research was conducted.

Conflicts of Interest: The authors declare no conflict of interest.

References

1. ITE/FHWA. *Traffic Calming: State of the Practice*; Federal Highway Administration: Washington, DC, USA; Institute for Transportation Engineers: Washington, DC, USA, 1999.
2. Brindle, R.E. *Australia's Contribution to Traffic Calming*; Pearson: London, UK; PTRC: Kernersville, NC, USA, 1992.
3. Bureau of Public Roads. *Traffic Assignment Manual*; Department of Commerce, Urban Planning Division: Washington, DC, USA, 1964.
4. GPS Accuracy. 2018. Available online: https://www.gps.gov/systems/gps/performance/accuracy/ (accessed on 18 March 2018).
5. Barbosa, H.M. *Impacts of Traffic Calming Measures on Speeds on Urban Roads*; University of Leeds: Leeds, UK, 1995.
6. Kucharski, R. Makroskopowy Model Przepływu Ruchu w Sieci Drugiego Rzędu–Alternatywny Opis Stanu Sieci. In Proceedings of the Poznań-Rosnówko, Problemy Komunikacyjne Miast w Warunkach Zatłoczenia Motoryzacyjnego, Poznań-Rosnówko, Poland, 19–21 June 2013.
7. Pipes, L.A. An operational analysis of traffic dynamics. *J. Appl. Phys.* **1953**, *24*, 274–281. [CrossRef]
8. Forbes, T.W.; Zagorski, H.J.; Holshouser, E.L.; Deterline, W.A. Measurement of driver reactions to tunnel conditions. In Proceedings of the Highway Research Board, Washington, DC, USA, 6–10 January 1958; Volume 37.
9. Gazis, D.C.; Herman, R.; Rothery, R.W. Nonlinear follow-the-leader models of traffic flow. *Oper. Res.* **1961**, *9*, 437–599. [CrossRef]

10. Newell, G.F. Nonlinear effects in the dynamics of car following. *Oper. Res.* **1961**, *9*, 209–229. [CrossRef]
11. Wiedemann, R. *Simulation des Straßenverkehrsflusses*, 1974th ed.; Institut für Verkehrswesen: Karlsruhe, Germany, 1974; Available online: https://trid.trb.org/view/596235 (accessed on 19 June 2021).
12. Bando, M.; Hasebe, K.; Nakanishi, K.; Nakayama, A. Analysis of optimal velocity model with explicit delay. *Phys. Rev. E* **1998**, *58*, 5429–5435. [CrossRef]
13. Treiber, M.; Kesting, A. Microscopic calibration and validation of car-following models—A systematic approach. *Procedia Soc. Behav. Sci.* **2013**, *80*, 922–939. [CrossRef]
14. Raju, N.; Arkatkar, S.; Joshi, G. Evaluating performance of selected vehicle following models using trajectory data under mixed traffic conditions. *J. Intell. Trans.Sys.* **2020**, *24*, 617–634. [CrossRef]
15. PTV-AG. *PTV VISSIM 9 User Manual*; PTV-AG: Karlsruhe, Germany, 2017.
16. Greenshields, B. A Study of Traffic Capacity. *Highw. Res. Board Proc.* **1935**, *14*, 448–477.
17. Daganzo, G. An analytical approximation for the macroscopic fundamental diagram of urban traffic. *Transp. Res. Part B Methodol.* **2008**, *42*, 771–781. [CrossRef]
18. Helbing, D. Derivation of a fundamental diagram for urban traffic flow. *Eur. Phys. J. B* **2008**, *70*, 229–241. [CrossRef]
19. Kucharski, R.; Drabicki, A. Estimating Macroscopic Volume Delay Functions with the Traffic Density Derived from Measured Speeds and Flows. *J. Adv. Trans.* **2017**. [CrossRef]
20. Spiess, H. Technical note—Conical volume-delay functions. *Trans. Sci.* **1990**, *24*, 153–158. [CrossRef]
21. Akcelik, R. Travel time functions for transport planning purposes: Davidson's function, its time dependent form and alternative travel time function. *Aust. Road Res.* **1991**, *21*, 49–59.
22. Petrik, O.; Filipe, M.; de Abreu e Silva, J. The Influence of the Volume–Delay Function on Uncertainty Assessment for a Four-Step Model. *Adv. Intell. Syst. Comput.* **2014**, *262*, 293–306.
23. Cascetta, E. *Transportation Systems Engineering: Theory and Methods*, 49th ed.; Springer Science: Berlin, Germany, 2013. [CrossRef]
24. Manheim, M.L. *Fundamentals of Transportation Systems Analysis*; MIT Press: Cambridge, MA, USA, 1979.
25. Florian, M.; Gaudry, M.; Lardinois, C. A two-dimensional framework for the understanding of transportation planning models. *Transp. Res. B* **1988**, *22*, 411–419. [CrossRef]
26. Wardrop, J.G. Some Theoretical Aspects of Road Traffic Research. *Proc. Inst. Civil Eng.* **1952**, *1*, 325–362. [CrossRef]
27. Gentile, G.; Noekel, K. Linear User Cost Equilibrium: The new algorithm for traffic assignment in VISUM. In *Conference Papers 2009*; AET: Noordwijkerhout, The Netherlands, 2009.
28. Szarata's Team. Badania Zachowań Komunikacyjnych Mieszkańców Krakowskiego Obszaru Metropolitarnego; Kraków: Technical Report. 2014. Available online: https://www.bip.krakow.pl/?sub_dok_id=96964 (accessed on 19 June 2021).
29. Available online: http://datafromsky.com/ (accessed on 15 May 2019).
30. Richter, M.; Paszkowski, J. Modelling driver behaviour in traffic-calmed areas. *Czas. Tech.* **2018**, *8*, 111–124.

energies
MDPI

Article

Identification of the Determinants of the Effectiveness of On-Road Chicanes in the Village Transition Zones Subject to a 50 km/h Speed Limit

Alicja Barbara Sołowczuk * and Dominik Kacprzak

Road and Bridge Department, Faculty of Civil and Environmental Engineering, West Pomeranian University of Technology Szczecin, 71-311 Szczecin, Poland; dominik.kacprzak@zut.edu.pl
* Correspondence: alicja.solowczuk@zut.edu.pl; Tel.: +48-(091)-449-40-36

Abstract: In recent years, in which a considerable increase in the road traffic volumes has been witnessed, traffic calming has become one the key issues in the area of road engineering. This concerns, in particular, trunk roads passing through small villages with a population of up to 500 and the road section length within the village limits of ca. 1400–1700 m. A successful traffic calming scheme must involve primarily effective reduction in inbound traffic speed. A review of the data from various countries revealed that chicanes installed in the transition zones may have a determining effect on the success of the traffic calming project. The effectiveness of such chicanes depends mainly on the type of chicane, its location on the carriageway, its shape and the size of the lateral deflection imposed by the chicane on the inbound lane. The purpose of this study was to identify the speed reduction determinants in traffic calming schemes in village transition zones, based on a central island horizontally deflecting one lane of a two-lane two-way road with 50 km/h speed restriction. As part of the study, vehicle speeds were measured just before and after the chicanes under analysis. Furthermore, the inbound lane traffic volumes were measured in field and a number of factors were identified, including the applied traffic management scheme, road parameters, view of the road ahead and of the village skyline, isolated buildings, road infrastructure and adjacent roadside developments. The obtained data were analysed with a method employing tautologies of the selected 32 factors affecting the drivers' perception. A single aggregate parameter was proposed for assessing the coincidence of the influence of selected factors on speed reduction. The analysis of the existing schemes and the results of statistical analyses carried out in this study confirmed the authors' hypothesis that the combined selected factors produce a desirable effect and that they should be additionally enhanced by the application of solar powered devices.

Keywords: traffic calming; transition zone; chicane; speed restriction; speed reduction; solar cells

Citation: Sołowczuk, A.B.; Kacprzak, D. Identification of the Determinants of the Effectiveness of On-Road Chicanes in the Village Transition Zones Subject to a 50 km/h Speed Limit. *Energies* **2021**, *14*, 4002. https://doi.org/10.3390/en14134002

Academic Editors: Anna Granà, Elżbieta Macioszek, Margarida Coelho and Paulo Fernandes

Received: 26 May 2021
Accepted: 29 June 2021
Published: 2 July 2021

1. Introduction

An increase in the traffic volumes observed in the recent decades exacerbates transport-related problems in small towns and villages lying on busy through roads. Taking the above into account, in various countries, traffic calming schemes have started to be introduced in design guidelines for both transition zones and for the entire sections of the through roads within the village limits [1–4]. The ever-increasing volumes of traffic lead, in many cases, to an increase in the number of road traffic incidents [4–7], higher noise levels [8–11] and increased concentrations of exhaust fumes and pollutants [4,11,12] along the entire section of the road's passage through the village. In numerous instances, private plots of the villagers and a high density of buildings do not allow unrestricted expansion of the existing road system [13,14]. All these factors have an adverse effect on the acceptance and perception of the road section passing through the village and on the local residents' living conditions [15–19]. Rapid accumulation of these factors can be encountered, especially on the relatively short sections of roads carrying traffic through villages (according to data given in the Danish [1], British [2,19–21] and Polish design guidelines [22] and in the article on the design requirements for the arterial roads in Iran [23]). In the literature, this problem has been discussed for many years; however, the focus has mainly been on design requirements for roads carrying through traffic in built-up areas [1–4,13,24] while less attention has been paid, so far, to village transition zones [25–27]. The first traffic safety analyses on through roads crossing rural areas were conducted in the 1990s in Denmark [1] and the United Kingdom [2,19,20,23,28] and, later on, also in Canada [18], Germany [3], Sweden [26], Spain [29,30] and in the USA [19,31]. To prevent similar incidents from occurring on many through roads of lesser importance, the introduction of traffic calming schemes has been recommended in many countries, for application both in the transition zones and over the whole length of the road within the built-up area, e.g., in Austria [32], Denmark [1,33], France [34], the United Kingdom [2,35], Germany [3], Italy [36], Iran [37], Poland [13,22], Sweden [26], Spain [29,30], Switzerland [38] and in the USA [31,39–41]. These recommendations are confirmed in numerous publications (e.g., Danish [1,10], British [2,4,8,11], Belgian [9], American [12], and Italian [14] ones), in which the authors argue that it is a stochastic process, making the adverse impact of the road on the surrounding environment dependent on the actual traffic speeds. However, considering the regional road classes one should differentiate the impacts of traffic-calming measures on the reduction in speed depending on the speed limit applied on the approach to and within the village limits [42–44]. Based on the analyses of available data and the described results of experimental research projects conducted on the completed traffic calming schemes, it has been concluded that the implementation of traffic calming measures and various speed control schemes resulted in reducing the number of road traffic accidents by ca. 40–50% on average (according to [4]—40%, [20]—10–15%, [28]—54%, [32]—45%, [41]—40–50%, [45]—8–71%, [46]—40–50%).

The most frequently used traffic calming measures are based on horizontal or vertical deflection, depending on the speed limit applicable on the road and the main design parameters of the road in question [1,2,4]. Although horizontal deflections (due to use of chicanes) may bring speed reduction without causing discomfort of passing through vertical traffic calming measures, the design guidelines [1–3] recommend consideration of many additional aspects in the design process, such as the applied traffic control measures, lighting system, location along the approach section, etc. An important—in the authors' opinion—problem regarding the location of the chicane in the transition zone to the village has been more extensively described, for example in [42–44]. In [3,19,26], it is also stressed that the fundamental problem with using horizontal traffic calming measures, especially in rural areas, is the need to ensure the means of passage for oversized agricultural vehicles. However, the analysis of the above problems presented in German [3,27], Swedish [26], and American [19,21,28,47] publications showed that the suggested resultant speed reduction data obtained mainly from research conducted on test tracks or from simulation research fail to take into account a number of parameters related directly to the adjacent features in the transition zone and that have an unquestionable influence on the drivers' perception. These problems were addressed in [42,44] in reference to roads with 70 km/h speed limit or in publication [43], analysing the synergy of various traffic calming measures on the passage through a village lying on regional roads with 70 km/h, 50 km/h and 40 km/h speed limit zones. The current state of knowledge on the scale of speed reduction in the transition zones into villages is discussed in various road engineering circles, but it is still advisable to search for methods to describe the synergy of the numerous factors of the road surroundings that have an effect on the drivers' perception and to obtain more efficient scores of the expected speed reduction. This research area still has a great analytical potential and could be mandatorily implemented in many countries in village transition zones on regional roads.

This article deals with the evaluation of the effectiveness of a specific traffic calming measure, namely a chicane deflecting one lane, installed in the village transition zones on two-lane single carriageway regional roads with the speed limit of 50 km/h posted on the B-33 sign. In this article, the authors implemented the methods and conclusions from article [44] based on the data obtained on roads with 70 km/h speed limit, elaborating on them considerably and supplementing them with new factors observed on roads with a 50 km/h speed limit. The state of knowledge on the scale of speed reduction due to on-road chicanes, against the existing body of scientific research, is presented in Section 2. The relevant test sections and the assumptions adopted in the research conducted on roads with a 50 km/h speed limit are characterised in great detail in Section 3. In Section 4, the authors managed to demonstrate a high consistency between the analysed variables, i.e., between the chosen parameters of speed distribution and the proposed aggregate parameter representing the transition zone surroundings. This allowed them to state in the final conclusions that the proposed method of selecting the type of chicane and its placement in the transition zone to the village should be further developed and researched on other roads, or with other design options of horizontal traffic calming measures.

2. Background

The recommended distances on which speed restriction should be implemented vary between the currently applicable design guidelines and various scientific publications [1–3,27]. The lengths of the speed limit zones depend primarily on the hierarchy of the road and on the total length of the road's passage through the village. The process of designing a traffic calming scheme over the length of the passage of a road through a village is a complex task, involving experimental testing and analytical studies. Different options tend to be chosen in different countries based on the local empirical studies on the drivers' behaviour and habits, and on the resultant speed restrictions applied in the village approach zones. These options take into account the data on the achieved speed reductions, obtained from studies conducted in test track conditions [19,21,28,47] or speed reductions achieved during research conducted with the use of traffic simulators [12,24], and the results of research projects employing speed radar devices installed on existing roads [14,19,46], but also the results of research on the impact of the existing road lighting systems on drivers' perception, and on the use of solar technologies in road infrastructure components, conducted in the natural environment [48–52] or in an "artificial" environment with traffic simulators [53,54].

Still, most of the above-mentioned experimental research results refer to "artificial" environments lacking the natural surroundings of the road, with the only variables being class of the road, speed restriction, width of the traffic lane, size of horizontal deflection [1–3], and also the location and size of lateral obstructions installed directly on the chicane or on the edge of the bending lane [21,28,47]. The shape of the chicane, its physical length and the length of its influence on drivers, as well as the resultant speed reduction were additionally taken into account in [27,32]. However, a majority of scientific studies refer to horizontal or vertical traffic calming measures situated on a through roads in the central parts of the villages or on suburban streets in larger towns and cities. Studies addressing directly the issue of speed reduction in the village transition zones were carried out in several countries, including Germany [3,27], Austria [32], Italy [14] and the USA [19,55–57].

Summing up the above review of the results of the existing research, it can be stated that the variation in the amount of speed reduction is stochastic in nature and the basic qualitative and quantitative description thereof depends on many very diverse factors, as described above. A comparison of the data from the above-mentioned studies demonstrates the presence of fundamental differences between the respective results and numerous deficiencies in repeatability of Δv results obtained in the respective countries. Considering the above, it is advisable to search for the ways in which many other additional factors acting in combination impact the speed reduction process, to understand it better and thus make it more effective. In the authors' opinion, the above-mentioned studies lack an analysis of:

- existing traffic control devices (i.e., location of the upright signs related to the built-up area and the view of the village skyline or individual buildings;

- various road parameters and road infrastructure components related to the built-up area, visible or not visible from the driver's seat on the approach to the village;
- visibility of the road ahead and the village skyline or single buildings situated in the vicinity of the road;
- adjacent roadside development (open rural area, rural area including a tree row or small groves of trees or forest).

The analysis of the conclusions given in [42,44] showed that the design process should also include the aspects associated with the actual features adjacent to the road in the transition zone under analysis, i.e., the existing traffic management scheme, road design parameters, adjacent roadside development and visibility conditions. The research conclusions, due to demonstrated high consistency of the speed reduction data and the speed change index with the aggregated parameter of the combined effect of various factors characterising the road surroundings, as proposed by the authors have, in the author's opinion, a significant research potential, which can be utilised in subsequent studies. However, publications [42,44] refer to transition zones with a 70 km/h speed limit, in which chicanes horizontally deflecting one lane by 2 m were implemented. Still, no research results are available for road sections with a 50 km/h speed limit and various types of chicanes resulting in different amounts of horizontal deflection. These aspects were considered in the current study.

Figure 1 illustrates the state of knowledge contained in previous publications.

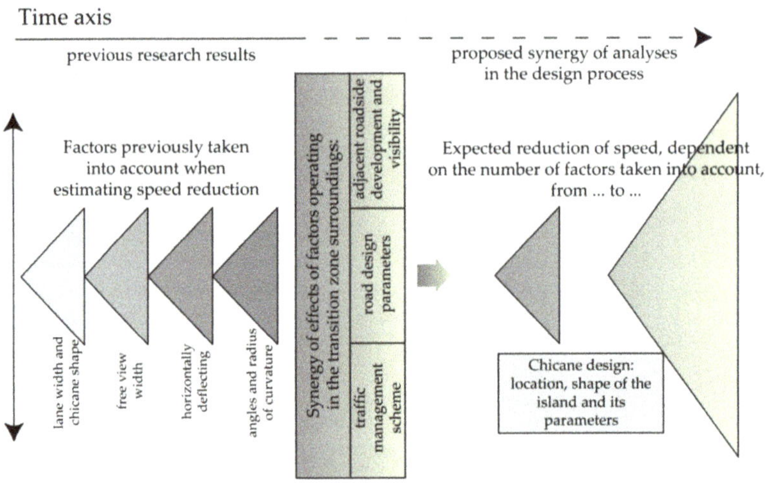

Figure 1. Visualisation of the state of knowledge on the effect of various factors on the reduction in speed based on the literature review.

3. Materials and Methods

3.1. Assumption of the Homogeneity of Parameters of the Road Sections under Comparison Adopted in the Research

The main condition for obtaining robust relationships is the homogeneity of most of the factors describing the studied stochastic object. To this end, purposeful sampling [58,59] should be used, which allows to ensure obtaining a homogenous set of data, that is a set of variables of one kind (i.e., binary and qualitative). This approach to statistical analyses allows the researchers to determine the influence of the selected independent variables on one dependent variable. In the studies described above, the factors characterising the independent variables were the factors defining the surroundings of the road in the transition zone, and the dependent variables were the speed distribution parameters. To assess the influence of the road surroundings on the reduction in speed, from nearly

100 analysed transition zones, the authors chose only regional road sections with 50 km/h speed limits (posted on the B-33 sign placed before the transition zone or just before the chicane). All the test sections were lit by lighting systems conventionally used in built-up areas, covering also the surroundings of the chicane or the area behind it.

All the chosen test sections were situated on renewed regional roads with the pavement in a very good condition, with drainage system in proper working order and traffic control devices clearly visible to the motorists. The traffic controls included raised pavement markers (RPM) or light emitting diode systems and solar powered upright signs. The width of the traffic lane on most of the test sections was 3.5 m. Only on two test sections with 2 m horizontal deflection, were the traffic lanes 3 m wide. Hard shoulders were present only on four sections including, a semi-circular island chicane imposing 6 m horizontal deflection.

On most of the test sections, comparable hourly traffic volumes ranging between 200 and 500 veh/h, and similar proportions of heavy vehicle traffic of 3.5–6% were recorded. Only four test sections carried higher traffic volumes in the region of 850 veh/h. The correlation coefficient of speed reduction for the noted hourly traffic volume level in both directions was $R = -0.29$, and $R = -0.18$ for the traffic volume level on the inbound lane alone. Taking this into account, the effect of traffic volumes on the extent of the speed reduction noted was ignored in further analyses.

All speed surveys were conducted in summer (June, July and August), on working days from 10:00 a.m. till 4:00 p.m. in sunny weather with dry road surfaces and under good natural lighting conditions.

3.2. Description of the Test Sections

For the purpose of the survey of the actual vehicle speeds on two-lane two-way through roads with a speed limit of 50 km/h posted on the B-33 traffic sign, the authors identified twelve test sections located in village transition zones. The test sections were situated in open rural areas or in rural areas including a row of trees lining the road or small groves in the roadside area. The chicanes affecting one lane only installed on the test sections differ considerably in terms of shape, length and width (Figure 2). On four sections horizontal deflection a = 6 m was applied, on one section a = 2.5 m, on three sections a = 2.15 m, and on two sections a = 2 m. On two sections, selected for comparison purposes, no chicanes were installed. Conventional traffic signs U-5, U-6 and C-9 (Figure 2a) were placed on the chicanes. Tapers on road markings before and after the chicane differed depending on the size of horizontal deflection and on local conditions associated with the presence of road structures or access points to agricultural fields. The taper ratios applied before and after the chicane are represented in Figure 2a. To show the taking of the lane and horizontal deflection of the vehicle paths, HGV envelopes a are presented in Figure 2b, parallel to the chicane layout plans. An in-depth analysis of the HGV envelope (drawn assuming that passage by the chicane kerb will be avoided) demonstrated that with "more aggressive" taper ratios of 1:7 and 1:5, on curved sections of the traffic lane, the vehicle envelope extended beyond the traffic lane, and the path of the outer wheels went beyond the roadway edge into the dirt shoulder. Only with the 1:10 taper and 2.5 m horizontal deflection did the envelope fit completely within the traffic lane width. When the lane was deflected by 6 m and on roads with paved shoulders, both the outer wheels of the tractor-trailer unit and the vehicle body envelope went beyond the traffic lane area into the paved shoulder. An analysis of the vehicle envelopes demonstrated their dependence the configuration of the following factors: shape of the chicane, traffic lane width, amount of horizontal deflection and the applied taper ratio.

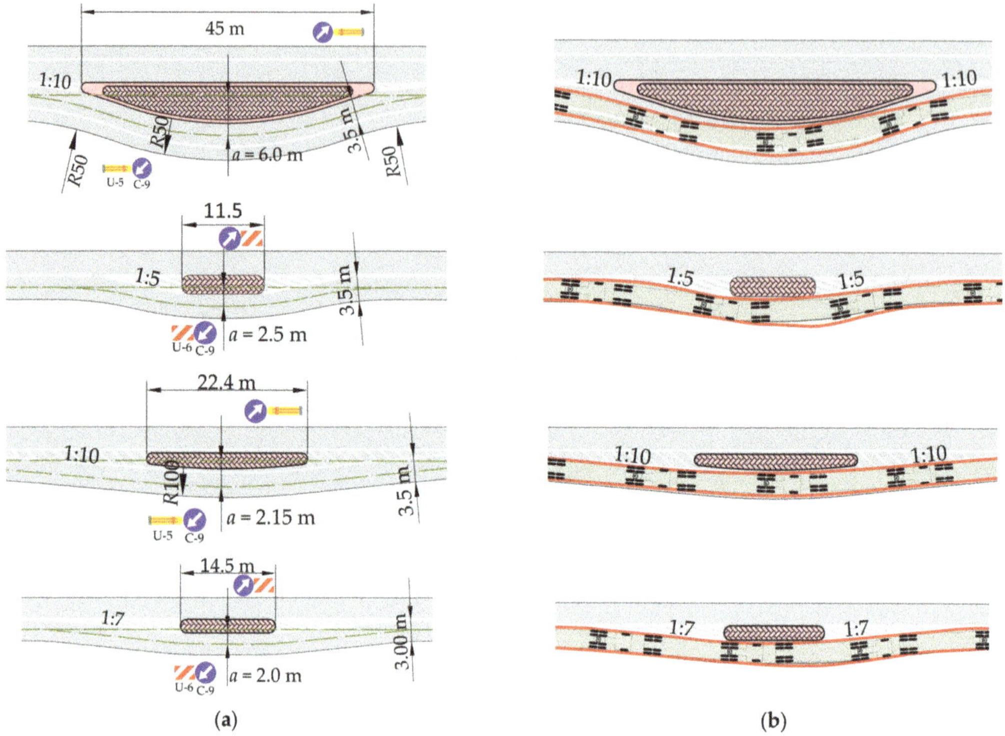

Figure 2. The shape and the main dimensions of chicanes affecting one lane only, installed on the analysed test sections imposing deflections of 6 m, 2.5 m, 2.15 m and 2 m: (**a**) basic layout and traffic control signs used; (**b**) HGV driving envelope and path.

Layout of the vertical traffic signs E-17 "entry sign" and D-42 "built-up area" was another important characteristic of the sections under analysis. In accordance with the vertical traffic signs design trends given in the design guidelines [1,2,13] and in publications [25,42], chicanes should be located after the upright sign E-17 placed at the administrative boundary and the built-up area sign D-42. However, in reality, chicanes were located both past these signs and also between them, and on approach to them. Furthermore, in a few cases of villages sprawling over very large areas with dispersed settlement patterns, they were not accompanied by these signs at all.

All the test sections were located in rural areas. Seven of them were situated in an open area, five of them run through an area overgrown with small groves of trees or tree rows lining the road. The test sections differed significantly in terms of: chicane parameters and factors related to traffic management scheme, road design parameters, adjacent roadside development in the transition zone and visibility conditions.

3.3. Adopted Measurement Methodology

The process of assessment of the effectiveness of chicanes located in the village transition zones with a 50 km/h speed limit, the same as on regional road sections with 70 km/h speed limit, started with an analysis of the vehicle speeds measured before and after the chicane. The measurement apparatus for simultaneous measurement of the traffic volume and vehicle speeds in both directions of traffic [60] was used, as was the case in the study reported in [44]. The measurements covered both traffic lanes in all cases. Using the prescribed measuring procedures that make it possible to immediately assess the number of free-flow vehicles, the surveys were stopped when the number of free-flow vehicles on the inbound lane exceeded 100. The speed survey duration was 2–3 h on average. Speed measurements were correlated one with another and they were conducted simultaneously at points located before and after the chicane, thus making it possible to determine the actual speed reductions achieved for the respective vehicles. The test stations were always located at the same distance from the taper end to avoiding measuring vehicle speeds on curves. As the apparatus can distinguish free-flow situations in either direction of traffic, values needed for the purpose of the performed analysis could be selected on completion of the survey. As mentioned before (in Section 3.1), only small differences in hourly traffic volumes were found on the sections under analysis, and the differences in the speed distribution parameters between free-flow and stable-flow situations were also small as a consequence. Taking this into account, only free-flow values were analysed in the further part hereof.

3.4. Selection of Appropriate Statistical Tests

Analysing the effect of various factors on the traffic speed and the extent of the reduction in the speed, the authors performed a number of statistical tests and adopted the conventional statistical inference methodology. In addition to the significance test, goodness-of-fit test and the test for homogeneity, i.e., the standard tests, the measured speeds were subjected also to nonparametric tests, i.e., the test of independence and the median test. Standard statistical tests used for speed results served, in the first instance, to verify whether the speed populations had normal distributions. The Kolmogorov and Pearson goodness-of-fit tests were used for this purpose. Considering that the speed populations have, as a rule, a continuous probability distribution, more reliable results were obtained in *the Kolmogorov goodness-of-fit test*. *The Pearson goodness-of-fit test* depends on variance tests results and, in the cases under analysis, the results of the tests of variance varied, hence the use *of the Pearson goodness-of-fit test* was considered questionable. Based on the test results, it was assumed that all the populations with a continuous distribution had a normal distribution.

Another *Kolmogorov–Smirnov goodness-of-fit test* was carried out in relation to the "before" and "after" speed populations to examine whether these populations vary in any way.

Since on a number of test sections a small difference in speeds $\Delta v = v^{before} - v^{after}$ was noted and the results of *Kolmogorov–Smirnov goodness-of-fit tests* varied, nonparametric tests were considered in further analyses. Similarly to the statistical inference methodology described in articles [42,44], the nonparametric tests allowed the authors to determine whether two analysed features (not necessarily measurable ones) were independent, i.e., whether the speed distributions depended on the test station location. *The nonparametric tests* (*of independence and median*) were also carried out for the successive sections ordered by the amount of reduction in the speed measured after the chicane. In both tests, positive results were obtained on most of the test sections, confirming the relevance of the test station location, meaning that a change in speed took place across the treatment.

3.5. Main Assumptions Taken in the Qualitative Analysis Concerning Determination of the Speed Reduction Determinants

The estimated speed range data for all the test sections are given in Figure 3. On two sections, No. 11 and No. 12, with no chicanes in place, test stations were located before and after the road sign communicating the boundary of the village (E-17) and the built-up area (D-42). Taking into consideration the fact that traffic calming schemes are implemented in the transition zones primarily to achieve reduction in the village entry speed, Figure 3 shows also the value of ${v_{85}}^{after}$ (85th percentile speed after the chicane).

Since different types of the chicanes were analysed, Figure 3 also shows symbolic representations of the chicanes, the amount of horizontal deflection a and the $l/2a$ ratio, taking into account the relationships given in the Austrian guidelines [32]. The shape of the chicane is related directly to its length and the length-to-width ratio, since, according to the conclusions from the research results published in [27,32], these parameters were critical to the extent of speed reduction Δv_{85}. However, the data in Figure 3 show that, in the analysed case, neither the lateral deflection of the traffic lane a, nor the $l/2a$ above-mentioned ratio was a factor critical to the extent of speed reduction and, what is more, no relationship of these factors with the vehicle speed was confirmed. The analysis of the estimated speed ranges data (Figure 3) showed a considerable variation in the speed values, which is independent of the shape of the chicanes, amount of lateral shift and speed reduction. Therefore, the coefficients of correlation of Δv_{85} and Δv_{av} speed reductions related to the amount of the horizontal shift of traffic lane a and the $l/2a$ ratio were compared in Table 1.

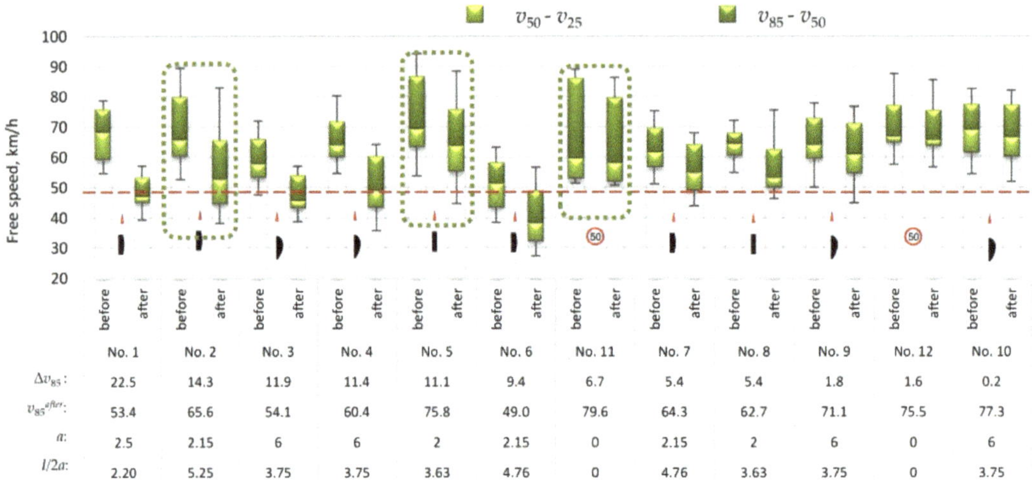

Figure 3. Distribution of free-flow speed ranges on the test sections according to the speed reduction value Δv_{85}.

Table 1. Coefficients of correlation between the vehicle speed after the chicane and reduction in speed Δv and the amount of the horizontal shift of traffic lane a and the $l/2a$ ratio in relation to the test section data.

Correlation Coefficients in Relation to Conclusions Given in [27]				Correlation Coefficients in Relation to Conclusions Given in [32]			
$v_{85} = f(a)$	$\Delta v_{85} = f(a)$	$v_{av} = f(a)$	$\Delta v_{av} = f(a)$	$v_{85} = f(l/2a)$	$\Delta v_{85} = f(l/2a)$	$v_{av} = f(l/2a)$	$\Delta v_{av} = f(l/2a)$
0.18	−0.36	0.17	−0.27	0.5	−0.33	−0.07	−0.29

The speed variations were particularly scattered on sections No. 2, No. 5 and No. 11. What is more, this dispersion is distinctive both before and after the chicane. When analysing the results of similar studies described in articles [42,44], the authors examined the conditions on the approach and through the treatment to find the cause of the demonstrated changes in the dispersion of speeds. Given the conclusions concerning the effect the adjacent roadside development and the visibility conditions have on traffic speed reduction and inbound speed to the village, visualisations of the relevant conditions on the approach and across the treatment on section No. 2 are presented in Figure 4. The visualisations presented below were developed on the basis of the resultant drivers' field of focal attention and based on the conclusions from earlier studies on drivers' central and peripheral vision field, as described in publications [16,18,61].

An analysis of the conditions on the approach and through the treatment shown in Figure 4a demonstrates that the adjacent roadside development on the approach to the chicane with various speeds fails to communicate the built-up area to drivers. They can see only the B-33 signs posting the 40 km/h speed limit and the E-17 sign placed directly before the chicane. A long straight section on the approach to the chicane also contributes to large speed dispersion before the chicane (Figure 4a). On the other hand, visibility conditions across the treatment, which are depicted in Figure 4b, indicate that the visible road infrastructure components (bridge, pedestrian pavements, bridge parapet) and a visible D-42 sign placed on approach impact drivers' perception and result in vehicle speed reduction. However, it needs to be stressed that although the speed reduction Δv_{85} = 14.3 km/h is achieved on this section, the village entry speed, despite the 40 km/h speed limit indicated on the traffic sign placed before the chicane (Figure 3), still significantly exceeds even the 50 km/h country-wide built-up area speed limit.

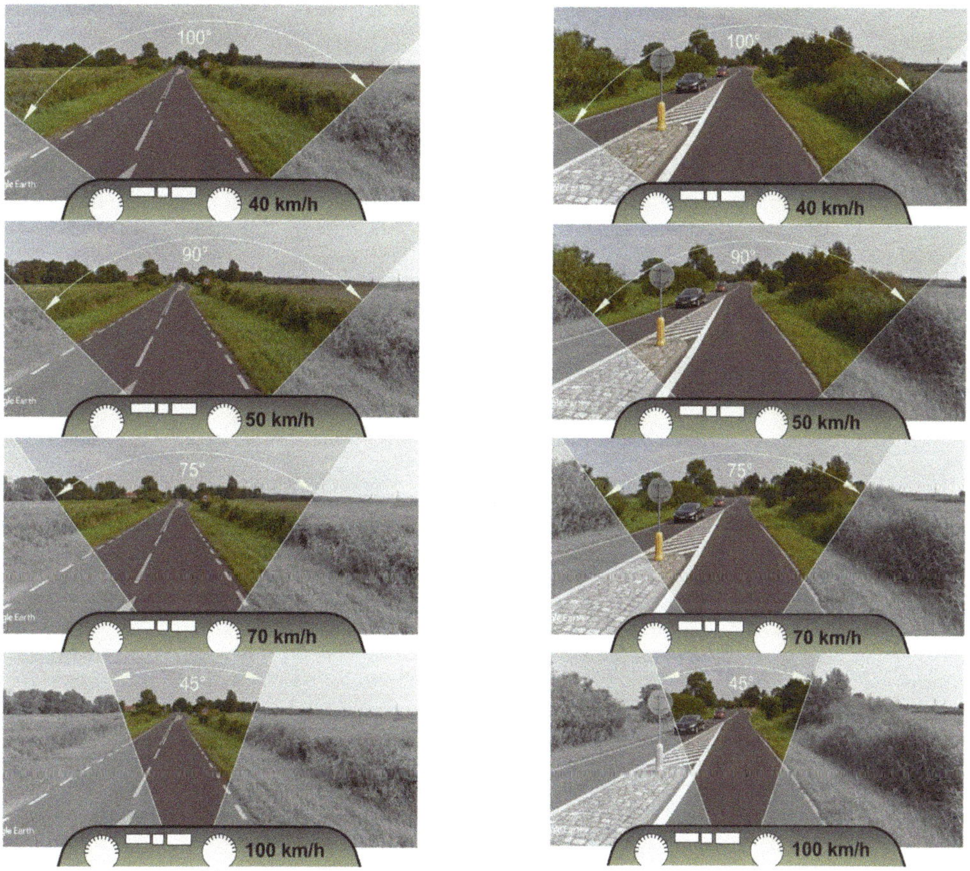

(a) (b)

Figure 4. A visualisation of the central and peripheral vision areas on section No. 2, speed reduction Δv_{85} = 14.3 km/h: (**a**) 100 m before chicane at v of: 40, 50, 70 and 100 km/h; (**b**) in the axis of the chicane at v of: 40, 50, 70 and 100 km/h. (Background photographs source: Google Earth [62]).

A similar analysis can be made for the two other sections, No. 5 and No. 11. The visibility conditions of transition zone surroundings and of the road ahead on these sections are presented in Figure 5 below.

Both on test section No. 5 and on test section No. 11, vehicles approach the village through a straight road section running through an open agricultural land. It is probably these very factors that are responsible for this large speed dispersion. Still, it needs to be stressed that it is probably the presence of the chicane and the road infrastructure components in view (pedestrian pavements, junctions, road curve further ahead) that contributed to the smaller speed dispersion after the chicane.

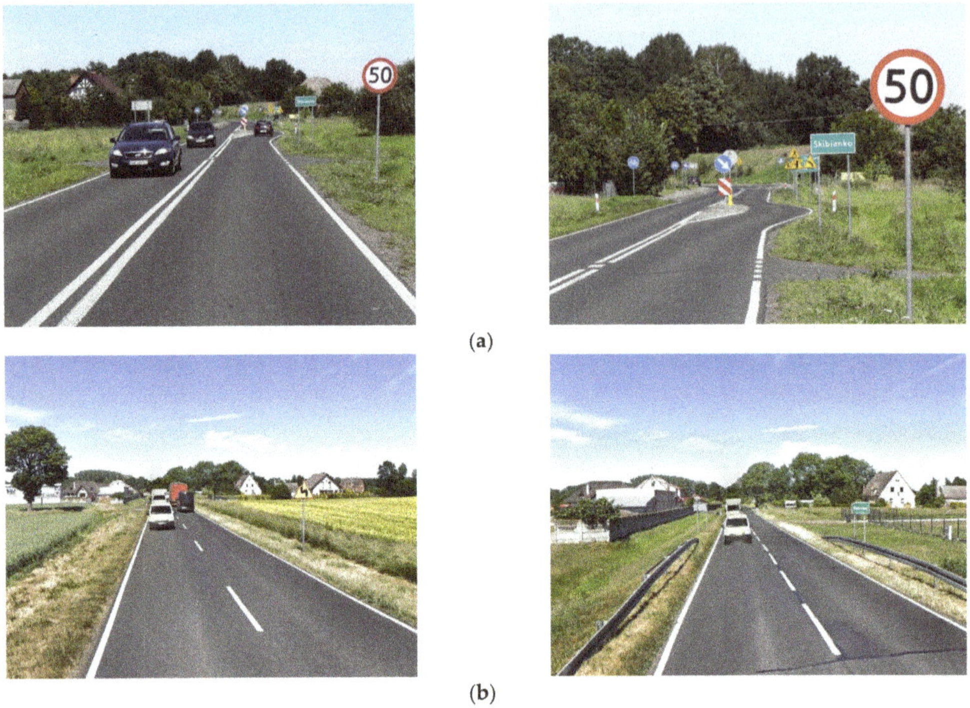

Figure 5. Visibility conditions on the approach to the village ca. 150 m before the chicane or E-17 sign and just at the chicane or the E-17 sign: (**a**) on test section No. 5 with a speed reduction Δv_{85} = 11.1 km/h; (**b**) on a comparative test section No. 11 with a speed reduction Δv_{85} = 6.7 km/h without a chicane in place (Source: Google Earth Street View [62]).

The smallest dispersions of speeds measured after the chicane were recorded on sections No. 1 and No. 3 (Figure 3), and the lowest speed values $v_{85}{}^{after}$ and $v_{av}{}^{aftter}$ were noted on section No. 6. Therefore, similar visualisations of the central and peripheral vision areas on sections No. 1 and No. 6 are presented in Figures 6 and 7. A comparative analysis of the speed dispersion and the resultant drivers' field of vision indicates an existence of some determinants that impact drivers' perception resulting in different village entry speeds. An analysis of drivers' field of vision shown in Figures 6 and 7 indicates that the driver approaching the village centre area can focus his/her attention on various elements of the road and its surroundings. In two sections, these are buildings, bridge parapets, pedestrian pavements, road curvature, etc. It is probably these elements present in the central vision field that, in combination, had an effect on the extent of the reduction in the speed. The two additional cases analysed, i.e., test sections No. 11 and No. 12 also had divergent inbound speed values (Figure 3), yet another piece of evidence supporting the thesis of the probable existence of additional roadside determinants, which influence the drivers' perception and the vehicle speeds as a consequence.

A thorough analysis of the speed dispersion data represented in Figure 3 indicates that the large variation in changes in the 85th percentile speed can be associated with various factors influencing the approach and departure speeds. These determinants, acting in combination, can influence, to a varying degree, the vehicle speeds and the extent of speed reduction.

Furthermore, it needs to be stressed that the 85th percentile speed data compiled in Figure 3 do not corroborate the thesis put forward in publication [27] that chicanes at the road centreline imposing horizontal deflection of one lane by:

- up to 3 m have in every case the same effect on the speed $v_{85}{}^{after} \approx 60$ km/h and its reduction $\Delta v_{85} \approx 10$ km/h;
- more than 3 m have in every case the same effect on the speed $v_{85}{}^{after} \approx 50$ km/h and its reduction $\Delta v_{85} \approx 20$ km/h.

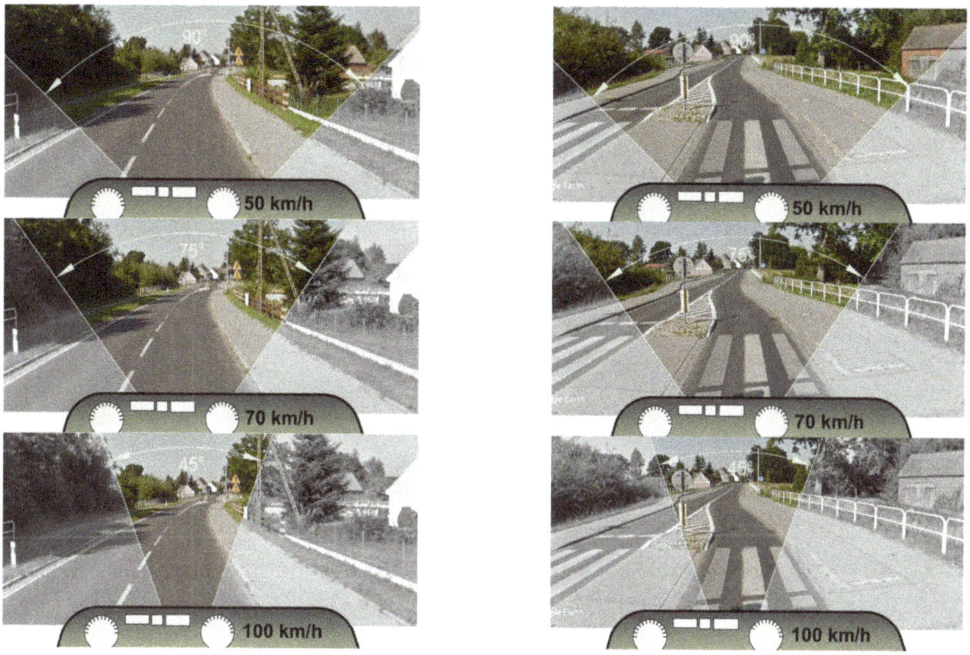

(**a**) (**b**)

Figure 6. A visualisation of the drivers' central and peripheral vision areas on section No. 1 with the greatest speed reduction Δv_{85} = 22.5 km/h. (**a**) 100 m before chicane at v: 50, 70 and 100 km/h; (**b**) in the axis of the chicane at v: 50, 70 and 100 km/h. (Background photographs source: Google Earth [62]).

In all the sections under analysis, the shape of the chicanes and the amount of horizontal deflection varied to a large extent, with the latter ranging from 2 to 6 m, and yet neither the village entry speeds nor the extent of the reduction in speed were proportional to the amount of deflection or to the shape of the chicane, as it was suggested in publications [27,32,47]. Based on these observations, the authors proposed to analyse various factors noted in the transition zones with a 50 km/h speed limit "specified in three criteria concerning: traffic management scheme, road design parameters and adjacent roadside development and visibility conditions", as they did on 70 km/h roads [44].

3.6. Genesis of the Aggregate Parameter z and Its Combined Effect on the Traffic Parameters

The analysis of the diversity and variation in the factors noted on the analysed test sections in relation to the above-mentioned criteria indicated a need for detailed analyses of the effect of the respective factors both on the vehicle speeds and the extent of the reduction in speed. To this end, the authors implemented the "innovative analytical approach" and introduced an "independent variable z to consider the criteria adopted in the analyses" (Equation (1)) [44]. The proposed independent variable z is, in essence, "an aggregate parameter, to quantify the overall impact of the adjacent roadside development on the drivers' perception, leading as a result to a reduction in the village entry speeds", introduced initially by the authors of article [44].

$$z_i = \Sigma(z_{o\,i}, z_{d\,i}, z_{zw\,i}) = \Sigma z_{o\,i,j} + \Sigma z_{d\,i,j} + \Sigma z_{zw\,i,j} \tag{1}$$

where: $z_{o\,I}$ is the total score for the traffic management scheme, $z_{d\,I}$ is the total score for the road design parameters, $z_{zw\,I}$ is the total score for the adjacent roadside development and visibility conditions, $z_{o\,i,j}$ is the score of factor j for the traffic management scheme, $z_{d\,i,j}$ is the score of factor j for road design parameters, and $z_{zw\,i,j}$ is the score of factor j for the surrounding landscape and visibility conditions [44].

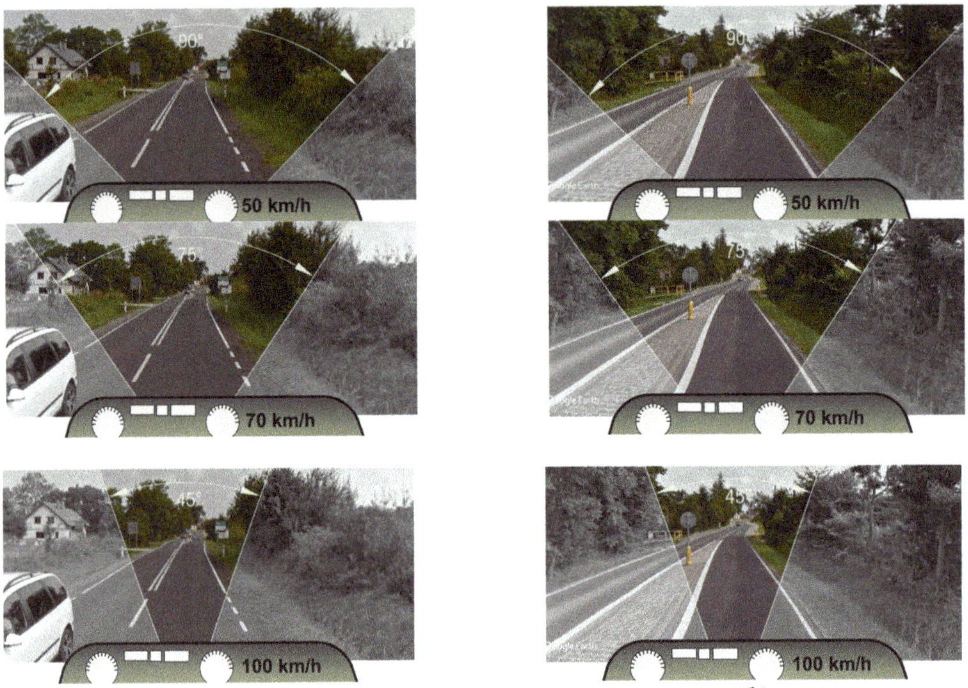

(a) (b)

Figure 7. A visualisation of the central and peripheral vision areas on section No. 6 with a speed reduction of Δv_{85} = 9.4 km/h: (**a**) 100 m before chicane at v: 50, 70 and 100 km/h; (**b**) in the axis of the chicane at v: 50, 70 and 100 km/h. (Background photographs source: Google Earth [62]).

As in article [44], the authors assumed in further analyses that the assumed factors will be characterised by binary "quantitative measures", which simultaneously testify to the presence of a given factor by logical tautology. For instance, as is the case in article [44], if a given factor was confirmed in field, it received the value 1. If, on the other hand, a given factor was not observed, then it received the value 0. Furthermore, the value of 0.5 was used in some specific cases. If, for example, "the village skyline was visible on the way through the treatment at a distance smaller than 100 m, then logical tautology was confirmed and when it was not visible the logical tautology was not confirmed. There was a need for an intermediate measure for the cases when the village skyline was visible from some other distance, for example 300–500 m" [44].

In a consistent manner, in the case of intermediate traffic management-related factors, the authors chose the factors mentioned in article [44] (Figure 8). An analysis of the existing traffic management scheme on the test sections located on roads with a 50 km/h speed limit demonstrated that two further auxiliary factors related to the additional B-33 signs posting speed limits of 40 km/h and 70 km/h, respectively, should be considered in the analyses. These two factors were not applied on roads subject to a 70 km/h speed limit, as there was no need to do so due to lack of speed limit zoning there [44]. Taking the above into consideration, the factors were supplemented with another factor related to the locations of an additional B-33 sign (imposing a 70 km/h speed limit) on the approach to the chicane and another B-33 sign after the chicane (denoting a 40 km/h or 50 km/h speed limit). Thus, the traffic management scheme criterion on roads with a 50 km/h speed limit covered ultimately a total of eleven factors.

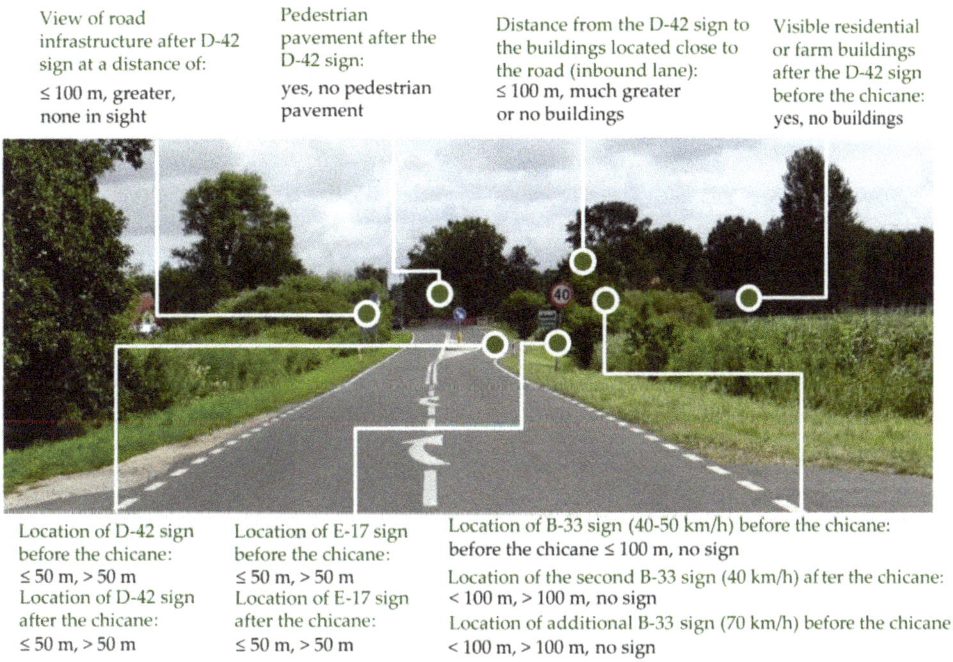

Figure 8. Factors selected for the analyses, along with the assigned quantitative measures related to traffic management scheme, against the background image of test section No. 2. Green font designates qualitative factors and black font designates the adopted interpretation of the quantitative measure.

As mentioned before (in Section 3.1), due to the statistically insignificant effect of the hourly traffic volumes on the reduction in speed on the sections under analysis, this factor was ignored in the criterion referring to road design parameters. One of the major factors in the criterion relating to the road design parameters, in accordance with the conclusions formulated in studies [1–3,27,32,47], was the highly variable amount of the lateral deflection of the inbound lane (2–6 m). However, taking into consideration the varied amount of horizontal deflection of the inbound lane, in relation to the sections considered in article [44], (with a uniform 2 m deflection), the quantitative measure records were changed accordingly. On the basis of site visits and Google Earth imagery [62], the qualitative and quantitative factors, as in article [44], were expanded with "elementary road data relative to the curved road sections, road infrastructure, accesses and junctions". The selected road design factors and their associated quantitative measures are presented in Figure 9. However, an analysis of the village transition zones on the road sections under analysis demonstrated the need to include additional factors related to: combining the chicane with a pedestrian crossing; visible buildings or road structures present on the inbound lane side, located on the horizontal curve after the chicane; and the presence of safety barriers along the drainage ditches installed on both sides of the road along the approach to the chicane, which had a significant effect on the approaching speeds. An analysis of the road geometry in the transition zones on roads subject to a 50 km/h speed limit made it possible to combine the two previously defined factors related to the curved road section into a single factor with confirmed logical tautology of the curved section and to the applied horizontal curvature radius. Summing up the above criterion relating to road design parameters, 9 parameters were finally distinguished (including one parameter combining the confirmation of the horizontal curvature and the size of its radius.

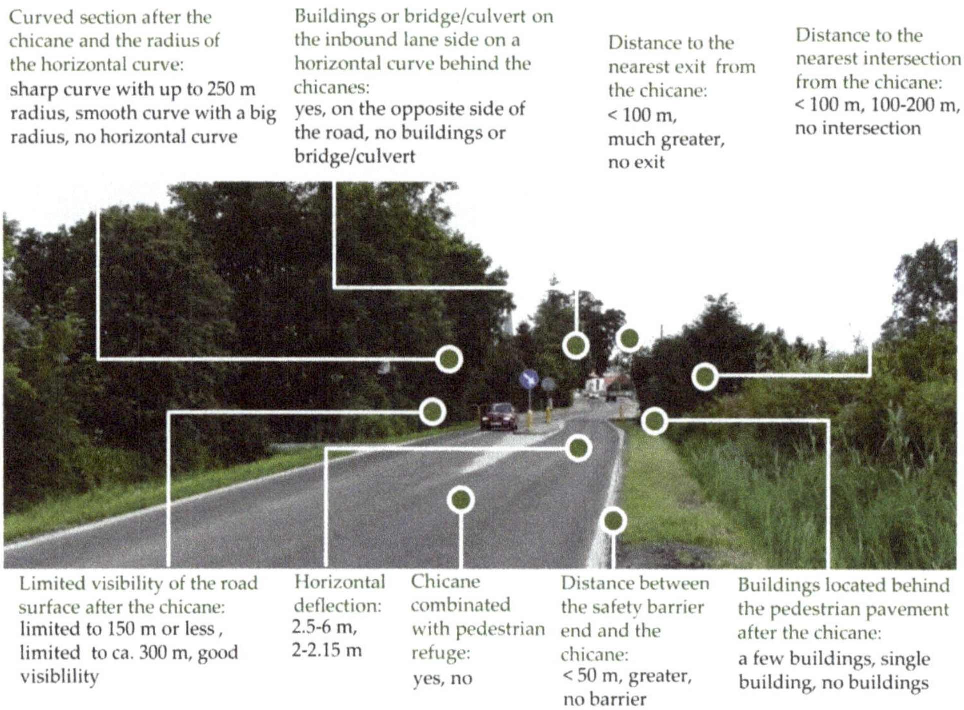

Figure 9. Factors selected for analyses, along with the assigned quantitative measures related to road design parameters. against the background image of test section No. 7. Green font designates qualitative factors and black font designates the adopted interpretation of the quantitative measure.

The last criterion concerning the adjacent roadside development and visibility conditions includes, as independent variables, various qualitative-quantitative factors defining the space in the close vicinity of the chicanes and various elements determining the visibility conditions (Figure 10), implemented based on the findings of article [44]. In the selection and description of the visibility conditions, the authors used the results of visual fixation points analysis in different driving conditions, as presented in publications [18,61]. In this case, it was appropriate to include five new factors related to the roadside development on the sections with large speed reductions, which probably also contributed to the obtained effect. These included, for instance, the factors related to various additional traffic calming measures visible after the chicane (such as safety barriers, additional traffic signs, traversable medians) or visible roadside obstructions having an impact on the perception of drivers. Thus, a total of 5 new factors were added to the 7 factors previously implemented from article [44], which means the initial assumption encompassed 12 factors from which the relevant quantitative measures were obtained, and logical tautology continued to be applied (Figure 10).

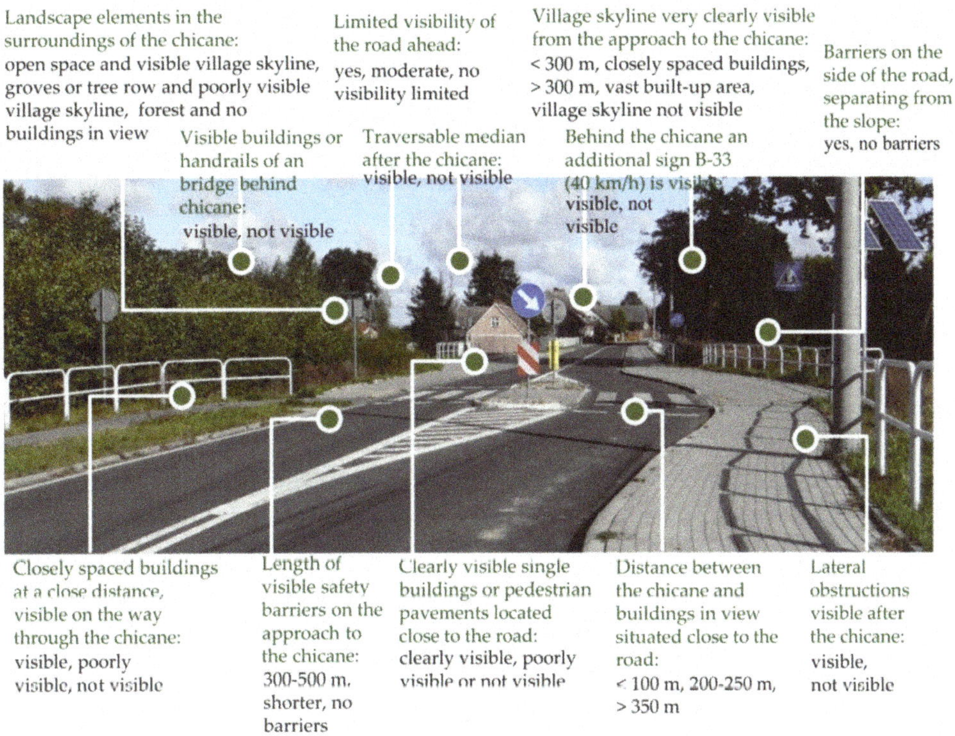

Figure 10. Factors selected for analyses, along with the assigned quantitative measures, related to the adjacent roadside development and visibility conditions, against the background image of test section No. 1. Green font designates qualitative factors and black font designates the adopted interpretation of the quantitative measure in question.

Summing up the above considerations, the authors—as was the case in article [44], in relation to the above-described three criteria—chose to perform statistical inference to determine the strength of correlation between the analysed variables describing the adjacent roadside development of the area and the speed-related parameters (v, Δv, w). The numerical value of the aggregate parameter z was determined using Equation (1). Next, the speed reduction indices $w\ (v_{85})$ (Equation (2)) and $w\ (v_{av})$ (Equation (3)), calculated with the following standard traffic engineering equations [44], were also analysed:

$$w\ (v_{85}) = (v_{85}{}^{before} - v_{85}{}^{after})\ /\ v_{85}{}^{before} \tag{2}$$

$$w\ (v_{av}) = (v_{av}{}^{before} - v_{av}{}^{after}) / v_{av}{}^{before} \tag{3}$$

4. Results

4.1. Outcomes of the Assessment of the Factors Associated with the Traffic Management Criterion

Similarly to article [44], quantitative measures were adopted in this study for the sake of clarity, using the logical tautology (Figure 7). Figure 11 shows the evaluated results recorded in a binary form, changing the colour of the quantitative measure equal to 1, i.e., confirmation of the presence of a given factor or of a corresponding small distance, to dark blue. The light blue colour was used to denote the quantification measure equal to 0, i.e., lack of a given factor, or a large distance to it. In the case of the additional test sections, No. 11 and No. 12 "which did not include chicanes, no weights were assigned to the traffic management related factors conditioned by the presence of a chicane (and hence they displayed no colours in the chart)" [44].

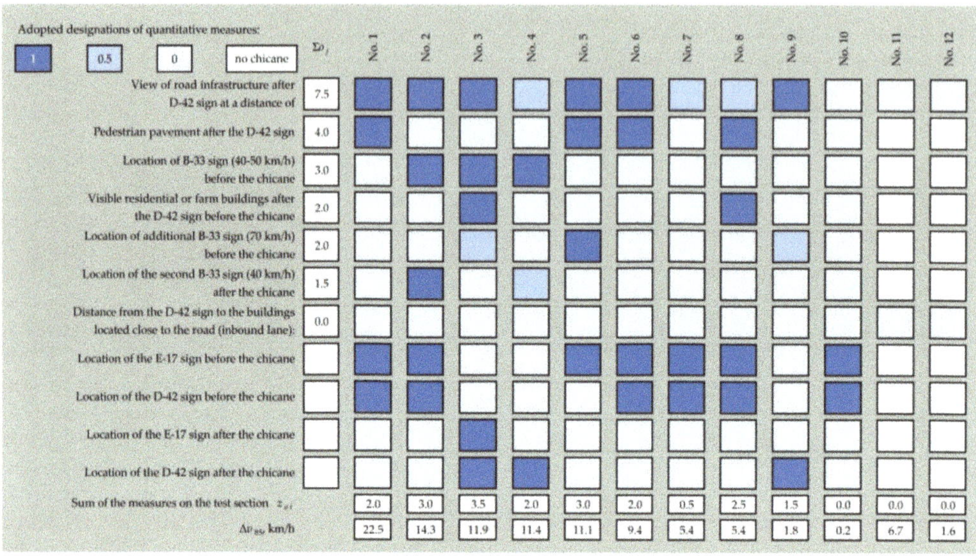

Figure 11. Outcomes of the assessment of the factors associated with the traffic management criterion. Where: o_j is a subsequent factor under consideration, Σo_j is the sum of confirmed logical tautologies on the test sections in relation to the considered j-th factor, and $z_{o\ i}$ is the sum of quantitative measures in all factors on the i-th test section.

An analysis of the data in Figure 11 demonstrates the consistency of some factors with the achieved speed reduction Δv_{85}, just as it was the case on 70 km/h roads [44]. To be specific, such consistency was obtained between speed reduction and "the factor describing the location of the B-33 sign in relation to the chicane axis, confirmation of the presence of buildings after the sign and confirmation of the presence of buildings in close proximity to the road edge after the D-42 sign". A comparison of the assessment of the traffic management related factors on 70 km/h roads [44] and a 50 km/h speed limit reveals large discrepancies in logical tautologies. As far as the selected factors are concerned, two or three logical tautologies were confirmed only on each of the sections with the speed reduction $\Delta v_{85} \geq 10$ km/h. As regards the factors related to the location of the E-17 and D-42 signs, their influence on the speed reduction was confirmed to be insignificant.

To sum it up, we can conclude that on 50 km/h roads, visible components of the road infrastructure, accompanied by the B-33 sign located immediately before the chicane, provide effective speed reduction in village transition zones.

4.2. Outcomes of the Assessment of the Factors Associated with the Road Design Parameters Criterion

The second criterion used for identification of speed reduction determinants is associated with the factors defining the road design parameters (Figure 12). In this criterion, various dependency relationships between the change in the 85th percentile speed Δv_{85} and the road design parameters were confirmed. However, it needs to be pointed out that no significant relationship between speed reduction and the amount of horizontal deflection and chicane shape, as formulated in the conclusions of publications [27,32] was confirmed herein.

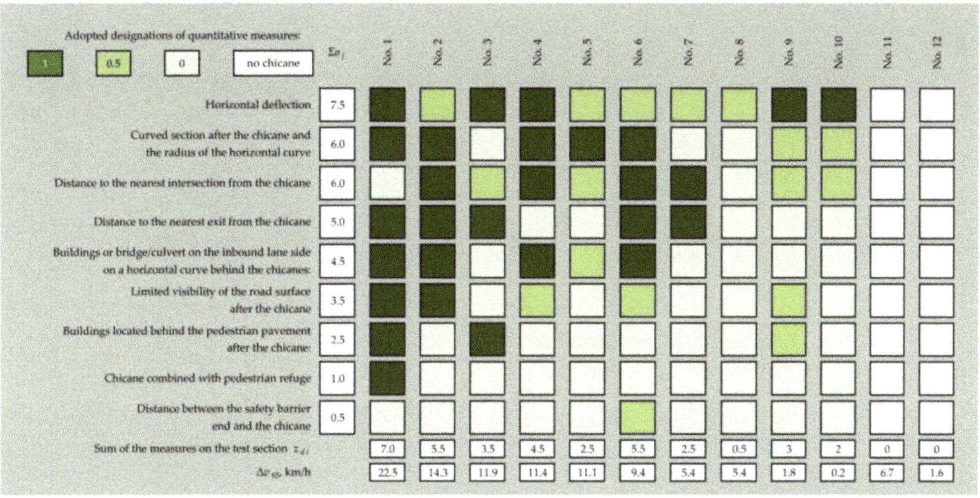

Figure 12. Outcomes of the assessment of the factors associated with the road design parameters criterion. Where: o_j is a subsequent factor under consideration, Σo_j is the sum of confirmed logical tautologies on the test sections in relation to the considered j-th factor, and $z_{d\,i}$ is the sum of quantitative measures in all factors on the i-th test section.

The road layout, and especially a curvature along with its parameters, an upcoming junction or exit in sight, buildings present by the inbound lane and limited visibility of the road ahead also had a diverse impact on speed reduction on the analysed test sections. However, a comparison of the assessment of the factors on the test sections under analysis with the results obtained on the 70 km/h roads [44] demonstrated that not all the factors selected in the road design parameters criterion have consistently confirmed logical tautologies, and the two new added factors were confirmed on only two test sections with different speed reductions. The adjacent roadside development on both of these sections differs significantly. On section No. 1, where the largest speed reduction was achieved (Figure 6a), there are no buildings on the approach to the chicane and it is primarily these conditions that determined the approach speed as high as $v^{before} \approx 76$ km/h. When approaching the chicane, the driver can see a built-up area and a large horizontal deflection of the lane ahead, which considerably influences the village entry speed ($v^{after} \approx 53$ km/h). On section No. 6, on the other hand, on the approach section, the driver sees roadside barriers along roadside ditches on both sides of the road over the length of 400 m and ending 150 m before the chicane, which has a significant effect on the reduction in the approach speed (${v_{85}}^{before} = 58$ km/h). At the chicane itself, the authors noted a few factors relevant to the village entry speed ${v_{85}}^{after} = 49$ km/h (Figure 7).

An analysis of the consistency of changes in the sums of the quantitative measures representing the road design parameters z_d and the specific values of Δv_{85} represented in Figure 12 confirmed the earlier hypotheses that the identified factors jointly contribute to reduction in village entry speeds, in line with the conclusions of the study reported in the article [44].

4.3. Outcomes of the Assessment of the Factors Associated with the Adjacent Roadside Development and Visibility Criterion

The last transition zone criterion, transferred from article [44] and related to visibility and adjacent roadside development, was supplemented with a few factors, represented in Figure 10. Additionally, in this case, significant dependency relationships of the analysed variables were confirmed on nearly all the test sections (Figure 13). The above observations are also consistent with the assessment results presented in article [44].

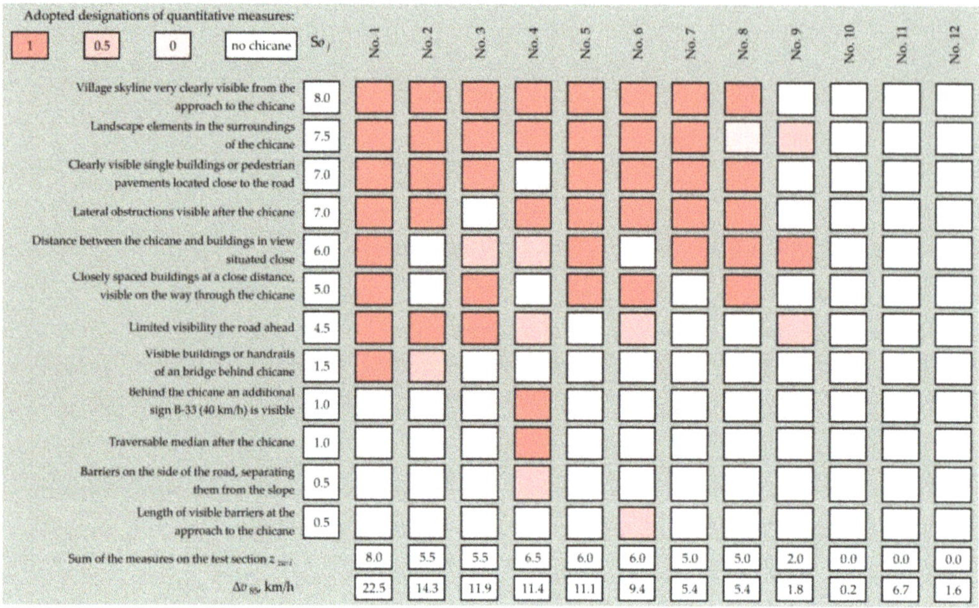

Figure 13. Outcomes of the assessment of the factors associated with the adjacent roadside development and visibility criterion. Where: o_j is a subsequent factor in consideration, Σo_j is the sum of confirmed logical tautologies on the test sections in relation to the considered j-th factor, and $z_{zw\ i}$ is the sum of quantitative measures in all factors on the i-th test section.

Nevertheless, it needs to be pointed out that confirmation of logical tautologies was not obtained for all the factors on all test sections. The obtained scores showed that if more than five logical tautologies were confirmed, a speed reduction by more than 5 km/h was also noted. It is also characteristic that with ca. six logical tautologies noted, speed reductions in the range of 10 to 14 km/h could be achieved. Moreover, it needs to be stressed that the four additional factors related to the additional traffic calming measures or road infrastructure proved to be of little significance. The above outcomes of the assessment in the criterion concerning the adjacent roadside development and visibility demonstrate consistency with an analogical assessment carried out on the 70 km/h roads [44].

4.4. Relationship between the Aggregated Parameter Z and the Speed Parameters

Summing up the above analyses, as in the case of the dependencies presented in article [44], Figure 14 below represents the probable consistency between the analysed parameters. The arrow used symbolically in Figure 14 indicates the direction of consistency between the analysed parameters. An analysis of variations in the analysed values demonstrates the greatest consistency between the speed reduction Δv_{85} and the speed reduction index w (v_{85}) on the one hand, and the proposed aggregate parameter z on the other.

4.5. Regression Analysis of the Outcomes of Assessment of the Combined Effect of the Selected Factors and Speed Distribution Parameters

As a result of the revealed dependencies, as shown in Figure 14, the authors carried out appropriate regression analyses, shown in Figures 15 and 16.

The detailed correlation coefficient data are given in Tables 2 and 3. In Tables 2 and 3, bold numerals denote cases when the correlation coefficient R was greater than 0.9, indicating a statistically significant relationship between the analysed variables [58,59].

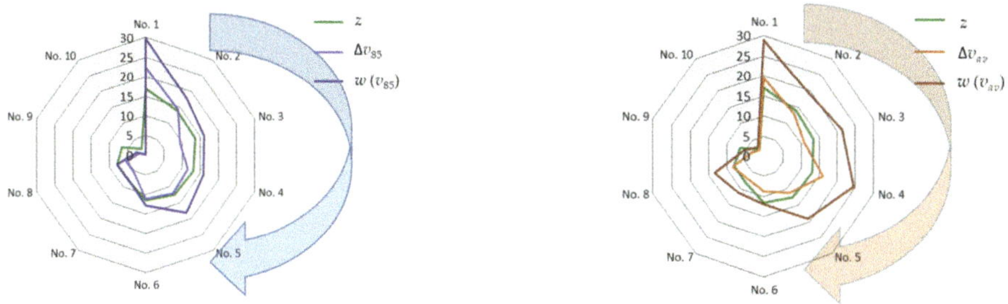

(a) (b)

Figure 14. Compilation of changes in the analysed variables: (**a**) consistency of variations in speed reduction Δv_{85} and Δv_{av}, and the aggregate parameter z (R = 0.94 and R = 0.96, respectively); (**b**) consistency of variations in speed reduction index w (v_{85}) and w (v_{av}) and aggregate parameter z (R = 0.92 and R = 0.93, respectively).

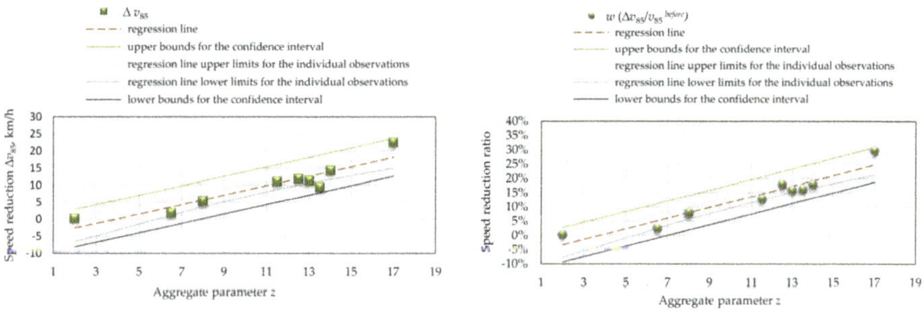

(a) (b)

Figure 15. Linear regression function defining the relationship between: (**a**) free-flow speed reduction Δv_{85} and the aggregate parameter z, (R = 0.94); (**b**) free-flow speed reduction index w (v_{85}) and the aggregate parameter z (R = 0.96).

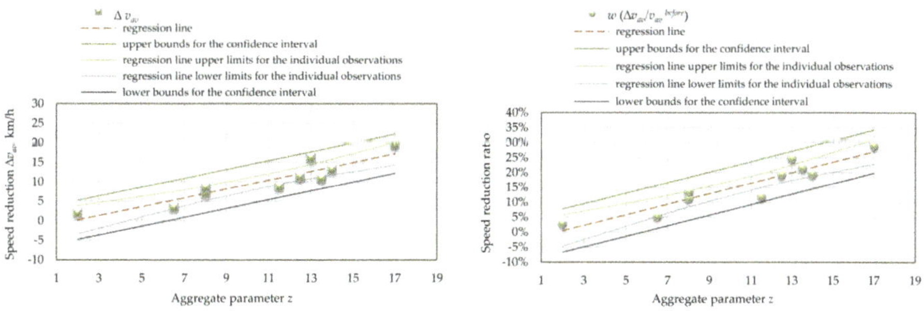

(a) (b)

Figure 16. Linear regression function defining the relationship between: (**a**) the free-flow speed reduction Δv_{av} and the aggregate parameter z, (R = 0.92); (**b**) the free-flow speed reduction index w (v_{av}) and the aggregate parameter z (R = 0.93).

The regression analyses represented in Figures 15 and 16 corroborate the hypotheses previously put forward by the authors that the reduction in speed in the village transition zones depends on a combination of several factors characterising the adjacent roadside development. Referring to Danish [1], British [2,4,8,35], German [3,27], Austrian [32], Italian [14], American [19,31] and Polish publications [13,22,25] on the subject, which, depending on the country, give various data regarding the percentage reduction in speed across chicanes, the authors confirm the validity of the statements contained therein.

However, there are also external factors related to the roadside development, which have a bearing on the extent of speed reduction Δv and the percentage values of the speed reduction index w on the analysed 50 km/h roads. The first studies identifying these factors were carried out by the authors, and are reported in this article and in article [44].

Table 2. Coefficients of correlation R between the quantitative measures and the chosen speed distribution parameters (including the additional sections No. 11 and No. 12).

	v_{85}^{before}	v_{av}^{before}	$v_{av}^{pp\ before}$	v_{85}^{after}	v_{av}^{after}	$v_{av}^{pp\ after}$	Δv_{85}	Δv_{av}	Δv_{av}^{pp}	$w\ (v_{85})$	$w\ (v_{av})$
z_{zw}	−0.35	−0.27	−0.33	−0.78	−0.78	−0.80	0.81	**0.92**	0.89	0.84	**0.92**
z_d	−0.31	−0.29	−0.35	−0.72	−0.74	−0.79	0.77	0.83	0.85	0.81	0.85
z_o	−0.23	−0.20	−0.33	−0.56	0.66	0.71	0.61	0.66	0.71	0.63	0.67
z	−0.35	−0.29	−0.37	−0.79	−0.80	−0.86	0.84	**0.93**	**0.93**	0.87	**0.94**

Table 3. Coefficients of correlation R between the quantitative measures and the chosen speed distribution parameters (excluding the additional sections No. 11 and No. 12, which had no chicanes installed).

	v_{85}^{before}	v_{av}^{before}	$v_{av}^{pp\ before}$	v_{85}^{after}	v_{av}^{after}	$v_{av}^{pp\ after}$	Δv_{85}	Δv_{av}	Δv_{av}^{pp}	$w\ (v_{85})$	$w\ (v_{av})$
z_{zw}	−0.10	−0.12	−0.10	−0.67	−0.67	−0.67	0.86	0.89	0.83	0.88	0.89
z_d	−0.06	−0.17	−0.16	−0.59	−0.62	−0.67	0.78	0.77	0.77	0.80	0.77
z_o	0.05	−0.05	−0.13	−0.35	−0.37	−0.49	0.56	0.60	0.55	0.56	0.50
z	−0.07	−0.15	−0.16	−0.70	−0.72	−0.77	**0.94**	**0.92**	**0.92**	**0.96**	**0.93**

5. Discussion

This article analyses 10 sections of roads passing through villages that include chicanes installed in the transition zones. The analysis of the obtained dispersion of speed data presented in Figure 3 showed that the obtained speed reductions are highly varied, and this variation does not depend solely on the amount of horizontal deflection or the shape of the chicane. A comparison of the speed reductions obtained in the sections including chicanes with the data obtained on the sections No. 11 and No. 12 with no chicanes demonstrates that also on the latter two, the speed reduction values did vary. Therefore, it is clear that some other determinants must have been involved.

In the course of the analyses, the authors formulated the total number of thirty two such determinants (Figures 11–13). Now, based on a detailed analysis of the variation in the values of aggregate parameter z on the test sections, in recapitulation, we can identify the main determinants, in a similar way to article [44]. For instance, in the traffic management criterion, the speed reduction was influenced predominantly by the roadside development past the D-42 sign (Figure 11), and in the road design criterion, the predominant influence was the parameters of the horizontal curvature at the transition zone (Figure 11). However, the greatest number of factors with confirmed logical tautology in the spatial surroundings and visibility criterion were related to the physical surroundings along the approach to the village and in the transition zone (Figure 12). On the test sections with the greatest speed reductions measured, the authors confirmed numerous logical tautologies, relating particularly to visibility conditions on the approach to and through the chicane, such as: village skyline, nearby single buildings or clearly visible road infrastructure associated with the built-up area (Figure 13). To recapitulate the above, we can say that the coincidence of the proposed factors in three criteria combined does have an impact on the anticipated speed reductions in the transition zone, and does corroborate the hypotheses regarding the aggregate parameter z put forward by the authors, that the higher the value of this parameter, the greater the reduction in village entry speed that can be expected (Tables 2 and 3, and Figures 15 and 16).

An analysis of the speed reductions and assessments of the factors related to the road design parameters criterion and the roadside development/visibility criterion demonstrates that the greatest speed reductions were obtained on the test sections running through open agricultural land. On these sections, logical tautologies were confirmed, which were related to a large extent to the view of: the village skyline on the approach to the transition zone, nearby buildings and road infrastructure elements such as pedestrian pavements, exits or private entries, junctions, visual obstructions such as bridge parapets, and also horizontal curvature located near the chicane, limiting the view of the road ahead and its immediate vicinity. Taking the above into consideration, in choosing the location for a chicane to be provided as part of a traffic calming scheme, road planners and designers should, in the first instance, consider the logical tautologies of the above given factors, along with their coincidence to the expected speed reductions. This confirms that the design process proposed in article [44] (Figure 21, p. 22) for chicanes located at the transition zones to villages is also valid for 50 km/h roads. Therefore, whenever logical tautologies are confirmed for several (more than eight) factors, the designers should position the chicane near the D-42 sign. With these conditions met, it can be expected that the desired speed reduction of ca. 10 km/h or more will be achieved within the transition zone on roads with a 50 km/h speed limit.

However, in the opinion of the authors, the process of designing chicanes in village transition zones involving mainly choosing the appropriate chicane location and shape should also include the selection of the most adequate from among the technological options which are currently gaining widespread application in the field of traffic engineering [49,51]. These new technologies relate primarily to traffic control elements (i.e., traffic signs and road markings) and to the improvement of their visibility by application of photovoltaic cells, which bring considerable energy savings. In recent years, we have witnessed an increasing use of various reflective elements and traffic signs powered by photovoltaic cells and solar batteries (Figure 17), as well as other technologies, such as LED lighting [51–54]. An analysis of the visibility conditions of chicanes on the existing roads (at night and in bad weather), revealed inadequacy of the conventional reflective elements based on omnidirectional reflectors (Figure 17a–c). Solar powered LED lighting systems—highly-efficient light sources, self-charging at daytime, even under low sunlight conditions—perform much better and are a much safer option in road applications (Figure 17d,e). LED elements offer long service life and reliability, do not require external power supply and contribute to energy conservation. Last but not least, they are not toxic or harmful to living organisms.

Taking into account the effect of the location of road edge marker posts installed 0.5 m beyond the road edge on the reduction in speed, as described in publication [47], it can be stated that they should be applied on road curves, both before and after the chicane where there are no village buildings in sight or there is no road lighting system in the village. Road edge marker posts or bollards should also be used when elements of the existing road infrastructure (e.g., pedestrian refuges, cycle paths, etc.) can be seen, yet with no buildings in view (Figure 18). Obligatory amounts of speed reduction recorded in the above-mentioned cases were not given in publication [47], since the test sections differed in terms of terrain and adjacent roadside development and speed data collection method. These elements of terrain and roadside development characteristics were taken into account by the authors in this study, in which a uniform speed data collection method was used.

Furthermore, in compliance with the conclusions of articles [48–51,53], the road edge marker posts should be fitted with reflective elements to enhance their visibility. If there is a need to use an additional lighting system due to specific local road design parameters associated with the surroundings of the road, then placement of edge marker posts equipped with solar-powered lighting elements (Figure 17e), widely used in numerous countries on roads running through agricultural land [50,51] should be considered. This option brings considerable conservation of energy and cost savings, as these devices can do without extension of the existing village road lighting system beyond the built-up area limits. Depending on the local conditions, either road edge marker posts (Figure 18a) or bollards separating shared pedestrian and cycle paths from the road (Figure 18b) can be used.

Figure 17. Example applications of auxiliary lighting elements: (**a**) conventional reflective PEO curb markers on the verge of the island; (**b**) conventional reflective elements placed on various elements of the chicane; (**c**) conventional surface mounted reflective road studs PEO; (**d**) SolarCAP Road LED lights mounted in the outer edge of the carriageway; (**e**) road edge marker posts including LED elements.

Figure 18. A visualisation of improved chicane surroundings due to the use of: the B-33 sign (50 km/h speed limit) placed close to the chicane, reflective road edge marker posts or chain fences, solar-powered vertical signs, additional kerb-mounted omnidirectional or LED SolarCAP type elements: (**a**) using the example of test section No. 7; (**b**) using the example of the test section No. 8.

The last problem refers to the application of cold plastic materials including retroreflective glass beads for road surface markings to ensure good visibility at night time and in bad weather [63]. Depending on the availability of land adjacent to the road and planned horizontal deflection, the road markings can be designed over the chicane length in two ways (Figure 19). As per the conclusions given in [47] regarding the application of road edge marker posts, the more "aggressive" the taper on the hatched markings (e.g., 1:5 taper ratio) with a solid line painted along the chicane and along the outer edge of the carriageway (Figure 19a), the greater the speed reduction that may be expected. In the opposite case, when only the markings along the road edge are used and the taper ratio is smaller (e.g., 1:7 or 1:10), considerably smaller speed reductions in the transition zone should be expected, even with the road edge marker posts in place (Figure 19b).

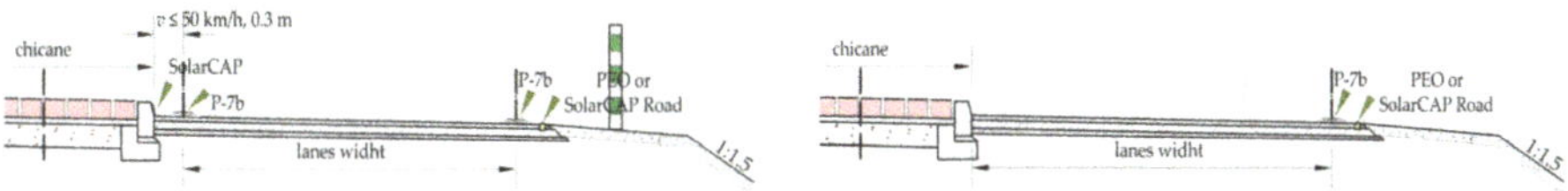

(**a**) (**b**)

Figure 19. A road cross section with different scheme of road surface markings P-7b "solid white line along the edge of the carriageway": (**a**) with two P-7b solid white lines (on both sides of the traffic lane) and 1:5 taper ratio; (**b**) with one P-7b solid white line and 1:7 or 1:10 taper ratio.

Summing up the discussion, the authors would like to state that there is a need for further analyses of data obtained on other roads or of the application of different chicane designs used in village transition zones.

6. Conclusions

Summing up the analyses, it can be concluded that the authors have obtained a confirmation of the influence of the identified factors characterising the adjacent roadside development on the effective speed reduction in the transition zones to villages on roads with a 50 km/h speed limit posted on the B-33 sign, which is consistent with the conclusions presented in article [44]. The important determinants that have a considerable effect on speed reduction are associated primarily with the view of the road surroundings on the approach section to the village and directly in the transition zone, i.e., the view of the village skyline or buildings or the elements of road infrastructure typical of built-up areas.

The amount of horizontal deflection and the shape of the chicane are other relevant factors, particularly important in the case of chicanes affecting one travel lane only, installed on the inbound lanes of roads passing through villages. The research results also indicate that it is more than one or two factors that have a bearing on the practicable speed reduction. The greatest speed reductions were achieved on test sections with a coincidence of many factors influencing drivers' perception of the surrounding environment (i.e., many confirmed logical tautologies of the identified determinants).

Author Contributions: Conceptualization, A.B.S. and D.K.; methodology, A.B.S.; formal analysis, A.B.S. and D.K.; data curation, D.K.; writing—original draft preparation, A.B.S.; writing—review and editing, A.B.S.; visualization, A.B.S.; supervision, A.B.S.; funding acquisition, D.K. All authors have read and agreed to the published version of the manuscript.

Funding: The researches included in the paper were partly financed by grants of WBiA ZUT: DKD Decision No. 517-02-033-6728/17.

Conflicts of Interest: The authors declare no conflict of interest.

References

1. *Urban Traffic Areas—Part 7—Speed Reducers*; Vejdirektoratet-Vejregeludvalget: Copenhagen, Denmark, 1991.
2. *Traffic Calming Guidelines*; Devon County Council Engineering & Planning Department: Devon, UK, 1992.
3. *Directives for the Design of Urban Roads. RASt 06*; Road and Transportation Research Association, Working Group Highway Design, FGSV: Köln, Germany, 2006.

4. Harvey, T. *A Review of Current Traffic Calming Techniques*; University of Leeds: Leeds, UK, 2013.
5. World Health Organization (WHO). *Global Status Report on Road Safety 2010*; World Health Organization: Geneva, Switzerland, 2010.
6. World Health Organization (WHO). *Global Status Report on Road Safety 2015*; World Health Organization: Geneva, Switzerland, 2015.
7. World Health Organization (WHO). *Global Status Report on Road Safety 2018*; World Health Organization: Geneva, Switzerland, 2018.
8. Paige, M. *Speed and Road Traffic Noise*; UK Noise Association: Chatham, UK, 2009. Available online: http://www.ukna.org.uk/uploads/4/1/4/5/41458009/speed_and_road_traffic_noise.pdf (accessed on 1 July 2019).
9. Ellebjerg, L. *Noise Reduction in Urban Areas from Traffic and Driver Management—A Toolkit for City Authorities*; Silence: Brussels, Belgium, 2008.
10. Ellebjerg, L. *Noise Control through Traffic Flow Measures—Effect and Benefits*; Report 151; Danish Road Institute: Hedehusene, Denmark, 2007.
11. King, R. Noise and Speed—A Guest Blog from UK Noise Association. 2019. Available online: http://www.20splenty.org/noise_and_speed (accessed on 18 May 2020).
12. Ghafghazi, G.; Hatzopoulou, M. Simulating the air quality impacts of traffic calming schemes in a dense urban neighbourhood. *Transp. Res. Part D Transp. Environ.* **2015**, *35*, 11–22. [CrossRef]
13. Krystek, R. *Principles of Traffic Calming on the Roads of the Pomorskie Region in Poland, Part 1: Street Layouts in Towns and Cities*; GAMBIT Pomorski: Gdańsk, Poland, 2008.
14. Lantieri, C.; Lamperti, R.; Simone, A.; Costa, M.; Vignali, V.; Sangiorgi, C.; Dondi, G. Gateway design assessment in the transition from high to low speed areas. *Transp. Res. Part F Traffic Psychol. Behav.* **2015**, *34*, 41–53. [CrossRef]
15. Babkov, V.F. *Road Design Parameters and the Safety of Traffic*; WKŁ: Warszawa, Poland, 1975.
16. Hobbs, F.D.; Richardson, B.D. *Traffic Engineering*; WKŁ: Warszawa, Poland, 1971.
17. Lunenfeld, H. *Evaluation of Traffic Operations, Safety, and Positive Guidance Project*; Federal Highway Administration, Office of Traffic Operations: Michigan, MI, USA, 1980.
18. Crevier, C. *Les Aménagements En Modération De La Circulation, Étude Et Applications*; École De Technologie Supérieure Université Du Québec: Montréal, QC, Canada, 2007.
19. Dixon, K.; Zhu, H.; Ogle, J.; Brooks, J.; Hein, C.; Aklluir, P.; Crisler, M. *Determining Effective Roadway Design Treatments for Transitioning from Rural Areas to Urban Areas on State Highways*; Final Report SPR 631; Federal Highway Administration: Washington, DC, USA, 2008.
20. *Roads Development Guide*; South Ayrshire Council, Strathclyde Roads: Ayrshire, UK, 1995.
21. Mackie, A.M.; Ward, H.A.; Walker, R.T. *Urban Safety Project, Part 3: Overall Evaluation of Area Wide Schemes*; TRRL Report 263; Transport and Road Research Laboratory: Crowthorne, UK, 1990.
22. Krystek, R. *Principles of Traffic Calming on the Roads of the Pomorskie Region in Poland, Part 2: Sections of Major Roads Through Towns and Villages*; GAMBIT Pomorski: Gdańsk, Poland, 2008.
23. Abdi, A.; Rad, H.B.; Azimi, E. *Simulation and Analysis of Traffic Flow for Traffic Calming*; Proceedings of the Institution of Civil Engineers-Municipal Engineer: London, UK, 2017; Volume 170, pp. 16–28. Available online: https://www.icevirtuallibrary.com/doi/abs/10.1680/jmuen.16.00005 (accessed on 17 May 2020).
24. Akgol, K.; Gunay, B.; Aydin, M.M. Geometric optimisation of chicanes using driving simulator trajectory data. In *Institution of Civil Engineers—Transport*; Transport ICE Publishing: London, UK, 2019.
25. Zalewski, A. *Traffic Calming as a Transport Engineering Problem*; Publishing House of the Technical University of Łódź: Łódź, Poland, 2011; Scholary Papers, issue No. 1104 habilitation monograph 414.
26. *Safe Road Design Manual*; Amendments to the WB Manual, Transport Rehabilitation Project ID PO75207; Consulting Services for Safe Road Design: Loan, Sweden, 2011.
27. *Wirksamkeit Geschwindigkeitsdämpfender Maßnahmen Außerorts*; Dezernat Verkehrssicherheit und Verkehrstechnik, Hessisches Landesamt für Straßen—und Verkehrswesen: Hessen, Germany, 1997.
28. Sayer, I.A.; Parry, D.I. *Speed Control Using Chicanes—A Trial at TRL*; TRL Project Report PR 102; Transport Research Laboratory: Crowthorne, UK, 1994.
29. González, D.D. Evaluación de las Zonas 30 en Europa y definición de una Zona 30 revisada. Ph.D. Dissertation, Infraestructura del Transporte y del Territorio (ITT), Universitat Politècnica de Catalunya, Barcelona, Spain, 2012.
30. Hernández, E.; Abadía, X.; París, A.C. *Criterios de Movilidad ZONAS 30*; Fundación RACC: Barcelona, Spain, 2007.
31. *Guidelines for Traffic Calming*; City of Sparks, Public Works, Traffic Division: Reno, NV, USA, 2007.
32. Berger, W.J.; Linauer, M. *Speed Reduction at City Limits by Using Raised Traffic Islands*; Institut fuer Verkehrswesen (Institute for Transport Studies), Universitaet fuer Bodenkultur A-1190: Vienna, Austria, 1998.
33. Prato, C.G.; Rasmussen, T.K.; Kaplan, S. Risk Factors Associated with Crash Severity on Low-Volume Rural Roads in Denmark. *J. Transp. Saf. Secur.* **2014**, *6*, 1–20. [CrossRef]
34. Vahl, H.G.; Giskes, J. *Traffic Calming through Integrated Urban Planning*; Amarcande: Paris, France, 1990.
35. *Local Transport Note 01/07*; Traffic Calming, Department for Transport, Department for Regional Development (Northern Ireland), Scottish Executive, Welsh Assembly Government: Scottish, UK, 2007.

36. Seneci, F.; Avesani, F.; Bonomi, I. *Piani Particolareggiati Per Mobilita' CICLABILE e pedonale e Sicurezza Stradale*; Comune di Bassano del Grappa: Verona, Italy, 2012.
37. Sadeghi-Bazargani, H.; Saadati, M. Speed Management Strategies; A Systematic Review. *Beat* **2016**, *4*, 126–133. [PubMed]
38. *Le Temps des Rues*; IREC, Federal Technical University of Lausanne: Lausanne, Switzerland, 1990.
39. Bahar, G.B. *Guidelines for the Design and Application of Speed Humps*; Institute of Transportation Engineers: Washington, DC, USA, 2007.
40. Hallmark, S.L.; Peterson, E.; Fitzsimmons, E.; Hawkins, N.; Resler, J.; Welch, T. *Evaluation of Gateway and Low-Cost Traffic-Calming Treatments for Major Routes in Small Rural Communities*; Institute for Transportation, Iowa State University: Iowa, IA, USA, 2007.
41. *City of Seattle Staff Directory, Streetscape Design Guidelines, Chapter 6*; City of Seattle Staff Directory: Seatle, DC, USA, 2020. Available online: http://www.seattle.gov/rowmanual/manual/pdf/08/chapter6.pdf (accessed on 18 July 2020).
42. Kacprzak, D.; Sołowczuk, A. Effectiveness of road chicanes in access zones to a village at 70 km/h speed limit. In Proceedings of the World Multidisciplinary Civil Engineering—Architecture—Urban Planning Symposium, WMCAUS 2018, Prague, Czech Republic, 18–22 June 2018; Volume 471, p. 062010. [CrossRef]
43. Kacprzak, D.; Sołowczuk, A. Synergy Effect of Speed Management and Development of Road Vicinity in Wrzosowo. In Proceedings of the World Multidisciplinary Civil Engineering—Architecture—Urban Planning Symposium, WMCAUS 2019, Prague, Czech Republic, 17–21 June 2019. [CrossRef]
44. Kacprzak, D.; Sołowczuk, A. Identification of the Determinants of the Effectiveness of On-Road Chicanes in Transition Zones to Villages Subject to a 70 km/h Speed Limit. *Energies* **2020**, *13*, 5244.
45. *Department of the Environment, Transport and the Regions, Traffic Advisory Leaflet 5/01*; Traffic Calming Bibliography, DETR: London, UK, 2001.
46. Jateikienė, L.; Andriejauskas, T.; Lingytė, I.; Jasiūnienė, V. Impact Assessment of Speed Calming Measures on Road Safety. *Transp. Res. Procedia* **2016**, *14*, 4228–4236. [CrossRef]
47. Sayer, I.A.; Parry, D.I.; Barker, J.K. *Traffic Calming—An Assessment of Selected On-Road Chicane Schemes*; TRL Report 313; Transport Research Laboratory: Crowthorne, UK, 1998.
48. Jägerbrand, A.K.; Sjöbergh, J. Speed responses of trucks to light and weather conditions. *J. Cogent Eng.* **2019**, *6*, 1685365. [CrossRef]
49. Jun, W.; Ha, J.; Jeon, B.; Lee, J.; Jeong, H. LED traffic sign detection with the fast radial symmetric transform and symmetric shape detection. In Proceedings of the IEEE Intelligent Vehicles Symposium (IV), Seoul, Korea, 28 June–1 July 2015. [CrossRef]
50. De Bellis, E.; Schulte-Mecklenbeck, M.; Brucks, W.; Herrmann, A.; Hertwig, R. Blind haste: As light decreases, speeding increases. *PLoS ONE* **2018**, *13*, e0188951. [CrossRef] [PubMed]
51. Shahar, A.; Brémond, R.; Villa, C. Can light emitting diode-based road studs improve vehicle control in curves at night? A driving simulator study. *Sage J.* **2016**, *50*, 266–281. [CrossRef]
52. Jägerbrand, A.K.; Sjöbergh, J. Route familiarity in road safety: A literature review and an identification proposal. *Transp. Res. Part F Traffic Psychol. Behav.* **2019**, *62*, 651–671. [CrossRef]
53. Nygardhs, S.; Lundkvist, S.O.; Andersson, J.; Dahlback, N. The effect of different delineator post configurations on driver speed in night-time traffic: A driving simulator study. *Accid. Anal. Prev.* **2014**, *72*, 341–350. [CrossRef] [PubMed]
54. Yang, Q.; Overton, R.; Han, L.D.; Yan, X.; Richards, S.H. Driver behaviours on rural highways with and without curbs—A driving simulator based study. *Int. J. Injury Control. Saf. Promot.* **2013**, *21*, 115–126. [CrossRef] [PubMed]
55. Afghari, A.P.; Haque, M.M.; Washington, S. Applying fractional split model to examine the effects of roadway geometric and traffic characteristics on speeding behawior. *Bus. Med. Eng. Traffic Inj. Prev.* **2018**, *19*, 860–866. [CrossRef] [PubMed]
56. Ariën, C.; Brijs, K.; Brijs, T.; Ceulemans, W.; Vanroelen, G.; Jongen, E.; Daniels, S.; Wets, G. Does the effect of traffic calming measures endure over time? A simulator study on the influence of gates. *Geogr. Psychol. Transp. Res. Part Ftraffic Psychol. Behav.* **2013**, *22*, 63–75. [CrossRef]
57. Theeuwes, J.; Horst, A.V.D.; Kuiken, M. *Designing Safe Road Systems: A Human Factors Perspective*; CRC Press: London, UK, 2017. [CrossRef]
58. Greń, J. *Mathematical Statistics*; Models and Problems; PWN: Warszawa, Poland, 1982.
59. Taylor, J.R. *An Introduction to Error Analysis, The Study of Uncertainties in Physical Measurements*, 2nd ed.; University Science Books Sausalito: California, CA, USA, 1997.
60. *Technical Documentation of a Device for Measurement of Speed and Vehicle Composition of Traffic—Manufactured by MART*; (Funded as Part of the Instrumentation Purchasing Program of the Polish State Committee for Scientific Research (KBN), Decision No. 1829/IA/108/96, Application No. IA/926/96); MART: Szczecin, Poland, 1998.
61. Qu'est-ce que l'angle mort? Ornikar 2014. Available online: https://www.ornikar.com/permis/conseils-conduite/controle-visuel/angle-mort (accessed on 3 September 2018).
62. Google Earth. Available online: http://www.earth.google.com.
63. Burghardt, T. High Durability-High Retroreflectivity Solution for a Structured Road Marking System. In Proceedings of the International Conference on Traffic and Transport Engineering (ICTTE 2018), Belgrade, Serbia, 27–28 September 2018. Available online: https://www.researchgate.net/publication/328065486_HIGH_DURABILITY-HIGH_RETROREFLECTIVITY_SOLUTION_FOR_A_STRUCTURED_ROAD_MARKING_SYSTEM (accessed on 18 November 2020).

energies

MDPI

Article

Identifying the Optimal Packing and Routing to Improve Last-Mile Delivery Using Cargo Bicycles

Vitalii Naumov * and Michał Pawluś

Transport Systems Department, Civil Engineering Faculty, Cracow University of Technology, str. Warszawska 24, 31-155 Kraków, Poland; michal.pawlus@doktorant.pk.edu.pl
* Correspondence: vitalii.naumov@pk.edu.pl

Abstract: Efficient vehicle routing is a major concern for any supply chain, especially when dealing with last-mile deliveries in highly urbanized areas. In this paper problems considering last-mile delivery in areas with the restrictions of motorized traffic are described and different types of cargo bikes are reviewed. The paper describes methods developed in order to solve a combination of problems for cargo bicycle logistics, including efficient packing, routing and load-dependent speed constraints. Proposed models apply mathematical descriptions of problems, including the Knapsack Problem, Traveling Salesman Problem and Traveling Thief Problem. Based on synthetically generated data, we study the efficiency of the proposed algorithms. Models described in this paper are implemented in Python programming language and will be further developed and used for solving the problems of electric cargo bikes' routing under real-world conditions.

Keywords: cargo bicycles; last-mile logistics; MTSP; CVRP

Citation: Naumov, V.; Pawluś, M. Identifying the Optimal Packing and Routing to Improve Last-Mile Delivery Using Cargo Bicycles. *Energies* **2021**, *14*, 4132. https://doi.org/10.3390/en14144132

Academic Editors: Elżbieta Macioszek, Anna Granà, Margarida Coelho and Paulo Fernandes

Received: 4 June 2021
Accepted: 7 July 2021
Published: 8 July 2021

Publisher's Note: MDPI stays neutral with regard to jurisdictional claims in published maps and institutional affiliations.

1. Introduction

In 2018, over 50% of the world population lived in cities and produced around 80% of global gross domestic product. This number is even higher in regions such as Europe, Latin America, the Caribbean and Northern America, reaching values around 70%. It is estimated, that by 2030 urban populations are going to increase all around the world to values exceeding 60% [1] and to 85% by the year 2100 [2]. It is projected that Africa and Asia that currently has around 40% of the population living in rural areas will increase to 60% by 2050 with mega-cities (with over 10 million inhabitants) will become more common. Urban freight transportation is one of the key aspects of every city's economics.

Freight transport creates problems in urban areas, due to high population density. Modern cities need solutions to reduce external costs such as congestion, pollution and others, which have increased in the last few years, especially due to the increase in the supply of goods. Online sales and globalization are leading to new trends in freight transport, and more goods are expected to be delivered in the near future. In this context, most of the goods delivered go to city centers. Last-mile logistics is the least efficient stage in the supply chain, accounting for up to 28% of the total cost of delivery [3]. Therefore, improving last-mile logistics and significantly reducing externalities are very important challenges for scientists. New technologies and means of transport, innovative techniques and organizational strategies allow for more effective last-mile deliveries in urban areas.

Challenges described above led to formulating the concept of "green logistics". This idea intends to replace currently used combustion engine vehicles with zero-emissions technologies such as electric vehicles, cargo bikes, hybrid vehicles, etc. The use of zero-emission technologies leads to several benefits for logistics service providers and cities involved such as lower maintenance and operational costs, reduced noise emissions, access to pedestrian-only zones or access to historical city centers often inaccessible or accessible in off-duty hours (off-hour delivery) by internal combustion vehicles [4,5]. Due to the high total cost of ownership electric trucks with a payload of over one ton are not competitive over their diesel engine counterparts [6], due to the high purchase cost, lower range in winter and battery degradation during vehicle lifetime. In consideration of all presented advantages and limitations, the use of cargo bicycles and cargo e-bikes is very appealing for last-mile delivery.

The last-mile logistics concept using cargo bikes have been described in [3,7–10]. Location and planning of micro hubs was addressed in [11–13]. Methods for routing cargo bicycles have been described in [14,15]. Technical requirements for implementing cargo bicycles were researched in [16]. Last-mile delivery using cargo bikes have been introduced in many cities around the world including Seoul [15], Vienna, Budapest, Copenhagen [9] and Rio the Janeiro [17]. Tools for improving bicycle safety, routing and user behavior was studied in [18]. The difference in travel time between cargo bicycles and cars have been studied in [19].

In this paper, we aim to develop a theoretical framework for the optimization of deliveries by cargo bikes under real-world conditions: obtaining the solution that can be treated as an answer to both the routing problem and packing problem is an extremely challenging task, as far as two NP-hard (nondeterministic polynomial) problems must be solved simultaneously.

Routing problem can be addressed by a Travelling Salesman Problem (TSP) [20] approach, where we aim to create a route that visits all given nodes, minimizing distance. Creating a feasible packing plan can be tackled by Knapsack Problem (KP) [21] approach. In knapsack problem we aim to find a packing plan that does not exceed maximum cargo bicycle payload. Two problems given above are connected by Travelling Thief Problem [22]. This benchmark problem tackles combination and interdependence of two sub problems namely TSP and KP. In this problem there are n cities, ant the distance matrix between them is given. There are m items, each with value and weight. There is also maximum weight constraint, and the travel speed is related to the knapsack current weight.

The paper has the following structure in the next part we depict the categories of cargo bikes as the means of sustainable transport; a brief review of publications related to the research problem and the existing tools are presented in the third part; the fourth part contains the problem description followed by the overview of the developed algorithms; the fifth section introduces a synthetically generated case study of solving the combined knapsack and traveling salesman problem; the last part offers brief conclusions and directions of future research.

2. Cargo Bike Categories

The definition, categorization and commercial use cases such as postal, courier services, package delivery and passenger transport have been described by Gruber et al. [7]. Cargo bikes can be divided into several categories, according to their frame build, cargo distribution on the bike, number of wheels and maximum load. Different types of cargo bikes and variations of them are described in Table 1. Additionally, every type of bike presented below can be fitted with an electric motor to assist riders (e-cargo bike), especially when accelerating or going uphill, which will be very useful when servicing densely populated areas with routes consisting of many short travels with many stops. Electric variants of cargo bikes mostly have a quickly changeable battery and can go over 100 km on a single charge [8]. This range is sufficient for servicing city centers.

There are multiple advantages to using cargo bikes instead of delivery trucks for last-mile delivery, especially in congested city centers. Cargo bikes or e-cargo bikes are considered no emission vehicles, as no greenhouse gases are emitted when operating this type of bike. Operation of cargo bikes is mostly noiseless, except for geared motors, which can produce a barely noticeable sound, but still quieter than combustion engines. E-Cargo bikes can also be ridden on non-motorized vehicle zones and historical city centers. They are treated as regular bicycles in the EU and numerous countries in the world. Due to their small size, they cause less traffic congestion and use much less parking space. They also do not require driver licenses in EU countries. Small size, being one of the biggest advantages of cargo bikes is also one of its biggest drawbacks. Cargo bike compartments are limited, especially when comes to their cubic capacity. For example, Fiat Ducato L3H3, being one of the most popular small delivery trucks in Europe has a payload of 1415 kg and a cargo area cubic capacity of 14.1 m^3 [9]. This gives around 100 kg/m^3 of cargo density. We will compare it to Yokler U [10], a cargo bike made especially for last-mile logistics with a payload up to 150 kg and 1 m^3 cubic capacity [10]. Yoklers cargo density is 150 kg/m^3. This comparison clearly states that optimizing cubic capacity is even more important for cargo bikes, than for conventional small trucks. Another disadvantage of cargo e-bikes is much lower range than trucks, so only a narrow area can be serviced on a single charge. In addition, due to regulatory limited maximum motor power (250 W in EU), cargo e-bike acceleration and top speed are highly dependent on vehicle load, so that also should be considered when planning cargo bike routes.

Table 1. Categorization of cargo bikes [23].

Name	Description	Cargo Space Location
Post bike/bakers' bike	Two wheeled bikes with a frame geometry similar to conventional bicycle. Cargo can be located in front of the bike and/or behind the saddle. Main disadvantage of this type of cargo bike is high center of mass, which may cause stability issue, especially at low speeds with heavy loads. The maximum transport weight is usually 50 to 70 kg	
Longtail	Two wheeled bikes with longer rear frame triangle compared to conventional bicycle. More stable than the post bike, but the cargo space needs to be split to accommodate the wheel. The maximum transport weight goes up to 50 kg	
Front-loader/Long John	The cargo compartment is placed in front of the cyclist, between the steering tube and the front wheel. This allows for lower center of mass thus improving stability and maneuverability, even in low speed and high load scenarios. This type of bikes can be equipped with two wheels on front axle improving stability even further, but at the cost of widening the vehicle and decreasing maneuverability. This type of bikes can usually carry up to 120 kg of cargo	
Rear-loader/Trike	In this case of rear loaded cargo bike the payload is placed behind the rider not to obstruct the view. This type of bike can accommodate 3 or 4 wheels and load cargos up to 500 kg in 4-wheel models. These bikes have a high degree of stability, at the cost of lower maneuverability than front loaders	
Bike trailer	Cargo bikes can also be equipped with additional trailer. This type of cargo compartment is easy to adapt to existing bicycles, although according to EN 15918:2011 their gross weight cannot exceed 60 kg	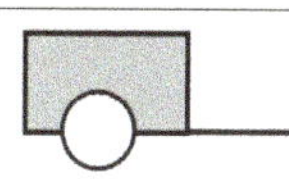

3. Overview about Related Problems and Corresponding Tools

As given in the introduction to this paper, solving the combined problem for packing and routing is not trivial. The software needs to calculate travel routes for multiple bikes with constraints according to maximum weight, cubic capacity and load-dependent speed constraints.

As this problem consists of multiple connected sub-problems, it is reasonable to try to inspect how different combination of solution algorithms affect the final outcomes. The sub problems consist of Knapsack Problem for finding the feasible packing plan, and the Travelling Salesman Problem for routing. These problems can also be combined by use of multiple Capacitated Vehicle Problem (CVRP). All of the problems given above are NP-hard, so optimally solving large instances of this problems is not possible due to computational complexity growing exponentially.

Routing problem can be addressed by a Travelling Salesman Problem approach. This mathematical description of the problem was formulated in 1930s by Karl Menger is one of the most studied combinatorial optimization problems. This problem and its generalizations and combinations such as Multiple Travelling Salesman Problem (MTSP) are widely adopted in real-life situations such as transport and delivery planning, agriculture, PCB drilling etc.

This problem can be formulated as follows: there are n cities and the distances between them are given by a distance matrix $D = \{d_{ij}\}$, where d_{ij} is the distance between city i and j. The salesman has to visit each city exactly once and minimize the distance of the complete tour. The objective function is given in Equation (1):

$$f(\overline{x}) = \sum_{i=1}^{n-1} \left(d_{x_i,x_{i+1}}\right),\ \overline{x} = (x_1, \ldots, x_n). \tag{1}$$

where $\overline{x}$ represents a tour, containing all of the cities exactly once. The aim is to find $\overline{x}$ which minimizes total tour distance $f(\overline{x})$.

Generalization of this problem is Multiple Salesman Problem, where multiple salesmen are needed to visit a given number of cities exactly once and return to the initial position with the minimum travelling cost. MTSP is a simplified version of VRP, by means of not considering the vehicle capacity or customer demands.

Recent approaches to solving the TSP and its generalizations include: Grey Wolf Optimizer [24], Genetic Algorithms [25,26], Swarm Optimization [27], Simulated Annealing [28] and hybrid approaches [29].

Knapsack Problem (KP) is a NP-hard optimization problem. The problem was first formulated in 1957 by George Dantzig. In this problem there are m items $I_1, \ldots, I_m$, which have a profit p_i and weight w_i. The knapsack is constrained by maximum weight W it can support. The aim of the problem is to pick items, maximizing total profit while their total weight does not exceed the maximum weight.

The problem is modelled as shown on:

$$maximize\ g(\overline{y}) = \sum_{i=1}^{m} p_i y_i,\ \overline{y} = (y_i, \ldots, y_m), \tag{2}$$

$$subject\ to\ \sum_{i=1}^{m} w_i \cdot y_i \leq W\ where\ y_1 = \begin{cases} 1 \text{ when item } i \text{ is picked,} \\ 0 \text{ when it is not.} \end{cases} \tag{3}$$

Vehicle Routing Problem was formulated by Danzig and Ramser in 1959 as a Truck Dispatching Problem [30]. The problem is defined as follows:

> "There is an undirected and complete graph of N locations (N − 1 customers and a depot) and a fleet of m vehicles. Each edge connecting two locations has a traverse cost (Euclidean distance). The goal is to visit each customer exactly once by a vehicle while minimizing the total cost of the routes. Each route must originate and terminate in the depot". [30]

Vehicle routing problems define a set of combinatorial optimization problems that allow optimizing planning for a fleet of vehicles, when vehicles operate trips that have multiple stops along the route.

There are multiple combinations and generalizations of this problem, including 3L-CVRP (Capacitated Vehicle Routing Problem with 3D Loading constraint) [31] consisting of CVRP with mass constraint and 3D packing constraint, SDVRP–Split Delivery Vehicle Routing Problem [32], 3L-SDVRP–3D Loading Split Delivery Vehicle Routing Problem [33], 3L-CVRPTWSDO–Split Delivery Vehicle Routing Problem with 3D Loading and Time Windows constraints [34].

Travelling Thief Problem [22] is an approach to create a new benchmark problem, to better approximate real-world problems. This benchmark problem tackles combination and interdependence of two sub problems namely TSP and KP. In this problem there are n

cities, ant the distance matrix between them is given. There are m items, each with value and weight. There is also maximum weight constraint, and the travel speed is related to the knapsack current weight. The aim of the problem is to find a tour that visits all of the cities exactly once and heads back to the starting city, optimizing objective function while the total weight of the knapsack is not exceeded.

4. Problem Description and Algorithms Overview

In comparison to conventional cargo trucks bikes have more restriction according to cubic capacity than trucks, that is why it is important to focus on dimension constraints and not only consider maximum weight as it is in classical Capacitated Vehicle Routing Problem (CVRP) formulations. It is also important that the overall bike weight (consisting of rider weight, bicycle weight and cargo weight) will affect acceleration times and top speed. This is similar to Travelling Thief Problem, but instead of adding weight during trip the vehicle leaves depot fully loaded, and then it reduces cargo weight at delivery locations thus increasing its speed.

The problem can be formulated as follows: There are n homogenous cargo bicycles with maximum cargo compartment dimensions and maximum cargo weight. There also are m consignments each with its weight, length, width, height and destination coordinates. Each bicycle must start and end its route in depot which coordinates are given. Bike maximum speed depends on the bicycle load. The goal of this problem is to find a set of routes, which allows for serving all consignments and minimizing total time needed for delivery.

The problem is represented by the set of following interdependent and connected parameters:

- Bin packing sub-problem:
    - Parameters:
        - Number of bicycles: n
        - Maximum bicycle cargo weight: $Wmax$
        - Cargo Bike compartment length: $lmax$
        - Cargo Bike compartment width: $wmax$
        - Cargo Bike compartment height: $hmax$
        - Number of consignments: m
        - Weight of each consignment: W_i
        - Length of each consignment: l_i
        - Width of each consignment: w_i
        - Height of each consignment: h_i
    - Solution:
        - The solution is a matrix of binary values called *packing plan* (X_{nm}) where: $x_{ij} = 1$ when consignment j is placed inside the bike i or 0 when its not
- TSP sub-problem:
    - Parameters:
        - Number of consignments: m which corresponds to number of nodes/cities for classical TSP formulation
        - Number of bicycles n
        - Coordinates' vectors $coordinatesX_{m+1}$ and $coordinatesY_{m+1}$, where: $coordinatesX_0$, $coordinatesY_0$ being depot coordinates and $coordinatesX_j$, $coordinatesY_j$ being target coordinates for consignment j. From these vectors distance matrix can be calculated
        - Velocity v_c

- Solution:
    - The solution is called routes, where single route $\bar{r}$ consists of visited nodes for each bicycle

In TSP sub-problem velocity is calculated depending on the current load of the bicycle. Velocity calculation is carried out in a similar matter as in TTP approaches [22]:

$$v_c = v_c max - load \cdot \frac{v_c max - v_c min}{Wmax}, \quad (4)$$

where $v_c max$ and $v_c min$ are maximal and minimal cargo bike velocities.

4.1. Bin-Pack-3D Algorythm

First method used called bin pack 3D is used as benchmark, because it can deterministically calculate routes, find all possible solutions and then choose the best one according to service times or total distances. It consists of two sub-problems tackled independently, with problem constraints distributed accordingly. First tackled problem is generating all possible knapsack combinations as a constraint programming problem. This problem can be formulated as:

There are n cargo bicycles with capacity c_i and maximum cubic capacity vol_i. There are also m packages, each having its weight w_j, length l_i, width w_i and height h_i. The objective of this algorithm is to find all feasible solutions x_{ij}, where:

$$x_{ij} = \begin{cases} 1 \text{ when package } j \text{ is placed in the bicycle } i, \\ 0 \text{ when package } j \text{ is not placed in the bicycle } i, \end{cases} \quad (5)$$

subject to constraints:

$$\sum\nolimits_{i=1}^{n} x_{ij} = 1, \quad (6)$$

$$\sum\nolimits_{j=1}^{m} w_j \cdot x_{ij} \le c_i, \forall i \in (1, \ldots, n), \quad (7)$$

$$\sum\nolimits_{j=1}^{m} v_j \cdot x_{ij} \le vol_i,\ \forall i \in (1, \ldots, n). \quad (8)$$

First constraint Equation (6) ensures that every package is packed on exactly one bike. Second constraint Equation (7) checks that no bike have exceeded its maximum cargo weight. Volume constraint Equation (8) is a simplification of 3D packing as this constraint is computationally faster, than checking 3D packing feasibility for every single package and bike. Routes fulfilling those constraints will be referenced to as 2D feasible.

After generating all feasible knapsack packing plans, all possible permutations of routes are calculated for every single bike packing plan obtained from feasible knapsack generator, with restriction that every route needs to start and end at the depot. In this case depot is set to be node no. 0 and it is a beginning of the Cartesian coordinates system used for routing. Every node coordinate is a X or Y distance from the depot. After calculating all possible route permutations, the route with lowest travel time is taken. This brute force approach is described in Algorithm 1.

In this deterministic approach all possible TSP routes are calculated. Each TSP route starts and ends at depot. For each TSP route total transport time is calculated.

After calculating all route times, the results are searched for a minimal total time solution. When solution is found it is checked for 3D feasibility according algorithm (packer) described in [21,35]. This algorithm verifies 3D feasibility according to cargo space and packages dimensions, not only volume. If the solution is found not 3D feasible, next best solution is found and 3d feasibility check is conducted, until feasible solution is found. Not every route is checked for 3d feasibility, because 3D packer algorithm is computationally expensive and there is no need for it especially when best solution has been found.

This algorithm may not be suitable for large data instances due to its data capacity and computation time. However, it is very helpful as a benchmark, as it checks all possible outcomes and can generate optimal solution. That is why it is used to compare CVRP and TSP-first approaches.

Algorithm 1: Bin-Pack-3D

Input:

1. *Consignments*–vector consignment number used for indexing
2. *Distance matrix*–matrix containing distances between consignment target locations
3. *Weight*–vector, which holds weight of every package
4. Maximum bike load, maximum velocity $v_c max$ and minimum velocity $v_c min$
5. *Packing plan* X_m is a vector containing Boolean values for a single bike obtained from knapsack generator

Output:
Vector containing route with lowest service time, total service time, total distance
Procedure:

1. Create temporary vector *nodes* used for storing
2. For every $x_j \neq 0$ in packing plan: append temporary vector $nodes_n$ with *Consignment* number for x_j
3. Acquire all permutations of nodes and store it in the *permutations* matrix
4. For each permutation in *permutations*:

i. Create variables $time = 0$, $distance = 0$ and vector *route*
ii. Create *dataset* containing routes, total times and total distances
iii. Calculate initial *load* for vehicle in route by summing weights of all nodes in *nodes*
iv. Append *route* with $(0,\ nodes[0], \ldots nodes[n], 0)$ as every route starts and ends at depot
v. For every node in *permutation*:

- Calculate $v_c = v_c max - load \cdot \frac{v_c max - v_c min}{W max}$
- Acquire *Distance matrix*[*node*][*previous node*] (if no previous node, use depot instead)
- Increase *time* by $distance / v_c$
- Decrease load by $Weight[node]$

vi. Append dataset with *route*, *time* and *distance*

5. Search dataset for lowest route time

Return route, distances and time for a route with lowest time

4.2. MTSP-First Algorithm

In this approach at first MTSP routes for all vehicles are calculated without any restrictions considering weight or cubic capacity using Cheapest Arc method as a first solution strategy and Guided Local Search as local search metaheuristic. The algorithm of this experimental approach is described in Algorithm 2.

After calculating possible MTSP routes, every route for every vehicle is verified for mass and cubic capacity feasibility and rejecting solutions that are not feasible.

Later remaining solutions are verified for 3D feasibility using packer algorithm. When all solutions are deemed feasible total distances and times for all vehicles in all routes is calculated. Next, the results are searched for a feasible solution with lowest time. For sake of documentation and result analysis, all routes were saved instead of removing them from the memory, which would improve algorithm memory usage for large datasets.

4.3. CVRP-First Algorithm

In this approach at first CVRP routes for all vehicles are calculated with restrictions considering weight or cubic capacity using Cheapest Arc method as a first solution strategy and Guided Local Search as local search metaheuristic. Next the 3D feasibility is checked the same way as in MTSP and Bin Pack approaches. The algorithm for this approach is shown on Algorithm 3. As in previous approaches non feasible data was not removed for research documentation purposes.

Algorithm 2: MTSP-First

Input:

1. *Consignments*–dataset containing consignment ID, target locations, weights, lengths, widths, heights and calculated volumes
2. *Data model*–dataset containing depot location, number of bikes, bike maximum capacity, cargo compartment length, width, height, maximum and minimum velocity

Output: Feasible route with best travel time
Procedure:

a. Create *DistanceMatrix* containing Euclidean distances between every node
b. Solve MTSP with time limit of 2 s and save every generated route
c. Check feasibility of generated routes according to weight and cubic capacity constraints and save KP feasible routes
d. For every KP feasible route check feasibility using *packer*() function and save *3D feasible routes*
e. For every *3D feasible route*:
    i. Create variables *time* = 0, distance = 0 and vector *route*
    ii. Calculate initial *load* for vehicle in route by summing weights of all nodes in the route from *Consignments*
    iii. For every node in *3D feasible route*:
        1. Calculate $v_c = v_c max - load \cdot \frac{v_c max - v_c min}{Wmax}$
        2. Acquire *DistanceMatrix*[*node*][*previous node*] (if no previous node, use depot instead)
        3. Increase *time* by $Distance/v_c$
        4. Decrease *load* by *Weight*[*node*]
    iv. Append *dataset* for 3D feasible routes with route, times and distances
f. Search *dataset* for lowest route time

Return route, distances for every vehicle, total distance, delivery time for a single vehicle and total time for a route with lowest time

Algorithm 3: CVRP-First

Input:

1. *Consignments*–dataset containing consignment ID, target locations, weights, lengths, widths, heights and calculated volumes
2. *Data model*–dataset containing depot location, number of bikes, bike maximum capacity, cargo compartment length, width, height, maximum and minimum velocity

Output: Feasible route with lowest travel time
Procedure:

a. Create *Distance matrix* containing Euclidean distances between every node
b. Solve CVRP with time limit of 2 s with constraints according to maximum cargo weight and volume. Save every generated route
c. For every CVRP feasible route check feasibility using packer () function and save *3D feasible routes*
d. For every *3D feasible route*:
    i. Create variables *time* = 0, distance = 0 and vector *route*
    ii. Calculate initial *load* for vehicle in route by summing weights of all nodes in the route from *Consignments*
    iii. For every node in *3D feasible route*:
        1. Calculate $v_c = v_c max - load \cdot \frac{v_c max - v_c min}{Wmax}$
        2. Acquire *Distance matrix[node][previous node]*(if no previous node, use depot instead)
        3. Increase *time* by Distance$/v_c$
        4. Decrease *load* by *Weight*[*node*]
    iv. Append *dataset* for 3D feasible routes with route, times and distances
e. Search *dataset* for lowest route time

Return route, distances for every vehicle, total distance, delivery time for a single vehicle and total time for a route with lowest time

5. Numerical Results and Discussion

In this chapter we will compare three approaches to the problem formulated in Section 4. These approaches were chosen to check the impact of interdependence between sub-problems. First one called Bin-Pack-3D is deterministic method used for generating all feasible solutions to the given problem, and it is used as a benchmark for comparing other combinations of sub-problems. Other two namely: MTSP-First and CVRP are using metaheuristic algorithms to find feasible solutions.

For experiments a simple dataset was created consisting of 4 bicycles and 7 consignments. Vehicle capacity was set to 100 kg, cargo compartment length is 800 mm, width was 500 mm and height was 400 mm. The depot was set as point of origin of Cartesian coordinates system with coordinates $x = 0$ and $y = 0$. Bicycle maximum velocity was set to 25 km/h as it is a legal maximum assisted speed for e-bike in the EU. Minimum velocity was set to 5 km/h.

Consignment data was set that the total weight and cubic capacity does not exceed maximum values for all cargo bikes, so that all the consignments can be delivered. The consignment coordinates x and y were selected randomly in the following range: $\{distance \, \epsilon \, \mathbb{Z} | -1000 \leq distance \leq 1000\}$. Cargo dimension was generated semi-randomly so that they fit the normal distribution with mean value 320 mm and standard deviation of 100 mm. Cargo weight was generated in a similar manner but with mean value of 15.0 kg and standard deviation of 5.0 kg. The generated data is presented in Table 2. Graphical representation of node locations and best route according to delivery time is shown on Figure 1.

Table 2. Experiment data.

Consignment	1	2	3	4	5	6	7	8	9	10
X coordinate [m]	322	−348	835	53	346	−340	−482	−696	956	−168
Y coordinate [m]	−524	−517	751	−425	529	35	−988	537	535	−602
Cargo weight [kg]	33.257	21.254	25.058	5.188	18.634	7.416	21.037	7.568	26.254	23.697
Cargo length [mm]	103	155	167	252	354	236	384	290	36	219
Cargo width [mm]	484	298	443	280	309	278	194	384	197	164
Cargo height [mm]	204	343	152	521	306	403	778	327	703	328

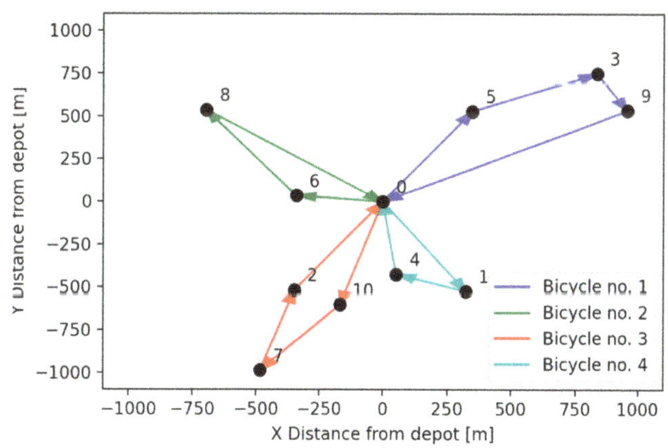

Figure 1. Node location and route with lowest total time.

For constraint programing solver and all metaheuristic approaches Google OR-Tools [36] libraries were used. The programming was carried out in Jupyter Lab using Python 3.9.4 as a kernel.

5.1. Bin-Pack-3D Results

Using this approach all feasible solutions were achieved. Due to large number of permutations, only best routes according to time were taken into consideration. The histograms for all 2D feasible, time-optimal routes are shown on Figure 2.

All TSP-optimal routes for all knapsacks have a mean time of 2780 s, maximum spread is 855 s and standard deviation of population is 164 s.

All 2D feasible solutions count was 827,772. They were calculated in 565 s. There were 41,253,456 route permutations, which were calculated in 298 s. Total algorithm runtime was 864 s. Average time of all 2D feasible routes was 2168 s with population standard deviation of 168 s and range of 1105 s. Mean distance of all 2D feasible routes was 11,325 m with population standard deviation of 991 m and range of 6290 m.

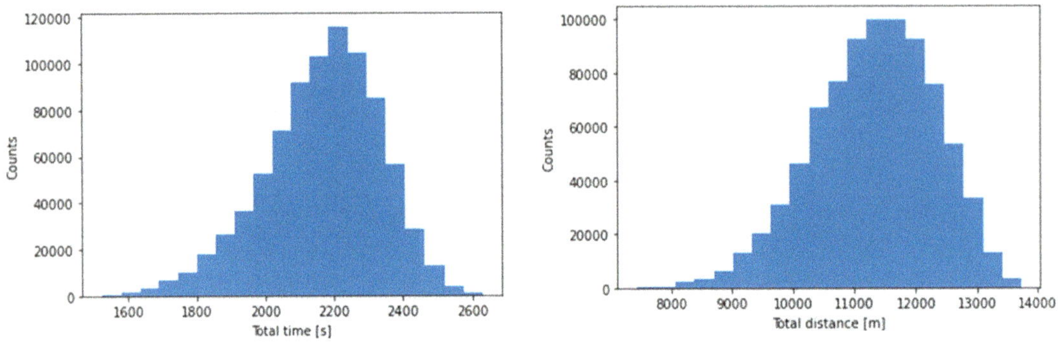

(**a**) (**b**)

Figure 2. Histograms for all Bin Pack routes according to (**a**) time and (**b**) distance.

This algorithm managed to find optimal route [[0, 0, 3, 5, 9], [0, 0, 6, 8], [0, 0, 2, 7, 10], [0, 0, 1, 4]] with total travel time of 1525 s and total travel distance of 7914 m. Distance of optimal solution was also the minimum of all feasible distances.

5.2. MTSP-First Results

After calculating possible MTSP routes, every route for every vehicle is verified for mass and cubic capacity feasibility and rejecting solutions that are not feasible.

Histograms of total times and distances for all generated routes is presented on Figure 3.

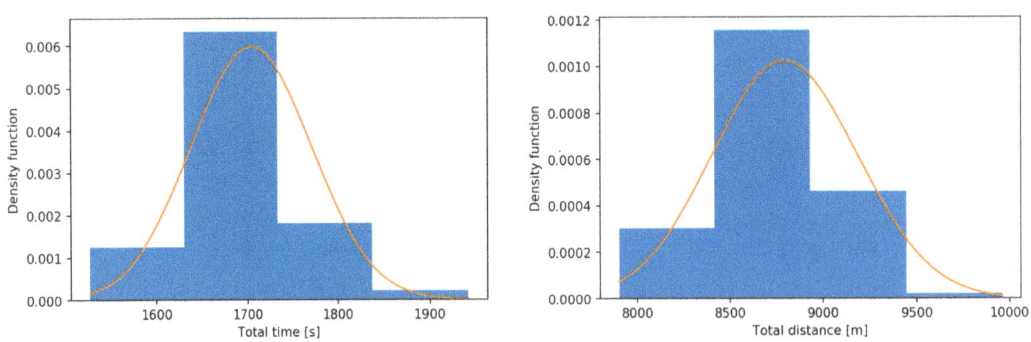

(**a**) (**b**)

Figure 3. Histograms for all MTSP routes according to time (**a**) and distance (**b**).

With MTSP solver being time-limited to 2 s total of 239 TSP solutions was found, 231 solutions were deemed feasible according to the weight and volume constraint. Total of 15 solutions were feasible according to packer algorithm thus being 3D feasible. 3D feasible routes histogram is shown on Figure 4.

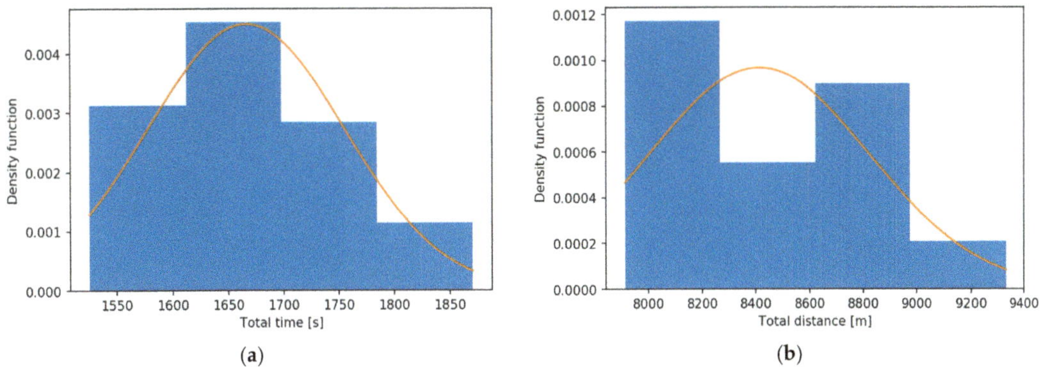

Figure 4. Histograms for 3D feasible MTSP routes according to time (**a**) and distance (**b**).

Mean total time for all 3D feasible solutions was 1652 s, with population standard deviation of 88 s and spread between maximum and minimum time was 305 s. Mean distance for all those routes was 8376 m with population standard deviation of 392 m and spread between maximum and minimum distances was 1416 m. All given values were rounded to a full second or meter accordingly.

Best found route, according to total travel time using MTSP-First algorithm was [[0, 5, 3, 9, 0], [0, 6, 8, 0], [0, 10, 7, 2, 0], [0, 1, 4, 0]] with total time of 1526 s and total distance of 7913 m. This route is shown on Figure 5.

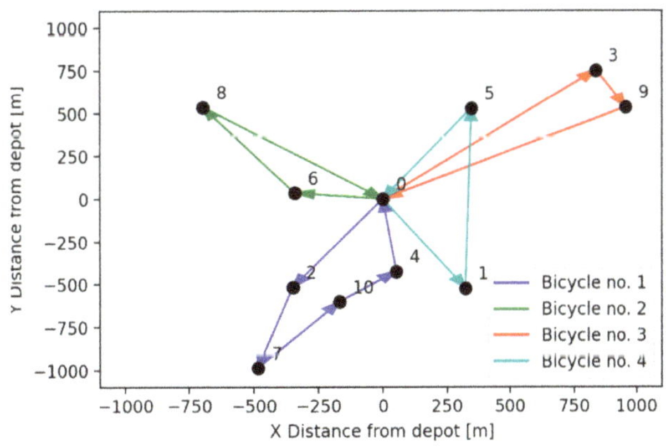

Figure 5. Route with best total travel time achieved from MTSP approach.

5.3. CVRP Approach Results

With CVRP solver runtime limited to 2 s it was able to generate 211 solutions, which were 2D feasible. Histograms of those values according to total travel time and distance are shown on Figure 6. 28 of those solutions were deemed 3D feasible.

Best found route according to total travel time was [[0, 10, 7, 2, 0], [0, 6, 8, 0], [0, 5, 3, 9, 0], [0, 1, 4, 0]], with travel time of 1525 s and distance of 7 914 m. Average total travel time of all 3D feasible solutions was 1705 s with population standard deviation of 70 s and range of 411 s. Mean distance of all 3D feasible routes was 8817 m with population standard deviation of 398 m and difference between maximum and minimum values of 2095 m. Histograms representing those data are shown on Figure 7.

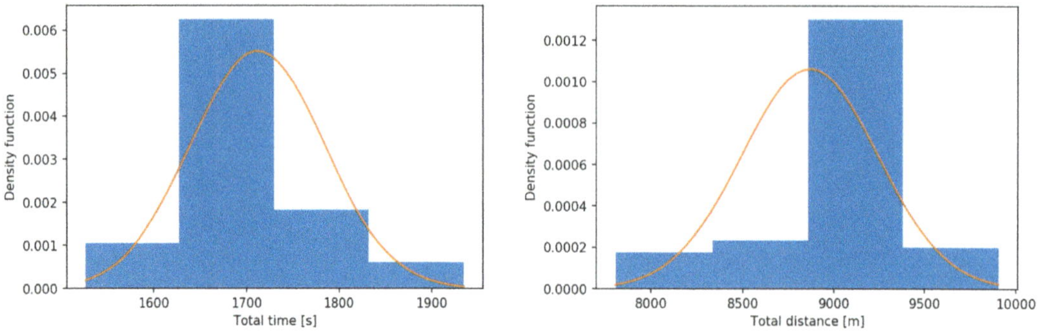

(**a**) (**b**)

Figure 6. Histograms for all CVRP routes according to time (**a**) and distance (**b**).

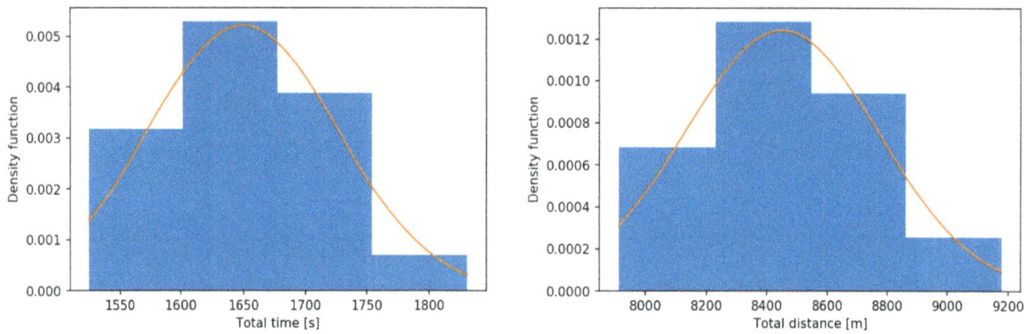

(**a**) (**b**)

Figure 7. Histograms for all 3D feasible CVRP routes according to time (**a**) and distance (**b**).

As can be seen on Figure 8 the bicycle routes with best total travel time had the same nodes as the route with minimum total travel distance, but the sequence of node visits was different.

5.4. Result Comparison

Results, achieved by metaheuristic methods (MTSP and CVRP) were similar to values calculated deterministically. The results are presented in Table 3.

Table 3. Experiment results.

Parameters	Used Algorithm		
	Bin Pack 3D	MTSP	CVRP
Total route time [s]	1525	1526	1525
Total route distance [m]	7914	7913	7914

As can be seen from the numbers in Table 3, for the synthetic dataset used in this research, the route parameters for the best solution generated by the corresponding algorithms are almost identical, although the shapes of the obtained routes differ insignificantly. Thus, the main criterion to choose the heuristic method is the time needed to achieve the result by running the corresponding algorithm: both proposed metaheuristics are characterized by the calculation time feasible under real-world conditions. However, the

CVRP approach in contrast with the MSTP heuristics was able to find an optimal solution, which is why it will be used for further research under real-life applications.

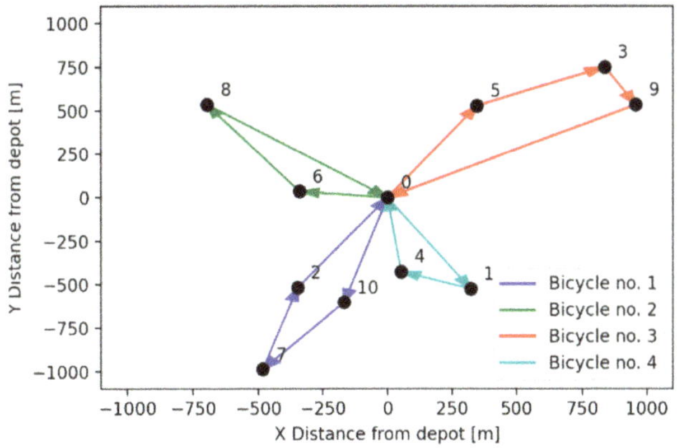

(a)

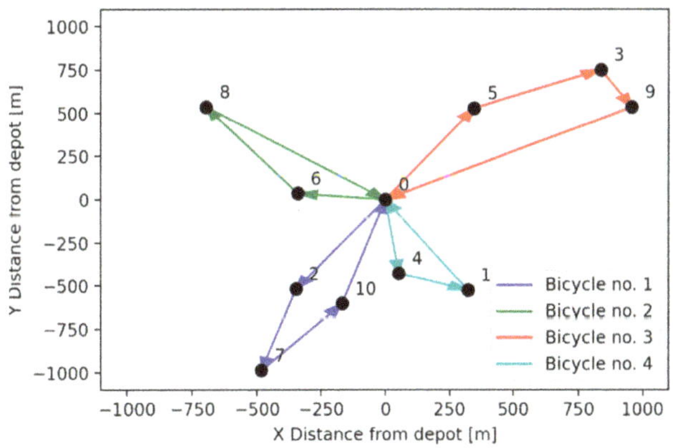

(b)

Figure 8. Diagram representing route for calculated routes with lowest time (**a**) and distance (**b**).

6. Conclusions

In this work, we deal with a combination vehicle routing problem with 3D loading constraints and load-dependent time. This approach is similar to CVRP formulated by Dantzig and Ramser except for additional constraints, namely, speed being dependent on the load carried by a vehicle as into traveling thief problem. This research is given by the problem of last-mile delivery with the use of cargo bicycles, which are powered by the strength of human muscles thus low power so they cannot achieve high speeds and those speeds will be even lower when the bicycle is fully loaded. Two wheeled cargo bicycles have capacity lower to their three or four wheeled counterparts, but with the lighter load it would allow for faster deliveries, due to being less cumbersome to ride in densely populated urban areas and being less susceptible to traffic congestions.

We have achieved an optimal solution for a given dataset by deterministically calculating all possible knapsack combinations and calculating the best route according to the total travel time for each knapsack combination. This deterministic approach was used as a benchmark for comparing two metaheuristic approaches, the first being CVRP with total load and cubic capacity and 3D packing feasibility restrictions. The second approach was made as an MTSP approach with a check for feasibility according to mass, volume and

3D packing constraints. Testing of algorithms proved that the CVRP approach was able to produce optimal solutions in 2 s for 4 bicycles and 10 consignments, which concludes that this approach is better suited for real-life applications with dynamically appearing consignments and will be further investigated. MTSP approach was not able to produce an optimal solution.

Future work should enable implementing efficient 3D packing plans with LIFO constraints so that packages can be removed without moving other consignments. In addition, a constraint programming approach to the whole problem will be attempted with various metaheuristic approaches.

Author Contributions: Conceptualization, V.N.; methodology, V.N. and M.P.; software, M.P.; validation, V.N. and M.P.; investigation, V.N. and M.P.; writing—original draft preparation, M.P.; writing—review and editing, V.N. All authors have read and agreed to the published version of the manuscript.

Funding: This research was funded by European Union's Horizon 2020 research and innovation program, grant agreement No. 769086 (CityChangerCargoBike project).

Institutional Review Board Statement: Not applicable.

Informed Consent Statement: Not applicable.

Data Availability Statement: The numeric results of the described simulations can be found at https://github.com/PawlusM/cargobike (accessed on 4 June 2021).

Conflicts of Interest: The authors declare no conflict of interest.

References

1. United Nations, Department of Economic and Social Affairs. The World 's Cities in 2018. In *The World's Cities in 2018—Data Booklet (ST/ESA/ SER.A/417)*; United Nations, Department of Economic and Social Affairs, Population Division: New York, NY, USA, 2018; p. 34.
2. United Nations. *World Population Prospects Key Findings & Advance Tables*; ESA/P/WP.241; United Nations: New York, NY, USA, 2015.
3. Ranieri, L.; Digiesi, S.; Silvestri, B.; Roccotelli, M. A Review of Last Mile Logistics Innovations in an Externalities Cost Reduction Vision. *Sustainability* **2018**, *10*, 782. [CrossRef]
4. Taefi, T.T.; Kreutzfeldt, J.; Held, T.; Fink, A. Supporting the adoption of electric vehicles in urban road freight transport—A multi-criteria analysis of policy measures in Germany. *Transp. Res. Part A Policy Pract.* **2016**, *91*, 61–79. [CrossRef]
5. Taefi, T.T.; Kreutzfeldt, J.; Held, T.; Konings, R.; Kotter, R.; Lilley, S.; Baster, H.; Green, N.; Laugesen, M.S.; Jacobsson, S.; et al. Comparative Analysis of European Examples of Freight Electric Vehicles Schemes—A Systematic Case Study Approach with Examples from Denmark, Germany, the Netherlands, Sweden and the UK. *Optim. Decis. Support Syst. Supply Chains* **2016**, 495–504. [CrossRef]
6. Taefi, T.T. Viability of electric vehicles in combined day and night delivery: A total cost of ownership example in Germany. *Eur. J. Transp. Infrastruct. Res.* **2016**, *16*, 512–553. [CrossRef]
7. Elbert, R.; Friedrich, C.; Boltze, M.; Pfohl, H.-C. *Urban Freight Transportation Systems: Current Trends and Prospects for the Future*; Published in Cooperation with WCTRS; Elsevier BV: Amsterdam, The Netherlands, 2020; pp. 265–276.
8. CIVITAS. Making urban freight logistics more sustainable. *Civ. Policy Note* **2015**, 1–63. Available online: http://www.eltis.org/ (accessed on 4 June 2021).
9. Anderluh, A.; Hemmelmayr, V.; Nolz, P.C. Sustainable Logistics with Cargo Bikes—Methods and Applications. *Sustain. Transport. Smart Logist.* **2019**, 207–232. [CrossRef]
10. Wrighton, S.; Reiter, K. CycleLogistics—Moving Europe Forward! *Transp. Res. Procedia* **2016**, *12*, 950–958. [CrossRef]
11. Baum, L. Planning of Cargo Bike Hubs. A Guide for Municipalities and Industry for the Planning of Transshipment Hubs for New Urban Logistics Concepts. 2020. Available online: https://cyclelogistics.eu/sites/default/files/downloads/Hub%20Planning%20Brochure_EN_Web_final.pdf (accessed on 4 June 2021).
12. Chiffi, C.; Reiter, K. Shared Micro-Hubs, Locker Systems and Cycle-Based Delivery Services for Achieving Zero Emission Urban Logistics. 2015. Available online: https://civitas.eu/sites/default/files/documents/cyclelogistics-civitas_final.pdf (accessed on 4 June 2021).
13. Naumov, V.; Starczewski, J. Choosing the Localisation of Loading Points for the Cargo Bicycles System in the Krakow Old Town. In *International Conference on Reliability and Statistics in Transportation and Communication*; Lecture Notes in Networks and Systems; Springer Science and Business Media LLC: Cham, Switzerland, 2019; pp. 353–362.
14. Enthoven, D.L.; Jargalsaikhan, B.; Roodbergen, K.J.; Broek, M.A.J.U.H.; Schrotenboer, A.H. The two-echelon vehicle routing problem with covering options: City logistics with cargo bikes and parcel lockers. *Comput. Oper. Res.* **2020**, *118*, 104919. [CrossRef]

15. Lee, K.; Chae, J.; Kim, J. A Courier Service with Electric Bicycles in an Urban Area: The Case in Seoul. *Sustainability* **2019**, *11*, 1255. [CrossRef]
16. Gruber, J.; Ehrler, V.; Lenz, B. Technical Potential and User Requirements for the Implementation of Electric Cargo Bikes in Courier Logistics Services. In Proceedings of the 13th World Conference on Transport Research (WCTR), Rio de Janeiro, Brazil, 15–18 July 2013; pp. 1–16.
17. De Mello Bandeira, R.A.; Goes, G.V.; Gonçalves, D.N.S.; de Almeida D'Agosto, M.; De Oliveira, C.M. Electric vehicles in the last mile of urban freight transportation: A sustainability assessment of postal deliveries in Rio de Janeiro-Brazil. *Transp. Res. Part D Transp. Environ.* **2019**, *67*, 491–502. [CrossRef]
18. Stamatiadis, N.; Pappalardo, G.; Cafiso, S. Use of technology to improve bicycle mobility in smart cities. In Proceedings of the 5th IEEE International Conference on Models and Technologies for Intelligent Transportation Systems, MT-ITS 2017, Naples, Italy, 26 June 2017; pp. 86–91.
19. Gruber, J.; Narayanan, S. Travel Time Differences between Cargo Cycles and Cars in Commercial Transport Operations. *Transp. Res. Rec. J. Transp. Res. Board* **2019**, *2673*, 623–637. [CrossRef]
20. Cheikhrouhou, O.; Khoufi, I. A comprehensive survey on the Multiple Traveling Salesman Problem: Applications, approaches and taxonomy. *Comput. Sci. Rev.* **2021**, *40*, 100369. [CrossRef]
21. Dube, E.; Kanavathy, L.R. Optimizing Three-Dimensional Bin Packing Through Simulation. MSO. 2006. Available online: https://www.academia.edu/35965688/OPTIMIZING_THREE-DIMENSIONAL_BIN_PACKING_THROUGH_SIMULATION (accessed on 4 June 2021).
22. Bonyadi, M.R.; Michalewicz, Z.; Barone, L. The travelling thief problem: The first step in the transition from theoretical problems to realistic problems. *IEEE Congr. Evolut. Comput.* **2013**, 1037–1044. [CrossRef]
23. Gruber, J.; Rudoph, C.; Lenz, B.P.; Liedtje, G.P.; Spath, C.; Wrighton, S. Untersuchung des Einsatzes von Fahrrädern im Wirtschaftsverkehr (WIV-RAD). 2016. Available online: https://elib.dlr.de/104273/1/WIV-RAD-Schlussbericht.pdf (accessed on 4 June 2021).
24. Shaheen, A.; Sleit, A.; Al-Sharaeh, S. A solution for traveling salesman problem using grey wolf optimizer algorithm. *J. Theor. Appl. Inf. Technol.* **2018**, *96*, 6256–6266.
25. Xu, J.; Pei, L.; Zhu, R.-Z. Application of a Genetic Algorithm with Random Crossover and Dynamic Mutation on the Travelling Salesman Problem. *Procedia Comput. Sci.* **2018**, *131*, 937–945. [CrossRef]
26. Hacizade, U.; Kaya, I. GA Based Traveling Salesman Problem Solution and its Application to Transport Routes Optimization. *IFAC PapersOnLine* **2018**, *51*, 620 625. [CrossRef]
27. Marinakis, Y.; Marinaki, M.; Dounias, G. A hybrid particle swarm optimization algorithm for the vehicle routing problem. *Eng. Appl. Artif. Intell.* **2010**, *23*, 463–472. [CrossRef]
28. Tatavarthy, A.; Rao, T.S.; Marimuthu, K.P. A Simulated Annealing and Nearest Neighborhood Approach to Solve a Vehicle Routing Problem in a FMCG Company. *Int. J. Mech. Prod. Eng. Res. Dev.* **2018**, *8*, 349–354. [CrossRef]
29. Mahi, M.; Baykan, O.K.; Kodaz, H. A new hybrid method based on Particle Swarm Optimization, Ant Colony Optimization and 3-Opt algorithms for Traveling Salesman Problem. *Appl. Soft Comput.* **2015**, *30*, 484–490. [CrossRef]
30. Dantzig, G.B.; Ramser, J.H. The truck dispatching problem. *Manag. Sci.* **1959**, *6*, 80–91. [CrossRef]
31. Gendreau, M.; Iori, M.; Laporte, G.; Martello, S. A Tabu Search Algorithm for a Routing and Container Loading Problem. *Transp. Sci.* **2006**, *40*, 342–350. [CrossRef]
32. Ozbaygin, G.; Karasan, O.; Yaman, H. New exact solution approaches for the split delivery vehicle routing problem. *EURO J. Comput. Optim.* **2017**, *6*, 85–115. [CrossRef]
33. Bortfeldt, A.; Yi, J. The Split Delivery Vehicle Routing Problem with three-dimensional loading constraints. *Eur. J. Oper. Res.* **2020**, *282*, 545–558. [CrossRef]
34. Chen, Z.; Yang, M.; Guo, Y.; Liang, Y.; Ding, Y.; Wang, L. The Split Delivery Vehicle Routing Problem with Three-Dimensional Loading and Time Windows Constraints. *Sustainability* **2020**, *12*, 6987. [CrossRef]
35. Li, X.; Zhao, Z.; Zhang, K. A Genetic Algorithm for the Three-Dimensional Bin Packing Problem with Heterogeneous Bins. 2014. Available online: https://www.researchgate.net/profile/Xueping-Li-4/publication/273121476_A_genetic_algorithm_for_the_three-dimensional_bin_packing_problem_with_heterogeneous_bins/links/54f742460cf2ccffe9dafbc2/A-genetic-algorithm-for-the-three-dimensional-bin-packing-problem-with-heterogeneous-bins.pdf (accessed on 4 June 2021).
36. Perron, L.; Furnon, V. OR-Tools. 2019. Available online: https://developers.google.com/optimization (accessed on 4 June 2021).

energies

Article

Investigating the Impact of Public Transport Service Disruptions upon Passenger Travel Behaviour—Results from Krakow City

Arkadiusz Adam Drabicki [1,*], Md Faqhrul Islam [2] and Andrzej Szarata [1]

[1] Department of Transportation Systems, Cracow University of Technology, 31-155 Kraków, Poland; aszarata@pk.edu.pl
[2] Transport Research Institute, Edinburgh Napier University, Edinburgh EH10 5DT, UK; faqhrul.islam@gmail.com
* Correspondence: adrabicki@pk.edu.pl

Abstract: Public transport (PT) service disruptions are common and unexpected events which often result in major impediment to passengers' typical travel routines. However, attitudes and behavioural responses to unexpected PT disruptions are still not fully understood in state-of-the-art research. The objective of this study is to examine how PT users adapt their travel choices and what travel information sources they utilize once they encounter sudden PT service disruptions. To this end, we conduct a passenger survey among PT users in the city of Kraków (Poland), consisting of a series of stated- and revealed-preference questions. Results show that passengers' reported choices during past PT disruptions mostly involve adjusting the current PT travel routine, exposing a certain bias with their stated choices (which tend to overestimate the probability of modal shifts). Factors influencing travel behaviour shifts include frequency and recency of PT disruption experience, as well as propensity to arrive on-time. With regards to travel information sources, staff announcement and personal experience play an important role in recognizing the emerging disruption, but real-time information (RTI) sources are the most useful in planning the onward journey afterwards. Based on these, we highlight the implications for future RTI policy during PT service disruptions; in particular, the provision of a reliable time estimate until normal service conditions are resumed. Such RTI content could foster passengers' tendency to use PT services in uncertain conditions, especially as their stated wait time tolerance often matches the actual duration of PT disruptions.

Keywords: public transport; travel behaviour; service disruption; real-time information; RTI

Citation: Drabicki, A.A.; Islam, M.F.; Szarata, A. Investigating the Impact of Public Transport Service Disruptions upon Passenger Travel Behaviour—Results from Krakow City. *Energies* **2021**, *14*, 4889. https://doi.org/10.3390/en14164889

Academic Editor: Andrea Mariscotti

Received: 29 June 2021
Accepted: 5 August 2021
Published: 10 August 2021

1. Introduction

Reliability (robustness) of public transport (PT) services is an important factor, shaping general passenger attitudes towards PT systems, and consequently affecting passengers' travel choices. Disruptions in PT services may cause various shift patterns in passengers' travelling strategies, depending on a wide range of aspects. This is especially the case for sudden (unexpected) service disruptions, which pose great journey disturbance and inconvenience. These invoke instantaneous shifts in the passengers' current travelling strategies and lead them to undertake much less-familiar travel options. Under such circumstances, availability of reliable travel information becomes even more crucial to the decision-making process, with the increasingly ubiquitous real-time information (RTI) sources offering a greater chance to minimise the resultant disturbance. Moreover, repeated experience of recurrent service disruptions may induce long-term shifts in typical trip patterns, including the risk of permanent modal shifts from PT systems. Understanding passengers' attitudes and responses to sudden PT disruptions is thus of paramount importance for defining such information and management strategies that will ultimately improve travel experience in uncertain PT service conditions.

1.1. Literature Review

Behavioural responses to both private and public transport travel disruptions are commonly acknowledged in research sources [1–3]. These comprise: adjusting the usual travel routine (i.e., route shifts or temporal shifts), switching to alternative transport modes, satisfying trip purposes at other destinations, consolidating trips (e.g., trip-chaining routines), changing the trip frequency, and cancelling the trip activity. The most common reactions comprise spatial and temporal shifts in existing trip patterns (i.e., change of travel route and travel time) [4], with remaining types of responses typically much smaller in magnitude, and depending on each individual case. As noted by monitoring studies [5], the majority of PT users regularly experience travel service disruptions and develop counteracting strategies. A summary of six city-wide surveys [2] reveals that ca. 10% of travellers encounter disruptions each week as part of their usual travel routine; ca. 50% are affected at least once a month. Experience of travel disruptions induces both short-term and long-term shifts in travel behaviour, which can be difficult to distinguish in practical observations [1] but might be substantially different from each other. Initial reactions tend to be exaggerated, especially in the event of sudden and unplanned service disruptions, after which the scale of trip pattern shifts decreases—a so-called settle-down effect [6]. Meanwhile, accumulated travel experience might facilitate other, permanent (and perhaps initially unforeseen) changes in travel behaviour.

Evidence summarised from multiple case studies [1] reveals the causes of various sudden, unplanned disruption causes and their impact on travel choices. Following a 1966 New York City public transport strike, 50% of travellers cancelled their trips on the first day—a figure which went down to 10% in following weeks, but in the longer run, the PT patronage only decreased by ca. 3%. However, for the 1986 Orange County strike, this figure was almost equal to 20%. After the 1994 earthquake in the Los Angeles area—another type of sudden disruption—31% of commuters changed their travel route and 30% of them adjusted their travel time. In comparison, other reactions were observed in much smaller magnitude, such as only a 6% rate of mode shift. Notably, the long-term propensity to change travel modes can be inhibited by the tolerable margin of increased travel time due to disruption—even as much as 10–20 min of extra delay before eventually switching from private to public transport. The scale of trip destination and trip cancelation changes is usually found to be lower, especially for more resilient commuting and business trips, affecting no more than 2–6% of total trips (excluding exceptional circumstances—explained below). For non-work trips, these rates are more variable, depending on trip motivation (e.g., shopping, leisure, health) and may reach up to 10–30% [2].

These behavioural responses induced by travel disruptions differ in type and magnitude depending on a wide range of factors, related to trip characteristics, travel experience, individual (socio-demographic) characteristics, local conditions, etc. In principle, changes in travel behaviour occur once a stimulus caused by service disruption exceeds a sufficient threshold—i.e., a certain tipping point of travel behaviour stability is achieved [7]. This stimulus can relate either to an isolated, high-impact disruption, or a series of low-impact disruptions which becomes frequent enough to spur changes in individual's travel patterns [8]. Travel shifts are more likely to occur with increased trip length [2] or time [9], as travellers are initially reluctant to rearrange their shorter-range trips. Among factors shaping individual's strategies, past travel experience is of seemingly high relevance [1], as it influences the traveller's perceptions and the resultant choice context. In this context, it should be noted that attitudes towards public transport are biased by negative travel experience and past worst-case situations [10]. Consequently, prior experience of service reliability in PT is relatively more important than in the case of private transport.

The importance of travel habits plays out especially in conjunction with another vital aspect, namely, familiarity with transport network [1]. Travellers who feel more familiar with their environment tend to remain more stable in their travel choices, or as formulated by [11]—reluctant to 'upset an ordered and well-understood routine'. Additionally, different underlying causes of travel disruptions lead to substantially variable adjustments in trip patterns, e.g., when comparing between the impact of natural catastrophes, mass events and technical failures [1]. Data provided by [2] indicates a cancelation rate for business and commuting trips of 6% for flooding-induced disruptions, but a rate of ca. 40% due to adverse winter weather conditions. These responses of PT users vary depending on the PT mode which they use: rail passengers tend to be more 'forgiving' towards service disruptions than bus passengers [12].

Finally, access to travel information is an important factor shaping passengers' decisions —especially in the event of sudden, unplanned service disruptions. Accurate, timely and relevant information allows travellers to tackle the feeling of being somewhat disconnected in first place and make more informed choices. UK surveys of bus passengers [5] outline the contents of information which are desirable during travel disruptions: cause (reason) of the on-going disruption, estimated impact on journey time, an idea of the scale of the problem, and possible travel alternatives. In terms of dissemination sources, a wide range of available sources are utilised to consult travel information. According to passenger surveys, the most popular source of information is electronic RTI displays at stops [5,13]—since they provide information relevant for current choice context, and their recurrently-updating nature is in itself reassuring [14]. Traditional sources of information (printed schedules), as well as interaction with staff (or the driver), with other travellers, or even relatives are also popular among travellers. Other modern-day sources, such as mobile apps and on-line planners are also increasingly widespread, though they are sometimes biased by unsatisfactory experience of past reliability and incur higher cognitive costs [9].

RTI provision is of vital importance; even though it does not always improve journey experience, it does reinforce the travellers' confidence (reassurance) in unusual circumstances [15]. Even under regular service conditions, passengers can experience certain frustration [16] associated with travel uncertainty, which will be even more important during disruptions. The consulted information may not therefore induce travel shifts in case of minor and short-lived service disruptions, as the propensity to change travel choices will only emerge beyond a certain disruption delay threshold. A survey in London [9] found that the majority of PT travellers would change their travel plans in the case of a 20-min disruption delay. Acquisition of RTI during disruptions was found to be influenced by issues such as prior travel experience, travel choice context, past experience of RTI reliability, and trip purpose [9,15]. It is also observable that on-time travel information is especially crucial in the recovery phase of disruption, once travellers seek for preventive measures to mitigate the disruption effects and recover their journey [8]. In general, RTI access induces different decision-making strategies among PT users [17] and increases the responsiveness in their choices to actual travel conditions. This has been demonstrated in simulation models [18,19] which observe additional passengers' travel benefits with access to RTI (as compared to "non-RTI" passengers), especially once disruptions occur. Experience of relevant and purposeful travel information thus contributes to an improved perception of the RTI system, and higher compliance with RTI content in future travel decisions [13].

1.2. Objectives and Contribution

Despite substantial advancements in recent works, certain gaps have not yet been thoroughly addressed in state-of-the-art research. We argue that further investigation is required to achieve a complete understanding of travel behaviour shifts and travel information usage during sudden PT service disruptions. This pertains to, among others, influence of travel experience upon ensuing strategies, consistency between passengers' attitudes and responses (i.e., stated vs. actual choices), as well as potentially different shifts in instantaneous vs. long-term travel patterns. Another aspect not yet fully explored concerns the utility of available travel information sources, either for recognizing the PT disruption itself and/or planning the onward PT journey afterwards. A further research question relates to whether passengers' decision-making processes in the face of sudden PT disruptions can be supported with an improved RTI provision policy.

With this study, we aim to address these gaps and contribute with comparable and enriching results to the state-of-the-art research. To this end, we conduct a series of stated- and revealed-preference surveys (SP/RP) among urban PT users in the city of Krakow (Poland). Its objective is to inspect the impact of sudden service disruptions—defined as short, unexpected service suspensions due to service failures, vehicle incidents, track obstructions, etc.—on passengers' ensuing travel strategies. We investigate the passengers' past (historical) experience of service disruptions, their output travel choices and shift patterns in travel choices with respect to relevant factors—such as: prior travel experience, trip purpose (i.e., time-criticality) and general socio-demographic characteristics. Moreover, we analyse the utilization of various travel information sources and their relevance for ensuing travel decisions, as well as preferences towards RTI content in the event of sudden service disruptions.

Our objective is to provide a valuable contribution to the developing stream of research on travel behaviour in public transport systems. The results of this study provide an evidence-based insight into stated and reported travel decisions due to sudden PT service disruptions. In particular, we observe the influence of passengers' recency of 'travel memory' and highlight potential discrepancies between their stated attitudes vs. actual behaviour. Moreover, our research investigation aims to underline how the RTI provision policy can be enhanced to mitigate the negative impacts of PT disruptions, by providing information that could meet passengers' expectations, and thus improve PT travel experience in such circumstances.

The remainder of this paper is organized as follows. Section 2 presents the methodology of our research investigation, including passenger survey design and data collection. Section 3 reports detailed results from the combined SP/RP passenger survey. We conclude with Section 4, which summarises the main findings of our study, discusses their research and practical implications and indicates fields for follow-up research.

2. Materials and Methods

2.1. Case Study—Public Transport (PT) System in Krakow

Surveys were conducted in the city of Krakow, the second-largest Polish city with ca. 750,000 inhabitants. The city PT system consists primarily of bus and tram services with ca. 130 permanent bus and tram routes operating on a typical weekday. As of March 2017, the RTI system was in operation in the form of electronic display signs (i.e., platform-level RTI) mounted at ca. 99% of tram stops in the city, providing passengers with information on estimated waiting times until the next tram departures. At the time, the RTI system in Krakow city was in its early development stages, and so far, its coverage has been extended only to the tram network; bus services are not yet included. Nevertheless, electronic RTI displays at tram stops are a popular source of real-time travel information for city PT users, and apart from remaining waiting times they also display notifications in case of both sudden service disruptions, as well as planned (long-term) line suspensions. However, the content of the displayed information is usually confined to a mere notification (report) of the on-going disruption, its location, and current diversion routes of affected tram services, without any time estimations or advisory travel information.

2.2. Data on Registered PT Service Disruptions

Although the city PT system has undergone substantial improvements in recent years, it still remains vulnerable to sudden service disruptions. As part of this study, we analysed data records obtained from the city transport management authority (ZIKiT Kraków), which covered all the notified service suspensions in the core part of the PT system in Krakow city (i.e., coverage of the full tram network and the majority of bus networks). The datasets were collected over a 6-month period between August 2015 and January 2016. It should be noted that such a time horizon was chosen on the grounds of data availability and representability. (Soon afterwards, a series of major investment works in the Krakow tram network has started, which significantly influenced the 2016–2017 data on PT disruptions' and thus undermined the overall data representativeness.) The obtained sample was carefully inspected to ensure that it yields a valid and characteristic picture of typical PT disruption events in the case-study PT system:

- A total of 577 unplanned service suspension events were registered for the 6-month analysis period, which gives a monthly average of 97 suspension events.
- The vast majority of service suspension events were short-term in their duration (Figure 1): 52% of suspensions took no longer than 15 min, and just 22% of them were longer than 30 min. Note that this only includes the time duration of suspension itself, excluding the subsequent service recovery period.
- No significant differences were observed in terms of temporal distribution. Disruption events tend to occur at fairly similar frequencies both throughout the daytime (with 78% of them taking place between 6 a.m. and 6 p.m.) and across the consecutive months (82 to 108 events registered per month).

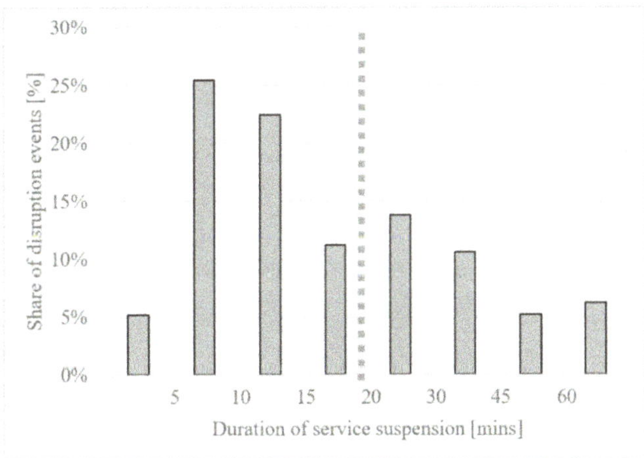

Figure 1. Share of registered service disruption events, depending on the suspension duration, in the PT system in Krakow (source: ZIKiT Kraków).

Importantly, the above-presented data includes only those disruption events which were registered by the city PT management authorities. It does not account for a vast number of short-term disruptions which quite often go unnotified, yet often cause enough disturbance to the passengers. The exact scale of these unregistered (and often recurrent) disruptions events remains unknown; as anecdotal evidence shows, in many cases it can be at least as high as the number of officially registered events.

2.3. *SP/RP Survey Setup and Design*

To investigate the impact of recurrent PT service disruptions on passengers' travel behaviour in the city of Krakow, a combined stated- and revealed-preference (SP/RP) survey was designed. The survey questionnaire consists of 15 questions in total, and its average completion time ranged between 3 to 4 min. It was carried out on electronic portable devices at (randomly selected) main bus and tram stops in the central area of Krakow city. Final passenger surveys took place over a 3-month period. i.e., April–June 2017 and additionally in September 2017. A total sample of ca. 450 fully completed questionnaires was collected, which was then further analysed, and the main findings are reported in the sections below. Before the main passenger survey was designed, focus-group discussions and pilot surveys were carried out. Their purpose was to avoid confirmation bias [20], observe the passengers' understandability of general survey purpose, gather their ideas and observations, and reformulate the questionnaire content. The final survey questionnaire can be subdivided into a few general sections as follows:

- The first (RP-based) part of the questionnaire focuses on their historical experience of travel disruptions, trip circumstances, information sources and reported travel choices in the event of the most recent disruption. In the case when a passenger did not experience any disruption at all, this survey part is omitted.
- In the second (SP-based) part, passengers are asked to state their maximum (tolerable) time thresholds and ensuing travel decisions for two hypothetical situations, i.e., service suspension for two distinct trip purposes: a time-critical journey and a non-time critical journey (i.e., depending on the propensity to arrive on-time).
- The final part contains questions on the judgement of the existing RTI system quality in the city of Krakow and desirable future RTI content during service suspensions.

Survey data collection was carried out with the prior consent of each respondent, ensuring their full anonymity and privacy protection. No personal data was stored thereafter, and all ethical standards were duly observed during this investigation.

3. Results

In the following section, we summarise the results of our survey investigation and elaborate on findings from each part of the survey.

3.1. General Experience of PT Disruptions

Overall results show that majority of respondents do indeed experience sudden, unplanned disruptions in their PT trips in Krakow city. In total, 68% of the interviewed passengers recall having encountered sudden disruptions, and 23% report a frequent experience of service disruptions (i.e., once a week or more often). Interestingly, a closer inspection reveals a valid discrepancy in obtained answers, depending on the data collection period (Table 1). For surveys carried out in May and June, about 9% of travellers recall not having experienced the PT disruption at all; for September surveys, this rate surges up to 56%; meanwhile, the shares of those who encounter frequent disruptions are 39% and 7% respectively. This indicates that, although the question objective refers to the travellers' past, long-term record travel experience, their travel memory seems to be in particular influenced by the most recent events and travel incidents. Consequently, travellers interviewed in September exhibit much lower service disruptions right after summer holiday period, compared to those questioned in the midst of typical commuting season. It should be noted that share of PT commuters is constant in both samples, comprising over 90% of all respondents (Table 1). In addition to that, city PT system returns to its typical service pattern in September, and (as reported earlier from the city registry data) the incidence rate of PT disruptions is rather constant over consecutive months.

Table 1. Frequency of encountered PT service disruptions. Results distinguished depending on survey period.

	Total Sample	Survey Period	
		May/June '17	September '17
More than once a week	10%	21%	2%
Ca. once a week	13%	18%	5%
Less than once a week	45%	53%	37%
Never experienced yet	**32%**	**9%**	**56%**
Share of regular PT users (i.e., 2 + PT trips per week)	92%	91%	91%

3.2. Revealed Impact of Past PT Disruptions

We investigated travel choices revealed (reported) by travellers for their most recent experience of journey disruption (Table 2). The most typical shifts in trip patterns once PT disruptions had been encountered either involved changing the initial PT trip route (i.e., using an alternative bus or tram connection—39%), waiting until the PT service has arrived (29%), and completing the rest of trip by walking to the destination (27%). Otherwise, remaining options such as mode shifts (4%) and trip cancelations (2%) are much less popular. The Chi-Square tests yield statistical significance of reported travel choices with a number of travellers' sociodemographic factors. Notably, this pertains to the propensity among elderly PT users (aged 60 and over) to remain and wait at their stop (63%) in the event of disruption; they were much less likely to consider other re-routing options. We also observe a certain increase in modal shift probability towards walking among students (32%) and respondents aged 18–35 (31%), as well as towards private transport (i.e., car, taxi, ride-sharing) among those employed (7%) and aged 26–45 (7%). Moreover, choice patterns are seemingly correlated with the stated frequency of past disruption experience. Frequent exposure to travel disruptions leads to a smaller likelihood of changing current trip itinerary, with 45% of travellers likely to wait further at the stop. Regardless of that, hardly anyone considers shifting to private transport.

Table 2. Revealed travel choice response during the most recent PT service disruption.

	Total Sample	Frequency of Experienced PT Disruptions		
		More Often	Ca. Once a Week	Less Often
Wait at the stop	29%	45%	29%	26%
Use an alternative PT route	39%	26%	39%	41%
Walk down to destination	27%	25%	20%	29%
Shift to private transport	4%	1%	8%	4%
Resign from travelling	2%	2%	3%	1%

Chi-square results: $x^2 = 21.74$, $p = 0.041$.

3.3. Long-Term Travel Behaviour Shifts Due to PT Disruptions

In the next stage of our survey, we ask travellers to report the long-term changes they made in trip patterns as a consequence of PT disruptions' experience, using a 1-to-5 Likert scale (where: 1—never, 5—always). The results presented in Figure 2 indicate that 77% of travellers acknowledge having made such long-term travel routine adjustments. Similar to previous questions, shifting to private transport seems to be the least likely option: 17% did so at least sometimes and only 4% did so on a regular basis (i.e., often or always). The remaining alternatives—changes in current PT route (different origin departure time, different alighting stop) or choosing an alternative PT route—exhibit much higher popularity, with respective figures ranging around 40–50% (at least sometimes) and 15–20% (on regular basis). Statistical tests reveal that travellers who frequently experience disruptions are more inclined towards changing their alighting stop (43%—at least sometimes) and especially towards adjusting their departure time from origin (52%—at least sometimes, 30%—on regular basis). Concurrently, shifting to an alternative PT route on a long-term basis is a popular option regardless of disruptions' experience frequency.

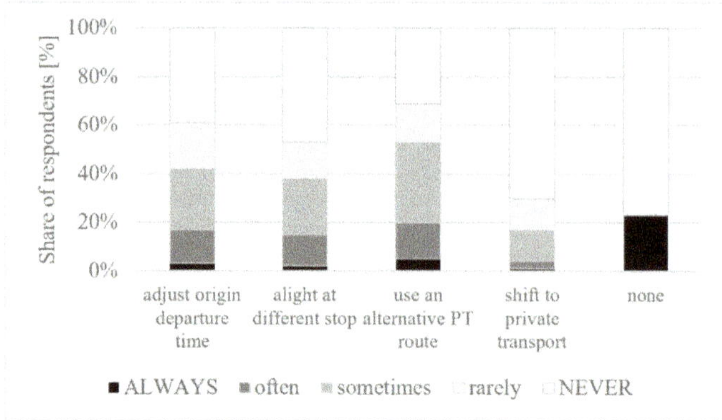

Figure 2. Reported long-term travel behaviour changes due to PT disruptions.

3.4. Stated Travel Choices during PT Disruptions

We then move to the next stage of our survey, where we ask travellers to report their preferred (stated) travel action in two hypothetical cases, assuming they have an exact information on remaining time of disruption duration. We distinguish here two scenarios depending on propensity to arrive on time: ***time-critical trips***—e.g., to work or study place (need to arrive on-time), vs. ***non-time-critical trips***—e.g., shopping or leisure purposes (need not be in a hurry). Respondents are asked to provide maximum acceptable waiting times and ensuing travel actions beyond this time limit for their typical PT trip context (to the workplace, shopping mall, etc.). As expected, resultant time tolerance thresholds are much lower for time-critical trips, with the vast majority of respondents (80%) reluctant to wait longer than 10 min (Figure 3). In contrast, for non-time-critical trips, results show that for a 10-min disruption just 35% of travellers would already consider taking alternative travel action.

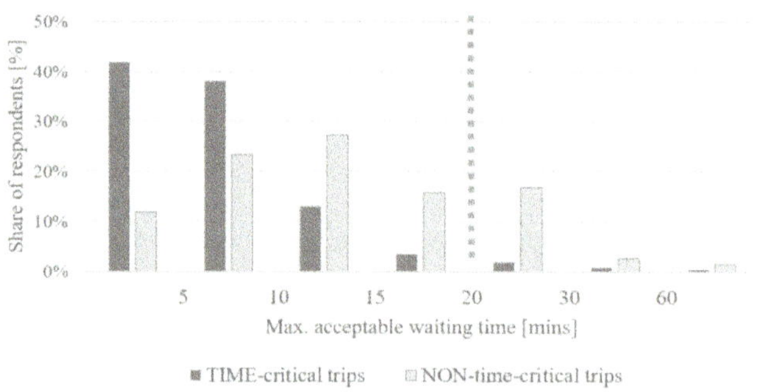

Figure 3. Stated max. acceptable waiting time before considering alternative travel action. Results are distinguished with respect to the propensity to arrive on-time (i.e., trip time-criticality).

In terms of ensuing travel action afterwards, choosing an alternative PT route is the most favourable option, with a rather uniform popularity rate in both scenarios (57–58%). Variability in results is exposed for other travel options, as shown in Table 3—with higher probability of potentially shifting to private transport for time-critical trips, walking down to destination for non-time-critical trips, and additionally—a potential 7% trip cancelation rate in the latter case. It should be emphasised though that these figures refer to passengers' stated-preference (SP) travel actions for hypothetical disruption scenarios. Validating these results against the revealed-preference (RP) choices for previous (past) disruption cases, reported also in Table 3, underlines an interesting discrepancy arising between SP vs. RP results. The most popular travel behaviour pattern pertains to using an alternative PT route, regardless of the time-criticality of trip. Its share is revealed to be similar to that of passengers' stated preferences (54% in RP choices vs. 58% in SP choices). However, important differences are traceable for remaining travel options: in contrast to stated results, walking is revealed to be much more popular (37% in RP choices). Likewise, the actual tendency to use private transport (i.e., car, taxi, ride-sharing) is over-exaggerated (6% in SP choices), as is the trip cancelation rate (ca. 2% in SP choices). These findings thus underpin a certain bias in passengers' answers to hypothetical stated-choice scenarios when contrasted with the actual choices that they reported for recent disruption.

Table 3. SP vs. RP travel choices due to PT disruption. Comparison between stated (potential scenario) vs. revealed (past experience) choices.

	Stated Choices Potential Trips		**Revealed Choices Past Trips**			
Need to arrive on-time?	yes	no	yes	no	yes	no
Wait at the stop	(n/a)		(excl. waiting):		25%	35%
Use an alternative PT route	58%	57%	54%	54%	40%	35%
Walk down to destination	22%	34%	38%	36%	29%	23%
Shift to private transport	19%	2%	5%	7%	3%	5%
Resign from travelling	~0%	7%	3%	1%	2%	1%

3.5. *Travel Information during PT Disruptions*

In the final part of our survey, we investigate impact of travel information sources and content during sudden PT disruptions. Firstly, we analyse which information sources are helpful in recognising past PT service disruptions. The results in Table 4 show that the highest (34%) share of respondents admit to having recognised the on-going disruption by themselves (i.e., without the need to consult other information sources). This is followed by 26% of travellers who became aware of the disruption by observing electronic RTI displays at stops. Traditional information sources are also found helpful in recognising service disruptions, i.e., driver or PT staff announcement (20%), and reactions of other passengers (13%). Moreover, a relevant correlation with the experienced disruption frequency (Table 4) is observable. Those who are used to frequent PT disruptions (more than once a week) are more likely to recognise them by themselves (52%), and less reliant on other sources such as RTI displays or passengers' reactions (i.e., collective behaviour).

Table 4. Revealed recognition source of the most recent PT service disruption.

	Total Sample	**Frequency of Experienced PT Disruptions**		
		More Often	**Ca. Once a Week**	**Less Often**
Driver or PT staff	20%	15%	12%	24%
Other passengers	13%	4%	17%	13%
Internet, social media	7%	9%	10%	6%
RTI displays at stops	26%	20%	20%	30%
Just notice them myself	34%	52%	41%	27%

Chi-square results: $x^2 = 20.25$, $p = 0.027$.

However, it is the dynamic and RTI devices which are the most frequently consulted travel information sources once service disruptions occur (Figure 4). Initially of little use in recognising of PT disruptions, smartphone apps (and on-line resources) become a popular means for the ensuing trip planning purposes—having been consulted at least sometimes by 75% of travellers, and by ca. 50% of them on regular basis (i.e., often or always). Similar figures are reported for electronic RTI displays at stops, which are the second most popular travel information source. On the other hand, social media alerts and feeds (e.g., local PT Facebook page) were yet utilised by a limited share of travellers (just 15% did so regularly).

This can be attributed to low awareness of social media travel alerts at the time of this research.

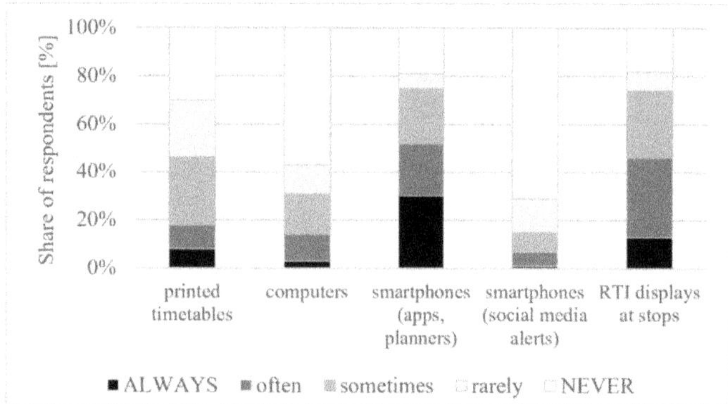

Figure 4. Consulted travel information sources during the PT service disruptions.

Finally, we turn our attention to the RTI system in the city of Krakow and expectations towards the future information provision policy during disruptions. In general, when asked to provide an assessment on a 1-to-5 Likert scale (1—poor, 5—excellent), current RTI reliability is reported on average equal to 3.9, and the RTI usefulness—equal to 4.5. Such results indicate a potential field for improving the perceived RTI accuracy and users' preference to obtain a more reliable (and extended) RTI content, especially during variable and disrupted travel conditions. This can be complemented by our next findings, i.e., future desirable RTI content during PT disruptions (Figure 5). Travellers are in particular interested in obtaining an accurate estimate of remaining time until normal conditions are resumed. Other information options, including real-time advice on alternative travel routes and cause of the on-going disruption, are also positively viewed, albeit evidently less important for survey respondents.

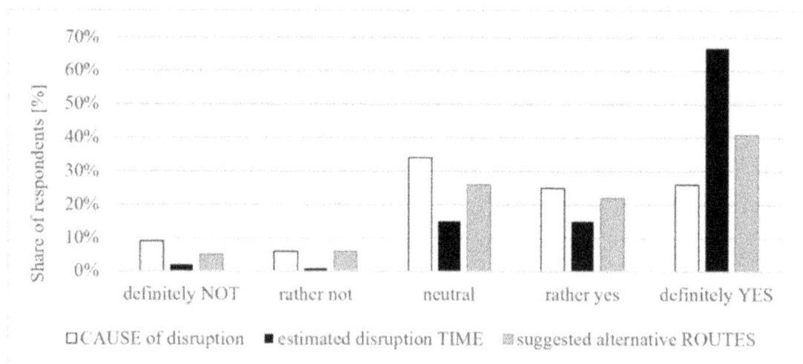

Figure 5. Preference towards future RTI content in the event of disruptions.

4. Discussion

The study presents results from combined stated- and revealed-preference (SP/RP) survey conducted among urban public transport (PT) users in the city of Krakow (Poland). The main objective was to investigate the impact of sudden PT disruptions upon resultant changes in passengers' travel behaviour. The secondary objective was to establish travel information sources and content relevant during sudden service disruptions.

The principal contribution of this study lies within an improved, evidence-based understanding of how passengers' travel behaviour is affected by experience of sudden PT service disruptions. Detailed results of our investigation (described in the previous section) shed more light on what adjustments in travel decisions may occur either in case of a single, sudden PT disruption, as well as a long-term consequence of repeated PT disruption experience. We examine the passengers' stated attitudes and waiting time tolerance and indicate how they match the actual (real-world) travel behaviour and duration of PT suspension events. As a secondary contribution, we investigate the popularity of travel information sources during the PT disruptions. Finally, as shown in the subsection below, we utilize our findings to indicate what travel information content should be particularly provided in real time to passengers in the event of such PT service disruptions.

The main findings of our study can be summarised as follows:

- Majority of surveyed PT users (68%) recall having experienced sudden, unplanned service disruptions in their PT trips and 23% of them report such experience on a frequent basis (at least once a week). Importantly, we observe that recency of experienced PT disruptions has significant influence upon passengers' travel memory. Only 9% of passengers interviewed in the midst of a typical commuting season (May–June), recall not having experienced any disruption before. However, the same survey conducted just after holiday break (September) yields a corresponding rate of 56% respondents. This is all the more remarkable, given that registered PT disruptions exhibit similar characteristics (frequency and duration) across consecutive months, and both samples are controlled against the steady share of regular PT users.
- Travel choices revealed by passengers during the latest (most recent) PT disruption primarily involve using an alternative PT (bus or tram) route (39% of travellers), followed by waiting at the stop (29%), walking towards the destination (27%), shifting to private transport (4%) and resigning from travelling (2%). Passengers more accustomed to frequent experience of PT disruptions are less likely to change their current PT travel routine and instead tend to wait further at the current stop.
- Furthermore, 77% of respondents admit to having made long-term adjustments in their travel behaviour as a consequence of recurrent disruption experience. These mostly involve using an alternative PT route or adjusting current PT trip itinerary, by changing the origin departure time or an alighting stop (40–50% of travellers). In contrast, increased frequency of car usage is reported only by ca. 20% of respondents.
- In the hypothetical (SP) disruption scenarios, trip time-criticality stands out as a major factor, influencing the stated choices and maximum acceptable wait time at the current PT stop before taking further action. For time-critical trips (e.g., work, study), only 7% of travellers would accept a max. waiting time longer than 15 min—whereas for non-time-critical trips (e.g., shopping, leisure) this rate increases to 37%.
- Our findings also expose relevant differences between passengers' preferred vs. actual choices during PT disruptions. Stated preferences (SP) vary with time-criticality, with higher propensity to use private transport for time-critical trips (19%), and to resign from travelling for non-time-critical trips (7%). Corresponding rates in the revealed preferences (RP), meanwhile, are equal to ca. 6% and 2% respectively, suggesting that SP answers overestimate the probability of modal shifts and trip cancellations. In contrast, RP answers exhibit uniform patterns regardless of trip time-criticality. Revealed travel choices primarily involve taking an alternative PT route (55%) or walking to a destination (38%—remarkably, also for time-critical trips).
- The main information sources which help travellers to recognise the emerging PT service disruptions are travellers' own observations (34%), and electronic RTI displays at stops (20%), followed by PT staff announcements or observations of other passengers. At the time of this research, on-line and internet resources were the least popular means of recognising the PT disruption. However, smartphone apps and RTI displays are the principal (most consulted) travel information sources for planning an onward journey once the PT disruption takes place, even despite limited information content utility.

The aim of our study is to deliver valuable insights that can enrich the state-of-the-art research on travel behaviour in modern-day PT networks. The findings of this study can help better understand the attitudes, factors and outcomes in passengers' decision-making process in uncertain, disrupted PT travel conditions. In addition to research implications, the outputs of our investigation can be of practical relevance for PT policymakers, planners and operators. As such, they can support the design and implementation of new analytical models, travel information and management strategies, aimed at mitigating the negative effects of sudden PT service disruptions.

Implications and Recommendations for RTI Policy in Krakow

Conclusions from our study indicate a few possible means of improving the current RTI provision policy in Krakow during sudden service disruptions. Firstly, in terms of desirable future RTI content, there is a clear passenger preference towards obtaining an accurate time estimation of disruption duration. Such a feature is found to be rated as most important (useful) among other travel advice options. This can be further supported by the fact that, based on PT operator data, over 50% of recorded disruptions tend to last no longer than 10–15 min. Once compared against the SP results, this is a time threshold that can match the stated waiting time tolerance, especially for the vast majority of non-time-critical trips (Figure 6). Providing a reliable disruption time estimation could therefore reinforce passengers' tendency to stay on their current PT route, particularly for the vast majority of trips which are not time critical (i.e., leisure trips, trips with arrival time flexibility). This will also help them make more informed choices and substantially decrease the travel disutility associated with service uncertainty and unreliability, improving the overall perception of PT service quality.

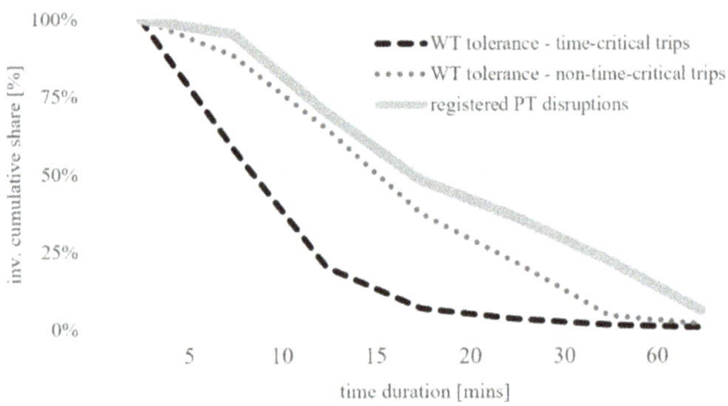

Figure 6. Results—revealed distribution of waiting time tolerance in the event of PT disruptions (acc. to surveyed passengers), plotted against the distribution of actual duration of PT suspension events (acc. to the PT operator's registry).

Secondly, another possible development relates to providing personalised RTI travel advice during disruptions, which would better match the individual travel preferences. Availability of the prescriptive O-D travel advice has prospective benefits for the majority of PT users, who exhibit constant travel (commuting) patterns even despite recurrent experience of service disruptions (Table 1). In this respect, customized, real-time O-D travel advice could decrease the uncertainty and cognitive effort associated with the urgent, unexpected trip re-planning process [9]. It should be noted that the awareness of RTI travel alerts and feeds was still relatively low among passengers (Figure 4) at the time of this research. Although RTI sources are already popular for en-route trip planning purposes—i.e., once the PT trip has already been disrupted—they are of more limited usefulness either in making pre-trip arrangements or recognising the on-going disruptions beforehand.

Thirdly, and as a final remark, passengers in general express a positive perception of present-state electronic RTI system in Krakow. RTI content is rated higher in terms of its usefulness than its reliability, which indicates a certain field for improving the RTI accuracy—especially crucial in terms of PT service irregularities and disruptions.

A certain limitation of this study can be attributed to the relative uniformity of the obtained sample: over 90% of respondents are regular PT users (commuters). Although surveys were conducted across multiple bus and tram stops in Krakow, on different weekdays and throughout the daytime, future surveys should take place also outside the central PT network and cover regional (suburban) trips as well. Additionally, follow-up surveys should be conducted in different urban and metropolitan areas to verify the potential influence of a wider range of aspects upon passengers' response to sudden PT disruptions such as transport network topology, demand levels, tariff policy, local specifications, etc. Further research will thus allow us to obtain more universal and transferable conclusions.

Author Contributions: Conceptualization, A.A.D. and M.F.I.; methodology, A.A.D. and M.F.I.; software, M.F.I. and A.S.; validation, A.A.D., M.F.I. and A.S.; formal analysis, A.A.D. and M.F.I.; investigation, A.A.D. and A.S.; resources, A.A.D.; data curation, A.A.D. and M.F.I.; writing—original draft preparation, A.A.D.; writing—review and editing, A.A.D., M.F.I. and A.S.; visualization, A.A.D.; supervision, A.S.; project administration, A.A.D.; funding acquisition, A.A.D. and A.S. All authors have read and agreed to the published version of the manuscript.

Funding: This research received no external funding.

Institutional Review Board Statement: This study did not involve humans or animals.

Informed Consent Statement: Not applicable.

Data Availability Statement: Not applicable.

Acknowledgments: Authors would like to express their sincere gratitude to Achille Fonzone from Edinburgh Napier University, whose support and advice contributed towards this work.

Conflicts of Interest: The authors declare no conflict of interest.

References

1. Zhu, S.; Levinson, D.M. Disruptions to transportation networks: A review. In *Network Reliability in Practice*; Levinson, D., Liu, H., Bell, M., Eds.; Springer: New York, NY, USA, 2012; pp. 5–20. [CrossRef]
2. Marsden, G.; Anable, J.; Shires, J.; Doherty, I. *Travel Behaviour Response to Major Transport System Disruptions: Implications for Smarter Resilience Planning*; OECD Discussion Paper: Paris, France, 2016; p. 9.
3. Ziółkowski, R.; Dziejma, Z. Investigations of the Dynamic Travel Time Information Impact on Drivers' Route Choice in an Urban Area—A Case Study Based on the City of Bialystok. *Energies* **2021**, *14*, 1645. [CrossRef]
4. Cairns, S.; Atkins, S.; Goodwin, P. Disappearing traffic? The story so far. *Proc. Inst. Civ. Eng. Munic. Eng.* **2003**, *151*, 13–22. [CrossRef]
5. Passenger Focus UK. *Bus Passengers' Experience of Delays and Disruption*; Research Report; Passenger Focus UK: London, UK, 2013.
6. Clegg, R. Empirical studies on road traffic responses to capacity reduction. In Proceedings of the International Symposium on Transportation and Traffic Theory (ISTTT17), London, UK, 13–15 July 2007.
7. Goodwin, P.B. *Enhancing the Effectiveness of Transport Policy by Better Understanding of Travel Choices*; Centre for Transport and Society, UWE Bristol: Bristol, UK, 2009.
8. Papangelis, K.; Corsar, D.; Sripada, S.; Beecroft, M.; Nelson, J.D.; Edwards, P.; Velaga, N.; Anable, J. Examining the effects of disruption on travel behaviour in rural areas. In Proceedings of the 13th World Conference on Transportation Research (WCTR), Rio de Janeiro, Brazil, 15–18 July 2013.
9. Maréchal, S. Modelling the acquisition of travel information and its influence on travel behaviour. In Proceedings of the 48th Universities' Transport Study Group (UTSG) Conference, Bristol, UK, 6–8 January 2016.
10. Guiver, J. Modal talk: Discourse analysis of how people talk about bus and car travel. *Transp. Res. Part A Policy Pract.* **2007**, *41*, 233–248. [CrossRef]
11. Goodwin, P. Habit and Hysteresis in Mode Choice. *Urban Stud.* **1977**, *14*, 95–98. [CrossRef]
12. Shires, J.D.; Cabral, M.; Marsden, G.; Wardman, M. The Impact of Disruption on Rail Demand. In Proceedings of the 44th European Transport Conference (ETC), Barcelona, Spain, 5–7 October 2016.
13. Islam, M.F.; Fonzone, A.; MacIver, A.; Dickinson, K. Modelling factors affecting the use of ubiquitous real-time bus passenger information. In Proceedings of the 5th IEEE International Conference on Models and Technologies in Intelligent Transport Systems, MT-ITS 2017, Naples, Italy, 26–28 June 2017. [CrossRef]
14. Islam, M.F.; Fonzone, A.; MacIver, A.; Dickinson, A. Use of ubiquitous real-time bus passenger information. *IET Intell. Transp. Syst.* **2019**, *14*, 139–147. [CrossRef]

15. Kattan, L.; de Barros, A.G.; Saleemi, H. Travel behavior changes and responses to advanced traveler information in prolonged and large-scale network disruptions: A case study of west LRT line construction in the city of Calgary. *Transp. Res. Part F Traffic Psychol. Behav.* **2013**, *21*, 90–102. [CrossRef]
16. Caulfield, B.; O'Mahony, M. A stated preference analysis of real-time public transit stop information. *J. Public Transp.* **2009**, *12*, 1–20. [CrossRef]
17. Fonzone, A. What do you do with your app? study of bus rider decision making with real-time passenger information. *Transp. Res. Rec.* **2015**, *2535*, 15–24. [CrossRef]
18. Cats, O.; Jenelius, E. Dynamic vulnerability analysis of public transport networks: Mitigation effects of real-time information. *Netw. Spat. Econ.* **2014**, *14*, 435–463. [CrossRef]
19. Fonzone, A.; Schmöcker, J.D. Effects of transit real-time information usage strategies. *Transp. Res. Rec.* **2014**, *2417*, 121–129. [CrossRef]
20. Evans, J.S. *Bias in Human Reasoning: Causes and Consequences*; Lawrence Erlbaum Associates, Inc.: Mahwah, NJ, USA, 1989.

energies

MDPI

Article

Effect of Traffic Calming in a Downtown District of Szczecin, Poland

Alicja Sołowczuk

Road and Bridge Department, Faculty of Civil and Environmental Engineering, West Pomeranian University of Technology in Szczecin, 71-311 Szczecin, Poland; alicja.solowczuk@zut.edu.pl; Tel.: +48-91-449-40-36

Abstract: The increasing use of road vehicles has caused a number of transport and environmental issues throughout the world. To cope with them, traffic calming schemes are being increasingly implemented in built-up areas. An example of such schemes are Tempo-30 zones. The traffic calming measures applied as part of this scheme must be carefully planned in terms of location and design details in order to obtain the desired reduction in speed, traffic volume and exhaust emissions and, last but foremost, to increase the safety and facilitate the movement of vulnerable road users. The coexistence and combined effect of these measures and their design details must also be taken into account. The purpose of this study was to investigate whether the applied traffic calming measures had a considerable bearing on the reduction in speed to the desired level, as assumed in the traffic calming plan. Three street sections starting and ending with different intersection types were chosen to examine the synergy of the applied traffic calming measures. The numbers and speeds of vehicles were measured in three day-long continuous surveys. As it was expected, the amount of speed reduction depended on the hourly traffic volume on a one-way street and various other traffic engineering aspects. The obtained results may be used to modify the existing speed profile models and can guide traffic engineers in choosing the most effective traffic calming measures.

Keywords: calming of traffic; Tempo-30 zone; speed reduction; traffic calming measures; choker; speed tables; raised pedestrian crossing

Citation: Sołowczuk, A. Effect of Traffic Calming in a Downtown District of Szczecin, Poland. *Energies* **2021**, *14*, 5838. https://doi.org/10.3390/en14185838

Academic Editors: Elżbieta Macioszek, Anna Granà, Margarida Coelho, Paulo Fernandes and Massimiliano Gobbi

Received: 26 July 2021
Accepted: 13 September 2021
Published: 15 September 2021

1. Introduction

The increasing use of road vehicles and the resulting high volumes of traffic bring about ever-increasing transport-related problems in urban areas. The first attempts to calm the street traffic in urban areas were made already in the 20th century. The problem of calming the traffic in downtown and uptown locations of cities has been dealt with by many researchers, urban planners, traffic modelling specialists and road engineers. The implementation of Tempo-30 zones in specific streets or whole urban blocks of congested uptown and downtown areas is one of the commonly applied solutions. Increased safety of traffic, volume and speed reduction (resulting in lower exhaust and noise emissions and fuel savings) and making the street more of a true public space rather than purely a transport route are the primary objectives of Tempo-30 zones. On the other hand, a reduction in speed may extend the time of travel and contribute to traffic jams at intersections. Thus, it is extremely important to apply appropriate design guidelines and be aware of the coincident effect of different traffic calming measures. One can look for such guidelines in the basic design manuals [1–3] and guidelines [4,5] published in the U.S. and in the U.K. [6–9] and German [10] guidelines, which describe the proposed applications of: raised intersections, small raised intersections between one-way streets, small raised intersections coupled with a neighbourhood gateway, gateways, traffic circles and the different traffic calming measures typically applied in Tempo-30 zones, i.e., curb extensions (so-called pinchpoints or chokers), curb extensions combined with pedestrian crossings or raised pedestrian crossings at corners of intersections, raised mid-block pedestrian crossings (speed tables) and traffic undulations: speed humps and bumps. This diversity in traffic calming measures is justified by a variety of local conditions, i.e., the traffic control system applied in the street, the route geometry, mid-block section length and different levels of pedestrian and vehicular traffic. The U.S. [4,5], U.K. [6–9] and German [10] guidelines and handbooks [2,3] provide different amounts of the 85th percentile speed reduction across the treatment, depending on the obtained results of studies and analyses. Additionally, the speed reduction ratio, i.e., the speed reduction divided by the initial speed ($w = \Delta v_{85}/{v_{85}}^{before}$), varies between these documents. With different parameters representing the effectiveness of traffic calming measures, designers can decide which of the available traffic calming measures will be the most appropriate for a given street or intersection. This is because this effectiveness depends, to a large extent, on the importance category of the street (2,3,6) and the applicable traffic calming speed reduction category [7,8]. The second parameter, which characterises the effectiveness of a given traffic calming measure, is the reduction in road incidents [4,6–8,11].

Based on the above-mentioned analyses and using the expertise developed during field studies, researchers have developed the principles of modelling traffic calming in traffic-calmed neighbourhoods. The method of modelling the speed profile for streets including traffic calming measures was first proposed in 1966 by Davidson [12] and modified in 1978 also by Davidson [13]. In 1991, Tisato [14] and Akçelik [15] introduced corrections to Davidson's traffic modelling function. In 1995, a further step was made by Barbosa [16], who added a number of new factors characterising the analysed road system to the previously applied ones (i.e., initial speed, amount of speed reduction, geometry of the street, length of mid-block sections and traffic volume). He carried out his research in the city of York on streets where various traffic calming measures had been installed, and the main outcome was a very comprehensive speed profile model. However, in [16], Barbosa concluded that "... the final effect depends on a number of design details applied in a given road system ... ". This conclusion was confirmed in subsequent research projects, and new factors were added, such as information on the applied traffic management system [17], behaviour of drivers waiting in queues before traffic lights [18], parameters of different behaviour patterns [19,20] and parameters of bike-sharing systems [21].

Many researchers have also analysed speed reductions, driving paths, decelerations and accelerations imposed by different traffic calming measures, for example, speed humps and speed bumps (Abdulmawjoud [22] and Baltrėnas [23]), different configurations and sequences of bulb-outs and chicanes (Akgol [24]) or speed tables, chicanes and road narrowings (Distefano [25,26]). The effects of speed humps and speed tables on the reduction in speed, exhaust emissions and fuel consumption were reported by Obregón-Biosca [27] and Lav [28]. Similar results concerning exhaust emissions and fuel consumption on a few Tempo-30-zoned streets were published by Int Panis [29] and Da Silva [30]. Additionally, the relationships between the reduction in pollution and fuel consumption and other factors were investigated and discussed for a wide range of speed changes: taking into account the behaviour of drivers and the resulting traffic jams (De Vlieger [31]), shifting gears on the way through the different traffic calming measures (Beckx [32]), speed control schemes, etc. [33–35]. Directly related to the reduction in speed in traffic-calmed neighbourhoods is the issue of noise reduction [36,37]. The speed reduction aspects directly related to air pollution, fuel consumption and traffic noise for different posted speed limits are regulated by an international code of practice [38].

Summing up the above literature review, we can state that the problems of traffic calming in cities and their outskirts have been extensively analysed and investigated. However, the detailed speed reduction data provided in the guidelines [4–10] for different traffic calming measures installed on selected streets do not cover the situations with a few measures installed at the same location or at a close distance from each other, as the case may be. This problem has not been dealt with in the articles published thus far, and this gap was noted already in 1995 by Barbosa [16].

Based on what has been noted in several articles [16–20], in the first place, we see a need for speed reduction analyses in Tempo-30 zones with a few traffic calming measures acting in combination, for example, curb extensions (further called chokers) combined with raised pedestrian crossings and raised intersections connected by short mid-block sections combined with a neighbourhood gateway. Since, as mentioned earlier, the reduction in pollution, fuel consumption and noise is strictly related to the amount of speed reduction, this article presents the results of a study on three different testing grounds at different traffic levels. Section 2 presents the subject of study and provides a detailed description of the testing grounds, taking into account different aspects of the chosen road system. Section 3 presents the data obtained in the traffic surveys and their analyses. This article ends with the discussion in Section 4 and final conclusions presented in Section 5.

2. Materials and Methods

2.1. Subject of Study and Testing Ground

The subject of study was an urban block located in downtown Szczecin, Poland, between two-way streets, in which Tempo-30 zones are being implemented (Figure 1).

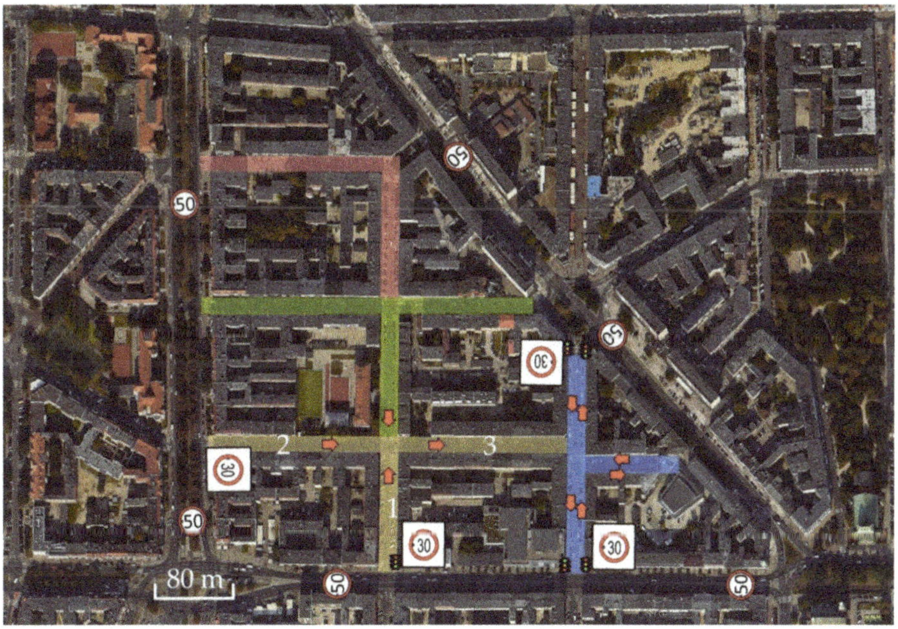

Figure 1. Analysed urban block and testing grounds. Legend: (1) testing ground No. 1; (2) testing ground No. 2; (3) testing ground No. 3. Source: own work of the author using Google Earth satellite imagery [39].

Figure 1 shows testing ground Nos. 1, 2 and 3, reconstructed in 2019 (marked yellow) and the two-way street reconstructed in 2016 (marked blue), i.e., streets with an implemented Tempo-30 zone. The arrows show the direction of traffic. Additionally, shown are the streets where reconstruction is currently taking place (marked green) or has been scheduled (marked red). Before reconstruction, the volume of traffic during peak hours was ca. 200–250 veh/h in both directions. Each of the above-mentioned streets had a 14 m-wide carriageway before reconstruction and carried traffic in both directions. Angled parking was allowed on the carriageway and footway on all these streets. The analysed urban block is bounded by main traffic arterials of Szczecin used by 2500–5500 veh/h during peak hours, including one dual carriageway with a tramway track running in the wide median between the two carriageways and three single-carriageway streets with four traffic lanes.

Each of the respective testing grounds included various traffic calming measures, i.e., mid-block chokers, speed tables, bulb-outs on raised intersections and also bollards blocking access or limiting the parking areas and trees planted on the carriageway between parking stalls (Figure 2).

(a)

(b)

Figure 2. Bollards limiting the parking spaces and greens planted on the carriageway between the parking spaces: (**a**) on on-street perpendicular parking lanes; (**b**) on on-street parallel parking lanes. Source: own work of the author.

The testing grounds were three one-way streets under reconstruction in which a Tempo-30 zone was implemented (Figure 3). The street selected for investigation of the combined effect of different traffic calming measures contained raised intersections at both ends and mid-block chokers combined with a raised pedestrian crossing (speed table). The main design data of the analysed streets were obtained from the plans developed by Maciej Sochanowski Design&Engineering company [40]. The primary difference between the testing grounds chosen to investigate the speed-reducing effect was the different types of intersections at the beginning and end of the mid-block sections. Common to all the testing grounds were raised pavements, chokers and bulb-outs.

There is a signalised intersection with a multilane main street located at the entry to testing ground No. 1. Testing ground No. 1 is a one-way street. Only bicycles may travel both ways. Testing ground No. 1 is left through a small raised intersection (Figure 3a).

Testing ground No. 2 is entered through a raised intersection with a dual carriageway featuring a neighbourhood gateway and ends with a small raised intersection (Figure 3b). As with testing ground No. 1, testing ground No. 2 is also a one-way street.

Testing ground No. 3 has raised intersections at both the entry and exit (Figure 3c). The exit intersection is with a two-way street, and traffic is controlled by signals in both directions (Figure 1). Being a staggered junction, at the outlet of testing ground No. 3, drivers can only turn right or left into the main two-way street (Figure 3c). These turning movements are, to a large extent, governed by the volume of traffic on the two-way street and by the lengths of queues before signals on both sides of the two-way transverse street (Figure 4).

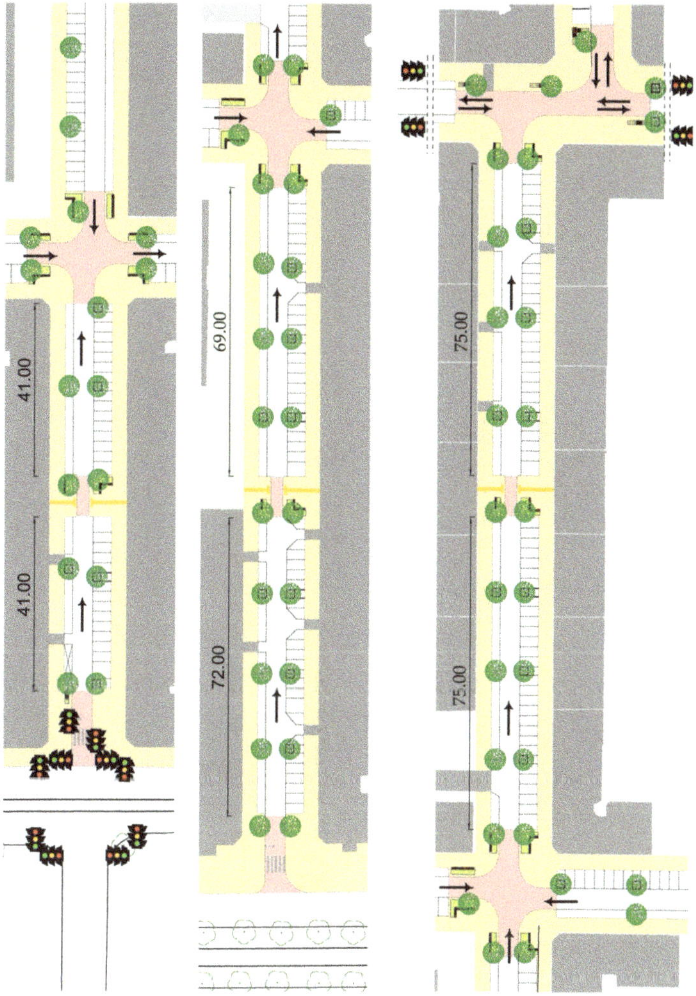

(a) (b) (c)

Figure 3. Analysed testing grounds: (**a**) testing ground No. 1; (**b**) testing ground No. 2; (**c**) testing ground No. 3. Arrows show the directions of traffic during the surveys. Source: own work of the author based on [40].

Figure 4. Queue of vehicles in testing ground No. 3, and two-way queue of vehicles on the transverse two-way street. Source: own work of the author.

Figure 5 shows test sections on the approach to and after chokers with raised pedestrian crossings in the respective testing grounds. Figure 6 shows the design details of the chokers and raised pedestrian crossings. After reconstruction, the carriageways in all the testing grounds are 5 m wide, narrowing down to 3 m at the chokers. The chokers are 10 m long. In testing ground No. 1, there is a planter on the footway extending towards the carriageway, spaced from the first kerb of the raised crossing by 5.5 m, while in the other testing grounds, there is a constant distance of 0.5 m between these elements. The planters are ca. 0.5 m high and are made of architectural concrete (Figures 5 and 6).

(a) (b)

Figure 5. Test sections before and after chokers with raised pedestrian crossings in testing ground No. 1, No. 2 and No. 3: (**a**) test section before choker; (**b**) test section after choker. Source: own work of the author.

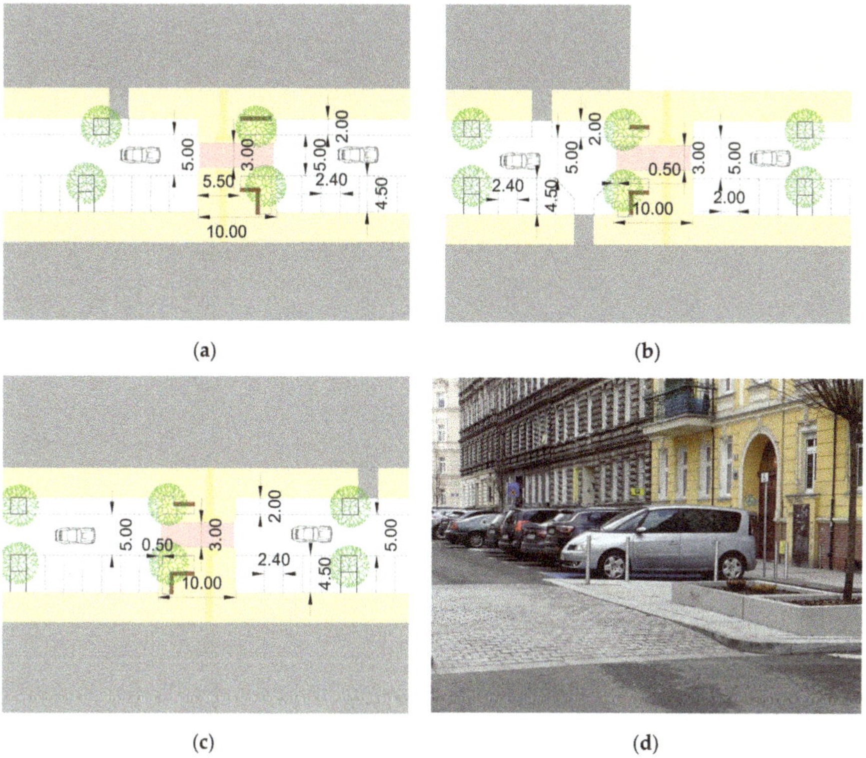

Figure 6. Treatments of chokers/raised pedestrian crossings: (**a**) testing ground No. 1; (**b**) testing ground No. 2; (**c**) testing ground No. 3; (**d**) design details of the sinusoidal approach ramp to the raised crossing and a view of a concrete planter. Source: own work of the author based on the design documentation [40].

The selected testing grounds differ in terms of the test section lengths on the approach to and after passing the choker (Figure 3), architecture (tenement houses with or without courtyard access), layout of on-street parallel and perpendicular parking stalls and a considerable fluctuation in demand for parking, related to the operation of street-level businesses. The main data characterising the public space in the analysed testing grounds are presented in Table 1 below.

Table 1. Public space features in the respective testing grounds.

Analysed Feature		No. 1	No. 2	No. 3
Length of approach to choker, m		41	72	75
Length of approach to intersection, m		41	69	75
Number of street-level businesses	on the left-hand side before choker	7	5	3
Number of access points		2 No.	2 No.	1 No.
Number of street-level businesses	on the right-hand side before choker	5	5	3
Number of access points		–	3 No.	–
Number of street-level businesses	on the left-hand side after choker	2	1	2
Number of access points		–	–	3 No.
Number of street-level businesses	on the right-hand side after choker	3	5	1
Number of access points		–	1 No.	1 No.
Number of trees within the carriageway on the left-hand side before choker		1	3	3
Number of trees within the carriageway on the right-hand side before choker		1	3	3
Number of trees within the carriageway on the left-hand side after choker		1	3	3
Number of trees within the carriageway on the right-hand side after choker		1	3	3
Location of the planter with tree on the choker from the approach edge, m		5.5	0.5	0.5

2.2. Measurement Method

In order to investigate the combined effect of different traffic calming measures installed in the Tempo-30 zone, 24 h vehicle speed and traffic count surveys were carried out for three days on selected streets, both before and after reconstruction. All measurements were performed during dry weather and with a dry pavement surface on Thursday, Friday and Saturday. Both free-flow and steady-flow speeds were measured simultaneously. For simultaneous measurement of driving speed and counting of traffic on the subsequent test sections, synchronised SR4 [41] electronic measuring devices were used. The devices were mounted on the posts of existing traffic signs. The siting of measurement points was based on the requirement of the study, i.e., on the inlets to and outlets from the testing grounds and on the approaches to and after the chokers. Nevertheless, only the data measured at the mid-block chokers were used for the purposes of this article.

The data were grouped in ranges at 25 veh/h intervals. This was conducted to be able to investigate the impact of the traffic volume on the effect of traffic calming. Interestingly enough, as it was established in the study, before reconstruction of streets in testing ground No. 1 and No. 2, the volume of traffic on one lane of a dual-lane carriageway was close to the traffic recorded on the one-way carriageway after implementation of the traffic calming scheme. The traffic volumes recorded after conversion to a one-way street testified to the attainment of one of the primary objectives of the Tempo-30 zone in question, namely, a substantial reduction in the traffic volume. Less traffic and lower speeds bring a reduction in noise and air pollution, i.e., the next objectives of the analysed Tempo-30 zone.

Conversely, in testing ground No. 3, the traffic recorded after reconstruction on the one-way carriageway was much greater than the traffic in both directions before the reconstruction. This was due to the traffic management system applied in the design [40] in which traffic flows from three one-way streets were collected along testing ground No. 3 (Figures 1 and 3c). Thus, a reduction in traffic was not achieved in testing ground No. 3.

We can therefore conclude that the implementation of Tempo-30 zones resulted in a considerable reduction in traffic, which is one of the traffic calming objectives, only in testing ground No. 1 and No. 2. In this way, the former arteries have become part of the public space, enhancing community activities, which is the second objective of traffic calming.

Due to a large variation of factors including traffic volumes, test section lengths (Figure 3) and public space features (Table 1), in order to ensure the consistency of the analysed data, each testing ground was analysed separately, and the effect of the volume of traffic on the recorded speed reductions was investigated. Considering a large proportion of vehicles travelling in linked small traffic flows, the next step of the analysis was to select the vehicles travelling in free-flow conditions, which was due to the following:

- Signalised intersection in testing ground No. 1 (Figure 1);
- Queue of cars waiting at the outlet of testing ground No. 3 (Figure 1) to enter the two-way transverse street with signalised intersections at both ends;
- High fluctuation of parked vehicles in all the testing grounds.

To this end, the functionality of the measuring system software [39] was employed which allowed the use of special procedures. These procedures were set up by the author after a detailed analysis of the initial speed measurements of vehicles parking into and driving out of the parking stalls, synchronised with video recordings of the analysed traffic conditions. Based on the analysis of traffic conditions obtained from three-day 24 h initial measurements, it was decided that the speed of driving into or out of the stall, i.e., less than 8 km/h, should be excluded from the continuous measurement database to be used in further analyses.

The experimental data were used to estimate the 85th percentile speed in the free movement condition at a given traffic volume. The share in free movement varied, as shown by the example in Figure 7, and depended primarily on the hourly traffic volume. Additionally, the road design and traffic conditions have a bearing on these variations. There are a few outliers, shown in Figure 7, which clearly indicate the effect of momentary traffic conditions on the share in the free movement traffic. This was probably due to the temporary impassability of the street section ahead (i.e., a number of cars waiting before traffic signals on the two-way street bounding testing ground No. 3). In summary, we can state that the obtained data validate the speed profile models of [14–19].

Since with the determination factor of 6%, the share in free movement in testing ground No. 1 before the choker does not depend on the traffic volume, the equation was not included in Figure 7a, in line with the principles of statistical analysis. The obtained equations are valid for a specific street only, and thus they should not be used in other analyses. The share in free movement also depends on the importance of the street under analysis, length of the test section between the intersections, number of main street businesses before and after the choker, architecture (multi-storey buildings located away from the footway, row or detached houses, detached public buildings such as schools or kindergartens), demand for parking before and after the choker, downtown or uptown location, etc.

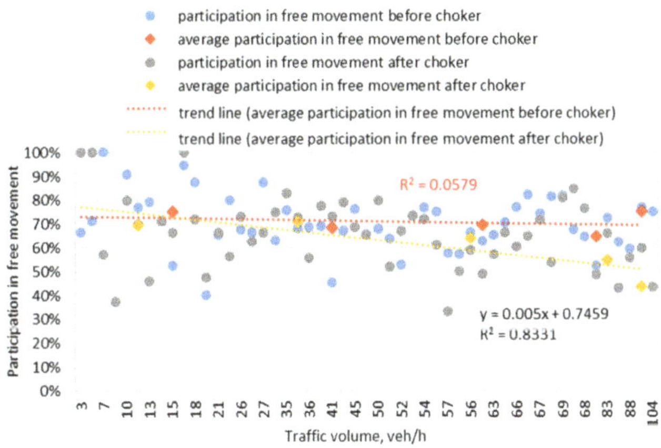

(a)

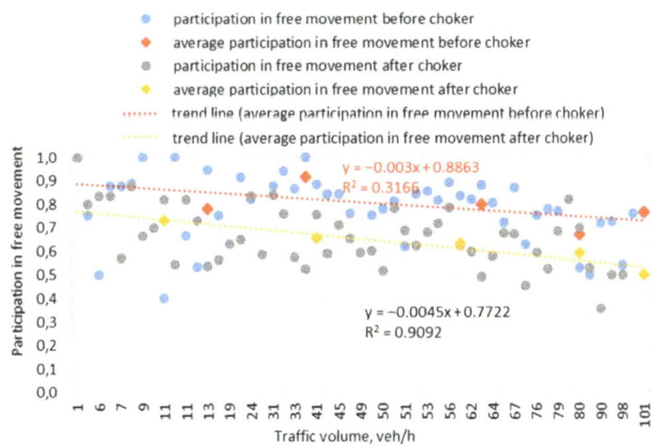

(b)

Figure 7. *Cont.*

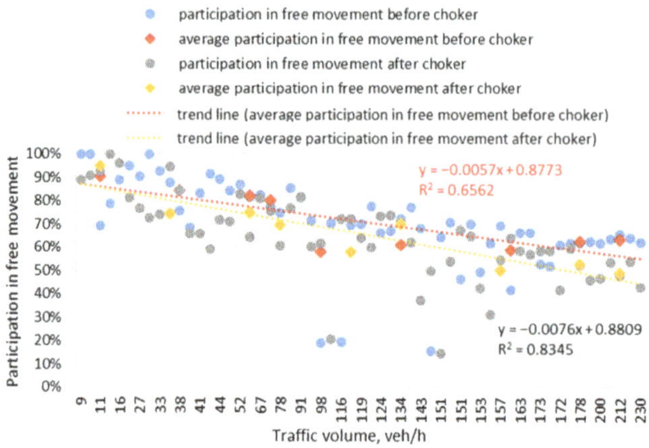

(c)

Figure 7. Variation in the share in free movement using the data obtained in the testing grounds: (**a**) No. 1; (**b**) No. 2; (**c**) No. 3. Source: own work of the author.

2.3. Research Methods

All free movement traffic data were subjected to a standard statistical analysis. A normal distribution of the measurement data was confirmed for all the cases under analysis (Appendix A). In addition, a few statistical tests were carried out (namely, goodness-of-fit, independence and median tests) to verify whether the driving speed depends on the traffic volume, location of the test section and public space features in the testing ground under analysis. The results of these statistical tests are presented in Appendix A. The results of the goodness-of-fit and median tests show statistically significant differences in speeds before and after the choker in all the testing grounds under analysis. The results of the independence test vary strongly only in testing ground No. 1 and No. 2. In testing ground No. 3, all values were positive, indicating a significant difference between the compared speed datasets. The results may be interpreted to indicate the relevance of other determinants in addition to the volume of traffic. These can include different public space characteristics in the respective testing grounds. The differences in this respect between the testing grounds are presented in Table 1. They include different numbers of access points to courtyards of three-storey tenement houses and public premises before and after the chokers, resulting in differing fluctuations in the parking demand. Most probably, the vehicles driving in and out of the parking considerably influenced the measured speeds. Summing up, the results of the statistical analyses performed as part of this study confirm the validity of Barbosa's conclusions [16] that speed profiles also depend on the traffic handling system in place.

The last analysis was an estimation of speed ranges in the respective testing grounds, supplemented with speed data before reconstruction. The obtained results are represented in Figures 8–10.

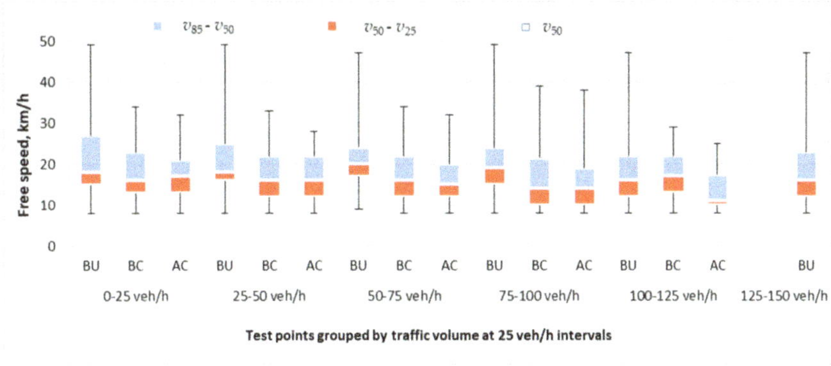

Figure 8. Summary of speed range analysis before and after upgrading, on the approach to and after the choker—data obtained in testing ground No. 1 (BU—before upgrading, BC—before choker, AC—after choker). Source: own work of the author.

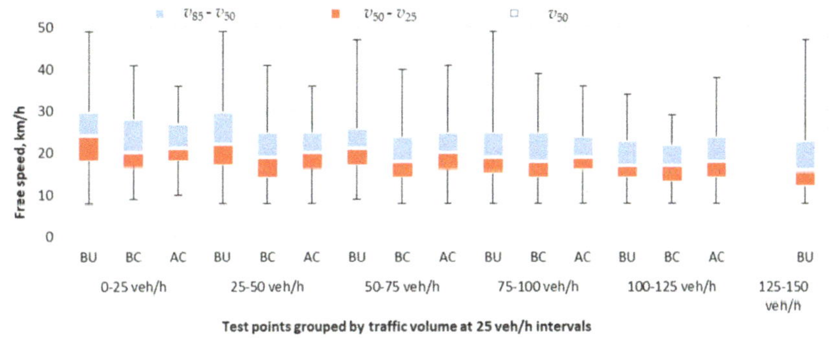

Figure 9. Summary of speed range analysis before and after upgrading, on the approach to and after the choker—data obtained in testing ground No. 2 (BU—before upgrading, BC—before choker, AC—after choker). Source: own work of the author.

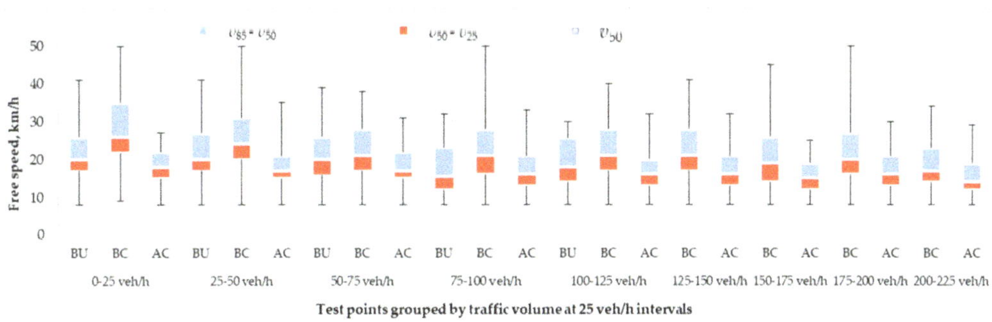

Figure 10. Summary of speed range analysis before and after upgrading, on the approach to and after the choker—data obtained in testing ground No. 3 (BU—before upgrading, BC—before choker, AC—after choker). Source: own work of the author.

The speed range analysis shown in Figures 8 and 9 demonstrated that in testing ground No. 1 and No. 2, the speeds measured before and after the choker/crossing treatment were lower than 30 km/h for almost all traffic volumes. This can be attributed to the short length of the test sections and a view on queues on the section after the small raised intersections located at the ends of both testing grounds (Figure 11), i.e., over the length of testing ground No. 3 (Figure 1). However, it must be clearly noted that isolated cases of higher speeds, up to 40 km/h, were also noted. Nevertheless, compared to the speeds before upgrading, a considerable speed reduction after implementation of the Tempo-30 zone was confirmed. Worth noting is also a higher traffic volume on one lane of the two-way street, as compared to the lower traffic volume on the reconstructed one-way carriageway. This also confirms the effectiveness of the applied traffic calming measures on the reduction in driving speeds and traffic volumes.

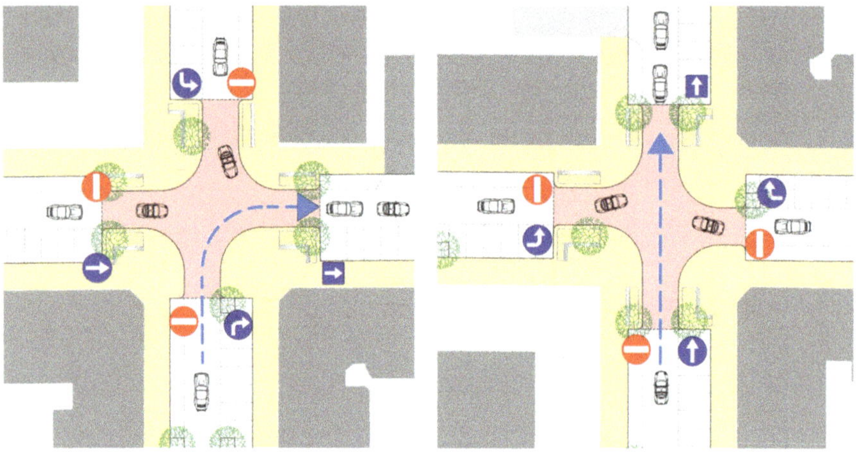

(a) (b)

Figure 11. Traffic conditions and existing traffic management on the raised intersection at the end of: (**a**) testing ground No. 1; (**b**) testing ground No. 2. Source: own work of the author based on [40].

Higher speeds were noted much more frequently in testing ground No. 3 with the longest test sections on the approach to and after the choker (Figure 10). This was caused, most probably, by the view of the unoccupied road ahead (which was the case with traffic volumes up to 100 veh/h) and no queues on the two-way transverse road bounding testing ground No. 3. As a result, the drivers tended to drive faster. In this case, the temporary traffic management implemented on the remaining streets in the block resulted in a considerable increase in the traffic volume on the one-way street after the implementation of the Tempo-30 zone in testing ground No. 3 up to 225 veh/h, as compared to the previous volume of 125 veh/h on one traffic lane of the two-way street. The traffic management and conditions on the raised intersection located at the end of testing ground No. 3 and on the two-way transverse street are presented in Figure 12.

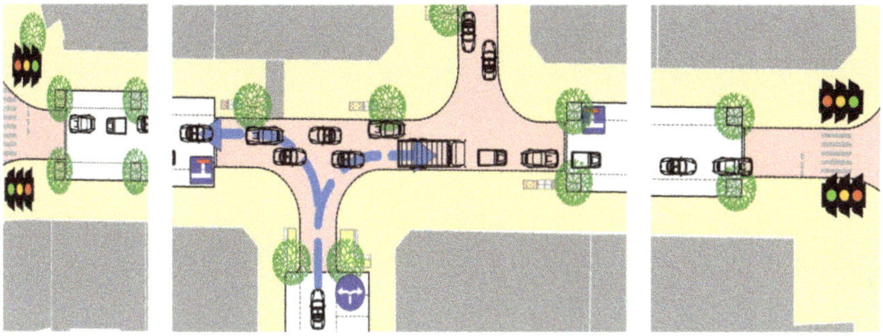

Figure 12. Traffic conditions and existing traffic management at the end of testing ground No. 3. Source: own work of the author based on [40].

3. Results

As the next step, the speed reductions obtained by the choker/crossing treatments were estimated. To this end, speed reductions were calculated for all the recorded traffic volumes. Next, average values were calculated for traffic volume ranges at 25 veh/h intervals. The total duration of the three-day 24 h continuous surveys provides 10–18 h of measurement per traffic volume range, which is considered a sufficient basis for further analyses. Additionally, the number of individual speed data in free movement of 300–1300 per traffic volume range is considered sufficient.

The speed change ratio was also calculated in accordance with the design guidelines [4–10] since the amount of speed reduction, the primary objective of the Tempo-30 zones, depends primarily on the initial speed, before the implementation of a traffic calming measure(s). The obtained results were subjected to regression analyses, separately for each testing ground. This is in line with Barbosa's suggestion [16] that the final speed reduction depends on various details concerning a specific traffic system. Therefore, the chosen testing grounds had different test section lengths, numbers of parking stalls, access points and types of inlet and outlet intersections.

The results of the regression analysis of the results obtained in the respective testing grounds and change in the 85th percentile speed, in accordance with the above assumptions, are presented in Figures 13–18. From the results of the regression analyses in Figures 13, 15 and 17, we see that, only in one case, only one result falls beyond the regression area (Figure 17). The confidence interval upper and lower bounds are also presented for a single observation, represented by dashed lines in Figures 13, 15 and 17. All the data under analysis fall within the area limited by these lines. As Barbosa's speed profile modelling [16] takes into account traffic handling and spatial layout parameters (Table 1), as well as traffic conditions, Figures 13, 15 and 17 also show regression equations for each of the testing grounds in turn.

Table 2 below compiles the main parameters associated with the regression analysis, i.e., coefficients of regression equations, correlation coefficient, coefficient of determination and also Guildford's and Pearson's interpretations of the magnitude of correlation.

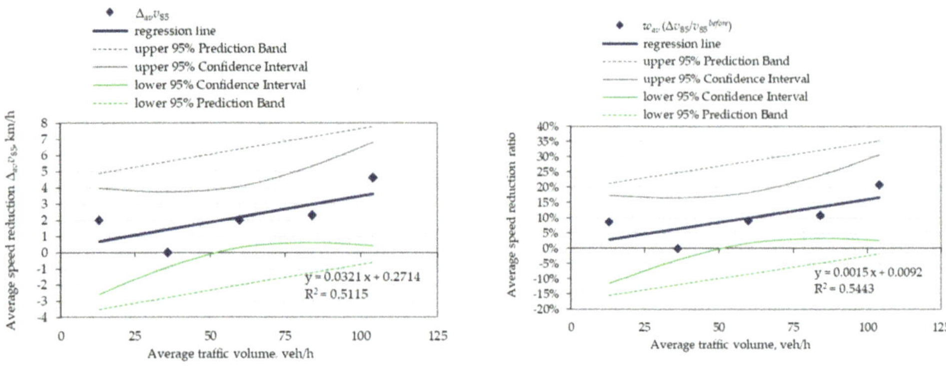

(a) (b)

Figure 13. Regression analyses (testing ground No. 1) for the following relationships: (**a**) average reduction in free-flow speed Δv_{85} and average traffic volume (R = 0.72); (**b**) speed reduction ratio and average traffic volume (R = 0.74). Source: own work of the author.

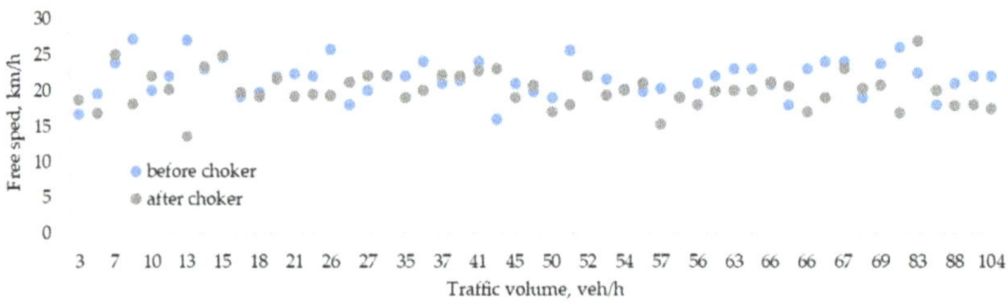

Figure 14. Summary of 85th percentile free-flow speed changes depending on hourly traffic (testing ground No. 1). Source: own work of the author.

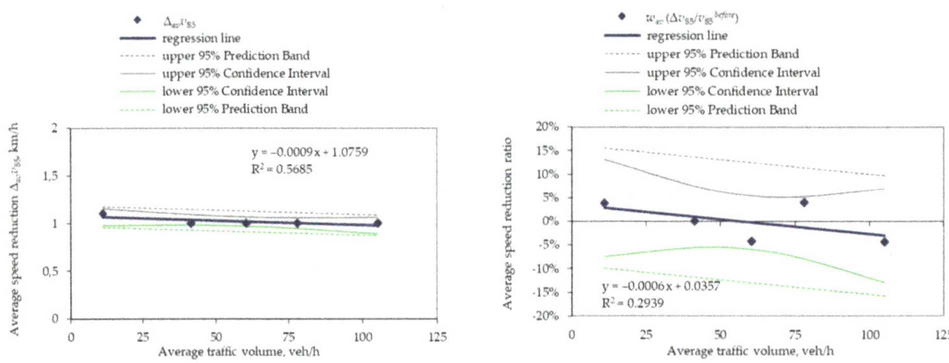

(a) (b)

Figure 15. Regression analyses (testing ground No. 2) for the following relationships: (**a**) average reduction in free-flow speed Δv_{85} and average traffic volume (R = −0.75); (**b**) speed reduction ratio and average traffic volume (R = −0.54). Source: own work of the author.

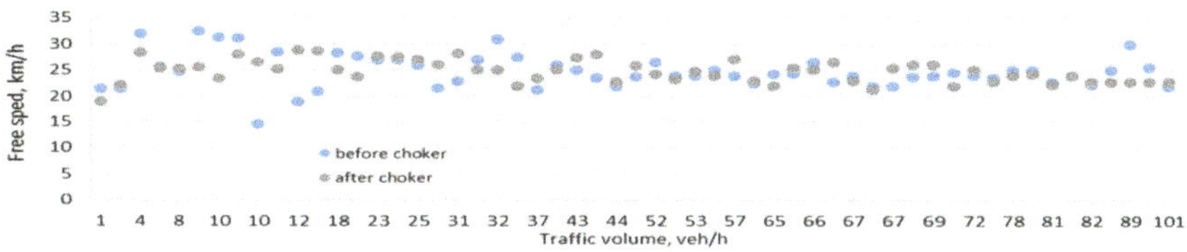

Figure 16. Summary of 85th percentile free-flow speed changes depending on hourly traffic (testing ground No. 2). Source: own work of the author.

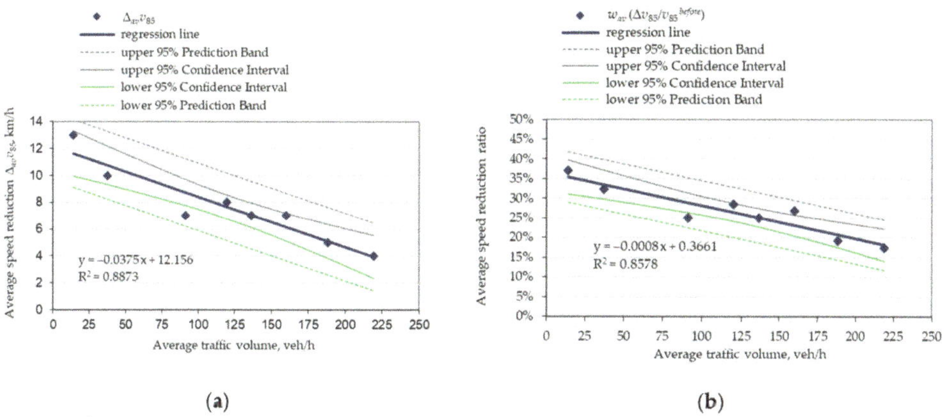

Figure 17. Regression analyses (testing ground No. 3) for the following relationships: (**a**) reduction in free-flow Δv_{85} and average traffic volume ($R = -0.94$); (**b**) speed reduction ratio and average traffic volume ($R = -0.93$). Source: own work of the author.

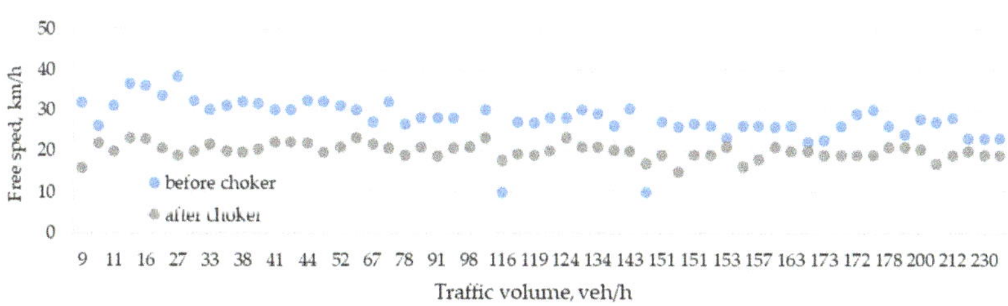

Figure 18. Summary of 85th percentile free-flow speed changes depending on hourly traffic (testing ground No. 3). Source: own work of the author.

Table 2. Regression analysis parameters.

Testing Ground	Relationship	Coefficients		R	R^2	Guilford's Interpretation of the Magnitude of Significant Correlations	Pearson's Interpretation of the Magnitude of Significant Correlations
		a	*b*				
No. 1		0.0321	0.271	0.72	51%	High correlation	High degree
No. 2	$\Delta_{av}v_{85} = f(N_{av})$	−0.0009	1.076	−0.75	57%	High correlation	High degree
No. 3		−0.0375	12.155	−0.94	89%	Very high correlation	Perfect
No. 1		0.0015	0.009	0.74	55%	High correlation	High degree
No. 2	$w = f(N_{av})$	−0.0006	0.036	−0.54	29%	Moderate correlation	High degree
No. 3		−0.0008	0.366	−0.93	86%	Very high correlation	Perfect

4. Discussion

As mentioned above, following the suggestions of Barbosa [16] and the factors included in the speed profile models found in [17,19], further analyses were carried out separately for the respective testing grounds.

The results of the regression analysis for the results obtained in testing ground No. 1 are represented in Figure 13. The value of the correlation coefficient (R = 0.72) indicates that speed reduction increases with the increase in the traffic volume, yet not exclusively. This relationship is also confirmed by the correlation coefficient obtained in relation to the speed change ratio (R = 0.74). It is worth noting that all the data considered in the regression analysis fall within the bounds of the confidence interval. Considering the low traffic volumes in the cases under analysis, only five ranges of traffic volume changes were used, and the relevant speed reductions ranged from 0 to 5 km/h. Taking into account the speeds recorded before upgrading, which are presented in Figure 8, we see a considerable reduction in speed for all the traffic volumes under analysis, both for these speeds and for the speeds measured on the chicane approach section.

The analysis of the obtained regression results also showed the relevance of the relationship between speed changes before and after the choker and the volume of traffic, as represented in Figure 14. From a simultaneous analysis of both results represented in Figures 13 and 14, it transpires that for smaller traffic volumes, the speed reduction is small and increases proportionally to the growth in traffic. The 85th percentile free-flow speeds on the approach to the choker are higher than the values of the 85th percentile free-flow speeds after the choker. The obtained increase in the speed reduction with the increase in the traffic volume depends not only on the traffic volume but also on the complex road system and traffic conditions noted in testing ground No. 1. Most probably, also relevant are the following: only 41 m-long test sections on the approach to and after the choker; the raised, signalised intersection at the entry to testing ground No. 1, splitting the traffic flow into small portions; and the small raised intersection at the end of the testing ground with a right turn being the only allowed movement (Figures 3a and 11). Additionally, worth noting are numerous street-level businesses (including boutiques) located in the testing ground, which increase the parking needs. In the free flow of traffic, on a short section after the choker, the drivers take into account the traffic conditions on the road ahead but also on the end intersection of testing ground No. 1 (Figure 5b).

The results of the analyses carried out as part of this study show the relevance, for the case under analysis, of the driver behaviour parameters, as described in [19]. Taking into account much smaller speeds and smaller traffic volumes, these results may be used to supplement the description of the time of travel models for traffic-calmed streets, as described in [19].

Figure 15 shows the regression analysis for the results obtained in testing ground No. 2. The value of the correlation coefficient ($R = -0.75$) indicates that speed reduction may depend on the increase in the traffic volume. However, the speed reductions of only 1 km/h are drastically different from the results obtained in testing ground No. 1. As regards the speed change ratio, $R = -0.54$, the relationship with the traffic volume was not confirmed with such small speed reduction values. Additionally, in testing ground No. 2, all the data fall within the bounds of the confidence interval, thus validating the obtained relationships. Setting the 0–1 km/h reductions next to the speed values presented in Figure 9, we see a substantial reduction in speed, compared to the situation before upgrading. That said, a statistically significant reduction in speed across the applied choker was not confirmed. Thus, the most probable primary determinants of speed reduction in testing ground No. 2 include the number of access points, parking demand fluctuation caused by numerous public premises located there and the fence along the school premises on the left-hand side of the street after the choker, which discourages pedestrians from crossing the street at this point.

The analysis of the obtained speed changes before and after the choker depending on the traffic volume, as shown in Figure 14, indicates completely different traffic conditions in testing ground No. 2. The free-flow speed changes at traffic volumes exceeding 25 veh/h indicate a 20–25 km/h stabilised speed both on the approach to and after the choker. This indicates that in testing ground No. 2, there were other parameters relevant to the obtained speed reductions. Most probably, much more important were the length of both test sections of ca. 70 m (Figure 3b) and also the uncongested entry into the Tempo-30 zone from the dual carriageway street through the raised intersection with a neighbourhood gate. In this case, stabilised speed values were recorded already on the initial test section on the approach to the choker. This can be attributed to the presence of street-level businesses and courtyard access points, i.e., the public space features. These features are missing after the choker, and the very stable speed, which is quite independent from the traffic volume, was due to the location of a primary school on the right-hand side of the street, with a chain link fence along the footway (Figures 1 and 3b). The above factors should also be introduced into the above-mentioned speed models [16,17,19]. Summing up the above discussion, it should be made clear that the Tempo-30 zone in testing ground No. 2 calmed the traffic, which is the primary objective of this scheme. This conclusion is supported by the data in Figure 9. Before the upgrading on one lane of the then two-way street, the traffic volumes and speeds were greater than on the one-way street after the implementation of the Tempo-30 zone. The variations in the 85th percentile free-flow speed were generally small in this case, with only isolated cases of ca. 40 km/h. Despite the small speed reductions, the results indicate a high effectiveness of the implemented traffic calming measures (choker, raised pedestrian crossing, sinusoidal approach and exit ramp, planters on extended footways discouraging drivers from making undesired manoeuvres to negotiate roadside obstacles and bollards). The obtained primary results, namely, the stabilisation of the speed over the whole length of testing ground No. 2, should be, most probably, attributed to the synergy of the above-mentioned traffic calming measures influencing the behaviour of drivers.

The results of the regression analysis for the results obtained in testing ground No. 2 are represented in Figure 17. The value of the correlation coefficient ($R = -0.94$) indicates that speed reduction depends primarily on the increase in the traffic volume. This relationship is also confirmed by the correlation coefficient obtained in relation to the speed change ratio ($R = -0.93$). In this case, all the data but one fall within the bounds of the confidence intervals. Overall, the speed drops from 4 to 12 km/h proportionally to the increase in the traffic volume. When the obtained relationships are paired with the speed values presented in Figure 10, similar to the previous testing grounds, we see that upgrading brought a large reduction in speed for all the traffic volumes in consideration. The relevant factors, besides the already mentioned raised intersection at the end of testing ground No. 3, include, most probably, the choker and reduced parking demand on this section.

However, attention is drawn to the specific nature of testing ground No. 3 in that the one-way street collects traffic from three access streets (Figures 1 and 3c), and this was, most probably, the primary factor that caused the increased volume of traffic on the analysed street after upgrading. From the chart in Figure 10, it transpires that the traffic volume on one traffic lane of the two-way street did not exceed 125 veh/h before upgrading, and after upgrading, twice the traffic volume was noted on the one-way street, i.e., up to 250 veh/h. It is also important to include this, in the future, in the speed profile models of the free-flow speed (Figure 7), since the traffic management at the entry to testing ground No. 3, represented in Figures 3c and 11, and at the end of testing ground No. 3, represented in Figure 12, largely influenced the traffic conditions in this area and the test results obtained in testing ground No. 3. A simultaneous analysis of the data in Figures 17 and 18 indicates that the amount of speed reduction decreases with the increase in traffic, i.e., it is inversely related to the growth in traffic. The changes in speed on the approach to and after the choker, represented in Figure 18, both for each of the respective traffic volumes separately and for the average values, point to the existence of a strong relationship between speed reduction and the volume of traffic in testing ground No. 3. However, it must be stressed that in this case, the strong relationship between speed reduction and traffic volume (confirmed by the high correlation coefficients) was due to the applied road system and traffic conditions in testing ground No. 3. The traffic conditions, i.e., combined movements from three streets and the queue of vehicles on the bounding street (Figure 4), were the main determinants of the measured speeds on the approach to and after the choker. Additionally, the test section lengths (75 m) most probably had a bearing on the measurement data. Worth noting are a few street-level businesses located in testing ground No. 3 on both sides of the one-way carriageway, which reduce the demand for parking in their area.

The drivers' behaviour, described in [19,20], had less effect on the final results in testing ground No. 3, since in this case, the speed depended on the queues in the bounding street or free carriageway without such queues visible to the driver, allowing turning movements without waiting. The results of the analysis confirm the necessity to take into account the traffic management parameters, as described in Cascetta's model [17].

Based on the results of this study carried out on traffic-calmed streets, we see the necessity to also include in the presently used speed profile models the above-mentioned factors which concern not only the road system but also the traffic conditions, locations of access points and street-level businesses and the applied traffic management.

Nevertheless, the results obtained in all the testing grounds clearly confirm the effectiveness of the applied traffic calming and reduction in free-flow speed as compared to the situation before upgrading. Since in the traffic-calmed neighbourhoods, the reductions in noise [36–39], exhaust emissions [27–30,38,42,43] and fuel consumption depend primarily on the actual speeds of vehicles, it is justified to state that the remaining objectives of the Tempo-30 zones were also attained on the analysed streets.

5. Conclusions

The primary objectives of the Tempo-30 zones include an improvement in traffic safety, reduction in speed and traffic volume and less noise, pollution and fuel consumption. Since no serious road incidents were noted in the analysed block before the upgrading [40], we can definitely declare that the remaining traffic calming objectives (reduction in speed, traffic volume, noise and air pollution and consumption of fuel, all of which are related to the driving speed) were attained on the upgraded streets.

The results of the Tempo-30 zone traffic calming measures show the effectiveness of the chokers applied combined with the mid-block crosswalks, bollards and concrete planters. The most impressive were the speed reductions obtained on streets featuring less access points to courtyards and less public premises. An effective speed reduction can also be obtained in streets featuring more access points and public premises. The amount of this speed reduction is smaller though. Beside the traffic volume, the speed reduction determinants include entry and exit intersections, the traffic management scheme and also street furniture along the street.

It must be noted that after the trial period with temporary traffic management to relieve testing ground No. 3 from excessive traffic, the responsible road engineer proposed that testing ground No. 2 should remain a one-way street, yet with the opposite direction of traffic. This change has actually relieved testing ground No. 3 from excessive traffic and long queues of cars. It is therefore postulated to pay particular attention when designing the directions of traffic on one-way streets in Tempo-30 zones, in order to avoid overcrowding other streets. Barbosa's modified traffic model can be used as a suitable tool.

In summary of the study results (elaborating on Barobosa's hypothesis), we can declare their suitability for the necessary modification of the existing speed profile models, allowing for a detailed description of the road system and traffic conditions in traffic-calmed neighbourhoods. The revisions that must be made to the existing Barbosa [16], Davidson [12,13] and Tisato [14] traffic models should concern the description of the parameters characterising different Tempo-30 zone sections, their lengths, number of access points, number of public premises, parking spaces and intersections at both ends of the analysed street section. Furthermore, the above-mentioned models should also consider speed reductions by chokers and mid-block crosswalks combined with safety barriers and street furniture for different traffic volumes.

In the near future, i.e., after the completion of the upgrading of all the streets in the analysed block, the author plans to repeat the traffic volume and speed reduction measurements with the purpose of supplementing the existing Tempo-30 zone speed profiles, if appropriate and required.

Funding: This research received no external funding.

Data Availability Statement: Not applicable.

Acknowledgments: The author would like to thank Maciej Sochanowski of Maciej Sochanowski Design&Engineering of Szczecin for providing details of the upgrading of the analysed city district.

Conflicts of Interest: The author declares no conflict of interest.

Appendix A

Table A1. Results of standard statistical goodness-of-fit tests.

Traffic Volume, veh/h	Goodness-of-Fit Test K-S λ H_0: $F(v) = F_0(v)$; H_1: $F(v) \neq F_0(v)$, $\lambda_{\alpha = 0.05} = 1.36$					
	Testing Ground No. 1		Testing Ground No. 2		Testing Ground No. 3	
	$\lambda^{(v \text{ before})}$	$\lambda^{(v\ after)}$	$\lambda^{(v \text{ before})}$	$\lambda^{(v\ after)}$	$\lambda^{(v \text{ before})}$	$\lambda^{(v\ after)}$
0–25	0.51	0.52	0.44	0.57	0.38	0.46
25–50	0.89	1.06	0.62	0.51	0.62	0.87
50–75	0.92	0.80	0.78	1.31	0.38	0.53
75–100	0.74	0.56	0.69	1.13	0.47	0.38
100–125	0.30	0.53	0.41	0.28	0.93	0.86
125–150	–	–	–	–	0.69	0.54
150–175	–	–	–	–	0.65	1.29
175–200	–	–	–	–	0.91	0.80
200–225	–			–	0.36	0.69

Appendix B

Table A2. Results of two-sample Kolmogorov–Smirnov test.

Traffic Volume, veh/h	Two-Sample Kolmogorov–Smirnov Test λ H_0: $F(v^{before}) = F(v^{after})$; H_1: $F(v^{before}) \neq F(v^{after})$					
	Testing Ground No. 1		Testing Ground No. 2		Testing Ground No. 3	
	λ	$\lambda_{\alpha}, {\alpha = 0.05}$	λ	$\lambda_{\alpha}, {\alpha = 0.05}$	λ	$\lambda_{\alpha}, {\alpha = 0.05}$
0–25	8.06	1.36	8.59	1.36	6.77	1.36
25–50	12.54	1.36	13.01	1.36	11.93	1.36
50–75	19.79	1.36	2.07	1.36	7.91	1.36
75–100	10.92	1.36	13.76	1.36	13.28	1.36
100–125	5.35	1.36	0.56	1.36	14.47	1.36
125–150	–	–	–	–	14.72	1.36
150–175	–	–	–	–	23.67	1.36
175–200	–	–	–	–	14.96	1.36
200–225	–	–	–	–	9.56	1.36

Appendix C

Table A3. Results of the test of independence and median test.

Traffic Volume, veh/h	Test of Independence H_0: $P\{X = v^{before}{}_i, Y = v^{after}{}_i\} = P\{X = v^{before}{}_i\}\, P\{Y = v^{after}{}_i\}$ H_1: $P\{X = v^{before}{}_i, Y = v^{after}{}_i\} \neq P\{X = v^{before}{}_i\}\, P\{Y = v^{after}{}_i\}$ $\chi_\alpha{}^2 = 3.84$, $\alpha = 0.05$ Testing Ground:			The Median Test H_0: $F_1(x) = F_2(x)$ H_1: $F_1(x) \neq F_2(x)$ $\chi_\alpha{}^2 = 3.84$, $\alpha = 0.05$ Testing Ground:		
	No. 1	**No. 2**	**No. 3**	**No. 1**	**No. 2**	**No. 3**
0–25	1.01	1.46	27.43	11.94	22.72	69.04
25–50	1.87	2.52	51.40	379	273	962
50–75	0.50	0.02	11.57	3731	850	164
75–100	0.02	4.62	27.18	162	508	1482
100–125	0.57	0.04	23.96	7.32	4.28	1599
125–150	–	–	25.87	–	–	1643
150–175	–	–	47.50	–	–	12917
175–200	–	–	19.77	–	–	964
200–225	–	–	4.86	–	–	639

References

1. *Urban Street Design Guide*; National Association of City Transportation Officials: Washington, DC, USA, 2013.
2. Krystek, R. *Principles of Traffic Calming on the Roads of the Pomorskie Region in Poland, Part 1: Street Layouts in Towns and Cities*; GAMBIT Pomorski: Gdańsk, Poland, 2008.
3. Krystek, R. *Principles of Traffic Calming on the Roads of the Pomorskie Region in Poland, Part 2: Sections of Major Roads through Towns and Villages*; GAMBIT Pomorski: Gdańsk, Poland, 2008.
4. *Guidelines for Traffic Calming*; City of Sparks Public Works Traffic Division, Sierra Transportation Engineers, Inc.: Reno, NV, USA, 2007.
5. *Urban Traffic Areas—Part 7—Speed Reducers*; Vejdirektoratet-Vejregeludvalget: Copenhagen, Denmark, 1991.
6. *Traffic Calming Guidelines*; Devon County Council Engineering & Planning Department: Devon, UK, 1992.
7. *Traffic Calming*; Local Transport Note 1/07; Department for Regional Development (Northern Ireland), Scottish Executives, Welsh Assembly Government: London, UK, 2007.
8. *Roads Development Guide*; East Ayrshire, Strathclyde Regional Council: London, UK, 2010.
9. Harvey, T.A. *Review of Current Traffic Calming Techniques*; University of Leeds: Leeds, UK, 2013.
10. *Directives for the Design of Urban Roads. RASt 06*; Road and Transportation Research Association (FGSV): Köln, Germany, 2006.
11. Zhu, H.; Almukdad, A.; Iryo-Asano, M.; Alhajyaseen, W.K.M.; Nakamura, H.; Xin Zhang, X. A novel agent-based framework for evaluating pedestrian safety at unsignalized mid-block crosswalks. *Accid. Anal. Prev.* **2021**, *159*, 106288. [CrossRef] [PubMed]
12. Davidson, K.B. A flow–travel time relationship for use in transportation planning. In Proceedings of the 3rd Australian Road Research Board (ARRB) Conference, Sydney, Australia; 1966; Volume 3, pp. 183–194. Available online: https://trid.trb.org/view/1209266 (accessed on 12 September 2021).
13. Davidson, K.B. The theoretical basis of a flow–travel time relationship for use in transportation planning. *Aust. Road Res.* **1978**, *8*, 32–35, Discussion, p. 45.
14. Tisato, P. Suggestions for an improved Davidson travel time function. *Aust. Road Res.* **1991**, *21*, 85–100.
15. Akcelik, R. Travel time functions for transport planning purposes: Davidson's function, its time dependent form and alternative travel time function. *Aust. Road Res.* **1991**, *21*, 49–59. Available online: https://www.researchgate.net/publication/242258239_Travel_time_functions_for_transport_planning_purposes_Davidson\T1\textquoterights_function_its_time-dependent_form_and_an_alternative_travel_time_function (accessed on 9 July 2021).
16. Barbosa, H.M. *Impacts of Traffic Calming Measures on Speeds on Urban Roads*; University of Leeds: Leeds, UK, 1995.
17. Cascetta, E. *Transportation Systems Engineering: Theory and Methods*, 49th ed.; Springer Science: Berlin, Germany, 2013.
18. Macioszek, E.; Iwanowicz, D. A Back-of-Queue Model of a Signal-Controlled Intersection Approach Developed Based on Analysis of Vehicle Driver Behavior. *Energies* **2021**, *14*, 1204. [CrossRef]
19. Richter, M.; Paszkowski, J. Modelling driver behaviour in traffic-calmed areas. *Czas. Tech.* **2018**, *8*, 111–124. [CrossRef]
20. Paszkowski, J.; Herrmann, M.; Matthias, R.; Szarata, A. Modelling the Effects of Traffic-Calming Introduction to Volume–Delay Functions and Traffic Assignment. *Energies* **2021**, *14*, 3726. [CrossRef]
21. Macioszek, E.; Świerk, P.; Kurek, A. The Bike-Sharing System as an Element of Enhancing Sustainable Mobility—A Case Study based on a City in Poland. *Sustainability* **2020**, *12*, 3285. [CrossRef]
22. Abdulmawjoud, A.A.; Jamel, M.G.; Al-Taei, A.A. Traffic flow parameters development modelling at traffic calming measures located on arterial roads. *Ain Shams Eng. J.* **2021**, *12*, 437–444. [CrossRef]

23. Baltrėnas, H.P.; Januševicius, T.; Chlebnikovas, A. Research into the impact of speed bumps on particulate matter air pollution. *Measurement* **2017**, *100*, 62–67. [CrossRef]
24. Akgol, K.; Gunay, B.; Aydin, M.M. Geometric optimisation of chicanes using driving simulator trajectory data. *Transp. Proc. Inst. Civ. Eng. Transp.* **2020**. [CrossRef]
25. Distefano, N.; Leonardi, S. Effects of speed table, chicane and road narrowing on vehicle speeds in urban areas. In Proceedings of the VI International Symposium, NEW HORIZONS 2017 of Transport and Communications", Sarajevo, Bosnia and Herzegovina, 17–18 November 2017. Available online: https://www.researchgate.net/publication/328738163_EFFECTS_OF_SPEED_TABLE_CHICANE_AND_ROAD_NARROWING_ON_VEHICLE_SPEEDS_IN_URBAN_AREAS (accessed on 12 June 2021).
26. Distefano, N.; Leonardi, S. Evaluation of the Benefits of Traffic Calming on Vehicle Speed Reduction. *Civ. Eng. Archit.* **2019**, *7*, 200–214. [CrossRef]
27. Obregón-Biosca, S.A. Speed humps and speed tables: Externalities on vehicle speed, pollutant emissions and fuel consumption. *Results Eng.* **2020**, *5*, 100089. [CrossRef]
28. Lav, A.H.; Bilgin, E.; Lav, A.H. A fundamental experimental approach for optimal design of speed bumps. *Accid. Anal. Prev.* **2018**, *116*, 53–68. [CrossRef]
29. Beckx, L.I.P.C.; Broekx, S. Impact of 30 km/h Zone Introduction on Vehicle Exhaust Emissions in Urban Areas. Available online: https://www.researchgate.net/publication/237327146 (accessed on 12 June 2021).
30. Da Silva, F.N.; Custódio, R.A.L.; Martins, H. Low Emission Zone: Lisbon's Experience. *J. Traffic Logist. Eng.* **2014**, *2*, 133–139. [CrossRef]
31. De Vlieger, I.; De Keukeleere, D.; Kretzschmar, J. Environmental effects of driving behaviour and congestion related to passenger cars. *Atmos. Environ.* **2000**, *34*, 4649–4655. [CrossRef]
32. Beckx, C.; Int Panis, L.; Debal, P.; Wets, G. Influence of gear changing behaviour on fuel-use and vehicular exhaust emissions. In Proceedings of the 8th International Symposium on Highways and the Urban Environment, Nicosia, Cyprus, 12–14 June 2006; Rauch, S., Morrisson, G., Eds.; Chalmers University: Göteborg, Denmark, 2006.
33. Liimatainen, H. Measures for Energy Efficient and Low Emission Private Mobility. In *Encyclopedia of the UN Sustainable Development Goals*; Springer International Publishing: Cham, Switzerland, 2020; pp. 1–12. [CrossRef]
34. Tang, J.; McNabola, A.; Mistear, B. The potential impacts of different traffic management strategies on air pollution and public health for a more sustainable city: A modelling case study from Dublin, Ireland. *Sustain. Cities Soc.* **2020**, *60*, 102229. [CrossRef]
35. Sun, L.L.; Liu, D.; Chen, T.; He, M.T. Road traffic safety: An analysis of the cross-effects of economic, road and population factors. *Chin. J. Traumatol.* **2019**, *22*, 290–295. [CrossRef]
36. Jeon, J.; Hong, J.; Kim, S.; Kim, K.-H. Noise Indicators for Size Distributions of Airborne Particles and Traffic Activities in Urban Areas. *Sustainability* **2018**, *10*, 4599. [CrossRef]
37. Bendtsen, H.; Haberl, J.; Litzka, J.; Pucher, E.; Sandberg, U.; Watts, G. *Traffic Management and Noise Reducing Pavements-recommendations on Additional Noise Reducing Measures*; Report 137; Road Directorate, Danish Road Institute: Copenhagen, Denmark, 2004.
38. Niebieska Książka. Sektor Transportu Publicznego w Miastach, Aglomeracjach i Regionach, Blue Book Road Infrastructure, Jaspers 2015. Available online: https://www.cupt.gov.pl/ (accessed on 30 November 2020).
39. Google Earth. Available online: http://www.earth.google.com (accessed on 12 August 2016).
40. Sochanowski, M. *Kompleksowa Modernizacja Chodników, Miejsc Postojowych i Nawierzchni Jezdni w Kwartale Ulic: Królowej Jadwigi, Małkowskiego, Bogusława X, Bohaterów Getta Warszawskiego, Ściegiennego*; Pracownia Projektowa Macieja Sochanowskiego: Szczecin, Poland, 2016.
41. *Speed Displays Traffic Detection, Radar, Detection, Software*; Vitronic: Kędzierzyn Koźle, Poland, 2015.
42. Künzler, P.; Dietiker, J.; Steiner, R. *Nachhaltige Gestaltung von Verkehrsräumen im Siedlungsbereich, Grundlagen für Planung, Bau und Reparatur von Verkehrsräumen*; Herausgegeben vom Bundesamt für Umwelt: Bern, Switzerland, 2011. Available online: https://www.bafu.admin.ch/dam/bafu/de/dokumente/luft/uw-umwelt-wissen/nachhaltige_gestaltungvonverkehrsraeumenimsiedlungsbereich.pdf (accessed on 12 August 2019).
43. Nina67, Consommation D'essence en Fonction de Vitesse et Rapport. Astuces-Pratiques 2015. Available online: https://www.astuces-pratiques.fr/auto-moto/consommation-d-essence-en-fonction-de-vitesse-et-rapport (accessed on 12 August 2019).

energies

MDPI

Article

Bicycle Traffic Model for Sustainable Urban Mobility Planning

Jacek Oskarbski *, Krystian Birr and Karol Żarski

Faculty of Civil and Environmental Engineering, Gdansk University of Technology, 80-233 Gdańsk, Poland; krystian.birr@pg.edu.pl (K.B.); karol.zarski@pg.edu.pl (K.Ż.)
* Correspondence: jacek.oskarbski@pg.edu.pl; Tel.: +48-604-475-876

Abstract: Modelling tools and transport models are required to assess the impact of measures for the effective planning of cycling routes in cities. This paper presents the methodology for developing a four-stage macroscopic model of bicycle traffic for the city of Gdynia, and its use in planning new bicycle routes, considering a modal shift. The model presented in this paper allows for the evaluation of the influence of the characteristics of the cycling infrastructure, along with the development of the cycling network based on the choice of cycling as an alternative to other modes of transport, by taking into account the modal shift. The model takes into account the influence of the longitudinal gradient, link, and surface type of cycling routes on the distribution and demand for bicycle traffic. The results of our research allow us to assess the impact of planned cycling routes on the reduction in the volume of car traffic, which is crucial for reducing energy consumption and negative environmental impacts. Experiences from the application of the model in Gdynia suggest that the model provides a strong basis to support mobility planning and monitoring processes in cities worldwide. Cities should take into account the methods proposed in this paper when planning the development of their transport systems.

Keywords: sustainable urban mobility planning; macroscopic transport model; bicycle traffic model; modal shift; modal split

Citation: Oskarbski, J.; Birr, K.; Żarski, K. Bicycle Traffic Model for Sustainable Urban Mobility Planning. *Energies* **2021**, *14*, 5970. https://doi.org/10.3390/en14185970

Academic Editors: Anna Granà, Margarida Coelho, Paulo Fernandes and Elżbieta Macioszek

Received: 7 August 2021
Accepted: 16 September 2021
Published: 20 September 2021

1. Introduction

Many cities (especially in Central and Eastern Europe) have developed their road transport networks with a focus on improving the efficiency of motor vehicle traffic and minimising congestion. Such activities usually aim to increase the capacity of elements of the road network. However, congestion—and the resulting increases in energy consumption and greenhouse gas emissions—can also be reduced more rationally by introducing measures to change the modal split and increase the share of cycling in daily trips. Therefore, effective development of the transport system requires prediction of the effectiveness of planned changes, and monitoring of the performance of implemented measures. Most often, cities do not use tools that allow for the consideration of cycling in the transport system and the possibility of changes in the modal split as a result of the implementation of infrastructural or organisational measures. Transport demand and transport network models can effectively support the process of mobility planning by taking into account active modes of transport, including cycling. This paper aims to review the state of the art in bicycle traffic modelling—taking into account both the modal shift, and measures contributing to the reduction in travel demand and the increase in the share of cycling trips—and to present the opportunities and benefits resulting from the application of such solutions in the example of the city of Gdynia.

Urban transport systems face major challenges due to the growth in private vehicle ownership and motorisation rates (number of private vehicles per number of inhabitants) [1,2]. There were more than 256 million passenger cars in EU countries in 2018 [3]. The World Bank's predictions indicate that by 2050 the number of vehicles on the road will double to 2 billion, and nearly 70% of the projected world population (approximately 5.4 billion people) will live in urban areas, tripling the number of urban trips [4,5]. According to recent estimates, passenger transport (in kilometres) in the United States will increase by 30–50% by 2100 [6], and in the EU by 42% by 2050 [7]. These estimates show that cities face a growing motorisation rate that needs to be addressed. The share of urban trips using public transport is decreasing, and the rate of cycling trips remains low, especially in Eastern European countries. Pedestrians and cyclists make between 13 and 51% of all trips in Western European countries [8]. However, in terms of kilometres travelled in the European Union, only 1% of passenger kilometres are travelled by bicycle, while 73% are travelled by car [9].

Many cities around the world are experiencing urban sprawl, rapidly increasing motorisation, inadequate public transport systems, a high share of car trips, high pollution, and poor infrastructure for pedestrians and cyclists [10]. The unsustainable growth of transport activities puts a strain on our planet's ecosystems and resources. Greenhouse gas emissions (GHGs) from energy production are one of the main causes of climate change. The excessively slow transition to alternative fuel sources and propulsion systems has led the transport sector to be increasingly blamed for the possible failure of individual countries to meet their commitments in international climate change agreements [11]. For this reason, transport remains at the centre of any debate on energy savings, due to its dependence on fossil fuels in car transport (passenger and freight), but also in rail, air, and maritime transport. In Europe, GHGs decreased between 1990 and 2017, except for the transport sector [12]. The transport sector is responsible for 30% of total energy consumption and 20% of total GHGs in the European Union (with road transport having the largest share, at 72%) [12–14]. The total external costs of transport (including accidents, congestion, noise, CO_2 emissions, and air quality) in urban areas of the EU amount to EUR 230 billion [15]. There are increasing indications that electrification in transport will not be fast enough or sufficient to meet the energy efficiency and low-carbon targets of the transport system [14,16–18]. These trends in the development of transport require actions aimed at reducing motorised traffic. Evaluation of the effectiveness of such measures is possible through the use of transport models that estimate changes in travel demand and the modal split, taking into account active mobility modes (including cycling).

There are very few reliable, integrated, and comprehensive tools for modelling the transport system that take into account bicycle traffic at the macroscopic level [19–22]. Most existing urban transport modelling tools inaccurately or incompletely take into account the influence of cycling on the functioning of the transport network. The tools used for transport modelling most often focus only on the performance of road transport, either excluding cyclists or treating them as a disruptive factor for motor vehicle traffic. Therefore, it is necessary to develop tools and models that allow assessment of the impact of planned actions for the implementation of cycling measures, taking into account the influence on the modal shift. Such analyses allow us to look differently at the possibility of improving the conditions of travel in the transport network through the reduction in car traffic. The model proposed in this paper contributes to filling the indicated gap.

The bicycle traffic model was developed by the authors of this paper to support the city of Gdynia in planning and decision making with regard to changes in the transport network, as well as in the development, implementation, and monitoring of the Sustainable Urban Mobility Plan (SUMP). The SUMP is a document proposed by the European Commission to establish a framework for urban mobility planning, and is the result of the sustainable urban mobility planning process [21,23]. The application of the model to planning the development of transport systems, monitoring the SUMP, consulting citizens and, ultimately, making decisions is a strong innovation of the presented solution.

The structure of the bicycle traffic model is also innovative (especially on a national scale). The data and variables used allow for a reliable representation of the existing cycling traffic and its predictions. Another contribution to the research is the consideration in the model of the influence of the longitudinal gradient on the speed of cyclists, and the consideration of the types of sections of the cycling network, and their surface, in the choice of transport mode and route. The research methodology includes both a survey of residents and direct measurements of bicycle and road traffic. The spatial approach to modelling mobility behaviour opens new opportunities for evidence-based planning and decision making in the broad area of cycling promotion and development, as well as for predicting and monitoring the effects of planned or implemented measures. The methodology for developing the model was tested and verified on the example of the city of Gdynia, but it can be an example of good practice for other cities.

This paper attempts to answer the following questions:

- Why is it important to consider a modal shift in urban transport modelling, and how can the experience from Gdynia help in the development of a bicycle traffic model that takes into account the modal shift?
- What data can be used to develop a bicycle traffic model, and what are the elements of the model to predict the choice of particular routes by cyclists?

Section 2 of this paper provides an overview of the solutions used in transport modelling, with a particular emphasis on cycling. Measures and factors that influence the modal split, the volume of bicycle traffic, cycling activity analysis techniques, and methods of estimating and predicting the volume of bicycle traffic are characterised. Section 3 presents the methodology for modelling bicycle traffic, with a particular emphasis on modelling the demand for cycling and the route choice. Section 4 presents the use of the modelling methodology in studies of the impact of cycling network development on transport demand and bicycle traffic assignment. A discussion and conclusions of the research and its results are included in Sections 5 and 6, respectively.

2. Modelling of Bicycle Traffic and Modal Shift in Cities: A Review

Transport models help to assess the effects of planned measures and decide on their introduction. Transport modelling allows the mapping of transport systems using mathematical tools that enable the simulation of transport processes that occur in reality [24–27]. The computer model is an abstract representation of the actual transport system, which allows researchers to study processes without having to experiment in the actual system environment. The transport model enables an analysis of the flow of people and goods in a network within an area with specific land use and socioeconomic characteristics [28]. The EU requires that the comprehensive implementation of transport measures be based on travel demand and transport network models in order to facilitate the calculation and comparison of economic indicators resulting from this implementation. Models help to supplement engineering knowledge with simulated results [29].

2.1. Measures Influencing Travel Demand and Modal Shift in Cycling

Motor vehicle congestion is considered one of the main problems in urban transport, as it contributes to longer travel times [30–33], increased energy consumption, and pollutant emissions [34,35] which, in turn, lead to a poorer quality of life in cities [36–38]. Therefore, it is necessary to promote infrastructure and technological development, regulatory instruments, and social change to reduce the impact of mobility demand on energy consumption and the environment [39]. Research on energy consumption and emissions in transport covers several areas of interest including, but not limited to, infrastructure measures, environmental impacts, analysis of transport policy, urban travel patterns, and lifestyles [30,40–42]. New alternative fuel technologies (e.g., electricity, natural gas, and hydrogen) are ways to break the dependence on fossil fuels in order to reduce energy consumption and traffic-related emissions. However, it is important to look for ways to change transport behaviour in the short term, as long-term technological solutions have not yet been fully realised [43–46]. Another way to reduce energy consumption and GHG emissions is to reduce VKT (vehicle kilometres travelled) and vehicle ownership rates. Due to the high level of energy consumption in the production phase, decreasing VKT alone is not sufficient—the number of vehicles owned must also be reduced [47,48]. Overcoming these negative trends is possible through the implementation of cycling, walking (active mobility or active travel), and smart mobility solutions that support them (e.g., intelligent transportation systems services, the Internet of Things, or Mobility-as-a-Service (MaaS), with active travel as one of the options) [39,49–52].

Walking and cycling can make a significant contributions to addressing urban challenges such as air pollution, health, and energy consumption [34,35,37,38,53–59]. Negative consequences of energy production and consumption are observed for both car use and cycling. The negative impact of motor vehicles arises from fuel consumption, the energy used to manufacture and maintain the vehicle, and the production and distribution of fuel. The energy consumed in the production and maintenance of a bicycle, and the increase in food demand (energy consumed in the production and distribution of food) resulting from the increased activity of a cyclist compared to a car driver, should also be taken into account in the case of bicycles [59,60]. The greatest energy savings can be achieved by reducing the fuel consumption of cars (non-renewable energy sources, dominating in the automotive industry) and by reducing the number of cars produced, which can be facilitated by increasing the demand for cycling [59]. Compared to private motorised transport, bicycles are healthier, cheaper, and more environmentally friendly, and can be more efficient in terms of travel time for short and medium distances in urban areas [33]. In recent years, we have seen an increase in the availability of electric bikes (e-bikes), which can increase the number of bicycle trips [61]. The e-bike can replace the car for distances that are considered too long for a conventional bicycle and, thus, can contribute to the development of sustainable mobility, both at the local and regional levels [62]. Surveys of commuters suggest that e-bikes can replace around 50% of car trips [63,64]. Furthermore, research shows a decrease in VKT per e-bike user of 20–28% [65,66], and a decrease in the distance travelled by car as a total modal share of around 10% [67]. Current developments in urban cycling include the introduction of adaptive systems for electric bikes to increase the number of users [68,69]. Such enhancements can improve cycling comfort through the use of innovative technologies that are usually found in individual motorised transport.

The European Commission, in the Green Paper [70], identified the main challenges in the field of urban transport development. The challenges of prioritising cycling are to promote active mobility, to promote eco-driving, to reduce motorised traffic, and to change transport behaviour and the ways in which urban communities perceive transport. Various strategies to reduce congestion were analysed in the research conducted by the OECD and the European Conference of Ministers of Transport [71]. The research identified the need to strengthen public transport and non-motorised modes of transport (walking and cycling), while implementing traffic management strategies to effectively reduce congestion. Research on the effectiveness and cost of congestion reduction strategies shows which types of measures are most successful in reducing congestion at the most cost-effective level [72]. Research shows that multimodal transport options, including cycling, are the most effective group of measures due to their lower implementation costs. Congestion, and its negative impacts on the environment and energy consumption, can be reduced by the modal shift from motor vehicle to bike. Cycling is widely considered an appropriate alternative to motorised traffic. Therefore, policymakers and city authorities aim to improve sustainable mobility by developing and promoting cycling [73,74]. State-of-the-art evidence suggests a positive correlation between comprehensive cycling promotion/development and a modal shift toward active mobility modes [75–79]. Changing transport behaviour is time-consuming and particularly difficult. Mobility means, among other things, the ability to move and make choices in transport; therefore, the role of mobility management in cities is to stimulate informed choices [80]. Changing lifestyles to adapt to new services and systems should aim to use alternative modes of transport instead of the car, and to reduce demand for transport [23,81–83]. A change in transport behaviour can be achieved, for example, by implementing the compact urban development concept, which aims to reduce the use of urban space for transport and reduce the demand for transport—especially motor vehicles [84–87]. There is also research showing that even the perception of "nice architectural design of residential, civic buildings and/or street furniture" can be positively related to cycling [88]. In addition to measures related to appropriate land use and urban road network planning, the following infrastructure and organisational measures should be recommended to cities in order to encourage modal shift and improve conditions for cycling:

- Development of bike lanes, paths, and bike highways [58,89–93];
- Development of traffic control strategies geared towards cyclists and pedestrians [94–97];
- Bike-sharing schemes [33,98–104];
- Improving cycling parking facilities [89];
- Traffic calming and introduction of lower speed areas [58,84];
- Mobility management measures that include educational campaigns, access restrictions (e.g., car-free areas), MaaS solutions, and congestion charges [84,105–109].

The most significant effects of changing the modal split can be achieved if infrastructural measures and those promoting active mobility are combined with measures that discourage car traffic, such as parking fees, congestion charges, or the removal of parking spaces.

2.2. Modelling of Transport in Cities

Models of transport demand and street networks can be developed at different levels of detail, i.e., macroscopic, mesoscopic, or microscopic [110]. Macroscopic models [19–21,111] are generally developed to analyse the flow of people and goods or vehicles over a larger area and at a lower level of detail (mainly due to limitations in research tools and data availability). Mesoscopic models allow us to simulate and study the movement of vehicle groups in road sections and intersections [112]. The third group is microscopic models, which are developed to take into account the behaviour of individual vehicle drivers [113]. Modelling of transport and traffic in heavily urbanised areas requires consideration of several factors [114]. These factors include a high density of buildings and dynamic changes in spatial development, a high density of the transport network, a high level of traffic volume, and a significant number of transport modes. The high risk of the occurrence of incidents that affect the level of reliability of the transport system should also be taken into account.

A typical model consists of several basic layers. The first element is a network model. The point (e.g., junctions, public transport stops) and linear (e.g., sections of streets, public transport routes, bicycle routes, walking routes) elements of the transport network are modelled using graph theory. The second element is the transport demand model. The demand model characterises the transport needs of particular groups of travellers in terms of the size of the needs, the origins and destinations of trips, and the motivation of traffic users. The issues of modelling demand using the classic four-step transport model are widely discussed in the literature, e.g., [115,116]. Further elements, such as transport interaction models, make it possible to take into account the relationship between the model parameters—for example, in the processes of traffic assignment or estimation of modal split and modal shift. Much research has led to the development of software packages that help to model transport demand and transport networks (supply) or their elements, and to forecast traffic or passenger trips (e.g., EMME, INTEGTARION, PTV VISUM, SATURN, TRIPS, AIMSUN, SUMO, etc.) [117,118].

Many cities in Europe, including some in Poland, have developed their models of transport networks and travel demand [119,120], but do not make full use of them. One reason for this is insufficient cooperation between the actors who use the models [29]. Many cities use models mainly at the macroscopic level, and even if they use models at the mesoscopic and microscopic levels, they do not integrate them in terms of data exchange between different models and transport authorities [121].

2.3. Modelling of Bicycle Traffic

Bicycle traffic models allow the resolution of problems related, among other things, to demand and supply [116,122], route choice [123], lane change, and queueing behaviour [124–126]. The demand for cycling, the choice of bicycle routes, and the choice of cycling modes were found to be influenced by many factors, including the built environment [127–130]; socioeconomic [131–134], psychological (habits, attitudes, norms, stress), and physical characteristics [135–139]; policies that promote cycling [46,59,84,138,140–142]; infrastructure for cyclists [39,143–145]; cost; effort; distance travelled; travel time; road safety; climate and weather; and travel motivation [146–150]. The frequency of commute to work, the time to cycle, and the length of the journey are important features of active commute behaviour. For example, these variables can be used to estimate the degree of physical activity that comes from commuting. The degree of physical activity was found to vary according to sex, age, and type of activity [151,152].

Several techniques for analysing cycling activity have been developed. Overlay mapping techniques or sketch-plan methods are useful for planning and prioritising [153,154], but are usually not calibrated based on actual bicycle counts. In the direct demand model, regression analysis is used to predict cycling traffic in the network [155–157]. Other methods use graph theory centrality estimates to distribute cycling volumes collected throughout the network [158,159]. In a study conducted in Southern California, prediction models of the number of bicycles were developed using linear regression, with the number of trips as a dependent variable, and measures of demographic and spatial development as independent variables. Four explanatory variables on land-use planning and transport systems (bus frequency, land-use mix, population density of inhabitants under 18 years of age, and proximity to the cycling network) were used to predict the number of bicycle trips at the junctions in Santa Monica [160]. The San Diego model was limited to two explanatory variables (employment density and length of nearby multifunctional paths) to predict the number of bicycle trips [161]. The study of bike modelling was used to explore the possibility of using space syntax in cycling modelling. A measure of space syntax representing direct paths in the street network was combined with land-use variables to predict the volume of cycling in the Cambridge network [132].

In [22,162], the results of agent-based modelling are presented to simulate cycling in the city of Salzburg. The model mainly aimed to test the role of the surrounding Salzburg area for inner-city cycle traffic. The results indicate that commuters from the surroundings have a significant impact on the spatial distribution of traffic in the city. Agent-based modelling proved to be a useful alternative to conventional transport modelling, as intuitive parameterisation allows for exploratory system analysis. MATSim agent modelling has been extended to include infrastructure attributes that influence cyclist route choice [163]. In the case of conventional (usually four-stage) models, most of the previous works on modelling cycling activity use transport zones as an analytical unit.

Due to the growing popularity of cycling in obligatory trips, some cities have decided to develop a model of cycling trips, which is a tool to support decisions at the stage of planning the development of a cycling system, similarly to other macroscopic conventional four-stage transport models. Some regional trip demand models have been adapted to estimate the share of cycling in a defined area [164,165]. Models can include different modes of transport with or without taking into account freight-related traffic. In both cases, cycling is one of the modes of transport. Research on modelling guides the planning of the development of cycling measures [166–168], and presents methods of bicycle traffic modelling [121,142,169–171].

As the analysis of approaches in the literature shows, the bicycle traffic model can be developed in two ways: by separating cycling at the stage of a modal split procedure, or by beginning travel demand modelling based on the results of surveys (not included in the modal split procedure). The second approach is used most frequently in Poland. Bicycle traffic models are usually independent of traffic and public transport models. Such a solution limits the possibility of analysing the impact of—for example—the development of the public transport system on the transport behaviour of the inhabitants in terms of changing the probability of choosing a bicycle to travel. The characteristics of the infrastructure of the cycling system, which may determine the choice of this travel mode as an alternative, are not taken into account.

The transport system models are usually estimated at the level of the traffic analysis zone (TAZ), and are not precise enough to capture the level of cycling activity at intersections. Models that study behaviour at an individual level are useful in identifying the factors that determine the choice of mode and route, but it is difficult to estimate the volume of bicycles. Most of these models are based on detailed data from household surveys, the collection of which can be expensive. Walking and cycling measures cannot be assessed under the conventional modelling procedure because transport demand and network forecasting models usually do not include walking and cycling trips. Furthermore, variables of pedestrian and bicycle links are rarely developed, or the attention necessary to accurately reflect walking and cycling routes is not given to them. Most demand models are not detailed enough to reliably take short trips into account. Models of transport networks usually do not include many local streets. Cycling routes are often located on local streets and alongside other public areas (seashores, rivers, etc.) that are not important from the traditional travel demand point of view.

It is very difficult to model infrastructure improvements that affect walking and cycling when these routes and connections are not part of the modelling network. Therefore, it is necessary to introduce local connections into the cycling network model, and to parameterise the connections available to cyclists. One of the most developed cycling models in Poland so far is the Warsaw Bicycle Traffic Model [172]. This model includes dedicated generation and distribution travel models as elements of the four-stage model, in which the modal shift is omitted due to the limitation of the model to one mode of transport. The network model includes routes dedicated to cycling (e.g., cycle paths and bicycle contraflow lanes), as well as road sections on which bicycle traffic is present according to general principles and shared walking and cycling routes.

3. Methodology of the Model of Bicycle Traffic Development

The model was developed based on the Multilevel Transport System Model (referred to by its Polish acronym, MST) for the city of Gdynia. The bicycle traffic model allows the application of the MST in the planning of new bicycle connections, taking into account modal shifts. The structure of the MST (including the macroscopic model that is the basis for the development of the bicycle traffic model) is discussed in more detail in the article describing its previous applications [23], and in the report by Oskarbski et al. [173]. The bicycle traffic model was developed within the FLOW project (Furthering Less Congestion by Creating Opportunities for More Walking and Cycling) [174] to extend the four-stage macroscopic model of the MST with a cycling layer. Therefore, the condition for the development of a bicycle traffic model, in this case, is the prior development of a four-stage macroscopic model of motorised traffic and public transport. Cities that do not have a macroscopic model should take into account the development of a comprehensive model for public transport and private motorised traffic, including the cycling part of the model in the planning stage. Figure 1 shows the main elements of the model, indicating which elements come from the MST and which elements were developed within the scope of the bicycle traffic model (elements within the scope of the cycling model are marked in blue). In terms of input data, only the data for the modelling of bicycle traffic are presented in Figure 1. Sociodemographic data and data used in the model calibration process (traffic volumes) were also updated as part of the model's development. The elements of the introduced bicycle traffic model are characterised in detail in Sections 3.1–3.4 and 4, using the example of the city of Gdynia. Speed models and stochastic impedance function parameters have been developed based on data measured in the Gdynia road network. Therefore, references to the example of Gdynia are presented in the description of the methodology to better illustrate the individual elements of the model. If the model is developed for another city, it will be necessary to verify the elements of the Gdynia model in terms of speed and stochastic assignment parameters. The transport network, including bicycles, was modelled with the use of PTV VISUM software [175], which is an additional constraint for cities that do not have access to dedicated software.

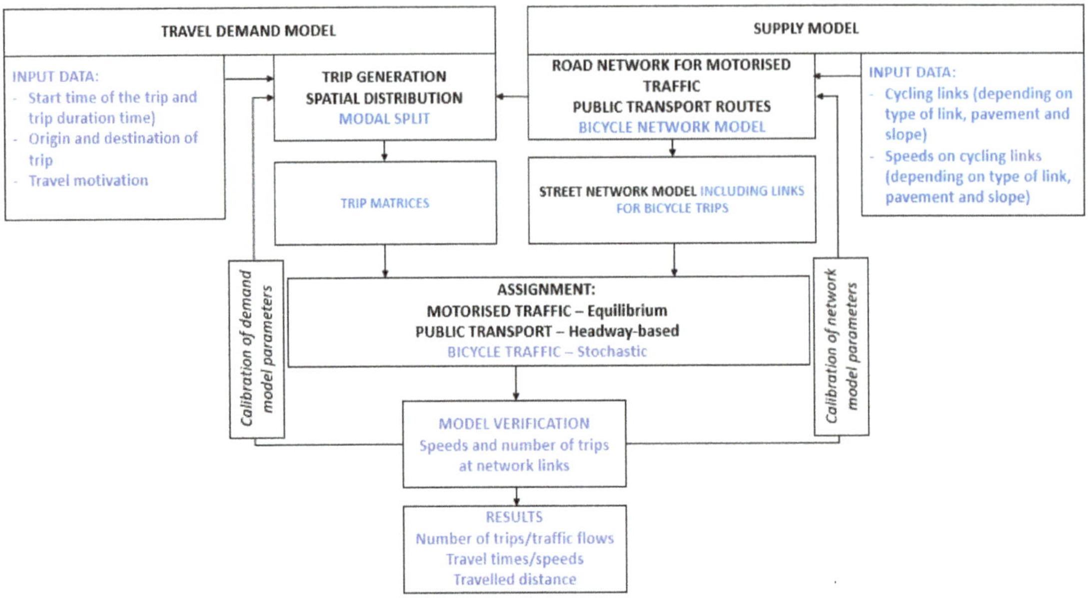

Figure 1. Integration of the bicycle traffic model into the MST.

3.1. Bicycle Network Model

The road network model in the macroscopic transport model was extended to include the modelling of important cycling links. Different types of links were developed to represent particular types of route in the bicycle network regarding the typical speed of cyclists and their separation from other traffic users, including vehicles or pedestrians. The types of links are also related to the technical class of the road along which they are located. The model also includes parameters that characterise the surface and slope of the bicycle routes. The following new attributes were developed for selected sections of the transport network, and different values of the following parameters were assigned to them (discussed in more detail later in this section):

- Longitudinal gradient;
- Available cycling infrastructure with a specification of link types: bicycle path, walking and cycling route (pavement), asphalt road on which bicycle traffic is present, together with motorised vehicles, according to general principles;
- The surface of the bicycle route, which is divided into the following types: asphalt, concrete blocks, and flagstones.

Longitudinal gradients were calculated based on the data from a numeric map provided by the Geodesy Department of the Gdynia City Hall. Each link between two nodes for which the altitude is known was assigned to one of the longitudinal slope classes. The longitudinal gradients of the cycling routes in Gdynia are shown in Figure 2. The longitudinal gradient is an important variable on which cycling speeds depend. Speeds and associated travel times were used as input parameters both in the modal split model and in the assignment of bicycle traffic presented later in this paper.

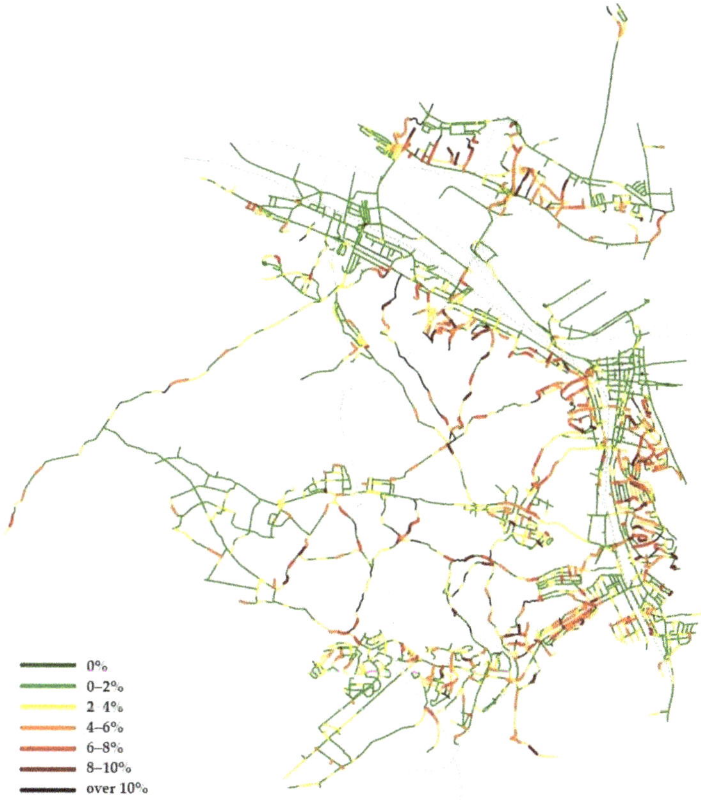

Figure 2. Slopes of bicycle routes in Gdynia.

The longitudinal gradient data were correlated with the average speed of sections with a given slope to determine the relationship between these variables. In addition, variables that describe the type of bicycle route and the type of road surface were also included (Section 4.2 shows the results of the survey). Based on the calculated speeds Vel_i for each bicycle route section i of length L_i (Equation (1)), the travel times TT0r along the bicycle routes between the transport regions (Equation (2)) were calculated to be used to determine the usefulness of bicycle transport. The estimated average speeds were mainly related to cyclists who regularly participate in the European Cycling Challenge (ECC), who represent a group of frequent cyclists. Similarly, bicycle speeds were studied using the Global Positioning System (GPS) and ECC data in Bologna [176]. Additional studies should be carried out shortly to make the results more reliable due to the possible influence of occasional cyclists but, currently, the majority of bicycle users are regular cyclists who influence the modal split on typical days of the week (these cyclists were taken into account in the development of the models).

$$Vel_i = 31 - \frac{25.5}{1 + e^{0.22 - 0.4 \cdot Slope_i - 0.07 \cdot Type_i - 0.077 \cdot Surf_i}} \quad (1)$$

Vel_i—average speed of link i;
$Slope_i$—slope of link i;
$Type_i$—type of road of link i;
$Surf_i$—type of surface of link i.

$$TT0r = \sum_{i=1}^{n} \frac{L_i}{Vel_i} \quad (2)$$

$TT0r$—travel time on bicycle routes between transport zones;
L_i—length of bicycle route link i;
Vel_i—average speed at link i.

The function Vel_i (Equation (1)) was developed by the authors of this paper, and allows the presentation of the dependence of the average speed of cycling on the longitudinal slope and the selected types of cycling link (Figure 3). Before applying the Vel_i function to the bicycle traffic model, the speed distribution functions were determined for the bicycle sections at 5 km/h intervals (in the range of up to 40 km/h). In the validation process, the sampling rate distributions in each speed interval were compared for the model dataset and the control observation dataset. The mean absolute percentage error (MAPE) for the mean speed calculated from the observation data from the control group and the model results did not exceed 12% (values in the upper error range were observed for steeper gradients for downhill cycling). The longitudinal gradient was the most decisive factor in the value of the average speed. The results were found to be satisfactory for planning purposes, as confirmed by a subsequent verification of the model presented in Section 4.5.

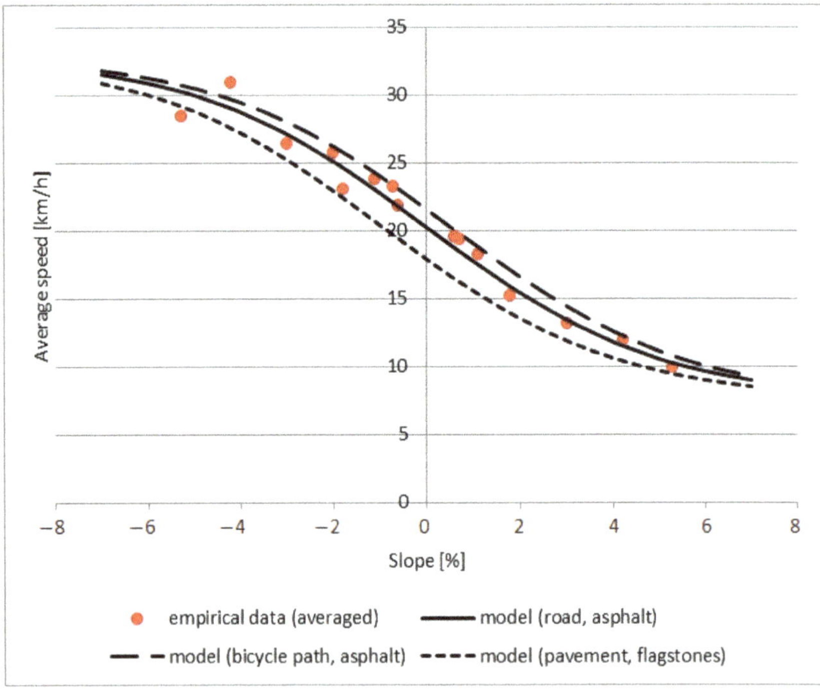

Figure 3. The relation between the average speed Vel_i of cycling on the longitudinal slope and the characteristics of the route.

In future studies, the presented model should be further developed by taking into account different user groups, e.g., by age or cycling experience.

3.2. Modelling of Demand for Bicycle Trips

The data collected (in the surveys listed in Section 4.1) were processed to develop an origin–destination matrix for bicycle traffic. All bicycle trips in the database are connected with attributes that describe:

- The start time of the trip and trip duration time;
- Origin and destination transport zones (TAZs), connected to the network through centroid connectors;
- Travel motivation;
- Modes of transport chosen by the respondents (who are also divided into groups, e.g., by age).

Based on the above data, cycling matrices were developed for the day, and separately for the peak hours of 7:00–8:00 (morning peak) and 15:00–16:00 (afternoon peak), in 2016 and 2017 (according to the biannual transport survey and traffic measurements, the hours with the most trips and the highest traffic volumes in Gdynia). The matrices were used to calibrate the MST model [173], which was developed with the use of PTV VISUM software.

The stages of generation and spatial distribution of trips in the MST presented in the report by Oskarbski et al. [173] were left unchanged when the model was extended to include cycling. Models of trip generation include explanatory variables (e.g., number of inhabitants, total number of jobs, number of people over 6 years of age, number of places of education in primary schools, junior high schools and secondary schools, area of service and commercial buildings, number of jobs in the service sector) in particular transport zones. The spatial distribution (trip matrices, including cycling mode) was developed using a gravity model. The share of morning and afternoon peaks in cycling traffic was determined based on traffic analyses for individual motivations. O–D (origin–destination) matrices were calculated for all zones in the MST based on sociodemographic data, compared and calibrated with O–D matrices estimated based on survey results. The bicycle modal split was calculated based on the survey results in each zone. Then, the modal split procedure was applied based on the travel times for each mode, as described in Section 3.3.

3.3. Modelling of Modal Split

Development of the MST to include cycling was carried out in the field of changing the modal split and shift models, changing the parameters, and extending the transport network model by sections important in the assignment of cycling traffic to the network.

The basic model based on which dependencies were built is a logit model. A utility function was defined for each of the transport modes. The utility of a given transport mode should be understood as the degree of its attractiveness, defined by any number of variables. This utility is determined for each alternative as a function whose variables are factors that significantly affect the attractiveness of travel by a given mode of transport (e.g., travel time). Due to the later assumptions of the logit model, utility is expressed as a component of two elements: the measurable utility V_{ij}, and the random part ε_{ij}, according to the formula given in Equation (4) [116]. This approach first allows us to grasp the different preferences of users and their subjective, perceptible attraction to the alternative (e.g., not every change from one mode of transport to another is as burdensome). Second, it allows users to take into account user errors in utility assessment (underestimation, overestimation). Third, it allows for the capture of all directly unmeasurable random factors influencing the choice of modes of transport (e.g., comfort, habits). The utility functions described by the selected variables were used in models with discrete selection. Discrete logit selection models allow a utility to be determined for each alternative separately, and the probability of selecting a particular mode of transport is determined by the difference between them. Thus, according to the above, each of the transport modes m has been described by a utility function $V^{m}{}_{ij}$ for the travel relationship between the origin of the trip i and the destination of the trip j. The logit model assumes that the user chooses the option with the highest utility from the available alternatives. The utility functions characterised in this way were used in the discrete choice model, whose parameters were adjusted using the generalised extreme value (GEV) model, which is consistent with the principle of maximisation of the stochastically defined utility function. The logit model was used to estimate the probability of choosing individual modes of transport (Equation (3)). The following available variables were used in the model: travel distance between i and j, travel time by bicycle, travel time by private transport, and perceived travel time by public transport [116,177,178]:

$$P_{gn} = \frac{e^{\mu_n U_{gn}}}{\sum_{g \in n} e^{\mu_n U_{gn}}} \cdot \frac{e^{\mu_g [\frac{1}{\mu_n} log(\sum_{g \in n} e^{\mu_n U_{gn}})]}}{\sum_{n=1}^{m} e^{\mu_g [\frac{1}{\mu_n} log(\sum_{g \in n} e^{\mu_n U_{gn}})]}} \quad (3)$$

P_{gn}—the probability of choosing the nth mode of transport belonging to the gth group of transport modes;
U_{gn}—utility of the nth mode of transport belonging to the gth group of transport modes;
μ_n, μ_g—model scaling coefficients.

$$U_{ij} = V_{ij} + \varepsilon_{ij} \quad (4)$$

${V^m}_{ij}$—measurable utility of transport modes m;
${\varepsilon^m}_{ij}$ —random variable with a logistical distribution reflecting values not included in the utility ${V^m}_{ij}$.

For each mode of transport (p—walk trip; R—bicycle trip; TI—private transport trip; TZ—public transport trip), the authors of this paper developed a nested logit model with the following utility functions:

$$\begin{gathered} V_P = \beta_{10} + \beta_{11} \cdot DIS \\ V_R = \beta_{20} + \beta_{21} \cdot TT0r \\ V_{TI} = \beta_{30} + \beta_{32} \cdot \frac{TTC}{PJT} \\ V_{TZ} = 0 \end{gathered} \quad (5)$$

DIS—travel distance (km);
$TT0r$—calculated time of cycling trip (min);
TTC—calculated travel time by private transport (min);
PJT—perceived travel time by public transport (min);
β_i—equation coefficients.

3.4. Bicycle Traffic Assignment

The calculation of the traffic flow was performed using stochastic assignment. Each link was characterised by attributes that describe the type of surface and the slope class, leading to a better representation of cyclist behaviour in the modal split and the route choice in the Gdynia bicycle traffic model. Important elements determining the assignment of bicycle traffic were to take into account factors that are random or not random. Random factors were taken into account in the stochastic assignment to calibrate the model. Factors that influence the choice of the route include the following characteristics and behaviours of cyclists:

- Traffic participants do not have complete knowledge of the transport network, which means that they do not always choose the route rationally according to their preferences [179,180];
- Different routes are sometimes chosen according to the need for variety [181,182];
- Cycling times on different routes can vary from day to day due to the influence of other traffic participants, slope, traffic signals, congestion, weather conditions, etc. [127,182–187];
- Different cycling habits make certain routes more attractive to certain groups of cyclists [135–139,186];
- Cyclists may have different preferences on the choice of more attractive routes, for example, due to the quality of the surface or the surrounding landscape, and different risk awareness [181,182,188,189];

- Cyclists have different levels of tolerance to weather conditions [89,149,150].

Stochastic assignment is by far more appropriate for bicycle trips, since it reflects the process of individual discrete choice. The variety of individual choices is more important for cyclists than volume-dependent travel times. Stochastic assignment assumes that the selection of the route is caused by changes in the subjective perception of the inconvenience of cycling in a given sequence of sections of the transport network. Stochastic assignment procedures assume that traffic participants, in principle, select the best route, but evaluate individual routes differently due to incomplete and different information. It is assumed that the resistance (impedance) of the route sections consists of two parts: deterministic, and random. Each link in the transport network model is characterised by information on the type of surface and the longitudinal slope (included in the Vel_i function, which characterises the individual sections of the cycling network), which affect the comfort of travel. Slopes directly influence the average cycling speed, which means greater resistance to uphill sections with high slopes. These factors greatly determine the choice of the route, and their application in the model aims to better represent the behaviour of cyclists in the transport network.

The stochastic assignment was performed using PTV VISUM software. To define the impedance of each section in the bicycle network, the authors of this paper calculated the travel time for each section of the bicycle network based on the Vel_i function (i.e., taking into account the type of section of the bicycle network, the longitudinal gradient, and the type of surface). The impedance of the routes is due to the impedance of individual sections of the cycling network. The results of the research, presented in Section 4.2, were used to develop the Vel_i function (Section 3.1) and parameterise the impedance of the section in the bicycle traffic model with travel time value.

According to the assignment procedure, not only is the shortest route found, but many alternative (but similar) routes are also found using a multiple best path search and a variation in the impedances of individual sections of the cycling network. The stochastic assignment procedure involves 5 steps: (1) route search considering the entire cycling network for the current impedance; (2) calculation of route independence (or commonality) based on the overlap of all routes for an origin–destination pair; (3) distribution of demand over the routes of each origin–destination pair, considering route independence (or commonality); (4) repeating the distribution of demand until demand for all origin–destination pairs is in equilibrium; (5) repeating steps 1–4 until no new routes are found, or the change in the number of connections between the two iteration steps is very small. The procedure is divided into external and internal iterations. The external (global) iteration is used to find routes. The internal iteration is used to assign the traffic volume to routes. This loop is repeated until the impedance deviations on the network elements and the bicycle traffic volume deviations on the routes between the two iteration steps are very small. The enhanced stochastic assignment procedure used in PTV VISUM is described in more detail by Szabo et al. [190].

The model was calibrated to achieve better goodness of fit between the model output data and the observed data by adding time penalties at major intersections that cause delays for cyclists (extensive intersections with traffic signals). The results of model verification are presented in Section 4.5.

4. Testing the Model on the Gdynia Case Study

Gdynia is a young and dynamically developing city, founded in 1926 and located on the Baltic Sea coast in the northern part of Poland; it has a population of almost 250,000, and is mainly made up of people of working age (151,000). The increasing number of motor vehicles is a potential threat to the attractiveness of the city centre, which is why the city is currently taking measures to promote sustainable transport development. Public spaces need to be designed efficiently and well. The number of inhabitants in Gdynia is decreasing; many of them move to neighbouring communes, where the population is rapidly increasing. At the same time, most people in these communities work or study in Gdynia. The average distance between home and work or study is steadily increasing which, together with the higher affordability of private cars, results in a systematic increase in the share of private car trips. The factors mentioned above (related to land use planning and the growth of motorization) can cause increased traffic congestion, travel time, accident costs, noise pollution, and exhaust emissions, as well as increasing the number of areas designated for cars (roads and parking spaces).

Regular monitoring of the transport preferences and behaviour of the inhabitants of Gdynia has been carried out for more than 20 years by the Gdynia Management of Public Transport (ZKM). Research is conducted every 2–3 years through an individualised interview survey, standardised on a representative sample of 1% of the population. The survey conducted in 2018 allowed us to estimate the following modal split for urban trips: 11.4% of respondents declared a pedestrian trip, 48.9% by car, 37.1% by public transport, 2.1% by bicycle, and 0.6 % others, including motorcycles (surveys were not conducted in 2020 due to COVID-19). The share of public transport in travel does not increase significantly: in 2013, for non-pedestrian trips, it amounted to 45.7%; in 2015 this figure decreased to 39.8%, and in 2018 it increased to 41.8%. We can also observe an increase in the share of cycling in non-pedestrian trips (2013: 0.8%; 2015: 1.8%; 2018: 2.4%) [191,192]. The development of cycling and walking could be a solution to the increase in trips made via private motorised transport. Therefore, it is crucial to increase the attractiveness of public transport and non-motorised modes of transport. The city has taken a series of actions to meet this challenge. First, these actions are related to the improvement of pedestrian and bicycle infrastructure (aiming at achieving continuity of bicycle routes, improving accessibility). Second, traffic organisation aims to prioritise pedestrians and cyclists over private motorised traffic within the city centre (traffic restrictions, traffic signal settings). The city has also implemented measures aimed at creating attractive spaces for pedestrians and cyclists, with facilities for them (including pilot shared-space projects).

The basis for informed and effective actions is the analysis of planned and implemented measures in the transport system using tools such as integrated transport demand and network models. The aim of Gdynia's participation in the FLOW project-(Furthering Less Congestion by Creating Opportunities for More Walking and Cycling) (HORIZON 2020) was, among other things, to develop tools used in the analysis of bicycle traffic. Gdynia uses a three-level transport model (MST) to model motorised private and public transport. The MST was developed as part of the CIVITAS DYN@MO project—"DYNamic citizens @ctive for sustainable mObility" (Seventh Framework Programme of the European Commission). Participation in the FLOW project allowed the extension of the transport model to include cycling, so that it could be used as a tool for planning, making decisions, and analysing the impact of cycling on traffic conditions. The results of the analyses carried out with the use of the MST updated with the bicycle traffic model help the municipal services in public consultations and the selection of effective measures to increase the share of cycling in daily trips. The transport network, including bicycles, was modelled with the use of PTV VISUM software [175]. The cycling model has been integrated into the existing macroscopic model, which is part of the MST. The model includes the same transport zones, but the network has been extended with important cycling parameters. The model includes the transport network of the city of Gdynia, with 173 TAZs. Each TAZ is characterised by variables that determine the generation of trips. The transport network model includes more than 5500 links (characterised by technical class, cross-section, capacity, and free-flow speed) and around 2100 nodes. The functions developed based on comprehensive traffic surveys conducted in Gdańsk in 2009 and the Gdynia study of the preferences and transport behaviour of the inhabitants in 2013 [191] were used to calculate the demand for trips and determine the transport behaviour of the inhabitants in a four-stage model [116,173].

4.1. Data Sources

Due to the different availability of data and significant costs of their acquisition in various cities, suggested data that may be used in the development of a bicycle traffic model are presented in the example of the city of Gdynia. When planning the development of a cycling model, it is advisable, in the first step, to diagnose the availability of data and to plan additional field research or the acquisition of data from other sources (e.g., mobile phone data, data from navigation systems, data from ITS services), taking into account the costs of data acquisition. The macroscopic model of Gdynia's bicycle traffic was developed using the following data (cycling data were collected in May, June, and September):

- The biannual transport survey (conducted by the local public transport authority) was used to calibrate and verify the sum of all trips and the modal split [191];
- Data collected during the European Cycling Challenge 2016 and 2017 (Global Positioning System track of 20,136 trips made within the competition, with speed estimation) were used to verify the spatial distribution of bicycle trips and bicycle traffic assignment, calibration, and speed verification;
- Data collected during the local cycling competition for companies in 2016 and 2017 (general information about all trips made within the competition)—necessary for the selection of transport system improvement scenarios;
- Surveys conducted amongst participants in a local cycling competition for companies in 2016 (1146 bicycle trips) were used to parameterise sections and select improvement scenarios;
- Cyclist traffic research (bicycle traffic volumes) in key city locations (at bicycle route sections and street intersections), along with surveys dedicated to FLOW in 2016 (1208 bicycle trips), were used for calibration and travel matrix verification, travel motivation, bicycle traffic volumes, and parameterisation of sections;
- Public and private transport and MST traffic parameters (VISUM, VISSIM, SATURN) [173], supported by data from intelligent transportation system services that form the TRISTAR system in Gdynia) [193];
- Statistics for individual transport areas in 2016 (e.g., number of inhabitants in particular transport areas, total number of workplaces, number of people aged 6 and over, number of study places in schools, area of service and commercial buildings, number of workplaces in the service sector) were used to update the transport demand model.

4.2. Survey Results

Based on the surveys (conducted amongst participants of a local cycling competition for companies in 2016—1146 bicycle trips and surveys dedicated to FLOW in 2016—1208 bicycle trips), it can be concluded that frequent cyclists travel not only for daily activities, but also for recreational purposes (obligatory trips with home–work or work–home motivation—40%; with motivation to educational places—5%; recreation—38%; other trips can be connected with shopping, for example). Modal split models do not take into account recreational trips, which are likely to have little impact on the modal share of daily trips during typical weekday peak hours.

Cycling surveys in key locations in the city showed that the largest group of cyclists was between the ages of 35 and 50 (32%); the second-largest group was 25–34 years old (25%). The joint share of the above age groups in cycling was more than half of the share of all users combined. The group over 50 years of age amounted to 25%. The least numerous group was children under 12 years of age, whose share in cycling did not reach 1%. In the breakdown by gender, the cyclists were mostly men; women constituted a much smaller group. The gender distribution in cycling can be expressed in a numerical ratio of 3:2.

The results of the research were used to develop the Vel_i function (Section 3.1), and to parametrise the impedance of the section in the bicycle traffic model with travel time value (according to the stochastic assignment procedure described in Section 3.4) in terms of the section type (Figure 4) and surface type (Figure 5), in a possible range of evaluation from 1 to 5.

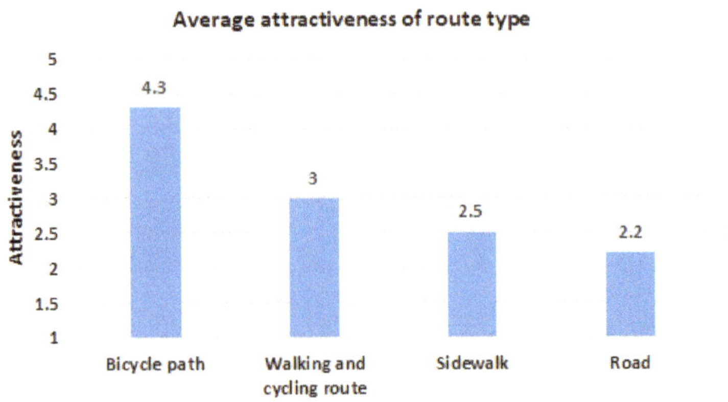

Figure 4. Attractiveness of the bicycle route to the user in terms of the route section type.

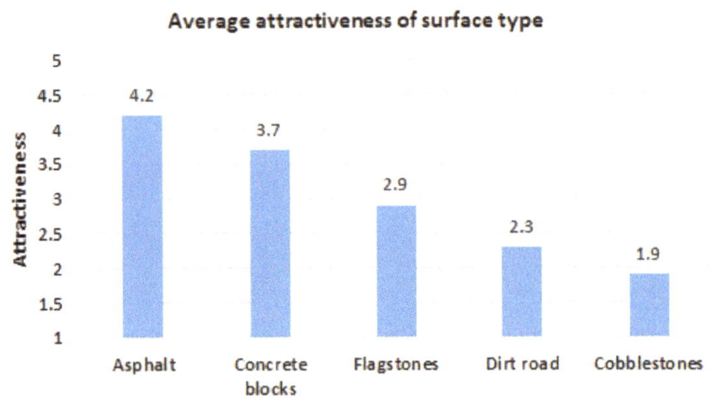

Figure 5. Attractiveness of the bicycle route to the user in terms of the route surface type.

The presented results of the surveys confirm previous findings that cyclists prefer separated on-street infrastructure or off-street paths [127,182–185]. To a lesser extent, cyclists indicated the attractiveness of mixed-traffic sections (marked walking and cycling routes, and sidewalks mostly made of flagstones that cyclists share with pedestrians, as well as road sections shared with motorised vehicles). According to the answers of the respondents, the pavement that cyclists indicated as the most attractive is a smooth surface without irregularities; these are made from asphalt and, to a lesser extent, smooth concrete blocks. Cobblestones (regular stone pavement) and dirt roads (most often forest roads) were indicated as the least attractive and most discouraging to cyclists.

4.3. Scenarios of the Development of the Bicycle Network in Gdynia

The bicycle traffic model was used to analyse various scenarios of the development of the bicycle network in Gdynia. The macroscopic model was used to estimate the effectiveness of the following six planning scenarios for the development of the urban cycling network compared to the baseline scenario (Scenario 0—without implementation of measures) (Figure 6):

- Scenario 1—bicycle overpass between Hutnicza and Wisniewskiego streets;
- Scenario 2—bicycle connection between the city centre and Oksywie across the harbour channel;

- Scenario 3—bicycle path along Wielkopolska and Zwyciestwa streets;
- Scenario 4—seaside bicycle path in Orlowo district;
- Scenario 5—17 December Avenue, a new path in the city centre;
- Scenario 6—completion of bicycle paths in the city centre.

Figure 6. Scenarios for the development of the Gdynia urban cycling network.

4.4. Results of Research Carried Out Using the Bicycle Traffic Model

The results of the simulation include data on bicycle traffic flow, travel time, speed, and the share of bicycle traffic on particular streets. The selected results of the simulation research are presented in Table 1. The results are also useful for assessing the travel conditions for each mode of transport in the network.

The implementation of each of these scenarios leads to a reduction in the average distance travelled, as well as the average and total travel time for cyclists, while at the same time the demand for cycling increases (Table 1). Figures 7 and 8 show example results for Scenario 2. The implementation of Scenarios 1, 2, and 5 has the greatest impact on the increase in cycling utility, although the increase in the number of cycling trips is quite small due to minor changes in the utility of other means of transport. These scenarios assume the development of new cycling routes, which significantly shortens the distance between important destinations in the city. In addition, they provide reliable and fast bicycle connections that are competitive with congested traffic routes during peak hours. These new cycling routes will reduce the average cycling time in the entire network by only ~3%, but will save time between origins and destinations that will be connected by the planned routes by up to 30%.

Table 1. Results of simulations for different scenarios of bicycle network development–afternoon peak hour.

Scenario	Traffic Demand (Cyclists/ Hour)	Bicycle Traffic Demand Increase (%)	Bicycle Network Performance	
			Total Travel Time (Pers-h)	**Total Travelled Distance (Pers-km)**
0—base	1689	–	556.0	12,381.1
1	1710	1.24	549.6	12,370.6
2	1704	0.89	543.4	12,115.0
3	1700	0.65	554.6	12,395.3
4	1693	0.24	553.0	11,490.7
5	1703	0.83	543.6	11,910.2
6	1702	0.77	552.6	12,375.8

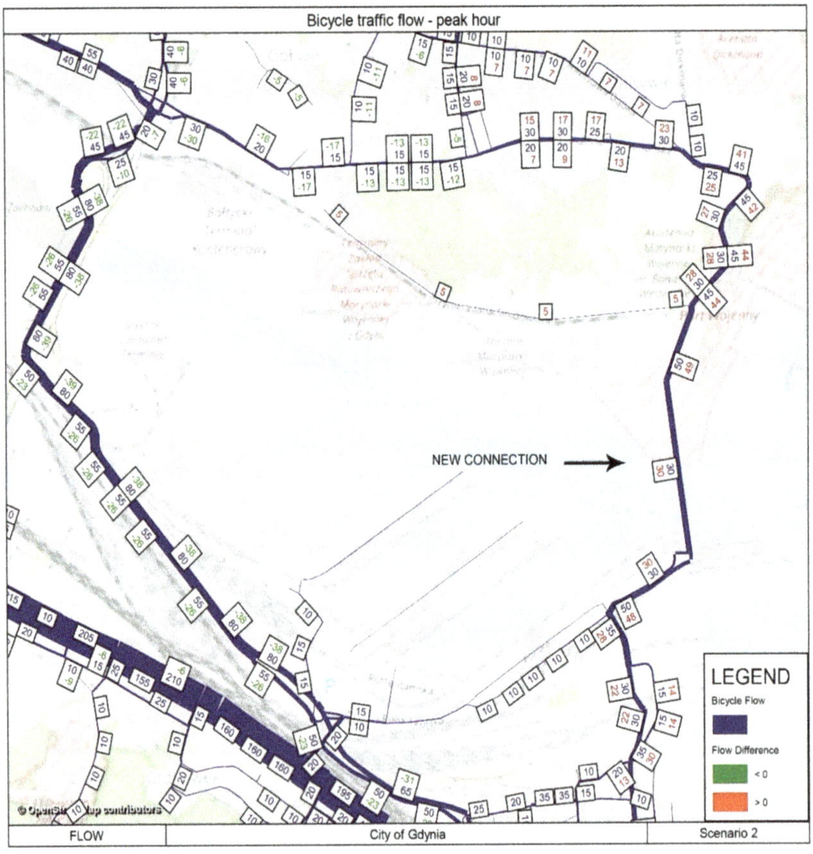

Figure 7. Bicycle traffic volumes in Scenario 2, and comparison of values with the baseline scenario.

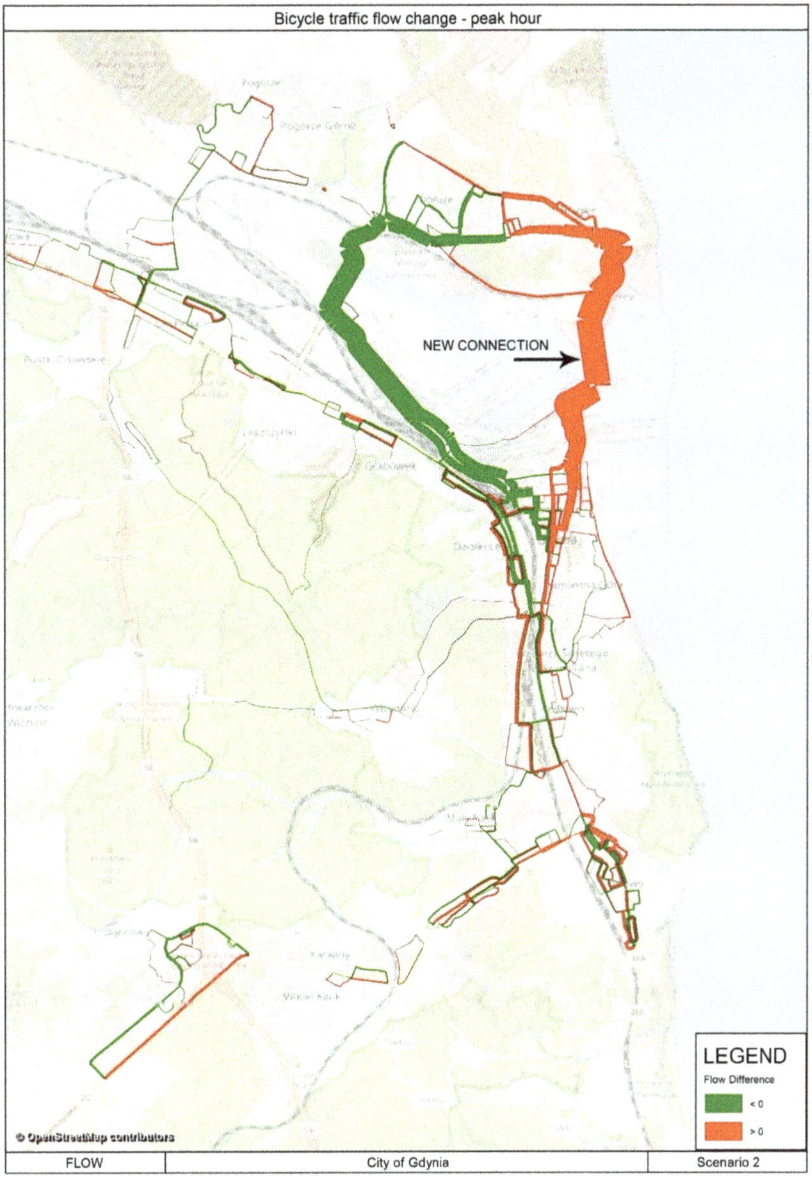

Figure 8. Change in bicycle traffic volumes in Scenario 2 compared to the baseline scenario.

The development of a section of a new bicycle path along the main arteries of the city was planned in Scenario 3. There is a noticeable gap in the network of bicycle paths along the main arteries in the south of the city today, which encourages cyclists to use the surrounding secondary roads. The evaluation carried out using the model showed that the completion of the main bicycle path network will attract cycling, and ~30% more cyclists will use a slightly longer but more comfortable route despite the longer distance.

Although the implementation of Scenario 4 has less impact on the change in cycling demand, the results show that cyclists would be more likely to use a comfortable bicycle path near the sea coast. It is noticeable that when choosing a route, the cyclist prefers to be separated from the car traffic, and not to have to cross the road within the intersections. A similar conclusion can be drawn from the results of Scenario 5, which includes the transformation of a park path in the city centre into a bicycle path. The development of a new high-speed bicycle connection through the city contributes to a significant reduction in travel time, and improves the efficiency of the transport network by reducing the distance travelled by bicycle.

The implementation of Scenario 6, including filling in gaps in the bicycle network in the city centre, has little impact on the change in travel time. On the other hand, a coherent and comfortable cycling network in the city centre encourages cycling.

4.5. Model Verification

The baseline scenario of the bicycle traffic model was verified, taking into account the average speed (Equation (1)), characterised by the described variables, such as the occurrence and type of separation of bicycle routes, slopes, and the type of road surface (Figure 9). For a separate control sample of data from GPS traces not used earlier to develop the model, a coefficient of determination of R2 = 0.94 was obtained.

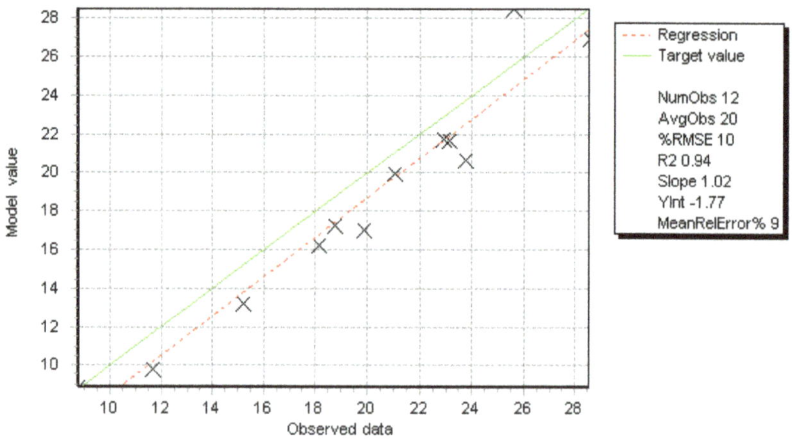

Figure 9. Goodness of fit of the model output data to the observed data from another measurement period for average bicycle speeds.

The model was also verified for the volume of bicycle traffic in particular sections of the transport network. A good fit (R^2 = 0.77) of the volume of bicycle traffic was obtained (Figure 10).

The application of the described approach makes it possible to consider the cycling mode of transport. The attractiveness of cycling is compared to the attractiveness of other modes. This approach makes it possible to determine the impact of road, cycling, or public transport investments or measures on the modal split.

The results show the need to improve the cycling network. The development and completion of the cycling network give the potential to improve the competitiveness of cycling with other modes of transport. The key factor that determines the modal split is the travel time of the different modes of transport. The impact of educational and promotional activities has not been directly analysed, but it may also affect the choice of mode of transport. At present, there is little interest in cycling as a mode of transport in Gdynia. As a result, the increase in the demand for cycling calculated based on the model is also insignificant. Educational and promotional activities that encourage cycling, as well as the introduction of the city bicycle system, can increase the number of cycling trips. These issues will be the subject of separate research. However, the proposed analytical methodology is useful in the planning and decision-making process to select the most effective solutions.

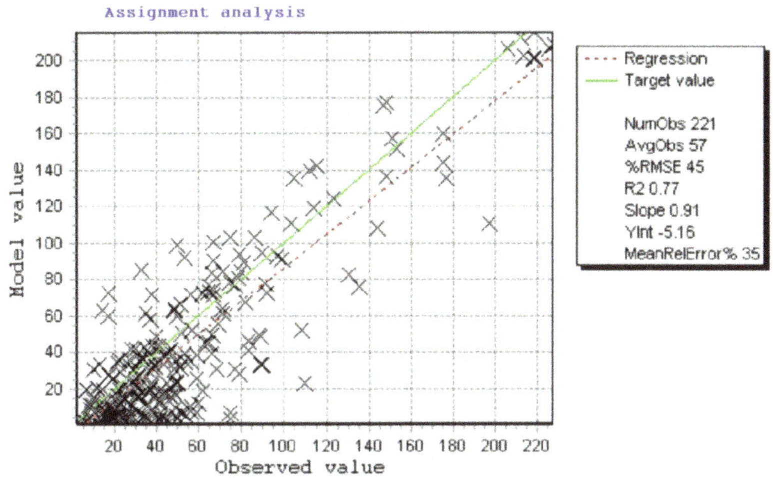

Figure 10. Goodness of fit of the bicycle traffic assignment to the Gdynia transport network.

5. Discussion

Many European cities implement their mobility planning processes following the EC guidelines [21], with the participation of decision-makers, stakeholders, and residents. Demand and transport network models can effectively support the mobility planning and decision-making process, taking into account active travel modes. Demand and transport network models can effectively support mobility planning and decision making. However, the inclusion of active modes of travel (such as cycling) in mobility planning requires knowledge about the impact of such measures on the performance of the transport system, taking into account the modal shift.

5.1. Considering the Modal Shift towards Cycling in Mobility Planning in the Example of Gdynia

It is essential to answer the question of why the consideration of a modal shift in urban transport modelling is important, and how the experience from Gdynia can help in the development of a bicycle traffic model, taking into account modal shift. Transport demand and transport network models have been key tools in shaping transport policy [29], but cities do not use them to their full potential, especially with regard to cycling and its impact on other modes of transport. The reasons for this may include insufficient cooperation between urban planning and transport management authorities, or different stakeholder perceptions of the importance of the models [29]. The perception of cycling in reducing the demand for other modes of transport and reducing congestion in the transport network is rather low. Thus far, cities use transport models to a very limited extent, and very rarely—especially when modelling active travel and its influence on motor vehicle trips and trips made via public transport. One of the reasons for this is the scarcity of reliable, integrated, and comprehensive tools to model the transport system while taking into account cycling traffic at the macroscopic level [19–22]. The tools used for transport modelling most often include the functioning of motor vehicle road transport and public transport, excluding cyclists or treating them as a factor disrupting traffic conditions.

Transport planning strategies consider the promotion and development of cycling as an appropriate way to address the problems associated with increased car traffic, such as energy reduction, pollution, congestion, and limited space. The modal shift towards cycling contributes to reducing the above problems. Cycling influences the demand for motor vehicles and public transport, as presented in the paper, and should be taken into account in feasibility studies of investments regarding the development of infrastructure and organisational measures dedicated to all modes of transport, as well as in the decision-making process. For this reason, it is necessary to develop and use tools and models that allow the evaluation of the impact of planned actions concerning the modal shift.

The model used in the planning and decision-making process provides quantitative results. As a result, an ex ante assessment of the effects of the decision-making process can be made. Effectiveness evaluations of planned measures are based to a greater extent on assumptions or subjective assessments if demand and transport network models are not applied. If an evaluation can be justified by modelled results, the conclusions are easier for residents to understand, and can significantly improve public participation as a complement to other decision support tools and methods [194–196]. The use of integrated models should be recommended for cities that are starting or participating in mobility planning. Such tools must first be developed or adapted to take into account multiple modes of transport, which requires planning and spending, but brings significant benefits in the process of transport planning and management.

The model presented in this paper was developed to support the city of Gdynia in developing and monitoring the Sustainable Urban Mobility Plan (SUMP), as well as in planning and decision making regarding changes in the transport network. Lockdown policies around the world—including travel bans, border closures, restrictions on public activities, and the closure or restriction of businesses, offices, and schools [197]—have changed transport behaviour and travel growth, with the fear of infection in public transport also playing a role. The COVID-19 pandemic allowed the authorities to take several measures to improve the sustainability and resilience of the transport system [198]. In Polish cities (including Gdynia), particular attention has been paid to investments in physical infrastructure, including the expansion of the cycling network, along with traffic calming and the introduction of lower speed areas to stimulate active travel. The application of the model to planning for the development of transport systems, SUMP monitoring, citizen consultation, and decision making is a strong innovation of the presented solution.

In the decision-making process of choosing the most effective solutions, it is justified to take into account the impact of types of cycling sections on road traffic safety, energy consumption, and the environment. The model presented in the paper can be used as a basis for estimating energy consumption and exhaust emissions. This approach has been used, for example, in research in which the transport model was applied to estimate exhaust emissions in the Polish city of Bielsko-Biała [199]. The main barrier to the application of such an approach is the difficulty of obtaining data on vehicle traffic at the local level (in Polish cities). However, in future studies, these aspects will be developed using the model, so an analysis of cycling network development scenarios resulting in a change in the modal split will allow assessment of the impact of such measures on the reduction in emissions and energy consumption.

An important element of the citizen consultation process, but also of improving the work of municipal departments, is to enable the acquisition, geographical visualisation, and analysis of data. In the case of the Gdynia model, the simulation results are presented to the citizens during meetings or through a dedicated website [23]. However, the data calculated using the model may in the future feed advanced solutions such as cyber–physical platforms for real-time management of smart cities. Such platforms could use artificial intelligence methods to manage data and replicate transport model databases [200,201].

5.2. *Aspects of Bicycle Traffic Modelling*

The method of analysis and the results give a different perspective on the possibility of improving travel conditions in the transport network, taking into account the reduction in car traffic and the change in demand for travel via public transport. The model and the methodology of its development, as well as the manner of using the available data, contribute to the filling of the indicated gap. However, in most cases, the evidence base for optimal cycling development, planning, and promotion are weak, and there is not enough information about cycling patterns with sufficient resolution.

Therefore, the next part of the discussion concerns the questions of which data can be used to develop a bicycle traffic model, and which elements of the model allow the prediction of the choice of particular routes by cyclists.

5.2.1. Data

Data on where and when people cycle within the urban transport network are still scarce [202]. Despite advances in measurement methods and sensor technology, cycling mobility data are still difficult to obtain and process. This is often the reason for deficiencies in data representativeness and integrability [203,204], which leads to the planning and implementation of cycling infrastructure in the absence of reliable data while, at the same time, the effects of infrastructural, organisational, and promotional measures of cycling cannot be properly monitored. Due to the considerable costs of data collection, different cities collect data useful in transport modelling at different scales. When planning the development of a bicycle traffic model, it is advisable in the first step to diagnose the availability of data and to plan additional field research or to collect data from other sources (e.g., mobile phone data, data from navigation systems, data from ITS services, MaaS solutions, or the Internet of Things), taking into account the costs of their collection. This paper presents the use of data available in the city of Gdynia, which may be used for the development of a bicycle traffic model. The methodology presented in this paper includes the use of the survey of inhabitants, as well as direct measurements of cycling (also using GPS tracking). This paper presents how to make full use of the data held by a city, or data that can be obtained at no or low cost, to build a transport model, showing that it is possible to use data from different sources to achieve the modelling objective. The methodology and structure of the model are also innovative (especially on a national scale). The data and variables used allow for a reliable representation of existing and predicted bicycle traffic. Multimodal modelling of transport, including cycling, requires the use of a very wide range of different data.

It is also very important to note that the surveys presented in this paper were conducted before the COVID-19 pandemic. Due to changes in transport behaviour and travel patterns, a study of bicycle use will be necessary in the post-COVID-19-pandemic period. In times of social distance, people travel less, try to avoid crowded public transport, and aim for more physical activity [205,206], which can be achieved through active modes of transport.

In addition, a more complex analysis using econometric techniques would be required in order to statistically identify the key variables that determine bicycle use. Finally, future research is needed in order to quantify the carbon footprint and impact of electric bicycles for certain components, such as batteries.

5.2.2. Method of Model Development

A four-stage macroscopic model was used to develop the bicycle traffic model [115,116]. The presented methodology of development of the model allows the analysis of the influence of the development of the public transport system, or individual motorised transport, on the transport behaviour of inhabitants in terms of a change in the probability of choosing bicycles for travel; it also takes into account the characteristics of the infrastructure of the transport system, which can determine the choice of cycling as an alternative to other modes of transport. An alternative approach to comprehensive modelling of travel demand and transport networks could be agent-based modelling [22,162]; however, the range of data available for modelling led the authors of this paper to use a four-stage approach. The availability of data can determine the choice of the modelling method. An additional constraint for cities is the availability of analytical tools and previously developed models. In the case of the model presented in this paper, a prerequisite for its development was the prior development of a four-stage macroscopic model of motorised traffic and public transport. Cities that do not have a four-stage macroscopic model should, in the planning stage, take into account the development of a comprehensive model of public transport and individual motorised traffic, including the cycling part of the model. A better representation of cycling can be achieved by taking into account factors in trip generation models related to facilities for pedestrians and cyclists, or land-use functions that encourage cycling. The Urban Transport Modelling System (UTMS) uses additional models to overcome the disadvantages described above [164]. In the case of the model presented in this paper, additional models (described in Section 3) were also developed to enable a more reliable representation of bicycle traffic.

The use of artificial intelligence (AI) methods, including machine learning, is worth considering [207]. The use of machine learning models can be considered for estimating O–D matrices (especially using data from satellite navigation, mobile phones, or the Internet of Things). However, such data are not widely available, or are expensive to acquire. The use of such methods may also raise problems in identifying users or groups of users travelling on the transport network. Identifying users is essential to predict changes in demand and travel routes, taking into account future changes in land use, especially over long time horizons. However, with the development of technology, mobile applications, and algorithms based on artificial intelligence, and the increasing availability of open data, AI methods—especially in the field of machine learning—are worth considering for use in transport demand modelling and route choice modelling, especially for short-term prediction.

The choice of a four-stage approach modelling method has met the expectations in terms of modelling cycling traffic for planning purposes.

5.2.3. Bicycle Network

A model with an accurate representation of the cycling network (including the type of pavement, the type of cycle path, the slope of the road, and the perceived attractiveness of each section of the cycle route) can be useful to predict the modal split and route choice. Relatively large transport zones limit the reliable replication of shorter routes for non-motorised modes (e.g., walking, cycling). Moreover, the length of links in networks is sometimes estimated manually by the modeller. Such methods often lead to imprecise calculations. Geographic information system (GIS) mapping systems provide a more accurate calculation of the length of links. These networks should also be improved with digital height models to ensure that topographic effects are included in distance and speed estimates (slope effects are essential in estimating link resistance for cycling travel mode). Other network inaccuracies may also cause problems in the reliable replication of the transport system. If links or intersection attributes (e.g., cycling speed) are inaccurate or misrepresented, this affects the estimation of speed and travel time. These inaccuracies can reduce the sensitivity to choosing a bike as a mode of transport to avoid congested road sections.

Networks of regional or urban models usually do not describe in detail the layout of local streets and the land-use mix. Such models cannot take into account street sections, local travel speeds, and the facilities available to pedestrians and cyclists due to the generalisation of local conditions. Large TAZs present a challenge to modelling local conditions, including facilities available to pedestrians and cyclists.

The model presented in this paper reliably represents the cycling network, taking into account longitudinal gradients, type of pavement, and type of cycling link by defining speeds Vel_i on particular sections of the cycling network based on these variables. In future studies, the presented model should be further developed by taking into account different user groups, e.g., by age or cycling experience.

A further contribution of this research is the inclusion in the model of the influence of longitudinal gradient on the speed of cyclists, using the GIS environment. The adopted number of transport regions and the level of network mapping are sufficient to reliably represent trips in the primary transport network, as confirmed by the model verification results presented in Section 4.5.

5.2.4. Modal Split

The basic model on which the relationships are based is the logit model. The choice of the model was dictated primarily by a four-stage approach, but also by the possibility of including the modal shift in the model results. The model takes into account the utility of individual modes of transport, expressed as a component of two elements: the measurable utility, and the random part. This approach allows us (1) to capture the different preferences of the users and their subjective perceived attraction to the alternatives, (2) to take into account user error in the evaluation of utility, and (3) to capture all directly non-measurable random factors influencing the choice of means of transport (e.g., convenience, habits). The logit model assumes that the user chooses the option with the highest utility among the available alternatives. The model uses cycling journey times calculated based on cycling speeds in particular sections of the cycle network.

The key factor that determines the modal split is the travel time of the different modes of transport. However, it should be noted that delays (affecting travel times) affect not only car drivers, but also other travellers, i.e., pedestrians, cyclists, and public transport passengers. Most road users perceive delays as the most important effect of congestion. Delay plays an important role in determining travel behaviour, but is only one factor of interest to travellers [208]. Delay occurs when the actual travel time exceeds a threshold acceptable to the traffic participant. Travellers who choose different modes of transport perceive delays differently. The perception of delay also depends, for example, on the motivation for the trip and the relationship between delay and total travel time, and is not linear [209,210]. In future research, issues related to travel time perception and the reliability of transport network performance should be developed.

More research and improvement of modal split (and traffic assignment) models are required, which should take into account, for example, the impact of weather conditions or the seasonality of cycling. Other factors to be considered in further research are the impact of educational and promotional activities, and the impact of cycling's accessibility (e.g., implementation of a city bike-sharing scheme) as a mode of transport, on bicycle traffic.

The Tri-City agglomeration (of which Gdynia is a part) is currently in the process of implementing a bike-sharing scheme. This system will mainly include electric bikes. This change will undoubtedly influence the modal split, as demonstrated by the research conducted so far on both bicycle-sharing systems [33,98–103] and the use of electric bikes [62–67]. Choosing an e-bike instead of a car for travel can reduce energy consumption, but the positive impact on the health of users may be less than if a conventional bicycle were used. Furthermore, a tool such as a bicycle traffic model can also be a data source for the study of the feasibility of this type of project, and the analysis of the effects of its development direction.

Another element that should be taken into account in further research is the integration of MaaS solutions into the cycling model, together with shared mobility services (including bicycle sharing, which is currently being implemented in the Tri-City agglomeration), and the possibility of changing the mode during the journey, to encourage modal shift and improve conditions for cycling. Shared mobility services, such as bike sharing, car sharing, ride sourcing, etc., influence travel behaviour by changing mobility patterns and modes and, thus, compete with more traditional modes of transport [4,33,46,51,101,105,106,108,211]. Shared mobility services have a positive impact on urban transport, as they can reduce the use of private vehicles and, thus, alleviate the current problems of public space scarcity, congestion, and negative environmental impacts. However, it is important to note that bicycles (including bike-sharing schemes) can also capture the demand for public transport [212]. Therefore, shared mobility services, in their current form, have an unclear impact on urban sustainability [213]. This fact underscores the importance of integrating cycling with public transport to promote its complementary use (including through multimodal trips, e.g., cycling to feed the public transport system in less accessible areas as a last-mile mode of transport).

5.2.5. Traffic Assignment

The calculation of the cycling flow was carried out using a stochastic assignment. Compared to deterministic distributions, the element that influences cycling assignment is the inclusion of random factors. In the case of cycling, the randomness of the assignment is influenced by incomplete knowledge of the transport network, which means that cyclists do not always choose a route rationally according to their preferences [179,180], that the choice of routes may be due to the need for a variety [181,182], and that cyclists may have different preferences to choose more attractive routes due to the quality of the road surface, sharing the route with other road users (cars or pedestrians), or different awareness of the risk of being involved in a road accident [181,182,188,189].

With the stochastic assignment, compared to equilibrium deterministic assignment, even in a less loaded network (such as the bicycle network), more routes are loaded, because some demand is also assigned to alternative but similar routes. This property is closer to reality than a strict application of Wardrop's first principle (as found by Daganzo [179]).

Stochastic assignment is more appropriate for bicycle travel, as it reflects the process of individual discrete choice. The variety of individual choices is more important to cyclists than volume-dependent travel times. Factors that influence the choice of a route to a large extent (information about the type of surface, type of section of the cycling network, and longitudinal gradient is included in the function characterising individual sections of the cycling network) influence the comfort of travel, and are particularly important for cyclists. According to the stochastic procedure, not only the shortest route is found, but also many alternative (but similar) routes, using multiple searches for the best path and the impedance variability of the different sections of the cycle network, which allows a more reliable distribution of cycling traffic, and the selection of routes to which cycling traffic would not be allocated if strictly deterministic distribution methods were used.

It can be observed in the evidence from Gdynia that, when choosing a route, cyclists preferred to be separated from motorised traffic, and to not need to cross the road at junctions (using separated on-street infrastructure or off-street paths), confirming previous research [127,182–185]. The model was calibrated by adding time penalties at major intersections that cause cyclist delays, but more detailed research in this area will be required.

Cyclists also prefer flat longitudinal gradients on bicycle routes, as confirmed by previous research [181,188,214]. From the opinions of the respondents presented in Section 4.2, it also appears that the pavement indicated by cyclists as the most attractive is a smooth, even surface (asphalt and, to a lesser extent, smooth concrete blocks). Many previous studies indicate that the type of surface is a factor that determines the choice of route by a cyclist [22,129,130,163,215,216]; however, it is difficult to find direct research results on cyclist preferences with regard to this issue [217,218]. One of the few works that unambiguously states that the quality of the surface is an important decision factor for cyclists when choosing a cycling route is the study by Landis et al. [219]. The results presented in this paper are an additional contribution to the knowledge in this field.

The factor concerning traffic safety was indirectly taken into account when assessing the attractiveness of particular types of bicycle routes. However, a more detailed approach will be required in future research. To update and improve the quality of the bicycle traffic model, it is necessary to conduct regular travel surveys with appropriate sample selection in smaller transport areas. This would allow for increasing the accuracy of the estimation of the origin–destination trip matrices. Cycling speed measurements must be carried out more comprehensively, taking into account, among other things, the type of journey, the experience of cyclists, and other personal characteristics, which will help to represent the behaviour of travellers with greater precision. As the volume of cycling increases, it will also be necessary to consider the capacity of the elements of the cycling network to address cycling bottlenecks.

5.2.6. Results of the Gdynia Case Study

The results of the calculations presented in Section 4 show that each planned element of the development of the cycling network will improve the efficiency of the entire Gdynia transport system. The development of the bicycle network leads to a reduction in the average travelled distance and the average and total travel time for cyclists, while the demand for cycling increases. Cycling has the potential to reduce car traffic, leading in turn to reductions in congestion, energy consumption, and emissions. The model, with a precise representation of the cycling route network that includes the type of pavement, the separation from other types of traffic, and the slope of the sections of the route, can be useful in predicting route selection and modal split due to the utility of each mode of transport based on travel times.

The results show that the greatest benefits for cycling in Gdynia can be achieved by filling in the gaps in the bicycle network—in particular, by connecting the city centre with the northern parts of the city. The most efficient connections compete with congested traffic routes during peak hours, and significantly shorten the distance and travel time between important points in the city. The implementation of bicycle routes separated from motorised traffic (bicycle paths), with flat longitudinal gradients and an asphalt surface, also contributes to the attractiveness of these routes.

6. Conclusions

Transport demand and network models that take into account active modes of travel can effectively support the mobility planning process, as demonstrated in this paper. Models should be used to evaluate measures implemented in the transport system to determine the impact of such measures on reducing traffic and congestion on the road network, and the resulting reductions in energy consumption and pollutant emissions. The models provide a better understanding of the changes in the functioning of the transport system and the behaviour of travellers in different scenarios for the development of this system. However, to allow for a reliable evaluation, it is necessary to use models that allow for the estimation of a modal shift towards cycling.

The model presented in this paper allows the analysis of the influence of the development of the public transport system or road infrastructure on the transport behaviour of inhabitants in terms of the change in the probability of choosing bicycles for travel. The model also allows us to assess the influence of the characteristics of the cycling infrastructure and the development of the cycling network on the choice of cycling as an alternative to other modes of transport. The application of the presented methodology and tools for transport planning provides a solid basis for the selection of the most effective solutions in the planning process.

As data on cycling times and routes within the urban transport network are still scarce, in addition to a survey of residents, direct measurements of cycling—including GPS tracking and data from ITS services—were used in the modelling. It is important to use data fusion from multiple available sources to reduce research costs.

A further contribution of this research is the inclusion in the model of the influence of longitudinal gradient on the speed of cyclists, using the GIS environment, and taking into account the types of cycling network sections and their surface in the choice of the mode of transport and the route of travel.

The model presented in this paper reliably represents the cycling network, taking into account longitudinal gradients, the type of pavement, and the type of cycling link by defining speeds on particular sections of the cycling network, based on these variables.

The key factor that determines the modal split is the travel time of the different transport modes. The logit model was applied to model modal split. The model takes into account the utility of individual modes of transport, expressed as a component of two elements: the measurable utility, and the random part. The model uses cycling journey times calculated based on cycling speeds in particular sections of the cycling network. The proposed approach makes it possible to model and estimate modal shifts between different modes of transport, taking into account random factors. This feature of the model is important for analyses that aim to determine the degree of reduction in congestion, emissions, and energy consumption.

The calculation of the cycling traffic flow was carried out using stochastic assignment, taking into account random factors. Stochastic assignment is more appropriate for bicycle trip assignment because it reflects the process of individual discrete choice. The variety of individual choices is more important to cyclists than volume-dependent travel times. When choosing a route, cyclists prefer separation from motorised traffic, flat longitudinal gradients, and smooth surfaces without bumps (asphalt and, to a lesser extent, smooth concrete blocks) in the cycle routes. The longitudinal gradient of a cycling route is the most important factor in determining the choice of route.

The simulation analysis carried out using the macroscopic model showed that well-thought-out and appropriate investments in the development of a network of bicycle transport are the most favourable in terms of potential changes in the transportation behaviour of the inhabitants. The results show that the greatest benefits for cycling in Gdynia can be achieved by filling in the gaps in the bicycle network.

The verification of the model confirmed its usefulness in planning the bicycle network, but some issues should be taken into account in future research related to modal split and route choice: future development of transport systems (bicycle-sharing systems and other MaaS solutions); travel time perception; and the reliability of transport network performance, including cyclist behaviour at intersections, different user groups—e.g., by age—cycling experience, and modelling of other personal characteristics, post-pandemic transportation behaviour of travellers, the impact of weather conditions or the seasonality of cycling, and the impact of educational and promotional activities on cycling travel preferences and behaviour.

The spatial approach presented to modelling cycling behaviour opens new opportunities for planning and evidence-based decision making in the broad area of cycling promotion and development, as well as for predicting and monitoring the effects of planned or implemented measures.

The methods proposed in this paper are worthwhile, and cities that plan the development of a bicycle traffic model should apply them in the mobility planning process. The example of Gdynia shows that the development of the MST with a bicycle traffic model can also be offered as a good practice for other cities. This may be important for cities with emerging economies and dynamic motorisation development, which will soon face the challenge of sustainable transport behaviour and transport needs.

Author Contributions: Conceptualization, J.O.; methodology, J.O.; software, K.B. and K.Ż.; validation, J.O., K.B., and K.Ż.; formal analysis, J.O.; investigation, J.O., K.B., and K.Ż.; resources, J.O., K.B., and K.Ż.; data curation, J.O., K.B., and K.Ż.; writing—original draft preparation, J.O.; writing—review and editing, J.O. and K.Ż.; visualization, K.B. and K.Ż.; supervision, J.O.; project administration, J.O.; funding acquisition, J.O. All authors have read and agreed to the published version of the manuscript.

Funding: The study was conducted under FLOW (Furthering Less Congestion by Creating Opportunities for More Walking and Cycling). This project has received funding from the European Union's Horizon 2020 research and innovation programme under grant agreement No 635998. No funding was obtained from the project to cover the costs of publication in open access.

Institutional Review Board Statement: Not applicable.

Informed Consent Statement: Not applicable.

Data Availability Statement: Not applicable.

Conflicts of Interest: The authors declare no conflict of interest. The funders had no role in the design of the study, in the collection, analyses, or interpretation of data, in the writing of the manuscript, or in the decision to publish the results.

References

1. Ferenchak, N.N.; Katirai, M. Pedestrian crossing behavior in relation to grouping and gender in a developing country context. *J. Glob. Epidemiol. Environ. Health* **2017**, *1*, 37–45. [CrossRef]
2. Lloyd, L.; Wallbank, C.; Broughton, J.; Cuerden, R. Estimating the potential impact of vehicle secondary safety regulations and consumer testing programs on road casualties in emerging markets. *J. Transp. Saf. Secur.* **2017**, *9*, 149–177. [CrossRef]
3. Eurostat. Passenger Cars, by Size and Type of Fuel Engin. 2021. Available online: http://ec.europa.eu/eurostat/data/database (accessed on 7 February 2021).
4. Machado, C.A.S.; de Salles Hue, N.P.M.; Berssaneti, F.T.; Quintanilha, J.A. An overview of shared mobility. *Sustainability* **2018**, *10*, 4342. [CrossRef]
5. Lerner, W. The Future of Urban Mobility. In *Towards Networked, Multimodal Cities of 2050*; 2011. Available online: https://robertoigarza.files.wordpress.com/2009/07/rep-the-future-of-urban-mobility-2050-little-2011.pdf (accessed on 7 August 2019).
6. Schäfer, A.W. Long-term trends in domestic US passenger travel: The past 100 years and the next 90. *Transportation* **2017**, *44*, 293–310. [CrossRef]
7. European Commission. Mobility and Transport Transport in the European Union Current Trends and Issues. 2019. Available online: https://ec.europa.eu/transport/sites/transport/files/2019-transport-in-the-eu-current-trends-and-issues.pdf (accessed on 20 March 2021).
8. Buehler, R.; Pucher, J. Walking and Cycling in Western Europe and the United States: Trends, Policies and Lessons. TR News 280. May–June 2012. 2012. Available online: http://onlinepubs.trb.org/onlinepubs/trnews/trnews280WesternEurope.pdf (accessed on 20 March 2021).
9. Vandenbulcke, G.; Dujardin, C.; Thomas, I.; de Geus, B.; Degraeuwe, B.; Meeusen, R.; Panis, L.I. Cycle commuting in Belgium: Spatial determinants and 're-cycling' strategies. *Transp. Res. Part A Policy Pract.* **2011**, *45*, 118–137. [CrossRef]
10. Pojani, D.; Stead, D. *The Urban Transport Crisis in Emerging Economies: An Introduction*; Springer: Cham, Switzerland, 2017.
11. Sims, R.; Schaeffer, R.; Creutzig, F.; Cruz-Núñez, X.; D'Agosto, M.; Dimitriu, D.; Meza, M.J.F.; Fulton, L.; Kobayashi, S.O.L.; McKinnon, A.; et al. Chapter 8: Transport. In *Climate Change 2014: Mitigation of Climate Change. Contribution of Working Group III to the Fifth Assessment Report of the Intergovernmental Panel on Climate Change*; Edenhofer, O., Pichs Madruga, R., Sokona, Y., Farahani, E., Kadner, S., Seyboth, K., Adler, A., Baum, I., Brunner, S., Eickemeier, P., et al., Eds.; Cambridge University Press: Cambridge, UK; New York, NY, USA, 2014; pp. 599–670.
12. European Environment Agency (EEA). Total Greenhouse Gas Emission Trends and Projections in Europe. 2021. Available online: https://www.eea.europa.eu/data-and-maps/indicators/greenhouse-gas-emission-trends-7/assessment (accessed on 7 May 2021).
13. Ajanovic, A.; Haas, R. The impact of energy policies in scenarios on GHG emission reduction in passenger car mobility in the EU-15. *Renew. Sustain. Energy Rev.* **2017**, *68*, 1088–1096. [CrossRef]
14. European Environment Agency (EEA). Primary and Final Energy Consumption in the European Union. 2021. Available online: https://www.eea.europa.eu/data-and-maps/indicators/final-energy-consumption-by-sector-11/assessment (accessed on 12 May 2021).
15. European Commission DG MOVE. *Study to Support an Impact Assessment of the Urban Mobility Package*; Activity 31 Sustainable Urban Mobility Plans Final Report; European Commission DG MOVE: Brussels, Belgium, 2013.
16. Hiselius, L.W.; Rosqvist, L.S. Mobility Management campaigns as part of the transition towards changing social norms on sustainable travel behavior. *J. Clean. Prod.* **2016**, *123*, 34–41. [CrossRef]
17. Masson-Delmotte, V.; Zhai, P.; Pörtner, H.-O.; Roberts, D.; Skea, J.; Shukla, P.R.; Pirani, A. *IPCC Report Global Warming of 1.5 °C*; 2018; Volume 2. Available online: https://www.ipcc.ch/sr15/ (accessed on 7 June 2021).
18. Creutzig, F.; Roy, J.; Lamb, W.F.; Azevedo, I.M.L.; Bruine De Bruin, W.; Dalkmann, H.; Edelenbosch, O.Y.; Geels, F.W.; Grubler, A.; Hepburn, C.; et al. Towards demand-side solutions for mitigating climate change. *Nat. Clim. Chang.* **2018**, *8*, 268–271. [CrossRef]
19. Gavanas, N.; Pozoukidou, G.; Verani, E. Integration of LUTI models into sustainable urban mobility plans (SUMPs). *Eur. J. Environ. Sci.* **2016**, *6*, 11–17. [CrossRef]
20. Alonso, A.; Monzón, A.; Wang, Y. Modelling Land Use and Transport Policies to Measure Their Contribution to Urban Challenges: The Case of Madrid. *Sustainability* **2017**, *9*, 378. [CrossRef]
21. Wefering, F.; Rupprecht, S.; Bührmann, S.; Böhler-Baedeker, S.; Granberg, M.; Vilkuna, J.; Saarinen, S.; Backhaus, W.; Laubenheimer, M. *Guidelines. Developing and Implementing a Sustainable Urban Mobility Plan*; European Union, 2013. Available online: https://www.eltis.org/sites/default/files/guidelines-developing-and-implementing-a-sump_final_web_jan2014b.pdf (accessed on 20 March 2021).
22. Kaziyeva, D.; Loidl, M.; Wallentin, G. Simulating Spatio-Temporal Patterns of Bicycle Flows with an Agent-Based Model. *ISPRS Int. J. Geo-Inf.* **2021**, *10*, 88. [CrossRef]
23. Okraszewska, R.; Romanowska, A.; Wołek, M.; Oskarbski, J.; Birr, K.; Jamroz, K. Integration of a multilevel transport system model into sustainable Urban mobility planning. *Sustainability* **2018**, *10*, 479. [CrossRef]

24. Daganzo, C.F. *Fundamentals of Transportation and Traffic Operations*, 1st ed.; Emerald Publishing: New York, NY, USA, 1 June 1997.
25. Jacyna, M. *Modelowanie i Ocena Systemów Transportowych*; Oficyna Wydawnicza Politechniki Warszawskiej: Warszawa, Poland, 2009.
26. Cambridge Systematics Inc. Travel Model Improvement Program. In *Travel Model Validation and Reasonability Checking Manual*, 2nd ed.; Cambridge Systematics: Medford, MA, USA, 24 September 2010.
27. Singh, R.; Dowling, R. Improved speed-flow relationships: Application to transportation planning models. In Proceedings of the Seventh TRB Conference on the Application of Transportation Planning Methods, Boston, MA, USA, 7–11 March 1999; pp. 340–349.
28. Sivakumar, A. *Modelling Transport: A Synthesis of Transport Modelling Methodologies*; Imperial College London: London, UK, 2007.
29. Givoni, M.; Beyazit, E.; Shiftan, Y. The use of state-of-the-art transport models by policymakers—Beauty in simplicity? *Plan. Theory Pract.* **2016**, *17*, 385–404. [CrossRef]
30. Wang, B.; Shao, C.; Ji, X. Influencing mechanism analysis of holiday activity-travel patterns on transportation energy consumption and emissions in China. *Energies* **2017**, *10*, 897. [CrossRef]
31. Neves, A.; Brand, C. Assessing the potential for carbon emissions savings from replacing short car trips with walking and cycling using a mixed GPS travel diary approach. *Transp. Res. Part A Policy Pract.* **2019**, *123*, 130–146. [CrossRef]
32. Zhou, Y.; Dey, K.C.; Chowdhury, M.; Wang, K.C. Process for evaluating the data transfer performance of wireless traffic sensors for real-time intelligent transportation systems applications. *IET Intell. Transp. Syst.* **2017**, *11*, 18–27. [CrossRef]
33. Faghih-Imani, A.; Anowar, S.; Miller, E.J.; Eluru, N. Hail a cab or ride a bike? A travel time comparison of taxi and bicycle-sharing systems in New York City. *Transp. Res. Part A Policy Pract.* **2017**, *101*, 11–21. [CrossRef]
34. Brand, C.; Götschi, T.; Dons, E.; Gerike, R.; Anaya-Boig, E.; Avila-Palencia, I.; de Nazelle, A.; Gascon, M.; Gaupp-Berghausen, M.; Iacorossi, F.; et al. The climate change mitigation impacts of active travel: Evidence from a longitudinal panel study in seven European cities. *Glob. Environ. Chang.* **2021**, *67*, 102224. [CrossRef]
35. Quarmby, S.; Santos, G.; Mathias, M. Air Quality Strategies and Technologies: A Rapid Review of the International Evidence. *Sustainability* **2019**, *11*, 2757. [CrossRef]
36. Horton, D. Environmentalism and the bicycle. *Env. Polit.* **2006**, *15*, 41–58. [CrossRef]
37. Amelung, D.; Fischer, H.; Herrmann, A.; Aall, C.; Louis, V.R.; Becher, H.; Wilkinson, P.; Sauerborn, R. Human health as a motivator for climate change mitigation: Results from four European high-income countries. *Glob. Environ. Chang.* **2019**, *57*, 101918. [CrossRef]
38. Woodcock, J.; Abbas, A.; Ullrich, A.; Tainio, M.; Lovelace, R.; Sá, T.H.; Westgate, K.; Goodman, A. Development of the Impacts of Cycling Tool (ICT): A modelling study and web tool for evaluating health and environmental impacts of cycling uptake. *PLoS Med.* **2018**, *15*, e1002622. [CrossRef]
39. Venturini, G.; Karlsson, K.; Münster, M. Impact and effectiveness of transport policy measures for a renewable-based energy system. *Energy Policy* **2019**, *133*, 110900. [CrossRef]
40. Chapman, L. Transport and climate change: A review. *J. Transp. Geogr.* **2007**, *15*, 354–367. [CrossRef]
41. Zhang, J.; Yu, B.; Cai, J.; Wei, Y.M. Impacts of household income change on CO_2 emissions: An empirical analysis of China. *J. Clean. Prod.* **2017**, *157*, 190–200. [CrossRef]
42. Loidl, M.; Werner, C.; Heym, L.; Kofler, P.; Innerebner, G. Lifestyles and cycling behavior—Data from a cross-sectional study. *Data* **2019**, *4*, 140. [CrossRef]
43. Thiel, C.; Perujo, A.; Mercier, A. Cost and CO_2 aspects of future vehicle options in Europe under new energy policy scenarios. *Energy Policy* **2010**, *38*, 7142–7151. [CrossRef]
44. Bozza, F.; De Bellis, V.; Malfi, E.; Teodosio, L.; Tufano, D. Optimal calibration strategy of a hybrid electric vehicle equipped with an ultra-lean pre-chamber SI engine for the minimization of CO2and pollutant emissions. *Energies* **2020**, *13*, 4008. [CrossRef]
45. Yang, D.; Timmermans, H. Households transport-home energy conservation strategies in response to energy price policies: A stated adaptation experiment based on portfolio choices and cross effects designs. *Int. J. Sustain. Transp.* **2017**, *11*, 133–147. [CrossRef]
46. Gallo, M.; Marinelli, M. Sustainable mobility: A review of possible actions and policies. *Sustainability* **2020**, *12*, 7499. [CrossRef]
47. Dillman, K.; Czepkiewicz, M.; Heinonen, J.; Fazeli, R.; Árnadóttir, Á.; Davíðsdóttir, B.; Shafiei, E. Decarbonization Scenarios for Reykjavik's passenger transport: The combined effects of behavioural changes and technological developments. *Sustain. Cities Soc.* **2020**, *65*, 102614. [CrossRef]
48. Heinonen, J.; Czepkiewicz, M.; Árnadóttir, Á.; Ottelin, J. Drivers of car ownership in a car-oriented city: A mixed-method study. *Sustainability* **2021**, *13*, 619. [CrossRef]
49. Butler, L.; Yigitcanlar, T.; Paz, A. How can smart mobility innovations alleviate transportation disadvantage? Assembling a conceptual framework through a systematic review. *Appl. Sci.* **2020**, *10*, 6306. [CrossRef]
50. Oliveira, F.; Nery, D.; Costa, D.G.; Silva, I.; Lima, L. A survey of technologies and recent developments for sustainable smart cycling. *Sustainability* **2021**, *13*, 3422. [CrossRef]
51. Tønnesen, A.; Knapskog, M.; Uteng, T.P.; Øksenholt, K.V. The integration of active travel and public transport in Norwegian policy packages: A study on 'access, egress and transfer' and their positioning in two multilevel contractual agreements. *Res. Transp. Bus. Manag.* **2020**, *40*, 100546. [CrossRef]
52. Gerboni, R.; Grosso, D.; Carpignano, A.; Dalla Chiara, B. Linking energy and transport models to support policy making. *Energy Policy* **2017**, *111*, 336–345. [CrossRef]

53. Edelenbosch, O.Y.; McCollum, D.L.; van Vuuren, D.P.; Bertram, C.; Carrara, S.; Daly, H.; Fujimori, S.; Kitous, A.; Kyle, P.; Broin, E.Ó.; et al. Decomposing passenger transport futures: Comparing results of global integrated assessment models. *Transp. Res. Part D Transp. Environ.* **2017**, *55*, 281–293. [CrossRef]
54. Castro, A.; Gaupp-Berhausen, M.; Dons, E.; Standaert, A.; Laeremans, M.; Clark, A.; Anaya, E.; Cole-Hunter, T.; Avila-Palencia, I.; Rojas-Rueda, D.; et al. Physical activity of electric bicycle users compared to conventional bicycle users and non-cyclists: Insights based on health and transport data from an online survey in seven European cities. *Transp. Res. Interdiscip. Perspect.* **2019**, *1*, 100017. [CrossRef]
55. Elliot, T.; McLaren, S.J.; Sims, R. Potential environmental impacts of electric bicycles replacing other transport modes in Wellington, New Zealand. *Sustain. Prod. Consum.* **2018**, *16*, 227–236. [CrossRef]
56. Woodcock, J.; Edwards, P.; Tonne, C.; Armstrong, B.G.; Ashiru, O.; Banister, D.; Beevers, S.; Chalabi, Z.; Chowdhury, Z.; Cohen, A.; et al. Public health benefits of strategies to reduce greenhouse-gas emissions: Urban land transport. *Lancet* **2009**, *374*, 1930–1943. [CrossRef]
57. Keall, M.D.; Shaw, C.; Chapman, R.; Howden-Chapman, P. Reductions in carbon dioxide emissions from an intervention to promote cycling and walking: A case study from New Zealand. *Transp. Res. Part D Transp. Environ.* **2018**, *65*, 687–696. [CrossRef]
58. Le Gouais, A.; Panter, J.R.; Cope, A.; Powell, J.E.; Bird, E.L.; Woodcock, J.; Ogilvie, D.; Foley, L. A natural experimental study of new walking and cycling infrastructure across the United Kingdom: The Connect2 programme. *J. Transp. Health* **2021**, *20*, 100968. [CrossRef]
59. Lovelace, R.; Beck, S.B.M.; Watson, M.; Wild, A. Assessing the energy implications of replacing car trips with bicycle trips in Sheffield, UK. *Energy Policy* **2011**, *39*, 2075–2087. [CrossRef]
60. Coley, D. Emission factors for human activity. *Energy Policy* **2002**, *30*, 3–5. [CrossRef]
61. Söderberg, A.; Adell, E.; Winslott Hiselius, L. What is the substitution effect of e-bikes? A randomised controlled trial. *Transp. Res. Part D Transp. Environ.* **2021**, *90*, 102648. [CrossRef]
62. Jones, T.; Harms, L.; Heinen, E. Motives, perceptions and experiences of electric bicycle owners and implications for health, wellbeing and mobility. *J. Transp. Geogr.* **2016**, *53*, 41–49. [CrossRef]
63. Haustein, S.; Møller, M. Age and attitude: Changes in cycling patterns of different e-bike user segments. *Int. J. Sustain. Transp.* **2016**, *10*, 836–846. [CrossRef]
64. Winslott Hiselius, L.; Svensson, Å. E-bike use in Sweden—CO_2 effects due to modal change and municipal promotion strategies. *J. Clean. Prod.* **2017**, *141*, 818–824. [CrossRef]
65. Cairns, S.; Behrendt, F.; Raffo, D.; Beaumont, C.; Kiefer, C. Electrically-assisted bikes: Potential impacts on travel behaviour. *Transp. Res. Part A Policy Pract.* **2017**, *103*, 327–342. [CrossRef]
66. Kroesen, M. To what extent do e-bikes substitute travel by other modes? Evidence from the Netherlands. *Transp. Res. Part D Transp. Environ.* **2017**, *53*, 377–387. [CrossRef]
67. Sun, Q.; Feng, T.; Kemperman, A.; Spahn, A. Modal shift implications of e-bike use in the Netherlands: Moving towards sustainability? *Transp. Res. Part D Transp. Environ.* **2020**, *78*, 102202. [CrossRef]
68. Lee, J.S.; Jiang, J.W. Enhanced fuzzy-logic-based power-assisted control with user-adaptive systems for human-electric bikes. *IET Intell. Transp. Syst.* **2019**, *13*, 1492–1498. [CrossRef]
69. Kim, D.H.; Lee, D.; Kim, Y.; Kim, S.; Shin, D. A power assistant algorithm based on human–robot interaction analysis for improving system efficiency and riding experience of e-bikes. *Sustainability* **2021**, *13*, 768. [CrossRef]
70. Commission of the European Communities. *Green Paper. Towards a New Culture for Urban Mobility*; Commission of the European Communities: Brussels, Belgium, 2007.
71. OECD/ECMT. *Managing Urban Traffic Congestion*; Paris, France, 2007. Available online: http://www.oecd-ilibrary.org/transport/managing-urban-traffic-congestion_9789282101506-en (accessed on 7 May 2021).
72. Litman, T. Smart Congestion Relief. In *Comprehensive Evaluation of Traffic Congestion Costs and Congestion Reduction Strategies*; Victoria Transport Policy Institute, 24 April 2014. Available online: www.vtpi.org/cong_relief.pdf (accessed on 20 March 2020).
73. ECF. *Cycling Strategy: Recommendations for Delivering Green Growth and an Effective Mobility in 2030*; European Transport Safety Council: Brussels, Belgium, 2017.
74. Karanikola, P.; Panagopoulos, T.; Tampakis, S.; Tsantopoulos, G. Cycling as a smart and green mode of transport in small touristic cities. *Sustainability* **2018**, *10*, 268. [CrossRef]
75. Pucher, J.; Dill, J.; Handy, S. Infrastructure, programs, and policies to increase bicycling: An international review. *Prev. Med.* **2010**, *50*, S106–S125. [CrossRef]
76. Sallis, J.F.; Cerin, E.; Conway, T.L.; Adams, M.A.; Frank, L.D.; Pratt, M.; Salvo, D.; Schipperijn, J.; Smith, G.; Cain, K.L.; et al. Physical activity in relation to urban environments in 14 cities worldwide: A cross-sectional study. *Lancet* **2016**, *387*, 2207–2217. [CrossRef]
77. Mertens, L.; Compernolle, S.; Deforche, B.; Mackenbach, J.D.; Lakerveld, J.; Brug, J.; Roda, C.; Feuillet, T.; Oppert, J.M.; Glonti, K.; et al. Built environmental correlates of cycling for transport across Europe. *Health Place* **2017**, *44*, 35–42. [CrossRef] [PubMed]
78. Christensen, H.R. Is the Kingdom of bicycles rising again? Cycling, gender, and class in postsocialist China. *Transfers* **2017**, *7*, 1–20. [CrossRef]

79. Maheshwari, M.; Jana, A.; Bandyopadhyay, S. Optimizing the Modal Split to Reduce Carbon Dioxide Emission for Resource-Constrained Societies. *Transp. Res. Procedia* **2020**, *48*, 2063–2073. [CrossRef]
80. Litman, T. Measuring Transportation—Traffic, Mobility and Accessibility, Institute of Transportation Engineers. *ITE J.* **2003**, *73*, 23–32.
81. Holden, E.; Høyer, K.G. The ecological footprints of fuels. *Transp. Res. Part D Transp. Environ.* **2005**, *10*, 395–403. [CrossRef]
82. Moriarty, P.; Honnery, D. Low-mobility: The future of transport. *Futures* **2008**, *40*, 865–872. [CrossRef]
83. Mallus, M.; Colistra, G.; Atzori, L.; Murroni, M.; Pilloni, V. Dynamic carpooling in urban areas: Design and experimentation with a multi-objective route matching algorithm. *Sustainability* **2017**, *9*, 254. [CrossRef]
84. de Koning, R.; Tan, W.G.Z.; Van Nes, A. Assessing spatial configurations and transport energy usage for planning sustainable communities. *Sustainability* **2020**, *12*, 8146. [CrossRef]
85. Hickman, R.; Hall, P.; Banister, D. Planning more for sustainable mobility. *J. Transp. Geogr.* **2013**, *33*, 210–219. [CrossRef]
86. Reisi, M.; Aye, L.; Rajabifard, A.; Ngo, T. Land-use planning: Implications for transport sustainability. *Land Use Policy* **2016**, *50*, 252–261. [CrossRef]
87. Organisation for Economic Co-Operation and Development (OECD). *Compact City Policies: A Comparative Assessment*; OECD: Paris, France, 2012.
88. Milakis, D.; Efthymiou, D.; Antoniou, C. Built environment, travel attitudes and travel behaviour: Quasi longitudinal analysis of links in the case of Greeks relocating FROM US to Greece. *Sustainability* **2017**, *9*, 1774. [CrossRef]
89. Kamargianni, M. Investigating next generation's cycling ridership to promote sustainable mobility in different types of cities. *Res. Transp. Econ.* **2015**, *53*, 45–55. [CrossRef]
90. Tiwari, G.; Jain, D.; Ramachandra Rao, K. Impact of public transport and non-motorized transport infrastructure on travel mode shares, energy, emissions and safety: Case of Indian cities. *Transp. Res. Part D Transp. Environ.* **2016**, *44*, 277–291. [CrossRef]
91. Harms, L.; Bertolini, L.; Brömmelstroet, M.T. Performance of municipal cycling policies in medium sized cities in the Netherlands since 2000. *Transp. Rev.* **2016**, *36*, 134–162. [CrossRef]
92. Schmid-Querg, J.; Keler, A.; Grigoropoulos, G. The munich bikeability index: A practical approach for measuring urban bikeability. *Sustainability* **2021**, *13*, 428. [CrossRef]
93. Jaszczak, A.; Morawiak, A.; Zukowska, J. Cycling as a sustainable transport alternative in polish cittaslow towns. *Sustainability* **2020**, *12*, 5049. [CrossRef]
94. Clark, S.D.; Page, M.W. Cycling and urban traffic management and control systems. *Transp. Res. Rec.* **2000**, *1705*, 77–84. [CrossRef]
95. Oskarbski, J.; Guminska, L.; Miszewski, M.; Oskarbska, I. Analysis of Signalized Intersections in the Context of Pedestrian Traffic. *Transp. Res. Procedia* **2016**, *14*, 2138–2147. [CrossRef]
96. Portilla, C.; Valencia, F.; Espinosa, J.; Núñez, A.; De Schutter, B. Model-based predictive control for bicycling in urban intersections. *Transp. Res. Part C Emerg. Technol.* **2016**, *70*, 27–41. [CrossRef]
97. Pucher, J.R.; Buehler, R. *City Cycling*; MIT Press: Cambridge, MA, USA, 2012.
98. Garcia-Palomares, J.C.; Gutierrez, J.; Latorre, M. Optimizing the location of stations in bike-sharing programs: A GIS approach. *Appl. Geogr.* **2012**, *35*, 235–246. [CrossRef]
99. Mora, R.; Moran, P. Public bike sharing programs under the prism of urban planning officials: The case of Santiago de Chile. *Sustainability* **2020**, *12*, 5720. [CrossRef]
100. Shaaban, K. Why don't people ride bicycles in high-income developing countries, and can bike-sharing be the solution? The case of qatar. *Sustainability* **2020**, *12*, 1693. [CrossRef]
101. Yao, Y.; Zhang, Y.; Tian, L.; Zhou, N.; Li, Z.; Wang, M. Analysis of network structure of urban bike-sharing system: A case study based on real-time data of a public bicycle system. *Sustainability* **2019**, *11*, 5425. [CrossRef]
102. Macioszek, E.; Świerk, P.; Kurek, A. The bike-sharing system as an element of enhancing sustainable mobility—A case study based on a city in Poland. *Sustainability* **2020**, *12*, 3285. [CrossRef]
103. Żochowska, R.; Jacyna, M.; Kłos, M.J.; Soczówka, P. A GIS-based method of the assessment of spatial integration of bike-sharing stations. *Sustainability* **2021**, *13*, 3894. [CrossRef]
104. Bąk, A.; Nosal Hoy, K.; Solecka, K. Multiple Criteria Evaluation of the Planned Bikesharing System in Jaworzno. In *Smart and Green Solutions for Transport Systems. TSTP 2019. Advances in Intelligent Systems and Computing*; Sierpiński, G., Ed.; Springer: Cham, Switzerland, 2020; Volume 1091, ISBN 978-3-030-35542-5. [CrossRef]
105. Blitz, A.; Lanzendorf, M. Mobility design as a means of promoting non-motorised travel behaviour? A literature review of concepts and findings on design functions. *J. Transp. Geogr.* **2020**, *87*, 102778. [CrossRef]
106. Börjesson, M.; Hamilton, C.J.; Näsman, P.; Papaix, C. Factors driving public support for road congestion reduction policies: Congestion charging, free public transport and more roads in Stockholm, Helsinki and Lyon. *Transp. Res. A Policy Pract.* **2015**, *78*, 452–462. [CrossRef]
107. Morton, C.; Lovelace, R.; Anable, J. Exploring the effect of local transport policies on the adoption of low emission vehicles: Evidence from the London Congestion Charge and Hybrid Electric Vehicles. *Transp. Policy* **2017**, *60*, 34–46. [CrossRef]
108. Piatkowski, D.P.; Marshall, W.E.; Krizek, K.J. Carrots versus sticks: Assessing intervention effectiveness and implementation challenges for active transport. *J. Plan. Educ. Res.* **2019**, *39*, 50–64. [CrossRef]
109. Scarinci, R.; Rast, F.; Bierlaire, M. Needed reduction in mobility energy consumption to meet the goal of a 2000-watt society. *Transp. Res. Part A Policy Pract.* **2017**, *101*, 133–148. [CrossRef]

110. Gerlough, D.L.; Huber, M.J. *Traffic Flow Theory*; TRB Special Report 165; Transportation Research Board National Research Council: Washington, DC, USA, 1975; ISBN 0309024595.
111. Williams, H.C.W.L. On the Formation of Travel Demand Models and Economic Evaluation Measures of User Benefit. *Environ. Plan. A Econ. Space* **1977**, *9*, 285–344. [CrossRef]
112. Jayakrisham, R.; Mahmassani, H.S.; Yu, T.Y. An evaluation tool for advanced traffic information and management systems in urban networks. *Transp. Res. Part C Emerg. Technol.* **1994**, *2*, 129–147. [CrossRef]
113. May, A.D. *Traffic Flow Fundamentals*; Prentice-Hall: Englewood Cliffs, NJ, USA, 1990.
114. Michalopoulos, P.G. Analysis of traffic flows at complex congested arterials. *Transp. Res Rec* **1988**, *1194*, 77–86.
115. Cascetta, E. *Transport Systems Analysis. Models and Applications*, 2nd ed.; Springer Optimization and Its Application: New York, NY, USA, 2009; ISBN 9780387758565.
116. de Ortuzar, J.D.; Willumsen, L.G.; Ortúzar, J.D.D.; Willumsen, L.G. *Modelling Transport*; John Wiley & Sons, Ltd: Hoboken, NJ, USA, April 2011; ISBN ISBN 9780470760390. [CrossRef]
117. Barcelo, J. *Fundamentals of Traffic Simulation (International Series in Operations Research & Management Science)*; Springer: New York, NY, USA; Dordrecht, The Netherlands; Berlin/Heidelberg, Germany; London, UK, 2010; ISBN 9781441961419.
118. Heller, B. *Deliverable 4.1. Exploring prediction perspectives. An open, Sustainable, Ubiquitous Data and Service Ecosystem for Efficient, Effective, Safe, Resilient Mobility in Metropolitan Areas—SETA*; European Commission: Sheffield, UK, 1 February 2016.
119. Kaparias, I.; Zavitsas, K.; Bell, M.G.H. *State-of-the-Art of Urban Traffic Management Policies and Technologies*; Imperial College London: London, UK, 2010.
120. May, A.D.; Shepherd, S.P.; Timms, P. Optimal transport strategies for European cities. *Transportation* **2000**, *27*, 285–315. [CrossRef]
121. Boile, M.P.; Ozbay, K. *The Future of Transportation Modeling*; New Jersey Department of Transportation, Federal Highway Administration: West Trenton, Ewing Township, NJ, USA, 2005.
122. Aziz, H.M.A.; Park, B.H.; Morton, A.; Stewart, R.N.; Hilliard, M.; Maness, M. A high resolution agent-based model to support walk-bicycle infrastructure investment decisions: A case study with New York City. *Transp. Res. Part C Emerg. Technol.* **2018**, *86*, 280–299. [CrossRef]
123. Rybarczyk, G. Simulating bicycle wayfinding mechanisms in an urban environment. *Urban Plan. Transp. Res.* **2014**, *2*, 89–104. [CrossRef]
124. Shan, X.; Li, Z.; Chen, X.; Ye, J. A Modified Cellular Automaton Approach for Mixed Bicycle Traffic Flow Modeling. *Discret. Dyn. Nat. Soc.* **2015**, *2015*, 213204. [CrossRef]
125. Agarwal, A.; Zilske, M.; Rao, K.R.; Nagel, K. An elegant and computationally efficient approach for heterogeneous traffic modelling using agent based simulation. *Procedia Comput. Sci.* **2015**, *52*, 962–967. [CrossRef]
126. Kucharski, R.; Drabicki, A.; Żyłka, K.; Szarata, A. Multichannel queueing behaviour in urban bicycle traffic. *Eur. J. Transp. Infrastruct. Res.* **2019**, *19*, 116–141. [CrossRef]
127. Nielsen, T.A.S.; Skov-Petersen, H. Bikeability—Urban structures supporting cycling. Effects of local, urban and regional scale urban form factors on cycling from home and workplace locations in Denmark. *J. Transp. Geogr.* **2018**, *69*, 36–44. [CrossRef]
128. Di Mascio, P.; Fusco, G.; Grappasonni, G.; Moretti, L.; Ragnoli, A. Geometrical and functional criteria as a methodological approach to implement a new cycle path in an existing Urban Road Network: A Case study in Rome. *Sustainability* **2018**, *10*, 2951. [CrossRef]
129. Shui, C.S.; Chan, W.L. Optimization of a bikeway network with selective nodes. *Sustainability* **2019**, *11*, 6531. [CrossRef]
130. Lin, S.J.; Shyu, G.S.; Fang, W.T.; Cheng, B.Y. Using multivariate statistical methods to analyze high-quality bicycle path service systems: A case study of popular bicycle paths in Taiwan. *Sustainability* **2020**, *12*, 7185. [CrossRef]
131. McCahill, C.; Garrick, N.W. The Applicability of Space Syntax to Bicycle Facility Planning. *Transp. Res. Rec. J. Transp. Res. Board* **2008**, *2074*, 46–51. [CrossRef]
132. Dill, J.; Voros, K. Factors Affecting Bicycling Demand: Initial Survey Findings from the Portland, Oregon, Region. *Transp. Res. Rec. J. Transp. Res. Board* **2007**, *2031*, 9–17. [CrossRef]
133. Xing, Y.; Handy, S.L.; Buehler, T.J. Factors Associated with Bicycle Ownership and Use: A Study of 6 Small U.S. Cities. In Proceedings of the 87th Annual Meeting of the Transportation Research Board, Washington, DC, USA, 13–17 January 2008.
134. Ryu, S. A bicycle origin-destination matrix estimation based on a two-stage procedure. *Sustainability* **2020**, *12*, 2951. [CrossRef]
135. Boettge, B.; Hall, D.M.; Crawford, T. Assessing the bicycle network in St. Louis: A place-based user-centered approach. *Sustainability* **2017**, *9*, 241. [CrossRef]
136. Furth, P.G.; Mekuria, M.C.; Nixon, H. Network Connectivity for Low-Stress Bicycling. *Transp. Res. Rec. J. Transp. Res. Board* **2016**, *2587*, 41–49. [CrossRef]
137. Useche, S.; Montoro, L.; Alonso, F.; Oviedo-Trespalacios, O. Infrastructural and human factors affecting safety outcomes of cyclists. *Sustainability* **2018**, *10*, 299. [CrossRef]
138. Westin, K.; Nordlund, A.; Jansson, J. Goal Framing as a Tool for Changing People's Car Travel Behavior in Sweden. *Sustainability* **2020**, *12*, 3695. [CrossRef]
139. Kim, J.; Choi, K.; Kim, S.; Fujii, S. How to promote sustainable public bike system from a psychological perspective? *Int. J. Sustain. Transp.* **2017**, *11*, 272–281. [CrossRef]
140. García, J.; Arroyo, R.; Mars, L.; Ruiz, T. The influence of attitudes towards cycling and walking on travel intentions and actual behavior. *Sustainability* **2019**, *11*, 2554. [CrossRef]

141. Pijoan, A.; Kamara-Esteban, O.; Alonso-Vicario, A.; Borges, C.E. Transport choice modeling for the evaluation of new transport policies. *Sustainability* **2018**, *10*, 1230. [CrossRef]
142. Santos, G.; Maoh, H.; Potoglou, D.; von Brunn, T. Factors influencing modal split of commuting journeys in medium-size European cities. *J. Transp. Geogr.* **2013**, *30*, 127–137. [CrossRef]
143. Aultman-Hall, L.; Hall, F.L.; Baetz, B.B. Analysis of Bicycle Commuter Routes Using Geographic Information Systems: Implications for Bicycle Planning. *Transp. Res. Rec. J. Transp. Res. Board* **1997**, *1578*, 102–110. [CrossRef]
144. Dill, J.; Gliebe, J. *Understanding and Measuring Bicycling Behavior: A Focus on Travel Time and Route Choice*; OTREC-RR-08-03; Oregon Transportation Research and Education Consortium: Portland, OR, USA, December 2008.
145. Nikitas, A. How to save bike-sharing: An evidence-based survival toolkit for policy-makers and mobility providers. *Sustainability* **2019**, *11*, 3206. [CrossRef]
146. Heinen, E.; van Wee, B.; Maat, K. Commuting by Bicycle: An Overview of the Literature. *Transp. Rev.* **2010**, *30*, 59–96. [CrossRef]
147. Miranda-Moreno, L.F.; Nosal, T. Weather or Not to Cycle: Temporal Trends and Impact of Weather on Cycling in an Urban Environment. *Transp. Res. Rec. J. Transp. Res. Board* **2011**, *2247*, 42–52. [CrossRef]
148. Holmgren, J.; Ivehammar, P. Mode choice in home-to-work travel in mid-size towns: The competitiveness of public transport when bicycling and walking are viable options. *Transp. Res. Procedia* **2020**, *48*, 1635–1643. [CrossRef]
149. Pogodzinska, S.; Kiec, M.; D'Agostino, C. Bicycle Traffic Volume Estimation Based on GPS Data. *Transp. Res. Procedia* **2020**, *45*, 874–881. [CrossRef]
150. Pazdan, S.; Kiec, M.; D'Agostino, C. Impact of environment on bicycle travel demand—Assessment using bikeshare system data. *Sustain. Cities Soc.* **2021**, *67*, 102724. [CrossRef]
151. Trost, S.G.; Owen, N.; Bauman, A.E.; Sallis, J.F.; Brown, W. Correlates of adults' participation in physical activity: Review and update. *Med. Sci. Sports Exerc.* **2002**, *34*, 1996–2001. [CrossRef]
152. Choi, J.; Lee, M.; Lee, J.K.; Kang, D.; Choi, J.Y. Correlates associated with participation in physical activity among adults: A systematic review of reviews and update. *BMC Public Health* **2017**, *17*, 356. [CrossRef]
153. Turner, S.; Shunk, G.; Hottenstein, A. *Development of a Methodology to Estimate Bicycle and Pedestrian Travel Demand*; Texas Transportation Institute, Research Project Number 0-1723, Report 1723-S; Texas Department of Transportation, Federal Highway Administration: Austin, TX, USA, September 1998.
154. Landis, B.W. The Bicycle System Performance Measures: The Intersection Hazard and Latent Demand Score Models. *ITE J.* **1996**, *66*, 18–26. [CrossRef]
155. Hankey, S.; Lindsey, G.; Wang, X.; Borah, J.; Hoff, K.; Utecht, B.; Xu, Z. Estimating use of non motorized infrastructure: Models of bicycle and pedestrian traffic in Minneapolis, MN. *Landsc. Urban Plan.* **2012**, *107*, 307–316. [CrossRef]
156. Fagnant, D.J.; Kockelman, K. A direct-demand model for bicycle counts: The impacts of level of service and other factors. *Environ. Plan. B Plan. Des.* **2015**, *43*, 93–107. [CrossRef]
157. Lu, T.; Mondschein, A.; Buehler, R.; Hankey, S. Adding temporal information to direct-demand models: Hourly estimation of bicycle and pedestrian traffic in Blacksburg, VA. *Transp. Res. Part D Transp. Environ.* **2018**, *63*, 244–260. [CrossRef]
158. McDaniel, S.; Lowry, M.B.; Dixon, M. Using Origin-Destination Centrality to Estimate Directional Bicycle Volumes. *Transp. Res. Rec. J. Transp. Res. Board* **2014**, *2430*, 12–19. [CrossRef]
159. Cooper, C.H.V. Using spatial network analysis to model pedal cycle flows, risk and mode choice. *J. Transp. Geogr.* **2017**, *58*, 157–165. [CrossRef]
160. Lindsey, G.; Nordback, K.; Figliozzi, M.A. Institutionalizing Bicycle and Pedestrian Monitoring Programs in Three States. *Transp. Res. Rec. J. Transp. Res. Board* **2014**, *2443*, 134–142. [CrossRef]
161. Jones, M.G.; Ryan, S.; Donlon, J.; Ledbetter, L.; Ragland, D.R.; Arnold, L.S. *Seamless Travel: Measuring Bicycle and Pedestrian Activity in San Diego County and Its Relationship to Land Use, Transportation, Safety, and Facility Type*; PATH Research Report: Berkeley, CA, USA, 1 February 2010.
162. Wallentin, G.; Loidl, M. Agent-based Bicycle Traffic Model for Salzburg City. *GI_Forum J. Geogr. Inf. Sci.* **2015**, *1*, 558–566. [CrossRef]
163. Ziemke, D.; Metzler, S.; Nagel, K. Modeling bicycle traffic in an agent-based transport simulation. *Procedia Comput. Sci.* **2017**, *109*, 923–928. [CrossRef]
164. DKS Associates; The University of California. *Assessment of Local Models and Tools for Analyzing Smart-Growth Strategies*; Final Report Prepared for the State of California Business, Transportation and Housing Agency, California Department; The State of California Business, Transportation and Housing Agency, California Department: Sacramento, CA, USA, 2007.
165. Transportation Research Board. *Smart Growth and Urban Goods Movement*; NCFRP Report 24; TRB Publication: Washington, DC, USA, 2013. [CrossRef]
166. Dimitriou, H.T.; Thompson, R. *Strategic Planning for Regional Development in UK*; Routledge: Abingdon, UK, 2008.
167. Banister, D.; Hickman, R. Transport futures: Thinking the unthinkable. *Transp. Policy* **2013**, *29*, 283–293. [CrossRef]
168. Bliemer, M.C.J.; Mulley, C.; Moutou, C.J. Handbook on Transport and Urban Planning in the Developed World. Edward Elgar Publishing Ltd: Cheltenham, UK; Northampton, MA, USA, February 2016.
169. Gudmundsson, H. Analysing Models as a Knowledge Technology in Transport Planning. *Transp. Rev.* **2011**, *31*, 145–159. [CrossRef]

170. Hanson, S.; Hanson, P. Problems in integrating bicycle travel into the urban transportation planning process. *Transp. Res. Rec. J. Transp. Res. Board* **1976**, *570*, 24–30.
171. Nosal, K.; Starowicz, W. Evaluation of influence of mobility management instruments implemented in separated areas of the city on the changes in modal split. *Arch. Transpor* **2015**, *35*, 41–52. [CrossRef]
172. Jacyna, M.; Wasiak, M.; Kłodawski, M.; Gołębiowski, P. Modelling of Bicycle Traffic in the Cities Using VISUM. *Procedia Eng.* **2017**, *187*, 435–441. [CrossRef]
173. Oskarbski, J.; Jamroz, K.; Budziszewski, T.; Birr, K.; Oskarbski, G.; Gumińska, L.; Oskarbska, I.; Michalski, L. *Report on the Traffic Model Development for SUMP Implementation Status Report G3.1, Civitas Dyn@mo Project*; Gdansk University of Technology: Gdańsk, Poland, 2016. Available online: https://civitas.eu/sites/default/files/d3.3_impl_stat_rep_g3.1_report_on_the_traffic_model_development_for_sump_final_0.pdf (accessed on 1 April 2021).
174. Project FLOW. HORIZON. 2020. Available online: http://h2020-flow.eu/ (accessed on 20 March 2021).
175. PTV Group. *VISUM Fundamentals*; PTV Group: Karlsruhe, Germany, 2012.
176. Rupi, F.; Poliziani, C.; Schweizer, J. Analysing the dynamic performances of a bicycle network with a temporal analysis of GPS traces. *Case Stud. Transp. Policy* **2020**, *8*, 770–777. [CrossRef]
177. Birr, K. Mode Choice Modeling for Urban Areas. Ph.D. Thesis, Cracow University of Technology, Kraków, Poland, 2018.
178. Birr, K. Mode Choice Modelling for Urban Areas. *Tech. Trans.* **2018**, *6*, 67–77. [CrossRef]
179. Daganzo, C.F.; Sheffi, Y. On Stochastic Models of Traffic Assignment. *Transp. Sci.* **1977**, *11*, 253–274. [CrossRef]
180. Kim, H.; Oh, J.S.; Jayakrishnan, R. Effects of user equilibrium assumptions on network traffic pattern. *KSCE J. Civ. Eng.* **2009**, *13*, 117–127. [CrossRef]
181. Winters, M.; Davidson, G.; Kao, D.; Teschke, K. Motivators and deterrents of bicycling: Comparing influences on decisions to ride. *Transportation* **2011**, *38*, 153–168. [CrossRef]
182. Chen, P.; Shen, Q.; Childress, S. A GPS data-based analysis of built environment influences on bicyclist route preferences. *Int. J. Sustain. Transp.* **2018**, *12*, 218–231. [CrossRef]
183. Misra, A.; Watkins, K. Modeling Cyclist Route Choice using Revealed Preference Data: An Age and Gender Perspective. *Transp. Res. Rec. J. Transp. Res. Board* **2018**, *2672*, 145–154. [CrossRef]
184. Veillette, M.-P.; Gris'e, E.; El-Geneidy, A. Does One Bicycle Facility Type Fit All? Evaluating the Stated Usage of Different Types of Bicycle Facilities among Cyclists in Quebec City, Canada. *Transp. Res. Rec. J. Transp. Res. Board* **2019**, *2673*, 650–663. [CrossRef]
185. Lu, W.; Scott, D.M.; Dalumpines, R. Understanding bike share cyclist route choice using GPS data: Comparing dominant routes and shortest paths. *J. Transp. Geogr.* **2018**, *71*, 172–181. [CrossRef]
186. Kaplan, S.; Luria, R.; Prato, C.G. The relation between cyclists' perceptions of drivers, self-concepts and their willingness to cycle in mixed traffic. *Transp. Res. Part F Traffic Psychol. Behav.* **2019**, *62*, 45–57. [CrossRef]
187. Alattar, M.A.; Cottrill, C.; Beecroft, M. Modelling cyclists' route choice using Strava and OSMnx: A case study of the City of Glasgow. *Transp. Res. Interdiscip. Perspect.* **2021**, *9*, 100301. [CrossRef]
188. Broach, J.; Dill, J.; Gliebe, J. Where do cyclists ride? A route choice model developed with revealed preference GPS data. *Transp. Res. Part A Policy Pract.* **2012**, *46*, 1730–1740. [CrossRef]
189. Ton, D.; Cats, O.; Duives, D.; Hoogendoorn, S. How Do People Cycle in Amsterdam, Netherlands?: Estimating Cyclists' Route Choice Determinants with GPS Data from an Urban Area. *Transp. Res. Rec. J. Transp. Res. Board* **2017**, *2662*, 75–82. [CrossRef]
190. Szabo, N.; Kretz, T.; Sielemann, S. *Extended Versions of PTV Visum, PTV Vissim and PTV Viswalk*; D2.2 Flow Report; PTV AG: Karlsruhe, Germany, 2016. Available online: https://ec.europa.eu/research/participants/documents/downloadPublic?documentIds=080166e5aff61994&appId=PPGMS (accessed on 5 June 2021).
191. ZKM Gdynia. *Transport Preferences and Behaviour of Residents of Gdynia*; Marketing Survey Report 2015; ZKM: Gdynia, Poland, 2015.
192. ZKM Gdynia. *Transport Preferences and Behaviour of Residents of Gdynia*; Marketing Survey Report 2018; ZKM: Gdynia, Poland, 2018.
193. Oskarbski, J.; Zawisza, M.; Miszewski, M. Information system for drivers within the integrated traffic management system—TRISTAR. In *Tools of Transport Telematics*; TST 2015; Communications in Computer and Information Science; Mikulski, J., Ed.; Springer: Cham, Switzerland, 2015; Volume 531, ISBN 9783319245768. [CrossRef]
194. Le Pira, M.; Marcucci, E.; Gatta, V.; Ignaccolo, M.; Inturri, G.; Pluchino, A. Towards a decision-support procedure to foster stakeholder involvement and acceptability of urban freight transport policies. *Eur. Transp. Res. Rev.* **2017**, *9*, 54. [CrossRef]
195. Zhong, T.; Young, R.K.; Lowry, M.; Rutherford, G.S. A model for public involvement in transportation improvement programming using participatory Geographic Information Systems. *Comput. Environ. Urban Syst.* **2008**, *32*, 123–133. [CrossRef]
196. Piantanakulchai, M.; Saengkhao, N. Evaluation of alternatives in transportation planning using multi-stakeholders multi-objectives AHP modelling. *Proc. East. Asia Soc. Transp. Stud.* **2003**, *4*, 1613–1628.
197. Dueñas, M.; Campi, M.; Olmos, L. Changes in mobility and socioeconomic conditions in Bogotá city during the COVID-19 outbreak. *Humanit. Soc. Sci. Commun.* **2021**, *8*, 101. [CrossRef]
198. Dias, G.; Arsenio, E.; Ribeiro, P. The Role of Shared E-Scooter Systems in Urban Sustainability and Resilience during the COVID-19 Mobility Restrictions. *Sustainability* **2021**, *13*, 7084. [CrossRef]
199. Jacyna, M.; Żochowska, R.; Sobota, A.; Wasiak, M. Scenario analyses of exhaust emissions reduction through the introduction of electric vehicles into the city. *Energies* **2021**, *14*, 2030. [CrossRef]

200. Corchado, J.M.; Chamoso, P.; Hernández, G.; San, A.; Gutierrez, R.; Camacho, A.R.; González-briones, A.; Pinto-santos, F.; Goyenechea, E.; Garcia-retuerta, D.; et al. Deepint.net: A Rapid Deployment Platform for Smart Territories. *Sensors* **2021**, *21*, 236. [CrossRef] [PubMed]
201. Jing, C.; Du, M.; Li, S.; Liu, S. Geospatial dashboards for monitoring smart city performance. *Sustainability* **2019**, *11*, 5648. [CrossRef]
202. Handy, S.; van Wee, B.; Kroesen, M. Promoting Cycling for Transport: Research Needs and Challenges. *Transp. Rev.* **2014**, *34*, 4–24. [CrossRef]
203. Shen, L.; Stopher, P.R. Review of GPS Travel Survey and GPS Data-Processing Methods. *Transp. Rev.* **2014**, *34*, 316–334. [CrossRef]
204. Loidl, M.; Stutz, P.; Fernandez Lapuente de Battre, M.D.; Schmied, C.; Reich, B.; Bohm, P.; Sedlacek, N.; Niebauer, J.; Niederseer, D. Merging self-reported with technically sensed data for tracking mobility behavior in a naturalistic intervention study. Insights from the GISMO study. *Scand. J. Med. Sci. Sports* **2020**, *30*, 41–49. [CrossRef]
205. Abdullah, M.; Dias, C.; Muley, D.; Shahin, M. Exploring the impacts of COVID-19 on travel behavior and mode preferences. *Transp. Res. Interdiscip. Perspect.* **2020**, *8*, 100255. [CrossRef]
206. De Vos, J. The effect of COVID-19 and subsequent social distancing on travel behavior. *Transp. Res. Interdiscip. Perspect.* **2020**, *5*, 100121. [CrossRef]
207. González-Briones, A.; Hernandez, G.; Corchado, J.M.; Omatu, S.; Mohamad, M.S. Machine Learning Models for Electricity Consumption Forecasting: A Review. In Proceedings of the 2019 2nd International Conference on Computer Applications & Information Security (ICCAIS), Riyadh, Saudi Arabia, 1–3 May 2019. [CrossRef]
208. Pooley, C.G.; Horton, D.; Scheldeman, G.; Mullen, C.; Jones, T.; Tight, M.; Jopson, A.; Chisholm, A. Policies for promoting walking and cycling in England: A view from the street. *Transp. Policy* **2013**, *27*, 66–72. [CrossRef]
209. Rudolph, F. *Analysing the Impact of Walking and Cycling on Urban Road Performance: A Conceptual Framework*; Project FLOW; European Commission: Brussels Belgium, January 2017. Available online: http://h2020-flow.eu/fileadmin/templates/documents/Deliverables/FLOW_Conceptual_Framework_FINAL_web.pdf (accessed on 11 August 2020).
210. Oskarbski, J.; Jamroz, K. Reliability and Safety As An Objective of Intelligent Transport Systems in Urban Areas. *J. Konbin* **2015**, *34*, 59. [CrossRef]
211. Shaheen, S.; Cohen, A.; Zohdy, I. *Shared Mobility: Current Practices and Guiding Principles*; Report No. FHWA-HOP-16-022; U.S. Department of Transportation, Federal Highway Administration: Washington, DC, USA, 2016.
212. Tyndall, J. Free-floating carsharing and extemporaneous public transit substitution. *Res. Transp. Econ.* **2019**, *74*, 21–27. [CrossRef]
213. Aguilera-García, Á.; Gomez, J.; Sobrino, N.; Vinagre Díaz, J.J. Moped Scooter Sharing: Citizens' Perceptions, Users' Behavior, and Implications for Urban Mobility. *Sustainability* **2021**, *13*, 6886. [CrossRef]
214. Scott, D.M.; Lu, W.; Brown, M.J. Route choice of bike share users: Leveraging GPS data to derive choice sets. *J. Transp. Geogr.* **2021**, *90*, 102903. [CrossRef]
215. Fitch, D.T.; Handy, S.L. Road environments and bicyclist route choice: The cases of Davis and San Francisco, CA. *J. Transp. Geogr.* **2020**, *85*, 102705. [CrossRef]
216. Rybarczyk, G.; Wu, C. Bicycle facility planning using GIS and multi-criteria decision analysis. *Appl. Geogr.* **2010**, *30*, 282–293. [CrossRef]
217. Bíl, M.; Andrášik, R.; Kubeček, J. How comfortable are your cycling tracks? A new method for objective bicycle vibration measurement. *Transp. Res. Part C Emerg. Technol.* **2015**, *56*, 415–425. [CrossRef]
218. Joo, S.; Oh, C.; Jeong, E.; Lee, G. Categorizing bicycling environments using GPS-based public bicycle speed data. *Transp. Res. Part C Emerg. Technol.* **2015**, *56*, 239–250. [CrossRef]
219. Landis, B.W.; Vattikuti, V.R.; Brannick, M.T. Real-time human perceptions: Toward a bicycle level of service. *Transp. Res. Rec.* **1997**, *1578*, 119–126. [CrossRef]

energies

Review

Evaluating the Environmental Impact of Using Cargo Bikes in Cities: A Comprehensive Review of Existing Approaches

Hanna Vasiutina [1,*], Andrzej Szarata [1] and Stanisław Rybicki [2]

[1] Faculty of Civil Engineering, Cracow University of Technology, 31-155 Kraków, Poland; aszarata@pk.edu.pl
[2] Faculty of Environmental Engineering, Cracow University of Technology, 31-155 Kraków, Poland; srybicki@pk.edu.pl
* Correspondence: hanna.vasiutina@doktorant.pk.edu.pl

Abstract: The impact of the use of cargo bicycles for delivery processes on the environment is undeniably positive: it leads to the reduction of pollutants, noise, and vibrations caused by traditional vehicles; decreases traffic jams; causes more effective use of public space; and others. But how should such an effect be measured? What tools should be used to justify the necessity for change to more sustainable means of transport? How can we improve the state of the environment considering the interests of logistics service providers? There is a large amount of scientific literature dedicated to this problem: by using different modeling approaches, authors attempt to address the issue of sustainable transport. This paper conducts a literature review in the field of green cargo deliveries, investigates the benefits and drawbacks of integrating cargo bikes in urban logistics schemes, and examines methodologies and techniques for evaluating the impact of using cargo bicycles on the environment. By providing an opportunity to get acquainted with the situation in the sphere of green deliveries, the authors aim to encourage a breakthrough in the field of sustainable transport that may be achieved by using cargo bikes in modern cities. We review the existing approaches and tools for modeling transport emissions and state the significant positive environmental consequences.

Keywords: environmental impact; cargo bikes; sustainable transport; transport modelling

Citation: Vasiutina, H.; Szarata, A.; Rybicki, S. Evaluating the Environmental Impact of Using Cargo Bikes in Cities: A Comprehensive Review of Existing Approaches. *Energies* **2021**, *14*, 6462. https://doi.org/10.3390/en14206462

Academic Editor: Bjørn H. Hjertager

Received: 22 September 2021
Accepted: 8 October 2021
Published: 9 October 2021

1. Introduction

The constant, alarming, worsening state of the environment is an issue of major concern to the European Union policy-making authorities for more than 50 years. The United Nations Framework Convention on Climate Change developed in 1992 is considered as the starting point of the international environmental-protection policymaking. This treaty was a base for the following and more famous Kyoto Protocol and the Paris Convention, which introduced the state parties' regulations for limiting and reducing greenhouse gas (GHG) emissions. Furthermore, the EU continues providing air-quality politics and ratifies with its member states legislative acts (starting with the Council Directive 96/62/EC) that establish the mechanism for acquiring and controlling ambient air quality. As well, all the European Union member states (that adopt the directives) are obliged for its implementation to conduct annual reports on the state of the air. Such reports are collected and inspected by the European Environment Agency (EEA), the leading organization for managing the cooperation with the EU members states, which gathers, verifies, and presents the information on Europe's environment.

While the polluting sectors (such as agriculture, industry, and others), which undergo the monitoring of EEA, show a steady downward trend in pollution emissions, the road transport segment indicates the growth of global emissions since 1990 [1]. Moreover, the EEA states that overall greenhouse gas emissions from transport reach up to a quarter of all GHG emissions in the EU. Road transport is denoted as the main producer of such emissions in 2018 (about 70%) [1]. Besides being responsible to a large extent for global warming, transport is accountable for significant social and environmental harm: it causes health issues, noise pollution, congestion, loss of public space, and negatively affects the overall livability of the city. On the other hand, transport plays the role of an engine in economic progress and recovery: it contributes to investment growth, brings employment, and widens markets. Therefore, the area of sustainable transportations is among the main targets of EU policymaking. Furthermore, a positive result could only be achieved through the collective interaction of global and local initiatives by mutual influence of bottom-up and top-down approaches [2].

Along with the rapid increase of digital technologies and demand for city logistics, the requirement for improvement of urban life quality also grows. As follows, the bottom-up approach is gaining momentum due to the growing awareness of the population about the threat of global warming and specifically about the impact of transport, one of the main contributors to climate change [3,4]. Whereas more citizens are concerned with the living condition of the neighborhood and appeal to city authorities for adequate reaction, the question of introducing the new, environmentally friendly business models and strategies arises. As a reaction to that prompt, many cities around the world introduced urban transport policies that aim to support the use of city space for pedestrians and non-motorized vehicles, limit speed, and restrict access of heavy diesel motor vehicles to the city centers (e.g., Cracow [5], Paris [6], London [7], Sao Paolo [8], Seoul [9], Rio de Janeiro [10], New York [11], Barcelona [12]). Moreover, officials in turn motivate residents to shift to more sustainable transport. For instance, purchasing of electric cars and cargo bicycles (CB) could be induced by funding and tax reduction; traveling by bicycles could be prompted by infrastructure improvement and encouragement through social initiatives; and use of public transport by increasing its efficiency and accessibility, and that is the representation of the top-down approach.

Consequently, there is increased attention for planning, adapting, and shifting to the new sustainable distribution solutions in the cities that could satisfy all parties interested in the delivery process. Not only does such a solution design eliminate negative environmental and social impacts from road transport on city inhabitants, but it also minimizes losses in efficiency and costs for urban logistic operators (ex. operating and investment costs). Inevitably, cycle logistics is gaining more attention in recent EU research and development projects, such as CityChangerCargoBike [13], Handshake [14], Park4SUMP [15], GreenCharge [16], etc.

Currently, the main concept for estimation of transport impact is to determine the level of pollution using the various transport emission models, which receive as an input the parameters' set of characteristics achieved from the traffic situation. Sometimes, such characteristics are oversimplified; in other cases, they are obtained from models of a transportation system (infrastructure of the study region, vehicle fleet characteristics). Furthermore, atmospheric dispersion models are applied for the evaluation of the behavior of pollutants in the atmosphere.

The great amount of scientific literature that analyzes alternative/ sustainable distribution scenarios for the last- (or first-) mile deliveries explores CB (and its variations) as an environmentally friendly means of transport and the main substitute for traditional vans (in last-mile deliveries). Primarily, the directions of such studies, which are the subject of interest of this article, could be outlined as:

(1) Analysis of business strategies. Commonly, in such an approach, the potential and the challenges of using CB as a green transport mode are investigated in the context of the development of strategies for logistics service providers (LSP) [12,17–19]. While developing the strategies, the economic impact area for CB use is assessed, its cost-effectiveness is evaluated, and the competitiveness of LSP is estimated.

(2) Transportation system modeling (TSM) is used to optimize the transport system parameters. For this approach, the routing models are usually implemented that integrate less-polluting vehicles (such as electric vehicles, CB, etc.) into a delivery system. These models could be used as a strategic and decision-making tool by city planners and authorities [20–25]. Additionally, the operations impact area and the

location of transshipment points are defined by the means of TSM [26–28]. Another promising direction of using TSM is the justification of the multimodal schemes with the incorporation of consolidation centers (for loading and unloading operations) on the edge of the city center with transshipment to the small, nonmotorized delivery vehicles, possibly CB [8,29,30].

(3) Estimation of transport impact. Transport emission and dispersion models are developed to evaluate amounts of different types of pollutants (CO_2, PM, SO_2, NOx, VOC, and CO) and their distribution in the atmosphere as well as to estimate the noise level [7,8,31,32].

(4) Combination and mixture of the approaches listed above [11,33–35].

Despite the great interest in the last years for the topic of ecological modes of transport, there is a lack of research with a special focus on the environmental impact of non-motorized solutions for goods' deliveries [6,36]. The authors of the paper [37] underline that the perspective in the exploration of new techniques and recommendations about using cargo bikes is vast.

This paper aims to identify the set of basic requirements that should be addressed by the contemporary methodology for evaluating the environmental impact of cargo bicycles. For these purposes, we explore the state of art in a topic of sustainable freight deliveries when CB are used as a mean of transport.

The paper is organized as follows: the next section is dedicated to the exploration of pluses and minuses of CB as a means of transport; the third part presents the review of existing studies on delivery schemes used by logistics service providers in real-world solutions; the fourth section contains a short analysis of the literature on using the transport systems modeling approach; methods for evaluation of transport impact on the environment are presented in the fifth section; and the last part provides conclusions and plans for the future research.

2. Cargo Bike as a Means of Transport: Advantages and Disadvantages

CB is a bicycle that serves for transportation of various freights (goods, passengers, etc.) and was specifically invented for such purposes nearly a century ago [2,38]. Depending on the design, the number of wheels, or their purpose, CBs are called freight, transport or box bikes, carrier cycles, tricycles (CT) and quadricycles, cycle-trucks, long-john, and others. Modern CBs are usually electrically assisted, and their models vastly vary from simple two-wheelers bikes equipped with boxes on the front or back wheel to more progressive longtails and long johns that can carry weights around 50–100 kg. The most advanced multi-wheelers or light electric vehicles (LEV) can transport cargo up to 500–700 kg [2,17,39,40]. The term "cycle logistics" commonly refers to any type of pedal bicycle [41].

Recent studies confirm the enormous potential of cargo cycles to serve as sustainable substitution of traditional delivery vehicles. At the same time, the readiness to transition to more ecological means of transport (powered by electricity) was detected in slightly more than 60% of studies dedicated to last-mile problems, whereas nearly 50% of scientific sources advocate for freight bicycle adoption [42]. Such great possibility of CB to replace conventional vans is conditioned, to a large extent, by the small or the medium size and light weight of the most cargoes and relatively short delivery distances in central areas of the cities. According to results of the CycleLogistics project, nearly half of all urban goods transportation in the EU could be carried by CB; furthermore, the average trip distances in the city centers do not exceed 7 km [43]. A similar figure was obtained by the authors of the paper [44]: they observed that the distance of most commercial delivery routes is less than 10 km. TNT-FEDEX data that was used in the research [8] show such average freight parameters, volume, and weight: 0.025 m^3 and 20 kg (maximum 30 kg), and those are medium packages of electronics and textiles. Moreover, the authors of the study [20] noted that 80% of all parcel flow are cargoes weighing less than 6 kg.

The following types of freights can be distinguished as the most frequently transported by CBs:

- small packages and boxes [17,45]:
    - food [6,7,11,17,39,40,46],
    - correspondences, documents [6,17,37],
    - pharmaceuticals [6];
- medium size: mail, retail [46], and
- service trips [39,46], home deliveries [37].

Especially, advantages of CB could be observed in the narrow streets of historical city centers, with one-way traffic management and a large number of pedestrian areas [12,27,41,45,47] (see Table 1). Such districts frequently have spatial and timing restrictions on cars entering and are highly congested due to the population density and high demand for goods delivery [17,48]. Moreover, most authors emphasize the effectiveness and suitability of CB for the last- (or first-) mile urban distribution [6,8,10–12,19,21–23,25,36,40,41,47–50].

Beneficial results of using CBs were indicated in:

- cost savings [6,8,17,20,21,36,48,50];
- decreasing congestions, CO_2 emissions, and local pollutants [6,8,9,11,19,23,27,51,52];
- improving safety on roads due to reduction of car-related accidents [6,38];
- presentation of attractive (green) image of a company amongst its customers [17,20,48]; and
- improving livability of the city [8].

Economic effectiveness could be achieved due to low charging, purchase, and maintenance costs, which are notably less than for traditional vans [50]. Furthermore, the increase in performance could be attained due to the reliability of CB and their agility in urban centers (possibility to ride on one-way streets, park on the sideways, etc.) [6]. By occupying less than 35% of the space that takes up the van, CBs contribute to eliminating the congestion delays [11]. The collaborative scheme of supplies, which includes CT and mobile depots (MD), indicates a significant reduction in GHG emissions [10]. The availability of ready-to-use bicycles and equipped transshipment hubs can ensure the elimination of delays in deliveries [53].

Nonetheless, along with the huge advantages of CB, their drawbacks and limitations must be considered. The most common disadvantages could be found in Table 1. As for ECB also, potential issues could arise in the areas of battery and engine malfunctions and, accordingly, the lowering of travel range and time. Additionally, researchers indicate the lack of regulations (e.g., parking policies) and cycling infrastructure (charging stations and transshipment points) [17,47,52].

Table 1. Advantages and disadvantages of using bikes for cargo deliveries.

Feature	Description	Source
	Advantages	
Compact	• needs little parking space • can use bike lanes • suitable for dense city center	[8,11,27,36,47,50]
Clean	• does not emit GHG • reduction of pollution and emission • noiseless	[6,8,11,12,20,27,36,40,47,51,54]
Cheap	• purchase cost • maintenance cost	[46,47]
Effectiveness	• congestion reduction • lesser driving distances • reduction of delivery time • energy savings	[6–8,11,12,17,20,21,27,40,46,50,51]
Cost savings	• fuel economy • operational/delivery cost	[8,36,49] [20,51]

Table 1. *Cont.*

Feature	Description	Source
Safe	• safer for pedestrians • lowering road accidents	[7,11] [6]
Attractiveness	• approval by society • improve life quality	[8,12,17,40]
	Disadvantages and limitations	
Capacity	• payload and volume capacity	[9,17,30,36,39,47]
Speed	• lower velocities	[9,29,36,39,47]
Range	• delivery distance limitations	[9,25,36,39]
Costs	• driver costs • loss of efficiency • high prices on bikes	[47] [20] [8]
Lack of infrastructure, topography features	• bike lanes • parking • transshipment points	[36,49]
Emissions	• CO_2 associated with electricity generation	[7]
Working conditions, risk factors (safety)	• riding behavior • traffic accidents • driver physique • winter conditions	[7] [49] [47] [29]

Special attention should be paid to the parameters, features of CB, and general recommendations when choosing an alternative vehicle. Table 1 summarizes the advantages and disadvantages of cycle logistics and provides a short description. Some disadvantages can be identified through a comparison of CB with conventional delivery vehicles.

In the literature on cycle logistics, much attention is paid to the problem of the implementation, concept testing, and determining the location of the urban consolidation centers (UCC), which are also called micro, mobile, delivery, transshipment center, or loading hubs, terminals and points, satellites, micro depots, micro-distribution platforms, and others. Usually, such centers are used for managing freight flows and represent temporal or permanent storage facilities to which cargo is delivered by one means of transport (conventional vans) and then is transported to the client by other means (cargo bikes) [5]. Authors of the paper [19] characterize a mobile depot as a trailer that is connected to a loading dock and serves as an office and a warehouse. The main consequence of the introduction of the UCC is the lowering of transport emissions due to the introduction of the more sustainable scheme of transport, where the route from the UCC to a customer was performed by CT [8]. In the presented study, the transshipment center (TC), besides just storing the freights, also serves as a garage for bikes. To use UCC in the form of containers and truck-trailers was offered as an option in one study [27]; the proposed solution led to a notable decrease in travel distances and annual emissions of local CO_2. Additionally, the authors of the manuscript [30] underlined the considerable role of UCC in pollution reduction; however, they pointed out that the large cost is the biggest obstacle for implementing such centers.

Besides focusing only on UCC, some studies are devoted to an exploration of the cycling infrastructure. For instance, a field experiment that was carried out using two CB was a subject of the research [40]. The authors observed bike deliveries performed on the pre-planned routes with aim of examining of quality and affordability of the bike lanes, road traffic threats, etc. As a result, attention was drawn to the poor bicycle infrastructure, the necessity of the thorough choice of the fleet, and the huge role of the policymakers in the sustainable shift. Similarly, by exploring a free CB-sharing system, the authors of the research [2] highlight the importance of extensive cycle infrastructure for the wider adoption of CBs. Likewise, the focus on bicycle paths as a basic need for the green mode shift was highlighted in the paper [37]. As a result of the literature review and surveys with representatives of the logistics industry, attention was drawn to the key role of interest groups and private businesses as well as the influence of municipalities on the sustainable development of the cities.

There are not many ways researchers can obtain data for the studies in the transportation area. The most common method is to conduct interviews and surveys; another more difficult procedure is gaining information from transport organizations (oftentimes, such data is protected by the confidentiality agreement and cannot be openly displayed). Furthermore, some perform field tests or observations or use GPS tracking data; others use knowledge from the literature or previously developed projects. The pie chart in Figure 1 shows the proportion of the data sources that have been used in the scientific papers cited in this article.

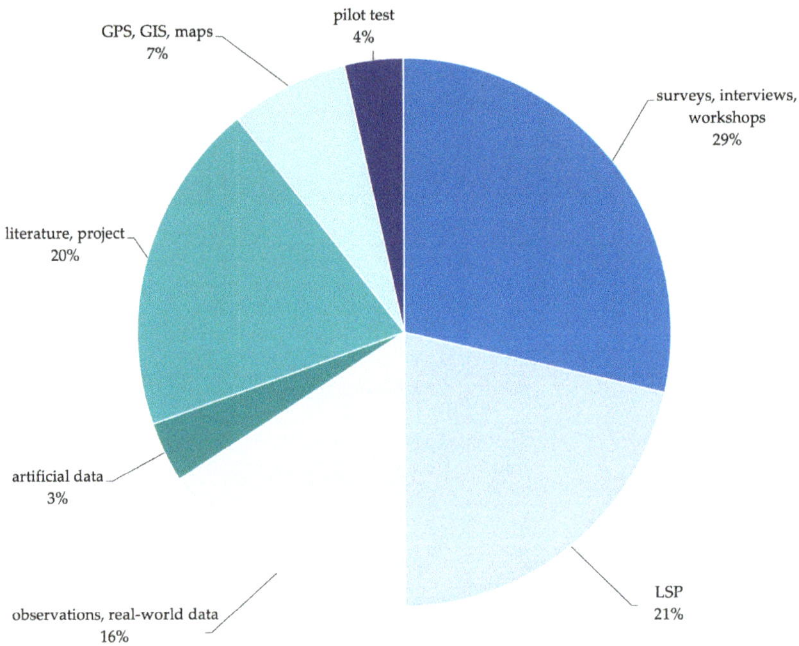

Figure 1. Rating of the most frequently used data sources.

The data represented on the graph illustrates that most researchers collect information from surveys, interviews, and workshops (nearly 30%). The second most popular position is shared by parcel delivery companies and scientific literature and projects (each of these sections gained approximately 20%). Slightly fewer papers refer to such sources as observations (a little more than 15%). However, GPS, GIS, and maps were used only by a small percentage of authors (about 7%), and the smallest segment is represented by studies for the implementation of which data from pilot projects were used (3.6%).

The described results demonstrate that data are mostly collected through questionnaires and surveys; thus, such information is often biased. Moreover, there is clearly a lack of works based on practical experiments and cooperation between the private sector and scientific institutions. On the other hand, we can observe the wide engagement of research and development projects that aim the popularization and expansion of sustainable mobility solutions in the urban areas.

Such projects demonstrate the potential, possibilities, and prospects of environmentally friendly transport to be adapted and implemented by the private sector, researched by scientists, and promoted among the public. Among recently implemented or started projects, the following ones should be listed:

- Ich ersetze ein Auto (I'm replacing a car) [55],
- SMILE [12],
- STRAIGHTSOL [12,19],
- Pro-E-bike [17],
- CycleLogistics [42],
- Ich entlaste Städte (Taking the load off cities) [44],
- Freie Lastenräder [2],
- LCL (Low carbon logistics) [40], and
- LEVV-LOGIC [39].

For the last years, the market of cycle logistics has grown rapidly. Large logistics companies adopt sustainable business models, take part in various pilot projects to test possibilities of green means of transport, cooperate with municipal authorities and non-governmental organizations, and provide data for the research. Some of these operators are TNT [8,17], EBC, GKC [17], DHL [17,50,51], GLS [17], UPS [18,50], FedEx [50], Hermes, Gnewt Cargo [41], CEP [8], and PonyZero [23].

Deliveries of freights by bicycles are performed in many urban areas around the world, although a study of the scientific literature of recent years has shown that most research on CB activities was conducted in the USA in such cities as Seattle, Austin, New York, Portland, and others. Germany comes second with six heavily researched cities. Slightly fewer articles were detected in Poland. The further ranking is shown in Figure 2.

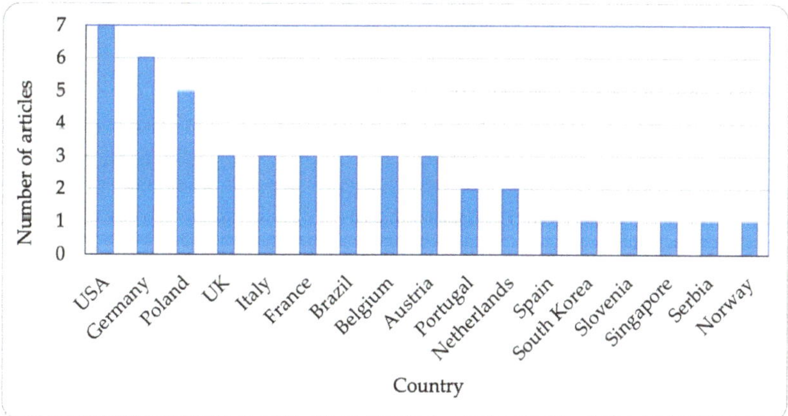

Figure 2. Geographical distribution of studies dedicated to CB issues.

In summary, most scientific papers attribute such main advantages of cargo bicycles as their environmental friendliness, compactness, accessibility, safety, and efficiency in urban conditions. However, it is necessary to pay attention to such features as their limited size, capacity, and speed, and, most importantly, to the working conditions and associated risks for the person operating such a vehicle.

3. Approaches to the Improvement of Business Strategies of Logistics Service Providers

Despite the obvious benefits of using zero-emission alternatives in city logistics, regardless of the eventual readiness of local authorities to put into action sustainable solutions, the decision-making process largely remains to the transportation service providers and their customers. Moreover, in many cases, such organizations show hesitance and a negative attitude to the introduction of green innovations [41,46,56]. Whereas the implementation of new strategies entails additional costs for the private sector, it is essential to consider such indicators as investment, operating costs, and other costs incurred by the company, although some researchers indicate such investment as beneficial, with return for the business [49]. However, authors of publications [6,7,17,37,41,57] emphasize the crucial role of the city authorities in influencing the implementation of effective green schemes. Such influence could be realized through policies, regulations (e.g., limitations of delivery hours, initiation of pedestrian zones, increase of penalties, etc.), and long-term planning. Additionally, the significance of pilot projects in the encouragement of the modal shift to environmentally friendly transport was pointed out in works [17,37].

For instance, city municipalities, transport operators, and eco-logistic and car-parking companies became the participants in the pilot project, which was examined by the authors of the work [12]. In the presented study, two electrically assisted tricycles along with transshipment points were used in two cities for performing parcel transportations for last-mile deliveries. Authors evaluate criteria concerning various spheres: economic, transport operations and energy, environmental, and social area. The outcome underlines the importance of contemplating the parameters of the study region and the crucial role of the cooperation between all the stakeholders of the transport system. Another pilot test was the subject of the study [17]. Through extensive research on big companies' (participants of the pilot project) operational practices, the major misconceptions about CB efficiency, load capacity, and reliability were identified. In addition, the environmental and economic effects of using CB were introduced through the evaluated savings of CO_2 emissions and energy costs.

Even though most research results indicate a significant positive economic, environmental, and social consequence from the implementation of cycle logistics, the authors of the paper [18] did not achieve any noteworthy outcome in reducing costs and emissions. The presented work describes the analyzes of the pilot test, in which one of the delivery variants included ECT. However, the general authors' remarks regarding the use of alternatives were positive.

Another research that was based on a cargo cycle trial project presents a model that could be practically used by decision makers [44]. The purpose of the experiment was a comparison of the travel times between vans and CB that perform distribution of goods in a public space in a mixed-fleet scheme. The estimated results for the option with substitution of 50% of traditional vehicles by bikes pointed out that expected delays in deliveries by CB would not exceed 10 min. Moreover, the replacement of 90% of motor vehicles would not result in delays of more than 20 min. Such promising outcomes could contribute to companies' decision to move to sustainable transport.

One more work, which was based on a real-life project, provides information on the competitiveness of clean transport mode in urban logistics [55]. A thorough socio-demographic overview was carried out on one of the main representatives of the decision-making group, individual messengers. The research provides a glimpse on indicators that could prompt the adoption of new technologies, such as the raise of awareness through arranging advertisement campaigns, field tests, and other forms of social informing.

Searching for an effective solution, numerous studies analyze the benefits and implications of different cooperation strategies that could be adopted for sustainable urban transportations, such as the estimation of a green delivery concept of using MD in combination with ET [19]. The authors applied the methodology multi-actor multi-criteria analysis (MAMCA) for the decision-making evaluation of six scenarios. Besides, the environmental, social, transport, and economic impacts were examined. As a result, the most significant effect was obtained in decreasing of emission level. However, the delivery time slightly dropped, and operational costs turned out to be twice higher than during the traditional transportation process. However, the confirmation of the greater cost-effectiveness of EACB over trucks was achieved in the cost-function comparison model for four transportation scenarios, developed by the authors [50]. Beneficial results showed the scenario in which the deliveries were performed by electrically assisted CB (ECB) over short distances from the distribution center. The research on the integration of green solutions into the traditional transportation business model is presented in [20]. Authors indicate the lack of studies dedicated to the adaption of zero-emission schemes by the main participants of the delivery system, international courier delivery service providers. The GUEST (Go, Uniform, Evaluate, Solve, Test) business methodology was used for the analysis of the supply system from the business and operational angles. Further, the decision-support system (DSS) was developed for the evaluation of different supply scenarios. As a result, savings in CO_2 emissions were achieved in the green operational scheme. However, the loss of efficiency by traditional service providers emphasizes the need for a holistic approach while implementing sustainable concepts.

In addition, the effectiveness of transportations is largely influenced by the productivity of selected means of transport. For instance, to choose the most suitable electric bike on the market, the authors applied the multi-criteria decision-making model COMET [58]. Evaluating such criteria as battery and engine parameters, speed, driving range, gear characteristics, price, etc., the comparison of ten EB models was conducted considering the condition of incomplete knowledge. The proposed model could be freely used in practice. Furthermore, the efficiency of cycle deliveries varies seasonally. The dependency of the CB productivity on the cold season was analyzed by [29]. Results indicate a decrease in the bike's average speed by approximately 30%. Additionally, the authors highlight that overall effectiveness could be increased by the improvement of working conditions of bicyclists and winter-specific bicycle maintenance. No less important is the arrangement of the consolidation centers, which could provide the opportunity to warm up and get technical support.

The extensive research through the interviews and consultations with the business owners was conducted by the authors of the paper [46] for defining the most influential conditions for the suitability of LEV for innovative urban logistics. Among other things, the necessity of using transshipment points as well as thorough planning of a mixed delivery fleet was strongly emphasized in that study. Similar research on the ability of CT to compete with diesel vans was presented in the paper [47]. The results of the evaluation of logistics and cost minimization models clearly indicate the effectiveness and suitability and applicability of CB services in the sustainable goods movement.

4. Transport System Modelling Approach

Using TSM as the first phase of the analysis of transport impact on the environment is a common practice. In such an approach, at first, the transportation network is inspected: the characteristics of the research area are explored (infrastructure, terrain properties); fleet composition, traffic volume, and speed are examined. Afterward, the output parameters from the TSM can be used as input parameters for the transport emission modeling.

In many cases, advanced TSM use subsystems that describe the rational behavior of the transport system entities. Such subsystems are often based on the Vehicle Routing Problem (VRP) algorithms. For instance, in the paper [21], the Two-Echelon Capacitated Electric Vehicle Routing Problem with Time Windows and Partial Recharging (2E-EVRPTW-PR) was proposed for the last-mile logistics for exploring the possibilities of clean transport modes (ECB and e-vans). The authors emphasize the effectiveness of such a two-echelon scheme, where ECBs perform the deliveries inside the restricted city area after the cargos were delivered to the micro depot by the e-vans. Similarly, obtaining synchronization between CB and traditional vans with a transshipment point near the city center in the two-echelon supply scheme was a purpose of the research [22]. Through the analysis of different delivery scenarios, the authors presented the algorithm, based on the GRASP metaheuristics (Greedy Randomized Adaptive Search Procedure), which aims to assist in a decision-making process while determining the most optimal solution for the distribution process. Another decision-supporting system that combines the business management tool Odoo with the route optimization technique and that is based on the Vehicle Routing Problem with Time Window (VRPTW) was introduced in the study [23]. Analysis of the collaboration of zero-emission vehicles transporting cargo less than 5 kg with conventional means of transport shows reduction of CO_2 emissions up to 14 tons per year. The system that assists food deliveries by CB was designed by the authors of the research [24] for connecting producers, distributors, consumers, and carriers. As the result, the beneficial results of delegating fresh food transportations in urban zones to the CB with IoT were noted.

Determining the optimal composition of the transport fleet by analyzing various alternatives of distribution schemes is exactly the problem that could be resolved using simulation modeling. Thus, authors of the study [52] pursue the exploration of different delivery scenarios that include small-sized electric vehicles (SEV) that replace vans in varying proportions. The simulation procedure was implemented in the AIMSUNG program. The optimal attainable rate that leads to costs, energy, and emission savings would be the 10% scenario, due to which it is possible to achieve a 3.6% reduction of CO_2. However, the positive effects of using SEV are shaded by the unwanted private sector risks and additional costs. Another simulation system for comparison of two operational schemes (where deliveries are performed only by trucks or by CB with mobile hubs) is presented in the paper [59]. Considering the varying characteristics of demand, the evaluation of some key transport efficiency indicators was performed. As the outcome of the CB delivery variant, the notable reduction (by nearly 150 deliveries per km^2) in vehicle distances and times traveled was achieved.

Different route and supply characteristics were examined in the research [50] to ascertain the conditions for the most efficient performance of CB compared to conventional trucks. The scheme, located near where the UCC clients were served by ECB, turned out to be more cost-effective than the one with deliveries by trucks. Likewise, the supplies by the ECB over short distances (about 2 km) were evaluated by the authors of the study [36]. The calculation of the optimal number of emission-free vehicles, which could substitute traditional vans in the city districts without reducing the efficiency of the initial system, was conducted using the micro-simulation software AIMSUN and the fuel consumption Well-to-Wheel approach. The results reveal that replacing 10% of vans brings beneficial results for all the stakeholders of the innovative distribution solution.

The research [9] suggests the model for reduction of GHG emissions along with operational costs by combining delivery trucks with zero-emission vehicles, ECB. Applying Heterogeneous Fleet Vehicle Routing Problem (HFVRP) in conjunction with the Tier-1 method for emission evaluation, the authors seek to achieve the optimal proportion between bikes and trucks in the mixed delivery system. The obtained most favorable fleet scheme with 29 trucks and 9 ECB showed the costs lowering by nearly 14% and carbon emissions by 10% compared with the scheme with a fleet of only trucks.

Integrating CB into urban operational scenarios was the objective of the study [45]. A developed GIS-based simulation tool aims to support the planning process while designing alternative distribution patterns that include CB and TP. In addition to the route-simulation feature, implemented with help of AnyLogic software and Capacitated Vehicle Routing Problem with Multiple Depots (MDCVRP), the presented system has functionality for calculating the economic and environmental impact of the examined network. Similarly, the research on different variants of delivery scenarios using the simulation approach presents the evaluation of operational and external costs of different distribution concepts [56]. The outcome shows that a scheme that includes 10 to 25% of self-pickups plus CB deliveries (performed from DP) is beneficial and can notably reduce external costs of LSP (nearly by 30%).

Almost no modern, efficient urban logistics solution can function without UCC. Such concept of providing facilities close to the core of the city is especially useful in case of using small or light modes of transport for last-mile distribution. Therefore, to incorporate consolidation centers into the operational scheme, the whole set of methods must be implemented, such as localization and capacity planning, synchronization, scheduling, and others. Thus, the authors of the publications [5,26] used the computer simulation approach for determining the location of loading hubs near the city area with traffic restrictions. The proposed freight delivery model, which is based on the simple Facility Location Problem (FLP), was implemented in Python programming language, and considers the stochasticity of the demand for transport services. Likewise, the authors of the paper [28] were engaged with the problem of establishing positions for micro depots and developed DSS aiming to study different transportation scenarios to reduce total operating costs as well as negative impact on the environment. The question of optimization of the delivery schemes in a city center by effectively situating parcel depots and route planning was examined in the paper [27]. As the main instrument for the implementation of Pickup and Delivery Problem with Time Windows (PDPTW) and location-allocation tasks, the ArcGIS software was chosen.

The problem of minimization of overall transporting costs while considering the stochasticity of the process was a goal of the work [54]. The authors explored a 2-Echelon Vehicle Routing Problem (2EVRP) with synchronization, where deliveries on the first echelon were performed by vans, and the CB (as a second echelon vehicle) served customers starting from the depot points. Moreover, an interesting concept of transshipment points was proposed: using satellites for synchronization between vans and bikes for loading and unloading operations.

The transport network mathematical model that considers the stochasticity of the demand for transport services was suggested by the research [25]. The open-source library for performing simulations of the delivery system is designed with help of the Python programming language and could be used for analysis and optimization of alternative distribution variants and particularly for implementing CB delivery scenarios. A closely related problem of modeling a new delivery scheme that integrates CB with micro depots (MD) into an urban transport network was presented by the work [51]. The simulation of two supply variants (by traditional vehicles and by CB) was performed in the open-source framework MATSim. Results indicate the significant reduction in total transport costs as well as emissions due to the use of CB.

The complex problem was examined by the authors of the paper [30]: first, they search for the UCC localization, and further, the optimal routes and fleet composition are determined with help of the Multi Depot Vehicle Routing Problem with Heterogeneous Fleet (MDHFVRP) and Genetic Algorithm (GA). The research proposes using a UCC sharing system for minimizing operating costs and achieving a positive environmental impact through the adaptation of sustainable vehicles, CB, and electric vans.

Exploring a case study in which postal delivery vehicles were substituted by CB, authors attempt to determine the economic effect of such modal shift [48]. Analyzing three different areas of the city and several variants of zero-emission vehicles and considering consolidation centers, the model of transport network was implemented using a Capacitated Vehicle Routing Problem with Time Window (CVRPTW) and simulation programs MATSim and jsprit. The proposed model points up the competitiveness of cargo cycles in terms of cost-saving perspective for the LSP and could be used as a city planning tool.

Table 2 highlights algorithms and software that are often used for modeling delivery systems with CB as a means of transport.

Table 2. Approaches to modeling systems of delivery by means of CB.

Methodology	Used Tools	Features	Source
		Algorithms	
HFVRP	Java, Simulated Annealing	determining mixed-fleet size	[9]
2E-EVRPTW-PR	CPLEX 12.10	DSS: synchronization task	[21]
VRPTW	Monte Carlo simulations	DSS: module for geo-referencing the data	[20]
	Java	DSS: trips creation module	[23]
	Odoo	DSS: business management framework	
2EVRP	GRASP, C/C++	synchronization between CB and vans	[22,54]
PDPTW	ArcGIS	finding optimal routes from MD to client	[27]
MDHFVRP	GA	UCC localization, route, and fleet planning	[30]
MDCVRP	AnyLogic	multimodal delivery model using CB	[45]
CVRPTW	jsprit	delivery network modeling	[48]

Table 2. *Cont.*

Methodology	Used Tools	Features	Source
FLP	Python	defining the loading hub location	[26]
Clarke–Wright Savings algorithm	Python MATLAB -	determining delivery routes	[25] [45] [56]
		Software	
Android application	C#, MySQL	system for food delivery	[24]
Maps	Google Maps API	travel time estimation georeferencing the routes	[44] [20]
	Google Maps navigation API	DSS: clients management module	[23]
	Google Maps Distance Matrix API	trip distances	[7]
	OpenStreetMaps	infrastructure data, client's localizations visualization, distance-matrices simulation graph road network data trip duration	[51] [45] [33] [27] [56]
TSM software	MATSim	exploring different delivery scenarios simulation of transport system	[48] [51]
	MAINSIM	simulation of different traffic configurations, CO_2 emissions	[33]
	AIMSUN	simulation of different fleet variants	[36,52]
GIS-based	ArcGIS + Visual Basic	traffic performance measures calculation	[11]
	CyberGIS	coupling of MAINSIM and SCIPUFF models	[33]

The wide range of tools and software presented in Table 2 is evidence of the big number of problems to be solved when optimizing the delivery systems that use CB as a means of transport. On the other hand, here appear the key questions that should be answered to choose the proper model for estimations of CB impact:

- Which optimization problems should be solved while simulating the process of delivering goods by bikes so that the behavior of a transport operator would be adequately considered in the simulation model?
- How much would the consideration of the transport operator's behavior in the model affect the results of estimations of the CB impact on the environment?

The answers to the listed questions should be obtained at the stage of developing the structure of a simulation model to be used for assessing the impact of CB.

5. Methods for Evaluation of Sustainable Transport Impact on the Environment

The analysis of the scientific literature clearly indicates the lack of studies on the environmental impact of sustainable delivery systems. There are, however, many works that implement various transport emission or dispersion models for evaluation of pollution caused by activities of traditional means of transport [33,35,60,61]. Thereafter, by modeling alternative scenarios and comparing the results, the question of how different transportation concepts would affect the environment could be answered [8,11].

Another equally important indicator of sustainable city development that is affected by transport is noise. Road-traffic noise models aim for the evaluation, management, prediction, and, eventually, the reduction of the sound power and involve such parameters as area information and traffic characteristics (acceleration, speed, and volume). Among the different approaches to noise modeling can be found the regression analyzes [32] and such noise emission models as FHWA, NMPB, ASJ-RNT, Imagine [62], and others.

Vehicle-related pollutants are divided depending on their source: exhaust emissions are produced and discharged by the internal combustion engine and fuel evaporation, and non-exhaust emissions are related to vehicle clutch and breaks, tire, and road wear abrasion. The composition of the exhaust emissions mainly includes carbon dioxide (CO_2) and monoxide (CO), nitrogen oxides (NOx, NO, NO_2), volatile organic compounds (VOC), particulate matter (PM), nitrous oxide (N_2O), ammonia (NH_3), persistent organic pollutants (POP), and metals. Non-exhaust emissions mostly consist of PM.

Moreover, another pollutant associated with vehicles is mineral dust, which spreads from the roads and streets surfaces by traffic or wind. Mineral dust, which consists mainly of silica (SiO_2) and corundum (Al_2O_3), sucked in with the intake air by the motorized vehicle engine, causes accelerated wear of engine parts. However, the engine could be protected by using high-efficient air filters [63,64] that contain elements made of pleated filter material (cellulose, cellulose with polyester, or with a nanofiber layer). Such filters are characterized by high filtration efficiency, which is over 99.5% [65,66].

The evaluation of the road transport impact on the environment can be performed using many different tools and approaches. Transport emission models vastly vary from the most popular in the EU countries, COPERT methodology, applied for composing of yearly emission inventories, to EMISENS, IVE, MOVES, HBEFA, VERSIT+, PHEM, MOBILE, and others. Such models allow estimation of the level of discharged emissions near the source. Instead, transport and dispersion models (T&D), also called air quality or atmospheric dispersion models, determine the spread of pollutants in the atmosphere and their concentrations at different locations. For the most part, these models are based on Lagrangian, Eulerian, or Gaussian plum models, such as OSPM, SCIPUFF, CALINE, STREET 5, FLEXPART, EPISODE, and others. The extensive literature review on a sequential usage of models for determining the air and water pollution caused by road transport can be found in the publication [67].

The main concept for estimation of transport emission is the multiplication of emission factors with respective activity data for different types of vehicles. Emission factors are laboratory-obtained measurements (via comprehensive vehicle tests) that depend on the driving behavior (acceleration, speeding, and braking), vehicle categories, Euro class, road, and traffic characteristics. In some cases, the approximate approach for the evaluation of CO_2 levels could be applied [27]. According to such a method (passed by U.S. Environmental Protection Agency), a liter of burned diesel fuel is equal to 2.66 kg of CO_2.

Emission models differ depending on the input parameters:

- vehicle fuel consumption,
- traffic volume [34] and composition [61],
- meteorological data [34],
- vehicle driving characteristics (speed, acceleration) [35],
- fleet composition (types of vehicles: size, fuel, evaporation, and exhaust control systems) [35],
- infrastructure (road map, speed limits) [61], and
- emission factors [61].

The spread of pollution in the atmosphere is mostly affected by the intensity of the source and meteorological conditions, such as wind strength, temperature, and humidity.

Data for dispersion modelling may contain:

- source parameters [33,61],
- composition of transport network [33],
- terrain characteristics [33], and
- meteorological conditions [33,61].

There is the recent tendency to develop fully integrated models by joining TSM for simulating vehicle activities with EM or T&D for determining or predicting the environmental effect caused by traffic [34,35,67].

For instance, the integrated simulation model that combines mesoscopic TSM, created with the use of FlexSim software and the regression model for evaluation of NO_2 concentration was proposed by the authors of research [34]. Such a system predicts the amount of pollution in the atmosphere depending on seasonality and the fleet composition. This approach could be used for the examination of different sustainable traffic variants to improve the ecological condition of the city district. Another integrated traffic-emission computation system was presented in the research [35]. After connecting TSM, implemented in Paramix system, with relational MS Access database, authors then added the emission calculation module based on the IVE model. The presented approach was further tested and is recommended for the estimation of traffic-related pollution. Furthermore, the combination of several methodologies, such as road transport emission EMISENS and evaluation of pollutants dispersion, has been proposed in the paper [61]. Assessing different pollution-reduction scenarios for black carbon (BC) and NOx, the researchers indicate that eliminating 10% of the biggest emitters will lead to a significant reduction of the pollutants (nearly 35% of NOx and 16% of BC daily). The analyses of the impact of different compositions of the conventional transport fleet on the pollution dispersion in the atmosphere by combining several models were carried out by the researchers in [33]. Microscopic TSM, developed in the MAINSIM program, enables to determine the level of emitted by traffic emissions due to its built-in module. Further, a gas-dispersion module was implemented using the SCIPUFF model. The combination of both modules was performed using the CyberGIS framework. Authors present how to analyze and make predictions on the atmospheric pollution behavior depending on the traffic and meteorologic factors.

Investigation of the possibility of changing the transport system and the following possible environmental consequences is of interest to many authors. For example, the comparison of several operational scenarios to determine the difference in the resulting level of pollution was performed in the work [68]. The variant with transportation by ECT showed five times fewer emissions of CO_2 compared to the option with diesel vans.

The suggestion to reduce the movement of heavy and light foods vehicles through the adoption of new delivery strategies that implement UCC outside the city together with the substitution of main emitters with electric vehicles is presented in the research [31]. The transport network model for the study was developed in AIMSUN microsimulation software and further transport emissions were evaluated. The quantification results point to the main pollutants, heavy vans, responsible for 13.8% CO_2, 43.7% of NOx, and 9.2% of PM from the total amount of emissions. For the study [7], the delivery data of a large platform provider were used to determine operational characteristics and amount of GHG emissions from on-demand deliveries performed by car, moped, and bike. Such an approach could be taken into consideration for comparison of different fleet compositions and justification for the introduction of clean modes of transport into the meal-delivery industry. The substitution of conventional vans by the ECT was examined by the authors of the paper [8] using real-world data. Results indicate the crucial savings in CO_2e: more than 95% and nearly a third in operating costs savings, which confirms the competitiveness of clean freight-distribution strategies. Research [11] evaluates traffic parameters and their environmental impact for the comparative analysis of different good movement modes: CB and motor vehicles. Using ArcGIS and Visual Basic, the traffic performance characteristics were calculated. Furthermore, the GHG and vehicle emissions evaluation was implemented with help of the MOVES model. Consequently, the high competitiveness of CB in transporting light freights over short distances was denoted by the outcome.

Instead of focusing only on determining the levels of CO_2 emitted by transport, the authors of the manuscript [60] emphasized the necessity of evaluating the concentrations of local pollutants (CO, NO_2, hydrocarbons, and PM). Using the approach that is based on the Gaussian plum model and the data achieved from the radar detectors, researchers determined amounts of NOx and CO distributed near the road segment. The highest level of pollutants was observed at a distance of 20 m from the road.

The methodologies for determining the level of noise caused by transport have a similar approach: they mainly depend on studied area characteristics as well on traffic data. The significant impact on a noise level has the congestion, vehicle acceleration, average speed over time, and the number of stopped vehicles [62]. Besides standard input parameters (traffic speed, volume, and width of the road surface), authors of the research [69] included honking into their noise model, which was based on the graph-theoretic approach. The study [32] focuses on the prediction of the noise level in the city area: using the regression modeling technique, the authors analyzed the influence of such urban parameters as street geometry and location and traffic characteristics on the noise level.

Table 3 summarizes the approaches to the estimation of the sustainable transport impact used by the authors of recently published papers.

Table 3. Approaches to the estimation of the environmental impact of transport and corresponding tools.

Source	Main Concept	Description	Tools
[7]	Estimation of emissions based on analytical models	comparing emissions of CB, mopeds, and conventional cars in meal deliveries	analytical model
[11]	Traffic performance characteristics + emissions	comparing CB and motor vehicle traffic effectiveness	ArcGIS + Visual Basic + MOVES
[33]	Integrated model: TSM + emissions + dispersion	assessing impact of different traffic patterns on air pollution	MAINSIM + SCIPUF + CyberGIS
[34]	Integrated model: TSM + simplified air pollution	predict NO_2 concentrations	FlexSim + regression model
[35]	Integrated model: TSM + vehicle emission computation	computation of hourly vehicle emissions	Paramix + MS Access + IVE
[61]	Integrated model: traffic emission + dispersion	tests of several schemes of the emission reduction	EMISENS

As can be noted from the descriptions in Table 3, to estimate the emissions reduction as an effect of using CB, the commonly applied are combinations of different tools. That is explained by the lack of conventions on what to measure and how to estimate the results. The ambiguity of measures of the sustainable transport impact may be also observed in Table 4, which summarizes the results of the studies dedicated to the estimation of road transport's negative influence.

For various technological solutions listed in Table 4, the results are measured in terms of costs and emission savings. However, the lack of uniformity in the type of used indicators makes the comparison of achieved results practically impossible. As the indicator for costs reductions, operational costs are the most frequently used parameter; however, energy costs, external costs, or delivery costs are used by researchers. This ambiguity also refers to the emission savings indicator: although the carbon dioxide equivalent is the standard parameter to estimate the reduction of emissions, other indicators are also used additionally to characterize the impact.

Table 4. Parameters achieved by estimation of the various delivery concepts aiming the reduction of the environmental and operational impact.

Source	Technological Solution	Effects	
		Costs Reduction, Nearly	Emission Savings, Up to
[6]	Using CB instead of trucks	€0.76 M per year	CO_2: 1.7 tons/day
[8]	Replacing diesel vans by ECT	operating costs: 31%	CO_2e: 97% per year
[9]	Mixed fleet: CB + vans	average 14%	CO_2: 10%
[11]	Comparing CB replacing vans	-	CO_2: 11 metric tons; PM2.5: 0.5 kg

Table 4. *Cont.*

Source	Technological Solution	Effects	
		Costs Reduction, Nearly	Emission Savings, Up to
[17]	Comparing alternatives: van, e-van, and LEV	energy cost: €0.9–11.0 per day	CO_2: 1.7–21 kg/day.
[19]	Impact of scheme ECB + MD	punctuality drop on 7%	CO_2: 24%; PM2.5: 99%
[20]	Comparing 5 alternative operational scenarios	operational and environmental costs: 30%	14 tons per year
[23]	CB transporting cargos < 5 kg	-	CO_2: 14 tons per year
[27]	Logistics system: CB + ECB + MD	-	local CO_2: 20 tons per year
[36]	Replacing 10% of traditional fleet	external costs: 25%	CO_2: 73%
[51]	Deliveries for commercial clients by CB	delivery costs: 28%	22%
[52]	Replacing 10% of vans by LEV	operational costs: 90%	3–4% of energy and CO_2
[56]	Delivery concept: self-pick-up + CB + DP	operational costs: 30%	-
[61]	Eliminating 10% of emitters	-	BC: 39%; NOx: 16%
[68]	Comparing deliveries: by ECT vs by diesel vehicles	-	CO_2e: 51–72%

6. Conclusions

Cargo bikes are an environmentally friendly means of transport. However, their use in commercial delivery schemes, besides obvious advantages, has some restrictions related to relatively small capacity and low delivery speed. The conducted literature review shows that the most preferable use of CB is last-mile deliveries in the cities, especially in districts with traffic restrictions and a high density of the population.

Logistics service providers extensively use CB in real-world solutions. However, the implementation of this technology needs complex actions of the authorities; besides, they are often initiated by active citizens and non-governmental organizations. To implement the delivery scheme that uses CB, the methodology for transportation planning should be chosen and the appropriate zero-emission fleet should be selected, but also the effect of these changes should be substantiated in terms of operational costs (from the point of view of LSP) and emissions reduction (for city authorities and residents).

To estimate the effect of using CB for last-mile deliveries, simulations of the transportation process are usually performed. The completed review of algorithms and software used for the optimization and simulations of cargo deliveries by bikes shows that there is a variety of tools available that allow solving the specific problem (routing, optimizing hub location, scheduling, etc.). However, there is no dedicated instrument that estimates the resulting technological parameters of the process of goods deliveries by CB (e.g., the covered distance and operation time). Such a tool would be indispensable for transport planners and city authorities when substantiating the emissions reduction due to the use of CB and assessing the corresponding operational costs.

Existing approaches to the evaluation of transport impact on the environment allow assessing in detail numerous indicators. A tool that is dedicated to the assessment of emissions reduction for technological schemes with CB as a means of transport must provide the possibility to check the results for CB comparing to other alternative servicing technologies. For these, the aggregated indicator (e.g., CO_2e) reflecting the environmental impact of transport should be used to avoid ambiguous conclusions.

The above-mentioned features (i.e., the consideration of interests of all parties, the assessments made based on the set of technological indicators, and the use of the aggregated emissions indicator) should be addressed by the contemporary methodology for evaluating the environmental impact of cargo bicycles.

As the direction of future research, the development of the simulation model for the substantiation of positive environmental effects of CB must be mentioned. We plan to test this model within the CityChangerCargoBike project while estimating the impact of the CB use in the districts with restricted traffic of cities that are the project partners—Krakow (Poland), San Sebastian and Vitoria-Gasteiz (Spain), Lisboa (Portugal), and Dubrovnik (Croatia).

Author Contributions: Conceptualization, H.V., A.S. and S.R.; methodology, H.V., A.S. and S.R.; investigation, H.V.; writing—original draft preparation, H.V.; writing—review and editing, A.S. and S.R.; visualization, H.V.; supervision, A.S. and S.R. All authors have read and agreed to the published version of the manuscript.

Funding: The research was carried out as part of the project "ROAD TO EXCELLENCE—a comprehensive university support programme," implemented under the Operational Programme Knowledge Education Development 2014–2020, co-financed by the European Social Fund; agreement no. POWR.03.05.00-00-Z214/18.

Institutional Review Board Statement: Not applicable.

Informed Consent Statement: Not applicable.

Data Availability Statement: Not applicable.

Conflicts of Interest: The authors declare no conflict of interest.

References

1. *Annual European Union Greenhouse Gas Inventory 1990–2018 and Inventory Report 2020*; European Environment Agency (EEA): Copenhagen, Denmark, 2020. Available online: https://www.eea.europa.eu//publications/european-union-greenhouse-gas-inventory-2020 (accessed on 30 August 2021).
2. Becker, S.; Rudolf, C. Exploring the potential of free cargo-bike-sharing for sustainable mobility. *GAIA—Ecol. Perspect. Sci. Soc.* **2018**, *27*, 156–164. [CrossRef]
3. *The First and Last Mile—The Key to Sustainable Urban Transport*; Transport and Environment Report 2019; European Environment Agency (EEA): Copenhagen, Denmark, 2020. Available online: https://www.eea.europa.eu/publications/the-first-and-last-mile (accessed on 30 August 2021). [CrossRef]
4. Wang, S.; Ge, M. Everything You Need to Know about the Fastest-Growing Source of Global Emissions: Transport. World Resources Institute. Available online: https://www.wri.org/insights/everything-you-need-know-about-fastest-growing-source-global-emissions-transport (accessed on 30 August 2021).
5. Naumov, V.; Starczewski, J. Choosing the localisation of loading points for the cargo bicycles system in the Krakow Old Town. *Lect. Notes Netw. Syst.* **2019**, *68*, 353–362. [CrossRef]
6. Koning, M.; Conway, A. The good impacts of biking for goods: Lessons from Paris city. *Case Stud. Transp. Policy* **2016**, *4*, 4, 259–268. [CrossRef]
7. Allen, J.; Piecyk, M.; Cherrett, T.; Juhari, M.N.; McLeod, F.; Piotrowska, M.; Bates, O.; Bektas, T.; Cheliotis, K.; Friday, A.; et al. Understanding the transport and CO_2 impacts of on-demand meal deliveries: A London case study. *Cities* **2021**, *108*, 102973. [CrossRef]
8. Ormond Junior, P.A.; Telhada, J.; Paulo, A. Evaluating the economic and environmental impact of the urban goods distribution by cargo cycles—A case study in São Paulo City. In Proceedings of the 2018 Conference: World Conference on Transport Research—WCTR 2019, Mumbai, India, 26–31 May 2019.
9. Lee, K.; Chae, J.; Kim, J. A courier service with electric bicycles in an urban area: The case in Seoul. *Sustainability* **2019**, *11*, 1255. [CrossRef]
10. Marujo, L.G., Goes, G.V., D'Agosto, M.A., Ferreira, A.F., Winkenbach, M., Bandeira, R.A.M. Assessing the sustainability of mobile depots: The case of urban freight distribution in Rio de Janeiro. *Transp. Res. Part D Transp. Environ.* **2018**, *62*, 256–267. [CrossRef]
11. Conway, A.; Cheng, J.; Kamga, C.; Wan, D. Cargo cycles for local delivery in New York City: Performance and impacts. *Res. Transp. Bus. Manag.* **2017**, *24*, 90–100. [CrossRef]
12. Navarro, C.; Roca-Riu, M.; Furió, S.; Estrada, M. Designing new models for energy efficiency in urban freight transport for smart cities and its application to the Spanish case. *Transp. Res. Procedia* **2016**, *12*, 314–324. [CrossRef]
13. CityChangerCargoBike. Available online: http://www.cyclelogistics.eu (accessed on 29 August 2021).
14. Handshake. Available online: https://handshakecycling.eu (accessed on 30 August 2021).
15. Park4SUMP. Available online: https://park4sump.eu (accessed on 30 August 2021).
16. GreenCharge. Available online: https://www.greencharge2020.eu (accessed on 30 August 2021).
17. Nocerino, R.; Colorni, A.; Lia, F.; Luè, A. E-bikes and e-scooters for smart logistics: Environmental and economic sustainability in pro e-bike Italian pilots. *Transp. Res. Procedia* **2016**, *14*, 2362–2371. [CrossRef]
18. Athanassopoulos, T.; Dobers, K.; Clausen, U. Reducing the environmental impact of urban parcel distribution. *Lect. Notes Logist.* **2016**, *12*, 159–181. [CrossRef]
19. Verlinde, S.; Macharis, C.; Milan, L.; Kin, B. Does a mobile depot make urban deliveries faster, more sustainable and more economically viable? Results of a pilot test in Brussels. *Transp. Res. Procedia* **2014**, *4*, 361–373. [CrossRef]
20. Perboli, G.; Rosano, M. Parcel delivery in urban areas: Opportunities and threats for the mix of traditional and green business models. *Transp. Res. Part C Emerg. Technol.* **2019**, *99*, 19–36. [CrossRef]

21. Caggiani, L.; Colovic, A.; Prencipe, L.P.; Ottomanelli, M. A green logistics solution for last-mile deliveries considering e-vans and e-cargo bikes. *Transp. Res. Procedia* **2021**, *52*, 75–82. [CrossRef]
22. Anderluh, A.; Hemmelmayr, V.C.; Nolz, P.C. Synchronizing vans and cargo bikes in a city distribution network. *Cent. Eur. J. Oper. Res.* **2017**, *25*, 345–376. [CrossRef]
23. Perboli, G.; Rosano, M. A Decision support system for optimizing the last mile by mixing traditional and green logistics. *Lect. Notes Bus. Inf. Process.* **2018**, *262*, 28–46. [CrossRef]
24. Tegeltija, S.; Ostojić, G.; Stankovski, S.; Kukolj, D.; Tejić, B. Food delivery using cargo-bikes with IoT. *Lect. Notes Multidiscip. Ind. Eng.* **2020**, 483–491. [CrossRef]
25. Naumov, V.; Starczewski, J. Approach to simulations of goods deliveries with the use of cargo bicycles. *AIP Conf. Proc.* **2019**, *2078*, 020070. [CrossRef]
26. Naumov, V. Substantiation of loading hub location for electric cargo bikes servicing city areas with restricted traffic. *Energies* **2021**, *14*, 839. [CrossRef]
27. Niels, T.; Hof, M.T.; Bogenberger, K. Design and operation of an urban electric courier cargo bike system. In Proceedings of the 21st International Conference on Intelligent Transportation Systems (ITSC), Maui, HI, USA, 4–7 November 2018; pp. 2531–2537. [CrossRef]
28. Leyerer, M.; Sonneberg, M.O.; Heumann, M.; Breitner, M.H. Decision support for sustainable and resilience-oriented urban parcel delivery. *EURO J. Decis. Process.* **2019**, *7*, 267–300. [CrossRef]
29. Dybdalen, Å.; Ryeng, E.O. Understanding how to ensure efficient operation of cargo bikes on winter roads. *Res. Transp. Bus. Manag.* **2021**, 100652. [CrossRef]
30. Simoni, M.D.; Bujanovic, P.; Boyles, S.D.; Kutanoglu, E. Urban consolidation solutions for parcel delivery considering location, fleet and route choice. *Case Stud. Transp. Policy* **2018**, *6*, 112–124. [CrossRef]
31. Aditjandra, P.T.; Galatioto, F.; Bell, M.C.; Zunder, T.H. Evaluating the impacts of urban freight traffic: Application of micro-simulation at a large establishment. *Eur. J. Transp. Infrastruct. Res.* **2016**, *16*. [CrossRef]
32. Rey Gozalo, G.; Suárez, E.; Montenegro, A.L.; Arenas, J.P.; Barrigón Morillas, J.M.; Montes González, D. Noise estimation using road and urban features. *Sustainability* **2020**, *12*, 9217. [CrossRef]
33. Cervone, G.; Dallmeyer, J.; Lattner, A.D.; Franzese, P.; Waters, N. Coupling traffic and gas dispersion simulation for atmospheric pollution estimation. *GeoJ. Libr.* **2019**, *118*, 13–31. [CrossRef]
34. Mihăiţă, A.S.; Ortiz, M.B.; Camargo, M.; Cai, C. Predicting air quality by integrating a mesoscopic traffic simulation model and simplified air pollutant estimation models. *Int. J. Intell. Transp. Syst. Res.* **2019**, *17*, 125–141. [CrossRef]
35. Wei, Y.; Yu Ying, X.; Lifeng, H.; Wei, G.; Jianhua, W.; Cao, J. Vehicle emission computation through microscopic traffic simulation calibrated using genetic algorithm. *J. Artif. Intell. Soft Comput. Res.* **2019**, *9*, 1, 67–80. [CrossRef]
36. Melo, S.; Baptista, P. Evaluating the impacts of using cargo cycles on urban logistics: Integrating traffic, environmental and operational boundaries. *Eur. Transp. Res. Rev.* **2017**, *9*, 30. [CrossRef]
37. Rudolph, C.; Gruber, J. Cargo cycles in commercial transport: Potentials, constraints, and recommendations. *Res. Transp. Bus. Manag.* **2017**, *24*, 26–36. [CrossRef]
38. Riggs, W. Cargo bikes as a growth area for bicycle vs. auto trips: Exploring the potential for mode substitution behavior. *Transp. Res. Part F Traffic Psychol. Behav.* **2016**, *43*, 48–55. [CrossRef]
39. Balm, S.; Amstel, W.P.; Moolenburgh, E.; Anand, N. The potential of light electric vehicles for specific freight flows: Insights from The Netherlands. In Proceedings of the International City Logistics Conference, Phuket Island, Thailand, 14–16 June 2017; p. 10. [CrossRef]
40. Nürnberg, M. Analysis of using cargo bikes in urban logistics on the example of Stargard. *Transp. Res. Procedia* **2019**, *39*, 360–369. [CrossRef]
41. Schliwa, G.; Armitage, R.; Aziz, S.; Evans, J.; Rhoades, J. Sustainable city logistics—Making cargo cycles viable for urban freight transport. *Res. Transp. Bus. Manag.* **2015**, *15*, 50–57. [CrossRef]
42. Oliveira, C.M.; Mello Bandeira, R.A.; Vasconcelos Goes, G.; Schmitz Gonçalves, D.N.; D'Agosto, M.D.A. Sustainable vehicles-based alternatives in last mile distribution of urban freight transport: A systematic literature review. *Sustainability* **2017**, *9*, 1324. [CrossRef]
43. Wrighton, S.; Reiter, K. CycleLogistics—Moving Europe Forward! *Transp. Res. Procedia* **2016**, *12*, 950–958. [CrossRef]
44. Gruber, J.; Narayanan, S. Travel time differences between cargo cycles and cars in commercial transport operations. *Transp. Res. Rec.* **2019**, *2673*, 623–637. [CrossRef]
45. Hofmann, W.; Assmann, T.; Neghabadi, P.D.; Cung, V.D.; Tolujevs, J. A Simulation tool to assess the integration of cargo bikes into an urban distribution system. In Proceedings of the 5th International Workshop on Simulation for Energy, Sustainable Development & Environment (SESDE 2017), Barcelona, Spain, 2017; hal-01875988. Available online: https://hal.archives-ouvertes.fr/hal-01875988/document (accessed on 18 September 2021).
46. Moolenburgh, E.A.; van Duin, J.H.R.; Balm, S.; van Altenburg, M.; Ploos van Amstel, W. Logistics concepts for light electric freight vehicles: A multiple case study from the Netherlands. *Transp. Res. Procedia* **2020**, *46*, 301–308. [CrossRef]
47. Tipagornwong, C.; Figliozzi, M. Analysis of competitiveness of freight tricycle delivery services in urban areas. *Transp. Res. Rec.* **2014**, *2410*, 76–84. [CrossRef]

48. Choubassi, C.; Seedah, D.P.K.; Jiang, N.; Walton, C.M. Economic analysis of cargo cycles for urban mail delivery. *Transp. Res. Rec.* **2016**, *2547*, 102–110. [CrossRef]
49. Oliveira, C.; Nascimento, L.; Belcavello Rigatto, I.; Oliveira, L.K. Characterization and analysis of the economic viability of cycle logistics transport in Brazil. *Transp. Res. Procedia* **2020**, *46*, 189–196. [CrossRef]
50. Sheth, M.; Butrina, P.; Goodchild, A.; McCormack, E. Measuring delivery route cost trade-offs between electric-assist cargo bicycles and delivery trucks in dense urban areas. *Eur. Transp. Res. Rev.* **2019**, *11*, 11. [CrossRef]
51. Zhang, L.; Matteis, T.; Thaller, C.; Liedtke, G. Simulation-based assessment of cargo bicycle and pick-up point in urban parcel delivery. *Procedia Comput. Sci.* **2018**, *130*, 18–25. [CrossRef]
52. Melo, S.; Baptista, P.; Costa, Á. Comparing the Use of Small Sized Electric Vehicles with Diesel Vans on City Logistics. *Procedia-Soc. Behav. Sci.* **2014**, *111*, 1265–1274. [CrossRef]
53. Fikar, C.; Hirsch, P.; Gronalt, M. A decision support system to investigate dynamic last-mile distribution facilitating cargo-bikes. *Int. J. Logist. Res. Appl.* **2018**, *21*, 300–317. [CrossRef]
54. Anderluh, A.; Larsen, R.; Hemmelmayr, V.C.; Nolz, P.C. Impact of travel time uncertainties on the solution cost of a two-echelon vehicle routing problem with synchronization. *Flex. Serv. Manuf. J.* **2020**, *32*, 806–828. [CrossRef]
55. Gruber, J.; Kihm, A. Reject or embrace? Messengers and electric cargo bikes. *Transp. Res. Procedia* **2016**, *12*, 900–910. [CrossRef]
56. Arnold, F.; Cardenas, I.; Sörensen, K.; Dewulf, W. Simulation of B2C e-commerce distribution in Antwerp using cargo bikes and delivery points. *Eur. Transp. Res. Rev.* **2018**, *10*, 2. [CrossRef]
57. Buldeo Rai, H.; Verlinde, S.; Macharis, C. City logistics in an omnichannel environment. The case of Brussels. *Case Stud. Transp. Policy* **2019**, *7*, 310–317. [CrossRef]
58. Sałabun, W.; Palczewski, K.; Watróbski, J. Multicriteria approach to sustainable transport evaluation under incomplete knowledge: Electric bikes case study. *Sustainability* **2019**, *11*, 3314. [CrossRef]
59. Chiara, G.D.; Alho, A.R.; Cheng, C.; Ben-Akiva, M.; Cheah, L. Exploring benefits of cargo-cycles versus trucks for urban parcel delivery under different demand scenarios. *Transp. Res. Rec.* **2020**, *2674*, 553–562. [CrossRef]
60. Kijewska, K.; Konicki, W.; Iwan, S. Freight Transport Pollution Propagation at Urban Areas Based on Szczecin Example. *Transp. Res. Procedia* **2016**, *14*, 1543–1552. [CrossRef]
61. Ježek, I.; Blond, N.; Skupinski, G.; Močnik, G. The traffic emission-dispersion model for a Central-European city agrees with measured black carbon apportioned to traffic. *Atmos. Environ.* **2018**, *184*, 177–190. [CrossRef]
62. Can, A.; Aumond, P. Estimation of road traffic noise emissions: The influence of speed and acceleration. *Transp. Res. Part D Transp. Environ.* **2018**, *58*, 155–171. [CrossRef]
63. Dziubak, T.; Bąkała, L. Computational and Experimental Analysis of Axial Flow Cyclone Used for Intake Air Filtration in Internal Combustion Engines. *Energies* **2021**, *14*, 2285. [CrossRef]
64. Tian, X.; Ou, Q.; Liu, J.; Liang, Y.; Pui, D.Y.H. Particle Loading Characteristics of a Two-Stage Filtration System. *Sep. Purif. Technol.* **2019**, *215*, 351–359. [CrossRef]
65. Dziubak, T.; Dziubak, S.D. Experimental Study of Filtration Materials Used in the Car Air Intake. *Materials* **2020**, *13*, 3498. [CrossRef] [PubMed]
66. Liu, B.; Zhang, S.; Wang, X.; Yu, J.; Ding, B. Efficient and reusable polyamide-56 nanofiber/nets membrane with bimodal structures for air filtration. *J. Colloid Interface Sci.* **2015**, *457*, 203–211. [CrossRef] [PubMed]
67. Shorshani, M.F.; André, M.; Bonhomme, C.; Seigneur, C. Modelling chain for the effect of road traffic on air and water quality: Techniques, current status and future prospects. *Environ. Model. Softw.* **2015**, *64*, 102–123. [CrossRef]
68. Saenz, J.; Figliozzi, M.; Faulin, J. Assessment of the carbon footprint reductions of tricycle logistics services. *Transp. Res. Rec.* **2016**, *2570*, 48–56. [CrossRef]
69. Gilani, T.A.; Mir, M.S. Modelling road traffic noise under heterogeneous traffic conditions using the graph-theoretic approach. *Environ. Sci. Pollut. Res.* **2021**, *28*, 36651–36668. [CrossRef] [PubMed]

energies

MDPI

Article

Exploring the Sustainable Effects of Urban-Port Road System Reconstruction

Dariusz Bernacki [1,*] and Christian Lis [2]

[1] Institute of Transport Management, Faculty of Transport Engineering and Economics, Maritime University of Szczecin, 70-500 Szczecin, Poland

[2] Institute of Economics and Finance, Faculty of Economics, Finance and Management, University of Szczecin, 70-453 Szczecin, Poland; Christian.Lis@usz.edu.pl

* Correspondence: d.bernacki@am.szczecin.pl

Abstract: The aim of the research is to identify and quantify the direct sustainable effects resulting from the improved road infrastructure in the local urban-port transport system. This case study considers the city port of Szczecin (Poland). The effects are identified for the local road transport system by comparing freight road transport performance in two options: with the investment and without the investment. The sustainable effects are quantified in terms of money and physical units. Sustainable economic, social, and environmental effects concern generalized freight road transport cost, i.e., truck operating costs and costs of truck drivers' working time, as well as freight transit time, energy consumption, greenhouse gas emissions, and environmental savings. To capture effects, the forecast of truck traffic demand, as well as unit vehicle operating costs, values of time, and air pollution and climate change, values are elaborated and revealed in freight road transport. The investigations show that the primary effect of investment is the reduced traffic congestion, which enhance the velocity of trucks in the transport system. The increased trucks' speed affect freight road traffic performance, time of delivery, and environmental externalities.

Keywords: investment; impact; sustainability; energy savings; urban-port road system

Citation: Bernacki, D.; Lis, C. Exploring the Sustainable Effects of Urban-Port Road System Reconstruction. *Energies* **2021**, *14*, 6512. https://doi.org/10.3390/en14206512

Academic Editors: Elżbieta Macioszek, Anna Granà, Margarida Coelho and Paulo Fernandes

Received: 8 September 2021
Accepted: 4 October 2021
Published: 11 October 2021

1. Introduction

Relations between city and port are a dynamically evolving area of multifaceted studies. After analyzing megatrends in the development of cities and seaports, the competitiveness of port cities, and management of interactions at the interface between cities and ports [1,2], studies have focused on the relations between road freight transport, facility location, logistics, and urban form [3], as well as the interactions between seaports and urban logistics [4].

The interactions between ports and their cities are subject to significant and dynamic changes. It is no longer evident that well-functioning ports have a net positive impact on the port–city interface [5]. The rapid growth of international trade (and resulting container traffic) has imposed tremendous pressure on the intermodal transportation system. The bottleneck of such intermodal chains has shifted from the ship–port interface to the port–city interface [6]. New areas of tension in the relationship between ports and cities have emerged as a result of conflicts over the environment, congestion, pollution, and other impacts resulting from port-related traffic [7]. Transport congestion in port cities is a crucial problem and results both from port-related cargo traffic and from the fast-economic development of such cities [8]. The concentration and combination of port-related and city-related traffic flows result in considerable congestion at port–city interfaces.

City-related road traffic includes passenger transport and freight traffic supplying goods and serving businesses operating in the city and its surroundings, while goods traffic is generated by seaports located in urban spaces. As a result, the urban road system must cope with mixed passenger-freight traffic of high intensity and general road congestion.

From a port perspective, road congestion raises travel times and fuel costs, lowers the reliability of commercial truck operations, increases the risk of missed schedules, and hinders efficient usage of port assets. From a city perspective, traffic congestion results in reduced population mobility, loss of working time, and increased environmental costs, which, overall, have a negative impact on social welfare [2].

The hinterland traffic of most ports is dominated by trucks, and it is truck traffic that causes most of the congestion in and around port areas and generates most external costs [9].

There are numerous quantitative methods of evaluation for public projects, such as cost assessment analysis, standard cost model, cost-effectiveness analysis (ACE), cost–benefit analysis (CBA), risk assessment, risk-risk analysis, and multi-criterion analysis (MCA). To assess the sustainability of transport infrastructure projects, new integrated methods of sustainability appraisal have been proposed [10], while the approach defined as sustainability assessment with an incorporated set of predefined sustainable indicators is emerging [11].

However, in the transport sector, cost–benefit analysis (CBA) and multi-criterion analysis (MCA) predominate; the former is an advanced and standardized analytical tool widely applied for social and economic appraisal of transport (road) infrastructure projects [12].

A historical and cross-cutting review of the concept of sustainability [13] proves that this is a multifaceted and still advancing phenomenon, although it was originally developed based on the three pillars of social, economic, and environmental (or ecological) aspects and is commonly represented by three intersecting circles with overall sustainability at the center. In the absence of theoretical descriptions of the three pillars and solid foundations of conception, the operationalization of sustainability and the distinction of social, economic, and environmental aspects is the subject of pragmatic approaches and various solutions applied by scholars and politicians.

In the transport sector, these three dimensions (pillars) of sustainability, namely economic, environmental, and social, are the subject of elaboration in an overall economic appraisal of transport intervention. Rigorous elaboration of transport appraisal practice and advances in economic, environmental, and social impacts of transport sustainability are presented in research [14–16]. Quantified sustainable effects are assigned a monetary value and included in the cost–benefit analysis of transport investment projects. However, these three dimensions of sustainability are mutually dependent and overlap each other. Therefore, in the transport sector, the common distinction between economic, environmental, and social aspects is as follows. Economic effects are investigated through the transport user and transport operator benefits and changes in transport operating costs and travel times. Environmental effects relate typically to environmental externalities such as transport-related air pollution and greenhouse emission. The social dimension is mostly streamlined to issues of transport congestion, traffic noise, and transport accidents.

Reviews on methodologies and measurements of the transport economics sustainable effects are found in several studies [17–20], while issues of environmental and social transport developments are extensively elaborated in [21–24].

These effects can be expressed in monetary terms and/or measured in physical units. However, there are monetary values that allow for the coherent and consistently comprehensive appraisal of intervention and comparable assessment of magnitude and structure of sustainable effects and hence evaluation of sustainable gains of the intervention. The latter is of the utmost importance for policymakers in their decisions for public funding. While making decisions, public agents must take into consideration a proper balance between economic, social, and environmental outcomes of intervention. Additionally, they must consider transport policy strategies goals and recommendations for limiting the effects of global warming and reduction of emissions in the transport sector, sustainable development of freight transport, socially fair transition towards low-emission, competitive, and connected mobility [25–28]. While considering interrelations between seaports and cities from the sustainability viewpoint, it is underlined [29] that seaports have a considerable impact on surrounding cities. There is an urgent need for cooperation between ports and cities in all sustainable aspects, especially in the mitigation of road congestion [30,31].

Empirical findings on port-related road traffic flow developments are scarce. In [32], some aspects of the relationship between road accessibility and port performance are analyzed and the opportunities for freight flows bundling for neighboring EU ports and their hinterland regions are investigated. For each transport mode, academics calculated the generalized costs and focused on business models to identify bundling projects that would lower the direct, generalized, and environmental costs of the contestable hinterlands' connectivity.

In [33], a social cost–benefit research framework was developed for a large road project to improve the hinterland links of the port of Zeebruges. The study presents findings on economic benefits to the port region and the hinterland, and analysis concerned contestable, long-distance markets, road transport, and feasible modal shifts induced by enhanced hinterland access to the port.

The relations between road investment and the local urban-port road transport performance have not been sufficiently investigated. Additionally, the effects of investment in the port-captive hinterland have not been adequately addressed. There is a lack of evidence on the sustainable effects of intervention in the port–city interface.

This manuscript fills these research gaps and elaborates the link between the provision of road infrastructure and the sustainable effects in the port-related freight road traffic. With the use of the cost–benefit concept, economic, environmental, and social sustainable effects are measured in monetary terms and in absolute, physical units.

The main research problem addressed by this study is as follows.

What are the economic, environmental, and social effects in monetary and non-monetary terms induced by the enhanced capacity of the port-urban road system?

Our study elaborates on the local urban-port road transport system in the (Baltic) port city of Szczecin (Poland), and it concentrates on the captive market (first–last mile urban-port road system) and the road freight traffic. The empirical analysis is performed with the bottom-up approach. With the use of primary traffic measures, the long-term freight road traffic forecast is elaborated in terms of interaction with other traffic types. The effects of investment are identified by comparing transport performance in two scenarios—with the investment and without the investment—while the analysis addresses freight road transport performance as a result of the increased truck traffic velocity in the reconstructed road transport. Next, as results of congestion mitigation in the local road transport system, savings are quantified in terms of generalized freight road transport costs, in freight transit time and in externalities.

This manuscript is structured as follows. Section 2 briefly outlines the case study. Section 3 presents the materials and data used. Section 4 describes the research framework, while in Section 5, inputs and parameters are presented. Sections 6 and 7 contain measurements and results. Section 8 draws conclusions and discusses research limitations as well as further research avenues.

2. Brief Outline of Port–City Interface

Szczecin is a city covering an area of 300.6 km^2, located in northwestern Poland in the West Pomeranian Province. At the end of 2017, the population of Szczecin was 403.9 thousand, and the average population density of the city was 1347 inhabitants per km^2. The Odra River separates the city into two parts, namely the Left and the Right Banks. The basic transport routes in Szczecin are determined by the bisected layout of the city centers on the Left and the Right Banks, which are connected by State Road No. 10 (DK10). Due to the location of industry (jobs) and schools on the Left Bank of Szczecin, and the numerous housing estates on the Right Bank, there is a massive flow of people daily on the east–west axis. The urban transport system with a marked course of DK 10 is shown in Figure 1.

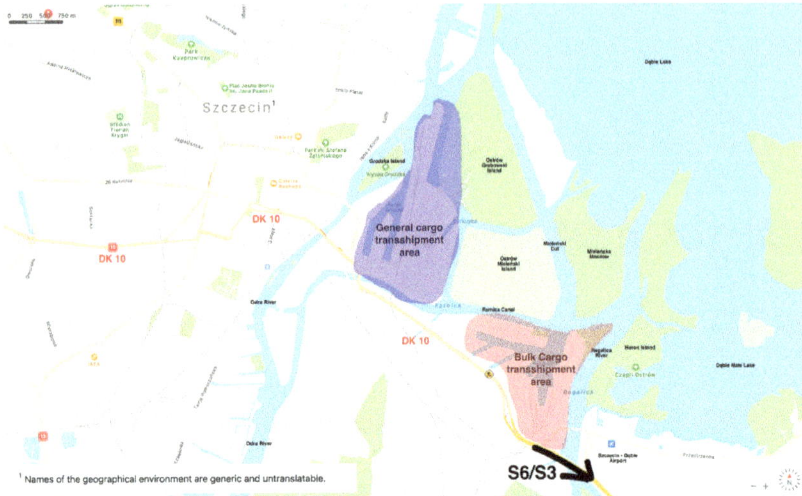

Figure 1. State Road DK10 and port areas in the city of Szczecin. Source: Own study based on www.openstreetmap.org (accessed on 22 June 2021).

The urban transport system is based on the main collector State Road No. 10 with an aggregate of high-intensity traffic, mainly comprising (1) intra-city passenger traffic, predominantly individual cars supported by public transport (buses, streetcars); (2) passenger cars and bus traffic connected with tourist transit; (3) truck traffic between Szczecin's port and its hinterland; and (4) truck traffic serving the needs of the city and businesses based on the Left (western) Bank of the Odra River. Simultaneously, the DK10 is the main road linking with the A6 freeway and the S3 expressway used by long-distance truck traffic to and from the city and the port hinterland. All entry roads to the port reloading areas are interconnected with the urban section of the DK10.

The port of Szczecin is located 65 km south of the Baltic Sea and is connected to it by a waterway and located in the city. It is a universal port and the cargo volume of 9 million tons per year qualifies it as a minor Baltic seaport. Poor maritime accessibility is the main deficiency of the port. Recently, investments are being made to improve nautical access to the port. This will enable the port to handle larger vessels and will also increase the cargo handling capacity.

The subject of research is 17 roads and road sections in the local transport system, with a total length of 9.812 km. The investment project consists of the reconstruction of the infrastructure in the existing urban-port system of roads. The length of the roads will remain unchanged, but they will be resurfaced, their axle load will be increased, some sections of road will be widened, entrance/exit lanes will be reconstructed, the road curvature will be reduced, entrance/exit ramps will be reconstructed, intersections will be reconstructed or built, and, lastly, the number of parking spaces for trucks will be increased. In Figure 2, the critical components of the urban-port road system planned for reconstruction are displayed.

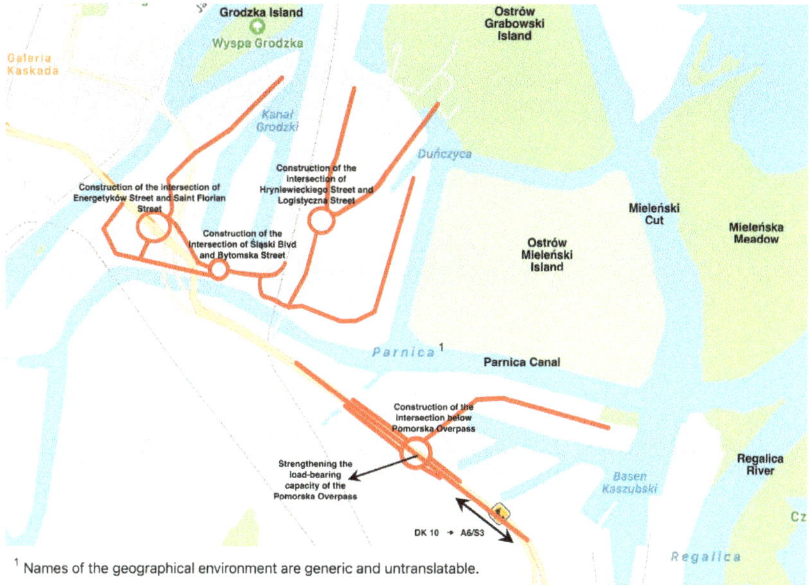

Figure 2. Scope of intervention in urban-port road system. Source: own study based on www.openstreetmap.org (accessed on 24 June 2021).

The modernized road system is scheduled to be operational from 2024, and the effects of its improved capacity have been set for 20 years (between 2024 and 2043).

3. Materials and Data Used

Primary road traffic flow measurements in the urban-port road system in the port city of Szczecin were made in line with the General Directorate for National Roads and Motorways methodology [34], and motor vehicles included cars, vans, light vehicles, trucks with and with out trailers, and buses.

Annual Average Daily Traffic (AADT) is the average number of vehicles passing a given section of a road per day. It is calculated according to the formula:

$$AADT_i = \frac{M_{Ri}\cdot N_1 + 0.85\cdot M_{Ri}\cdot N_2 + M_{Ni}\cdot N_3}{N} + R_{Ni}$$

where:

$AADT_i$—Annual Average Daily Traffic of motor vehicles at the i-th road;
MR_i—Annual Average Daily Traffic on working days (from Monday till Friday between 6 a.m. and 10 p.m.) on the i-th road;
$0.85\cdot MR_i$—Annual Average Daily Traffic on Saturdays and days before holidays (between 6 a.m. and 10 p.m.) on the i-th road;
MN_i—Annual Average Daily Traffic on Sundays and holidays (between 6 a.m. and 10 p.m.) on the i-th road;
RN_i—Annual Average Traffic in the night (between 10 p.m. and 6 a.m.) on the i-th road;
N_1—the number of working days within a year (in 2016, 252);
N_2—the number of Saturdays and days just before holidays (in 2016, 53);
N_3—the number of Sundays and holiday days within a year (in 2016, 61); and
N—the total number of days in a year (in 2016, 366).

The Average Annual Traffic (AAT) is a product of AADT multiplied by 365 days.

When calculating the individual transport effects, the engineering relations between the speed and traffic flow were applied. The engineering relationships between speed and (road) traffic flow was derived from the "Instruction for Assessing the Economic Efficiency of Road and Bridge Projects—Verification of the Research Method According to EU Recommendations and Update of Unit Prices as of 2007, Part II, Tables of Speed" [35].

The relationships include the vehicle traffic speed depending on the type of roads (urban double or single carriageways), their cross-section (width of the road), and their traffic flows measured by the number of vehicles passing urban road section per hour (Figures 3 and 4).

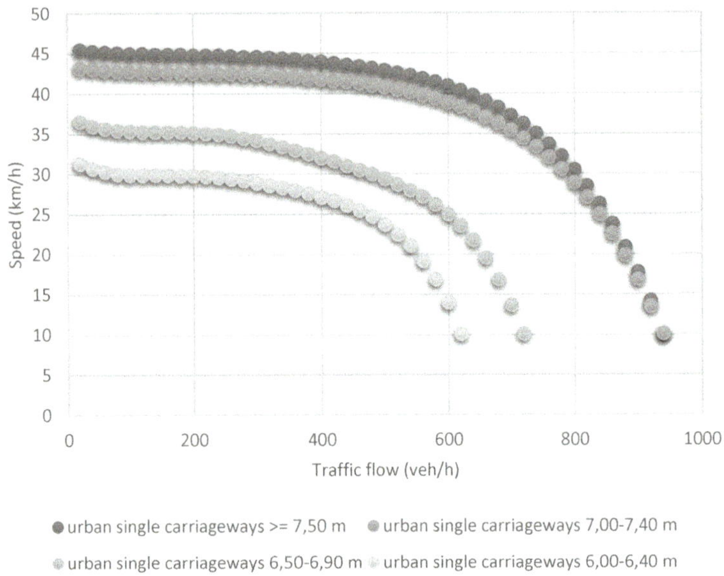

Figure 3. Speed-traffic flow curves for urban single carriageways. Source: Adapted from "Instruction for assessing the economic efficiency of road and bridge projects 2008".

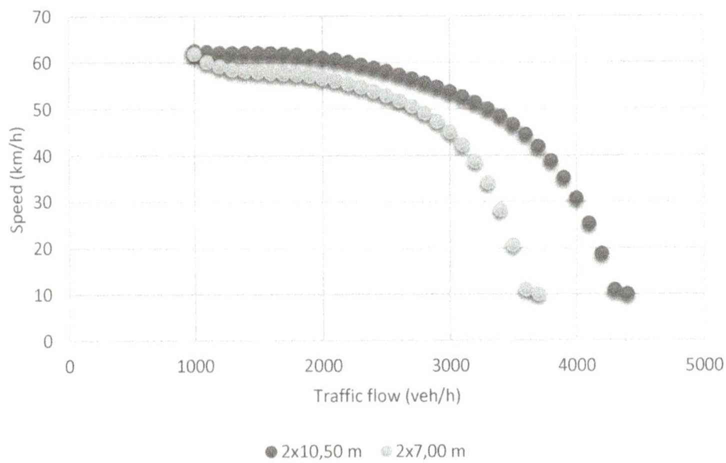

Figure 4. Speed-traffic flow curves for urban double carriageways. Source: Adapted from "Instruction for assessing the economic efficiency of road and bridge projects 2008".

In the city-port transport system, the speeds of trucks were measured on a road with a high level of technical wear and tear. The truck speed was measured on three road sections:

the 396.5 m long section No. 1, the 280.5 m long section No. 2, and the 394.5 m long section No, 3. A total of 34 truck transit time measurements were made, and then the average truck speed was calculated. The estimated average truck speed was 15.58 km/h, which corresponds to the speed on roads with a high degree of technical wear and tear.

As shown in the traffic flow tables (Tables A1 and A2), the average speed of vehicles on the roads in good technical conditions was 35.4 km/h. By comparing vehicle speed measured on roads with a high degree of wear and tear (15.58 km/h) with the average speed of vehicles derived form the traffic flow tables (35.4 km/h), the speed correction index (44.0%) was calculated for roads with high degree of technical wear and tear. The speed correction indices for medium and low levels of road technical wear and tear were obtained by enlarging the basic speed reduction index by 1/3 and 2/3, respectively, of the remaining part that completed the index to 100%. The values of truck speed correction indices depending on the degree of technical road wear and tear are depicted in Table 1.

Table 1. Truck speed correction indices depending on technical road wear and tear.

Average vehicle speed determined based on speed-traffic flow relations—road in good technical condition	35.4	km/h
Average vehicle speed determined based on statistical study—road in poor technical condition	15.58	km/h
Speed correction index for high level of road wear and tear	44.0%	
Speed correction index for medium level of road wear and tear	62.7%	
Speed correction index for low level of road wear and tear	81.3%	

Source: own study.

Additionally, degree of wear and tear for each of the road/road section within the city-port local transport system was estimated with corresponding truck speed correction indices (Table 2).

Table 2. Degree of roads wear and tear estimation with corresponding speed correction indices.

Nr of Road/Road Section	Road Type, Road Width (m)	Degree of Road Wear and Tear	Speed Correction Indices Due to Wear and Tear Degree
1	Dual carriageway 2 × 7.00 m	Low	81.3%
2	Single carriageway 7.00–7.40 m	High	44.0%
3	Single carriageway 7.00–7.40 m	High	44.0%
4,5	Single carriageway 6.50–6.90 m	High	44.0%
6,7	Single carriageway 7.00–7.40 m	High	44.0%
8	Single carriageway 6.00–6.40 m	High	44.0%
9	Single carriageway 7.00–7.40 m	Medium	62.7%
10	Single carriageway ≥7.50 m	Medium	62.7%
11	Single carriageway 6.50–6.90 m	High	44.0%
12	Single carriageway 6.00–6.40 m	High	44.0%
13	Single carriageway 7.00–7.40 m	Low	81.3%
14	Dual carriageway 2 × 7.00 m	Low	81.3%
15	Single carriageway 6.50–6.90 m	High	44.0%
16	Dual carriageway 2 × 10.50 m	Low, lowered bearing capacity	81.3%
17	Single carriageway 6.50–6.90 m	Medium	62.7%

Source: own study.

In the forecasting works, we used GDP forecasts of the Ministry of Development and Finance recommended for the Polish transport sector (Table A3). The GDP forecast for 2019–2043 is expressed in the annual average constant prices as growth indicators (calculated year-on-year, previous year = 100) and in corresponding values; however, the economic impact of the COVID-19 pandemic is not considered. The GDP forecast was applied for predicting truck traffic flow in the system.

Both the developed traffic forecasts' values of inputs and coefficients refer to the year 2019 as a base year for calculations. The sustainability effects are calculated for 20 years (2024–2043). Accounts are made with a discount rate of 4.5%, with fixed prices from the base year and without considering inflation in the analysis time horizon. Unit values of monetized benefits are presented in net terms (excluding VAT).

4. Research Framework

The research framework of the sustainable effects induced by the capacity improvement in the urban-port road system is depicted in Figure 5.

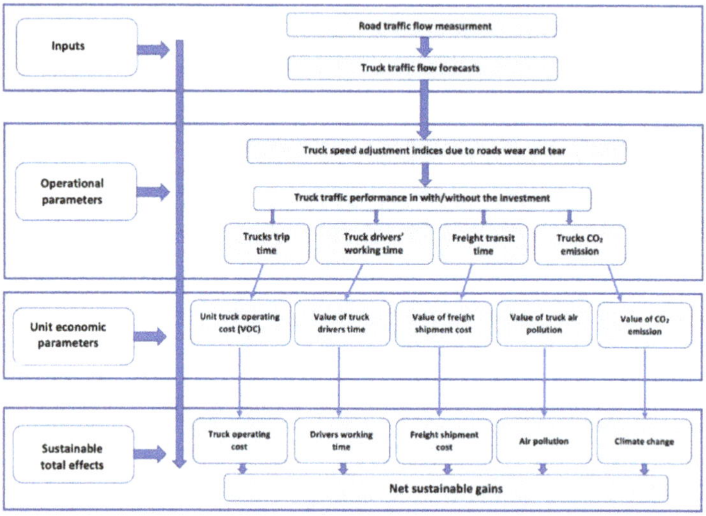

Figure 5. Research framework. Source: own study.

The analysis was carried out using the incremental method by calculating the net difference in transport effects that appear in the two options: with the reconstruction of the urban-port transport system (hereafter WI) and without the rebuilding of the urban-port transport system (hereafter W0), i.e., in the incremental calculus (WI-W0). This approach is commonly used when assessing the impacts of transport infrastructure projects [36–39].

The study concentrates on direct effects in the transport system as a consequence of intervention [40–42]. Road infrastructure investments reduce distances, travel time, and/or congestion. Direct benefits for transport operators and shippers are related to time (time savings and reliability) and vehicles' operating costs. Savings in operating and time costs and thereafter in related externalities increase the surplus of consumers and hence the welfare.

The main input in the research is the elaborated long-term demand for capacity services of the local road transport system. With the use of traffic primary surveys, the freight road traffic forecast is related to the growth of GDP, and prediction is made for truck flow, i.e., trucks/day passing in the urban-port road system.

The average truck speed, shown in the traffic flow tables for the WI option, varies depending on the road type and the change in traffic flow volume in each year of the analysis. Thus, the average speed of trucks in the W0 option in individual years of the analyzed period varies as well. Here, the average truck speed in option WI is multiplied by the speed correction index selected depending on the degree of technical road wear and tear.

Considering the interaction between different types of road users, traffic flow intensity, and technical wear and tear of roads, traffic congestion relief has been translated into an increase in trucks' speed.

Increases in trucks' speeds affect traffic performance in terms of truck trip time (vehicle-hours, v-hs), truck drivers' working time (working hours, w-hs), road transport time (tonnes-hours, t-hs).

In the economic calculus, we applied the notion of transport generalized cost, which is widely elaborated in transport economics [43–48]. Limited to the main components and related to freight road transport, the generalized transport costs are the sum of the vehicle operating costs (VOC) and the value of time (VOT), which is limited to the value of drivers' working time. Additionally, the valuation of externalities in terms of money has been accounted for, but it is limited to climate change and air pollution costs.

To capture economic effects, values of time and truck operating costs, as well as marginal external costs, have been validated in freight road transport. Next, as results of congestion mitigation in the local road transport system, savings in net terms (WI-W0) are quantified for trucks operating costs, drivers working time, freight transit time, and externalities.

5. Inputs and Parameters

5.1. Forecasted Truck Traffic

Road traffic forecasting has been a subject of numerous investigations. As per the literature [49–52], the main forecasting aspects relate to prediction methods (parametric or non-parametric), prediction horizon, prediction scale (single location, road segment, whole or part of network), prediction context (urban, rural, freeway), predicted variables as traffic flow (vehicles/hour), traffic density (vehicles/km), average speed, and travel time [53]. It was concluded that there is no universal method that fits every situation better than the rest [54]. In the context of our cost–benefit analysis, the freight road traffic forecast is related to the growth of GDP and prediction is made for truck flow, i.e., trucks/day passing in the urban-port road system. The long-term freight traffic forecast was elaborated since investments made in road infrastructure have a long economic lifetime and a long pay-back period.

The benefits of adding capacity to the road system may be reduced by the induced traffic of passenger cars and freight transit. Determining this effect of road capacity expansion is a complex problem and is concerned with interrelated system components: transportation supply system, land use, accessibility, and travel demand. Based on a review of the literature [55], we conclude that there are no empirical indications that added capacity to the road system generates a significant volume of induced traffic. Additionally, the benefits to port-related traffic may outweigh the negative effects of the induced commuters and freight transit traffic [56]. Because of the complexity of the problem, the trip-inducing effect of added road capacity is disregarded in the standard travel demand forecasting procedure [57,58].

Estimated Annual Average Daily Traffic (AADT) on road sections was a basis for a long-term prognosis of AADTs up to the year 2043. Forecasts of AADT were developed in the reference period for each road segment separately according to the following equation [59]:

$$AADT^k_{i,t+1} = AADT^k_{i,t} \cdot \left(\frac{GDP_{t+1} - GDP_t}{GDP_t} \cdot E^k(GDP) + 1 \right)$$

where:

$AADT^k{}_{i,t}$—Annual Average Daily Traffic for k-th category of vehicle in year t;

GDP_t, GDP_{t+1}—Gross Domestic Product in year t and $t + 1$ in constant prices from base period, wherein $\frac{GDP_{t+1} - GDP_t}{GDP_t}$ represents a relative increase in GDP in year $t + 1$ compared to year t (dynamics indicator); and

E^k (*GDP*)—elasticity coefficient for *k*-th vehicle category.

The annual average truck traffic forecast (AAT) was made as below:

$$AAT_{i,t} = 365 \cdot AADT_{i,t}$$

where:

$AADT_{i,t}$—Annual Average Daily Truck Traffic; traffic flow on the *i*-th road section in year *t* of forecast.

When analyzing the relationship between road freight transport demand and economic activity variables such as Gross Domestic Product, Gross Final Expenditure, index of industrial production, as well as commodity sectors are under investigation [60]. For the elasticity of freight transport demand in the UK with respect to the aggregate commodity sector, values in the range from 0.72 to 1.49 were found [61], while in another study [62], elasticity of freight road demand related to GDP was estimated at 0.66. In the research [63], the elasticity of road haulage with respect to the index of GDP (for 11 EU countries for the period from 1970–2010) amounted to between 1.0 and 1.2. As recommended [64], the elasticity coefficient of GDP for trucks with trailers for Poland amounts to 1.00, and this elasticity coefficient was used when predicting the truck traffic in the urban transport system.

The long-term prediction of truck traffic was made for every year in the forecasting period and for each road out of a total of 17 roads and sections of road in the transport system. Forecasting results limited to the first and last year of the prediction horizon are depicted in Table 3.

Table 3. Forecasted truck traffic on roads in urban-port road transport system in the selected years of the forecasting period (no. of trucks).

Number of Road/Road Section	Year	Average Annual Daily Traffic AADT	Average Annual Traffic AAT
1	2020 2043	8711 13,299	3,188,319 4,854,223
2	2020 2043	65,187 101,393	23,858,375 37,008,450
3	2020 2043	34,094 52,404	12,478,360 19,127,597
4	2020 2043	6203 9352	2,270,454 3,413,373
5	2020 2043	3608 5471	1,320,572 1,996,761
6	2020 2043	4054 6274	1,483,846 2,290,049
7	2020 2043	6973 10,858	2,552,014 3,963,040
8	2020 2043	7025 10,921	2,571,299 3,986,067
9	2020 2043	1083 1541	396,529 562,576
10	2020 2043	7014 10,459	2,567,032 3,827,948
11	2020 2043	7174 10,938	2,625,716 3,992,338
12	2020 2043	3014 4661	1,102,975 1,701,364

Table 3. *Cont.*

Number of Road/Road Section	Year	Average Annual Daily Traffic AADT	Average Annual Traffic AAT
13	2020 2043	3105 3524	1,136,283 1,286,144
14	2020 2043	467 707	170,949 258,066
15	2020 2043	5042 5783	1,845,429 2,110,824
16	2020 2043	906 1962	331,665 716,254
17	2020 2043	844 1152	308,970 420,639

Source: own study.

5.2. Unit Truck Operating Costs (VOC)

Vehicle Operating Cost (VOC) of trucks includes the following costs:

- Fuel costs: being a function of the road alignment and traffic conditions,
- Other costs: road quality affects the wear and tear of vehicles, including costs of oil, tires, and vehicle maintenance, as well as its depreciation.

VOC for trucks depends on travel speed, differentiating the type of terrain (flat, rolling, and mountainous) and road condition (pavement after reconstruction/rehabilitation or deteriorated pavement) and is estimated as the sum of fuel costs plus other costs. Applications of the calculation of unit truck operating costs are presented in Table A4.

It is assumed that there will not be any real increase in unit VOC over time as a potential increase in energy prices would be compensated by improved efficiency of vehicles [64].

5.3. Values of Travel Time (VOT) in Freight Road Transport

5.3.1. Unit Value of Truck Drivers' Time

The unit value of truck drivers′ time corresponds to the marginal gross cost of labor, including labor-related overhead of commercial drivers in Poland. Evolution over time is based on Polish GDP per capita growth, with an elasticity of 0.5 [65]. Therefore, unit values applied in research of truck drivers′ working time increase over time. In the base year of 2019, it amounts to EUR 21.83, while in the last year of calculation 2043, increases to EUR 30.93. Vehicle occupancy for trucks is assumed as equal to one driver.

5.3.2. Unit Value of Freight Shipments Time

The unit time values applied in the study come from the estimations made for the Netherlands [19,20] with the measure of willingness-to-pay (WTP) for freight time savings. In our calculations, converted into Polish values and indexed for 2019, the value of time for the average shipment in freight road transport amounts to EUR 0.88 per tonne-hour [65]. The time value of shipment is subject to escalation over time, with an elasticity of 0.5 to Polish GDP per capita growth, and in the last year of calculation, 2043, it increases to EUR 1.25.

5.4. Unit Costs of Trucks Air Pollution

The most important air pollutants related to transport are dust (PM10, PM2.5), nitrogen oxides (NOx), sulphur dioxide (SO_2), volatile organic compounds (VOCs), and ozone (O_3) as an indirect pollutant.

Costs of air pollution depend on speed, vehicle category, as well as condition, slope, and location (urban or non-urban area) of the road. Unit values of air pollution for trucks are based on the emissions evaluation method [65]. Applied in the calculation are unit trucks' air pollution costs, as presented in Table A5. Unit air pollution costs' evolution over time is based on GDP per capita growth with an elasticity of 0.8.

5.5. Unit Costs of Climate Change

The calculation method is in line with the approach described in the European Investment Bank Carbon Footprint Methodology [66].

Greenhouse gas emission (GHG), converted into emission factors of equivalent tonnes of CO_2 (tCO_2) depend on fuel consumption and therefore on the speed, on the vehicle category as well as on the pavement condition and geometry of the road, and these volumes of emissions are presented in Table A6. These emission factors are multiplied by a unit cost of CO_2, i.e., the economic ("shadow") cost of equivalent emission tonne of CO_2. In 2010 the unit costs of GHG emission amounted to EUR 25.0 [66], while the growth of CO_2 value applied in the calculation in the reference period are presented in Table A7.

6. Sustainable Effects Measured in Monetary Terms

6.1. Cost Savings as Result of Reduced Truck Operating Costs

Truck operating costs are represented by total operating costs of all trucks travelling in the relevant road system whereas costs savings are calculated for each option (W0 and WI) and each year throughout the entire reference period as in the formula:

$$VOC_t = 365 \cdot \sum_{i=1}^{17} AADT_{ti} \cdot L_i \cdot c(V,\ S, P)$$

where:

VOC_t—annual vehicle operating costs in EUR;
$AADT_{ti}$—annual average daily traffic on the i-th road section in vehicles/day;
L_i—length of the i-th road section in km;
c_t (V,S,P)—unit operating costs in the function of travel speed V, slope of terrain S, and technical condition of pavement P in EUR/veh km.

The total nominal cost savings of trucks operating costs for the period of 2024–2043 will amount to EUR million 8.31.

6.2. Cost Savings as Result of Reduced Truck Drivers' Working Time

The cost savings of drivers' working time are calculated for each option (W0 and WI), and each year throughout the entire reference period according to formula:

$$CoT_t = 365 \cdot \sum_{i=1}^{17} \frac{L_i}{V_i} \cdot AADT_{ti} \cdot utc_t$$

where:

CoT_t—annual cost of truck drivers' working time in EUR;
L_i—length of the i-th road section in km;
$AADT_{ti}$—annual average daily traffic on the i-th road section in vehicles/day;
utc_t—unit value of truck drivers' working time in EUR/h;
V_i—travelling speed on the i-th road section in km/h.

The total nominal cost savings of truck drivers' working time for the period of 2024–2043 will amount to EUR million 78.3.

6.3. Cost Savings as Result of Reduced Freight Shipment Time

Costs of freight shipment time are calculated for scenarios with and without the investment for each year of the reference period as follows:

$$CoFST_t = \sum_{l=1}^{17} FS_{ti} \cdot \frac{L_i}{V_i} \cdot cofst_t$$

where:

$CoFST$—annual cost of freight shipment time in EUR;
FS_{lj}—forecasted annual freight shipment on the i-th road section in tonnes;
L_i—length of the i-th road section in km;
V_i—travelling speed on the i-th road section in km/h. $cofst_t$—unit cost of freight shipment time in EUR/t km.

As a result of the investment, the total cost savings in freight shipment time in the years 2024–2043 will amount to EUR million 30.4.

6.4. Cost Savings as Result of Reduced Air Pollution

The costs savings of air pollution are calculated for each option (W0 and WI) and each year throughout the entire reference period according to formula

$$APC_t = 365 \cdot \sum_{i=1}^{17} L_i \cdot apc_t(V, S, P) \cdot AADT_{ti}$$

where:

APC_t—annual air pollution costs in EUR;
$apc_t(V,S,P)$—unit air pollution costs in the function of travel speed V, slope of terrain S and technical condition of pavement P in EUR/veh km;
$AADT_{ti}$—annual average daily traffic on the i-th road section in vehicles/day;
L_i—length of i-th road section in km.

As a result of the investment, the total cost savings in air pollutions in the years 2024–2043 will amount to EUR million 20.4.

6.5. Cost Savings as Result of Reduced Climate Change

Savings in greenhouse gas (GHG) emissions (equivalent to CO_2) are calculated for each option (W0 and WI), and each year throughout the entire reference period according to the formula:

$$GGE_t = 365 \cdot \sum_{i=1}^{17} AADT_{ti} \cdot L_i \cdot gge$$

where:

GGE_t—annual greenhouse gas emission in tonnes of CO_2;
$AADT_{ti}$—annual average daily traffic on the i-th road section in vehicle/day;
L_i—length of the i-th road section in the urban-port road system in km;
gge—unit greenhouse gas emission in tonnes of CO_2/t km.

Savings in costs of climate change are calculated following the formula below.

$$CCC_t = 365 \cdot \sum_{i=1}^{17} ccc_{ti}(V,\ S, P) \cdot L_i \cdot AADT_{ti}$$

where:

CCC_t—annual climate change costs in EUR;

ccc_{ti} (*V,S,P*)—unit climate change costs in the function of travel speed V_j, slope of terrain *S* and technical condition of pavement *P* of *i*-th road sections in the urban-port road system in EUR/veh km;
$AADT_{ti}$—annual average daily traffic on the *i*-th road section in vehicle/day;
L_i—length of *i*-th road section in km.

The total cost reduction resulting from the climate change will amount to EUR 1.85 million.

Total and discounted (with a discount rate of 4.5%) sustainable monetized effects induced by the investment are summarized in Table 4.

Table 4. Total and discounted sustainable monetized effects induced by the investment in 2024–2043.

Net Savings in:	Effects Induced by Investment Project Discounted Value (Million EUR)	Structure (%)
truck drivers working time	38.24	55.55
freight shipment time	14.80	21.48
trucks operating costs	4.33	6.29
air pollution costs	10.55	15.32
climate change costs	0.94	1.36
total	68.86	100.0

Source: own study.

Effects of the road infrastructure rehabilitation in the urban-port local market relates for the most to the savings in drivers working time as well as to savings in freight shipment time, and together they constitute 77.0% of the total discounted sustainable effects. Reduced truck operating costs amounts to 6.29%, while reduced environmental externalities (climate change and air pollution) constitutes 16.68% of the total sustainable monetized result of the investment.

The structure of monetized sustainable gains resulting from the reconstruction of the urban-port local road system is depicted in Figure 6.

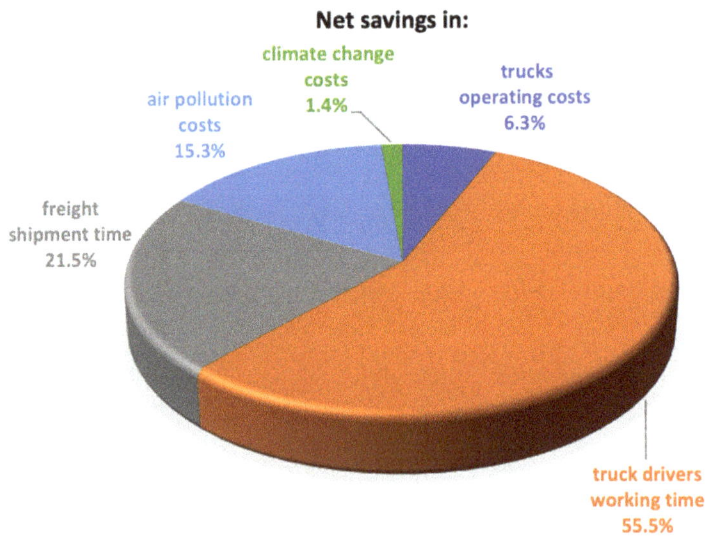

Figure 6. Structure of sustainable gains resulting from the reconstruction of the urban-port local road system. Source: own study.

7. Sustainable Effects Measured in Absolute Units

Applying the rearranged formulas from the former chapter, the results of the calculation for the transport and traffic performance induced by the intervention and accumulated in 2024–2043 are depicted in Table 5.

Table 5. Transport and traffic performance effects induced by the reconstruction of the urban-port road system.

Transport and Traffic Performance	WI	W0	WI-W0
in-service work (vehicle kilometers, v km)	95,007,071	95,007,071	0
truck trip time (Vehicle-hours, v-hs)	3,636,717	6,414,004	−2,777,287
truck drivers' working time (Working hours, w-hs),	3,636,717	6,414,004	−2,777,287
freight road traffic time (Tonne-hours, t-hs)	34,854,504	61,520,413	−26,665,909

Source: own study.

Because the forecast of truck traffic is an exogenous variable and the length of roads in the system remains unchanged, the projected in-service operation of trucks (measured in truck kilometers) in the port-city transport system will grow at the same rate regardless of investment options, and thus incrementally, the in-service operation of trucks will be 0.

However, improvement of the technical state of roads and the reconstruction or construction of new intersections, exit/entrance ramps, and lanes will increase the speed of trucks and reduce vehicle maneuvers. This leads to a reduction in trucks' trip time (measured in truck-hours) and truck drivers' working time (expressed in working hours). Since the number of trucks and the number of drivers is the same (1 truck = 1 driver), savings in travel time of trucks and in working time of truck drivers are the same in physical terms (i.e., 2,777,287 units). However, in economic terms, these quantities are different. In the former, there are the vehicle kilometers, while in the latter there are working hours of truck drivers. The monetary values of vehicle kilometers and the drivers' working hours vary, and this is reflected in the reduced vehicle operating and labor costs incurred by road transport operators. As a result of the increase in truck speed and smoother traffic flow, trucks' freight transit time will be reduced by 26,665,909 tonne-hours.

Savings in greenhouse gas (GHG) emissions (equivalent to CO_2 emissions) are calculated for each option (W0 and WI), and each year throughout the entire reference period according to formula:

$$GGE_t = 365 \cdot \sum_{i=1}^{17} AADT_{ti} \cdot L_i \cdot gge$$

where:

GGE_t—annual greenhouse gas emission in tonnes of CO_2;
$AADT_{ti}$—annual average daily traffic on the i-th road section in vehicle/day;
L_i—length of the i-th road section in the urban-port road system in km;
gge—unit greenhouse gas emission in tonnes of CO_2/t km.

Results of calculations of trucks' emission in equivalent tonnes of CO_2 in 2024–2043 are presented in Table 6.

Table 6. Trucks' traffic savings in the emission of CO_2 (in equivalent tonnes tCO_2).

Trucks' CO_2 Emission	WI	W0	WI-W0
(tonnes of CO_2, tCO_2)	75,378.2	106,977.2	−31,599.0

Source: own study.

Savings in equivalent CO_2 emissions in 2024–2043 will amount to 31,599 tonnes.

If one tonne of diesel produces 2.63 tonnes of CO_2 [67,68], gains in the consumption of fuel (diesel) will account for 12,014.8 tonnes or 14.1 million liters.

8. Conclusions

In this manuscript, the cost–benefit research concept has been applied to investigate the possible sustainable effects of the capacity improvement in the local urban-port road system. The sustainable effects are quantified in money terms and in physical units. Considering the interaction between different types of road users, traffic flow intensity and technical wear and tear of roads, traffic congestion release has been translated into an increase in trucks′ speed. The primary effect of investment is the reduced traffic congestion, which enhances the velocity of trucks in the transport system. The increased trucks′ speed affects freight road traffic performance, time of deliveries, and environmental externalities.

The investigated sustainable effects of the urban-port road system reconstruction are as follows.

Effects of economic sustainability; truck travel time in the system is reduced by 2,777,287 vehicle-hours, and the truck drivers' working time is reduced by 2,777,287 h. In terms of money, this corresponds to a reduction in generalized freight road transport costs: vehicles operating costs and drivers working time by EUR 4.33 and 38.24 million, respectively. The efficiency of resource use will increase, and cheaper transport services will be offered by the total EUR million 42.57, and it will be to the benefits of haulage operators and consumers.

Effects of social sustainability; the time of road freight shipments will be reduced by 26,665,909 tonne-hours, while monetized savings of time reduced deliveries by EUR 14.80 million. Time-efficient and reliable deliveries ensure better accessibility of consumers to goods and enable better availability of goods at lower costs/prices. These benefits will go to exporters and importers, as well as consumers. Social benefits originate from the reduced congestion in the urban-port road system.

Effects of environmental sustainability, in physical terms, corresponds to reduced emission of CO_2 with the total amount of 31,599 tonnes and savings in the consumption of fuel (diesel) of 12,014.8 tonnes. In monetary terms, it relates to a reduction in air pollution by EUR 10.55 million and greenhouse gases emission (GHG) by EUR 0.94 million. Through the reduction of truck-related GHG emissions and air pollution, the quality of societal well-being and health will be improved. The final beneficiary of these effects will be the community.

Research results are consistent with findings of intervention aimed at capacity expansion in the road infrastructure [12,39], where the prime effect is the release of traffic congestion. Transport congestion in transport is regarded as a socially sustainable effect and in freight transport, it enhances consumers availability of goods and reliability of deliveries/shipments. Furthermore, less congestion leads to time savings and reduction in transport generalized costs, which translates into a substantial increase in efficiency of transport services, the main economic sustainable effect of intervention. Reduced congestion impacts other social sustainable dimensions through the increased safety of transport users and reduced annoyance of transport noise. Increased velocity of vehicles in the road system and smoother traffic leads to savings in terms of fuel consumption, which in turn diminishes pollution and reduces a release of greenhouse gases. It is reflected in environmentally sustainable effects of intervention, limited air pollution, and reduced adverse climate change.

In this research, the sustainable impact of road capacity expansion is limited to the direct effects which occur in the road transportation system.

Moreover, because of a lack of data, the social effects related to the safety of transport users and noise annoyance nuisance are not specified.

Furthermore, research results depend heavily on forecasted truck traffic. Historical elasticities between truck traffic development and Polish GDP used in our research may change in the future with unknown magnitude and directions. Therefore, the question of the elaborated truck traffic forecasts' reliability, as well as estimated values of sustainable effects, remains valid.

The proposed further research relates to the extension of social and environmental effects to be incorporated in the evaluation of intervention in the transport sector. The recent advances [22] aimed at inclusion in appraisal such effects as well-to-tank emissions, habitat damage, soil and water pollution, externalities in sensitive areas, and separations in urban regions are of great importance.

Another research avenue is the elaboration of sustainable evaluation methods of intervention in the transport sector. In [69,70], the application of monetary methods (cost–benefit analysis) and the non-monetary method (multicriteria analysis) in the evaluation of intervention in transport (road) infrastructure is contemplated and validated.

Author Contributions: D.B., Conceptualization, Writing—original draft, Methodology; C.L., Data curation, Formal analysis, Review and editing; D.B. and C.L., Investigation and Resources. All authors have read and agreed to the published version of the manuscript.

Funding: This research was financed from a subsidy of the Polish Ministry of Science and Higher Education for statutory activities; Transport 4.0: 1/S/KGMiST/21 and Economics and Finance 503-0001-230000-ZS13.

Institutional Review Board Statement: Not applicable.

Informed Consent Statement: Not applicable.

Data Availability Statement: Not applicable.

Conflicts of Interest: The authors declare no conflict of interest.

Definitions and Abbreviations:

AADT, Average Annual Daily Traffic—the number of motor vehicles travelling through a given road cross-section within 24 consecutive hours; annual average is expressed in terms of the actual number of vehicles per day calculated according to relevant guidelines.

CUPT, The Centre for EU Transport Projects supports beneficiaries in the preparation and implementation of transport investments co-financed by the European Union.

Transport work—product of the number of kilometers travelled (the length of roads) and the number of vehicles (expressed in vehicle kilometers [veh km]) or product of the number of vehicles and the travel time (expressed in vehicle-hours [veh-h]).

Traffic Flow Capacity—the greatest number of units (vehicles) that can pass through a road section (street, intersection entry) during a given unit of time. Traffic flow capacity is expressed in terms of actual vehicles per hour [veh/h].

Road in flat terrain—a road on which the gradient is less than or equal to 2%. Road in rolling terrain—a road on which the gradient is between 2% and 6%.

Road in mountainous terrain—a road on which the gradient is bigger than 6%.

JASPERS, Joint Assistance to Support Projects in European Regions is an initiative operating within the structures of the European Investment Bank and aimed at improving the preparation of projects, including transport projects, by applying for EU funding.

Port hinterland presents the inland area surrounding a port from which the goods are either distributed or at which they are collected for shipping to other ports.

Captive hinterlands consist of all regions where one port has a substantial competitive advantage because of lower generalized transport costs to these regions. Consequently, this port handles most of all cargoes to/from these regions.

Contestable hinterlands consist of all regions where there is no single port with a clear cost advantage over competing ports. Therefore, various ports will have a share of the market.

Appendix A

Table A1. Vehicle travel speed (km/h) on urban dual carriageways depending on vehicle traffic flow N1 (vehicles/hour).

N1 No of Vehicles/Hour	Urban Dual Carriageway and Width of Way 2 × 10.50 m	2 × 7.50 m	2 × 7.00 m
1000	62.4	65.1 *	61.8
1100	62.2	63.3 *	60.1
1200	62.0	62.2 *	59.1
1300	62.0	61.5	58.4
1400	62.0	61.1	58.1
1500	62.0	60.9	57.9
1600	61.9	60.8	57.7
1700	61.8	60.6	57.5
1800	61.6	60.3	57.3
1900	61.3	60.0	57.0
2000	60.9	59.5	56.5
2100	60.5	58.9	56.0
2200	60.0	58.3	55.3
2300	59.4	57.5	54.6
2400	58.7	56.6	53.7
2500	58.0	55.6	52.8
2600	57.2	54.4	51.7
2700	56.4	53.1	50.5
2800	55.5	51.6	49.0
2900	54.6	49.7	47.2
3000	53.6	47.3	45.0
3100	52.5	44.4	42.2
3200	51.3	40.6	38.6
3300	49.9	35.8	34.0
3400	48.3	29.6	28.1
3500	46.5	21.6	20.6
3600	44.4	11.6	11.0
3700	41.8	10.0	10.0
3800	38.7		
3900	35.0		
4000	30.6		
4100	25.2		
4200	18.7		
4300	10.9		
4400	10.0		
4500			

Source: Instruction for assessing the economic efficiency of roads and bridges (2008, 84–86). * Unknown measurement errors.

Table A2. Vehicle travel speed (km/h) on urban single carriageways depending on vehicle traffic flow N1 (vehicles/hour).

N1 Number of Vehicles/hr.	Urban Single Carriageways ≥7.50 m	7.00–7.40 m	6.50–6.90 m	6.00–6.40 m	5.50–5.90 m
20	45.4	43.1	36.5	31.2	26.3
40	45.2	43.0	35.9	30.6	26.1
60	45.1	42.8	35.6	30.2	26.1
80	45.0	42.7	35.4	29.9	26.0
100	44.9	42.7	35.3	29.8	26.0
120	44.9	42.6	35.2	29.8	25.9
140	44.8	42.6	35.2	29.8	25.8
160	44.8	42.6	35.2	29.8	25.6
180	44.8	42.5	35.2	29.7	25.4

Table A2. *Cont.*

N1 Number of Vehicles/hr.	Urban Single Carriageways				
	≥7.50 m	7.00–7.40 m	6.50–6.90 m	6.00–6.40 m	5.50–5.90 m
200	44.7	42.5	35.1	29.7	25.0
220	44.7	42.5	35.0	29.6	24.6
240	44.7	42.4	34.9	29.5	24.0
260	44.6	42.4	34.7	29.3	23.2
280	44.5	42.3	34.4	29.1	22.1
300	44.5	42.3	34.1	28.8	20.6
320	44.4	42.2	33.7	28.5	18.5
340	44.3	42.1	33.3	28.2	15.7
360	44.2	42.0	32.9	27.9	11.8
380	44.0	41.8	32.4	27.5	10.0
400	43.9	41.7	31.9	27.0	
420	43.7	41.5	31.5	26.6	
440	43.5	41.4	30.9	26.0	
460	43.3	41.2	30.4	25.4	
480	43.1	40.9	29.9	24.6	
500	42.8	40.7	29.3	23.6	
520	42.5	40.4	28.6	22.4	
540	42.2	40.1	27.9	21.0	
560	41.8	39.7	27.1	19.1	
580	41.4	39.3	26.1	16.8	
600	40.9	38.8	24.9	13.9	
620	40.3	38.3	23.5	10.0	
640	39.7	37.7	21.7		
660	39.0	37.0	19.5		
680	38.2	36.3	16.8		
700	37.2	35.4	13.6		
720	36.2	34.4	10.0		
740	35.0	33.3			
760	33.6	32.0			
780	32.1	30.5			
800	30.4	28.8			
820	28.4	27.0			
840	26.2	24.9			
860	23.7	22.5			
880	20.8	19.8			
900	17.7	16.8			
920	14.1	13.4			
940	10.0	10.0			

Source: Instruction for assessing the economic efficiency of roads and bridges (2008, 84–86).

Table A3. GDP growth forecast for 2019–2043 (selected years only) expressed in GDP growth indicators (calculated year-on-year, previous year = 100) and in million PLN in constant annual average prices.

Year	Change in Polish GDP (In Annual Average Constant Prices)	Polish GDP Forecast (In Annual Average Constant Prices in PLN Millions)
2019	104.0	2,273,556.0
2020	103.7	2,357,677.6
2025	103.0	2,754,472.2
2030	102.7	3,165,373.7
2035	102.2	3,563,877.0
2040	102.0	3,946,388.7
2043	101.9	4,179,731.6

Source: Polish Ministry of Development and Finance; macroeconomic forecasts of May 2019.

Table A4. Unit truck operating costs (EUR/veh km), prices 2019.

Speed (veh/km)	Flat Terrain (Pavement after Rehabilitation/Construction)	Flat Terrain (Deteriorated Pavement)
0–10	0.530	0.630
11–20	0.506	0.588
21–30	0,487	0.556
31–40	0.474	0.532
41–50	0.467	0.517
51–60	0.465	0.511
61–70	0.468	0.514
71–80	0.477	0.526
81–90	0.492	0.547
91–100	0.512	0.576
101–110	0.538	0.614
111–120	0.538	0.614
121–130	0.538	0.614
131–140	0.538	0.614

Source: Tables of unit costs to use in cost–benefit analyses, July 2019, CUPT, www.cupt.gov.pl (accessed on 13 June 2021).

Table A5. Unit costs of truck air pollution (EUR/veh km), prices 2019.

Speed (veh/km)	Flat Terrain (Pavement after Rehabilitation/Construction)	Flat Terrain (Deteriorated Pavement)
0–10	0.495	0.695
11–20	0.447	0.612
21–30	0.409	0.547
31–40	0.382	0.499
41–50	0.366	0.469
51–60	0.362	0.456
61–70	0.368	0.461

Source: Own study based on Blue Book. Road Infrastructure, Jaspers, July 2015, and materials from Research Institute of Roads and Bridges (IBDiM).

Table A6. Climate change unit emission factors tCO_2/vehicle km, flat terrain: road surface after reconstruction/construction and degraded road surface.

Vehicle Travel Speed (km/h)	Climate Change Unit Emission Factors—tCO_2/Vehicle km—Road Surface after Reconstruction/Construction	Climate Change Unit Emission Factors—tCO_2/Vehicle km—Degraded Road Surface
	Heavy Goods Vehicles (HGV)	
0–10	0.000,999	0.001,399
11–20	0.000,900	0.001,232
21–30	0.000,825	0.001,101
31–40	0.000,772	0.001,006
41–50	0.000,741	0.000,946
51–60	0.000,732	0.000,921
61–70	0.000,746	0.000,933
71–80	0.000,783	0.000,980
81–90	0.000,842	0.001,063
91–100	0.000,923	0.001,181
101–110	0.001,027	0.001,335
111–120	0.001,154	

Source: Blue Book. Road Infrastructure, Jaspers, July 2015 and based on materials from the Institute of Roads and Bridges (IBDiM).

Table A7. Unit value of tCO_2 emission (EUR/tCO_2).

Years	Value of tCO_2 (EUR/tCO_2)
2019	40.63
2020	41.82
2021	43.02
2022	44.21
2023	45.41
2024	46.60
2025	47.80
2026	48.99
2027	50.19
2028	51.38
2029	52.58
2030	53.77
2031	54.97
2032	56.16
2033	57.36
2034	58.55
2035	59,75
2036	60.94
2037	62.13
2038	63.33
2039	64.52
2040	65.72
2041	66.91
2042	68.11
2043	69.30

Source: Tables of unit costs to use in cost–benefit analyses, July 2019, CUPT, www.cupt.gov.pl (accessed on 13 June 2021).

References

1. Merk, O. *The Competitiveness of Global Port-Cities: Synthesis Report*; OECD Publishing: Paris, France, 2014.
2. AIVP. *Port-City Governance*; Sefacil Foundation: Le Havre, France; AIVP: Le Havre, France, 2014.
3. Allen, J.; Browne, M.; Cherrett, T. Investigating relationships between road freight transport, facility location, logistics management and urban form. *J. Transp. Geogr.* **2012**, *24*, 45–57. [CrossRef]
4. Browne, M.; Woxenius, J.; Dablanc, L.; Cherrett, T.; Morganti, E. Port cities and urban logistics. In Proceedings of the 22nd Annual Conference of The Chartered Institute of Logistics and Transport, Logistics Research Network (LRN), Southampton, UK, 6–8 September 2017.
5. Merk, O.; Dang, T.-T. *The Effectiveness of Port-City Policies: A Comparative Approach*; OECD Regional Development Working Papers; OECD Publishing: Paris, France, 2013.
6. Heaver, T.D. The evolution and challenges of port economics. In *Port Economics. Research in Transportation Economics*; Cullinane, K., Talley, W., Eds.; Elsevier JAI Press: Oxford, UK, 2006; Volume 16.
7. Hall, P. *Traffic Planning in Port-Cities*; Discussion Paper; International Transport Forum: Paris, France, 2018.
8. Thompson, R.G.; Taniguchi, E. City logistics and freight transport. In *Handbook of Logistics and Supply-Chain Management*; Brewer, A.M., Button, K.J., Hensher, D.A., Eds.; Emerald Publishing Group: Bingley, UK, 2008; Volume 2, pp. 393–405.
9. Merk, O.; Notteboom, T. *Port Hinterland Connectivity*; Discussion Paper No 2015-13; International Transport Forum: Paris, France; OECD Publishing: Paris, France, 2015.
10. Sierra, L.A.; Pellicer, E.; Yepes, V. Method for estimating the social sustainability of infrastructure projects. *Environ. Impact Assess. Rev.* **2017**, *65*, 41–53. [CrossRef]
11. Suprayoga, G.B.; Bakker, M.; Witte, P.; Spit, T. A systematic review of indicators to assess the sustainability of road infrastructure projects. *Eur. Transp. Res. Rev.* **2020**, *12*, 19. [CrossRef]
12. Henke, I.; Carteni, A.; Molitierno, C.; Errico, A. Decision-making in the transport sector: A sustainable evaluation method for road infrastructure. *Sustainability* **2020**, *12*, 764. [CrossRef]
13. Purvis, B.; Mao, Y.; Robinson, D. Three pillars of sustainability: In search of conceptual origins. *Sustain. Sci.* **2018**, *14*, 681–695. [CrossRef]
14. Pearce, D.; Atkinson, G.; Mourato, S. *Cost-Benefit Analysis and the Environment. Recent Developments*; OECD Publishing: Paris, France, 2006.
15. Geurs, K.T.; Boon, W.; Van Wee, B. Social impacts of transport: Literature review and the state of the practice of transport appraisal in the Netherlands and the United Kingdom. *Transp. Rev.* **2009**, *29*, 69–90. [CrossRef]

16. Mardsen, G.R.; Kimble, M.; Nellthorp, J.; Kelly, C. Sustainability appraisal: The definition deficit. *Int. J. Sustain. Transp.* **2009**, *4*, 189–211.
17. Lautso, K.; Spiekermann, K.; Wegener, M.; Sheppard, L.; Steadman, P.; Martino, A.; Domingo, R.; Gayda, S. PROPOLIS. Planning and Research of Policies for Land Use and Transport for Increasing Urban Sustainability. 2004. Available online: http://www.spiekermann-wegener.de/pro/pdf/PROPOLIS_Final_Report.pdf (accessed on 20 June 2021).
18. Bickel, P.; Friedrich, R.; Burgess, A.; Fagiani, P.; Hunt, A.; De Jong, G.; Laird, J.; Lieb, C.; Lindberg, G.; Mackie, P.; et al. Developing Harmonised European Approaches for Transport Costing and Project Assessment (HEATCO). Deliverable 5: Proposal for Harmonised Guidelines. 2006. Available online: https://www.putevisrbije.rs/images/pdf/strategija/HEATCO_D5_eng.pdf (accessed on 3 October 2021).
19. de Jong, G.; Kouwenhoven, M.; Bates, J.; Koster, P.; Verhoef, E.; Tavasszy, L.; Warffemius, P. New SP-values of time and reliability for freight transport in The Netherlands. *Transp. Res. Part E Logist. Transp. Rev.* **2014**, *64*, 71–87. [CrossRef]
20. Significance. *Values of Time and Reliability in Passenger and Freight Transport in The Netherlands*; Report for the Ministry of Infrastructure and the Environment; VU University of Amsterdam: Amsterdam, The Netherlands, 2012.
21. Brons, M.; Christidis, P. External Cost Calculator for Marco Polo Freight Transport Project Proposals. 2013. Available online: https://op.europa.eu/en/publication-detail/-/publication/fb01b58c-96a5-4b00-a2b3-b4d234c954bf/language-en/format-PDF (accessed on 24 June 2021).
22. Schroten, A.H.; van Essen, L.; van Wijngaarden, D.; Sutter, R.; Parolin, D.; Fiorello, M.; Brambilla, S.; Mai, K.; Beyrouty, F.; Fermi, F.; et al. *Handbook on the External Costs of Transport*; European Union: Brussels, Belgium, 2019.
23. Korzhenevych, A.; Dehnen, N.; Bröcker, J.; Holtkamp, M.; Meier, H.; Gibson, G.; Varma, A.; Cox, V. Update of the Handbook on External Costs of Transport. 2014. Available online: http://ec.europa.eu/transport/themes/sustainable/studies/doc/2014-handbook-external-costs-transport.pdf (accessed on 30 June 2021).
24. Maibach, M.; Schreyer, C.; Sutter, D.; Essen, H.P.; Boon, B.H.; Smokers, R.; Schroten, A.; Doll, C.; Pawlowska, B.; Bak, M. Handbook on Estimation of External Cost in the Transport Sector. 2008. Available online: https://ec.europa.eu/transport/sites/transport/files/themes/sustainable/doc/2008_costs_handbook.pdf (accessed on 2 July 2021).
25. European Union. *The European Green Deal*; COM(2019) 640 Final; European Union: Brussels, Belgium, 2019.
26. European Commission. *A European Strategy for Low-Emission Mobility*; European Commission: Brussels, Belgium, 2016. Available online: https://eur-lex.europa.eu/resource.html?uri=cellar:e44d3c21-531e-11e6-89bd-01aa75ed71a1.0002.02/DOC_1&format=PDF (accessed on 3 July 2021).
27. ERTRAC. *Long Distance Freight Transport: A Roadmap for System Integration of Road Transport*; ERTRAC: Brussels, Belgium, 2019; Available online: https://www.ertrac.org/uploads/documentsearch/id56/ERTRAC-Long-duty-Freight-Transport-Roadmap-2019.pdf (accessed on 4 July 2021).
28. European Commission. *Europe on the Move. An Agenda for a Socially Fair Transition towards Clean, Competitive and Connected Mobility for All*; 31.5.2017 COM(2017) 283 Final; European Commission: Brussels, Belgium, 2017; Available online: https://eur-lex.europa.eu/legal-content/EN/TXT/?uri=CELEX:52017DC0283 (accessed on 7 June 2021).
29. Kotowska, I.; Mańkowska, M.; Pluciński, M. Inland shipping to serve the hinterland: The challenge for seaport authorities. *Sustainability* **2018**, *10*, 3468. [CrossRef]
30. Kotowska, I.; Mańkowska, M.; Pluciński, M. Socio-economic costs and benefits of seaport infrastructure development for a local environment. The case of the port and the city of Świnoujście. In *European Port Cities in Transition. Strategies for Sustainability*; Carpenter, A., Lozano, R., Eds.; Springer: Cham, Switzerland, 2020; pp. 327–345.
31. Fenton, P. Port-city redevelopment and sustainable development. In *European Port Cities in Transition. Strategies for Sustainability*; Carpenter, A., Lozano, R., Eds.; Springer: Cham, Switzerland, 2020; pp. 19–36.
32. Hintjens, J.; Van Hassel, E.; Vanelslander, T.; Van De Voorde, E. Port cooperation and bundling: A way to reduce the external costs of hinterland transport. *Sustainability* **2020**, *12*, 9983. [CrossRef]
33. Sys, C.; Vanelslander, T. Port hinterland relations: Lessons to be learned from a cost-benefit analysis of a large investment project. In *Ports and networks. Strategies, Operations and Perspectives*; Geerlings, H., Kuipers, B., Zuidwijk, R., Eds.; Routledge: London, UK, 2018.
34. General Directorate for National Roads and Motorways. *Method of Carrying out the General Traffic Measurement in 2015*; GTM 2015 Guidelines, attachment B.; General Directorate for National Roads and Motorways: Warsaw, Poland, 2015.
35. Instytut Badawczy Dróg i Mostów. *Instruction for Assessing the Economic Efficiency of Road and Bridge Projects—Verification of the Research Method according to EU Recommendations and Update of Unit Prices as of 2007*; Part II, Tables of Speed; Instytut Badawczy Dróg i Mostów: Warsaw, Poland, 2008; pp. 84–86. (In Polish)
36. Dekker, S.; Verhaeghe, R.J.; Pols, A.A.J. Economic impacts and public financing of port capacity investments: The case of rotterdam port expansion. *Transp. Res. Rec. J. Transp. Res. Board* **2003**, *1820*, 55–61. [CrossRef]
37. Directorate-General for Regional and Urban Policy. *Guide to Cost-Benefit Analysis of Investment Projects. Economic Appraisal Tool for Cohesion Policy 2014–2020*; European Commission: Brussels, Belgium, 2014.
38. van Exel, J.; Rienstra, S.; Gommers, M.; Pearman, A.; Tsamboulas, D. EU involvement in TEN development: Network effects and European value added. *Transp. Policy* **2002**, *9*, 299–311. [CrossRef]
39. De Langhe, K.; Sys, C.; Van de Voorde, E.; Vaneslander, T. *Analysis of a Cost-Benefit Study: Lessons to be Learned for Large Investment Projects in Port Areas*; Association for European Transport: Glasgow, UK, 2012.

40. Lakshmanan, T.R.; William, P. *Transportation Infrastructure, Freight Services Sector and Economic Growth*; U.S. Department of Transportation Federal Highway Administration: Washington, DC, USA, 2002; pp. 43–44. Available online: http://www.ncgia.ucsb.edu/stella/meetings/20020115/Lakshmanan.pdf (accessed on 11 July 2021).
41. Mackie, P.; Graham, D.; Laird, J. The direct and wider impacts of transport projects: A review. In *A Handbook of Transport Economics*; De Palma, A., Lindsey, R., Quinet, E., Vickerman, R., Eds.; Edward Elgar Publishing: Cheltenham, UK, 2013; pp. 501–527.
42. Borkowski, P. Bezpośrednie efekty ekonomiczne-podejście metodologiczne. In *Infrastruktura Transportu a Konkurencyjność Regionów w Unii Europejskiej*; Pawłowska, B., Ed.; Gdańsk University Press: Gdańsk, Poland, 2015; pp. 160–166.
43. Van Hassel, E.; Meersman, H.; Van De Voorde, E.; Vanelslander, T. Impact of scale increase of container ships on the generalised chain cost. *Marit. Policy Manag.* **2016**, *43*, 192–208. [CrossRef]
44. Button, K. *Transport Economics*; Edward Elgar Publishing: Cheltenham, UK, 2010; pp. 145–146.
45. Combes, P.-P.; Lafourcade, M. Transport costs: Measures, determinants, and regional policy implications for France. *J. Econ. Geogr.* **2005**, *5*, 319–349. [CrossRef]
46. Mackie, P.; Nellthorp, J. Cost-Benefit analysis in transport. In *Handbook of Transport Systems and Traffic Control*; Button, K.J., Hensher, D.A., Eds.; Emerald Group Publishing: Bingley, UK, 2009; p. 149.
47. Teodorovic, D.; Janic, M. *Transportation Engineering. Theory, Practice, and Modeling*; Elsevier: Amsterdam, The Netherlands, 2017; p. 642.
48. Hanssen, T.-E.S.; Mathisen, T.A.; Jørgensen, F. Generalized transport costs in intermodal freight transport. *Procedia-Soc. Behav. Sci.* **2012**, *54*, 189–200. [CrossRef]
49. van Arem, B.; Kirby, H.R.; Van Der Vlist, M.J.; Whittaker, J.C. Recent advances and applications in the field of short-term traffic forecasting. *Int. J. Forecast.* **1997**, *13*, 1–12. [CrossRef]
50. Vlahogianni, E.; Golias, J.C.; Karlaftis, M.G. Short-term traffic forecasting: Overview of objectives and methods. *Transp. Rev.* **2004**, *24*, 533–557. [CrossRef]
51. Vlahogianni, E.I.; Karlaftis, M.G.; Golias, J.C. Short-term traffic forecasting: Where we are and where we are going. *Transp. Res. Part C Emerg. Technol.* **2014**, *43*, 3–19. [CrossRef]
52. Hinsbergen, C.J.; van Lint, J.W.; Sanders, F.M. Short-term traffic prediction models. In Proceedings of the 14th World Congress on Intelligent Transportation Systems, Beijing, China, 9–13 October 2007.
53. Lana, I.; Del Ser, J.; Velez, M.; Vlahogianni, E.I. Road traffic forecasting: Recent advances and new challenges. *IEEE Intell. Transp. Syst. Mag.* **2018**, *10*, 93–109. [CrossRef]
54. van Lint, H.; Hinsbergen, C. Short-term and travel time prediction models. *Artif. Intell. Appl. Crit. Transp. Issues* **2012**, *22*, 22–41.
55. Kitamura, R. The effects of added transportation capacity on travel: A review of theoretical and empirical results. *Transportation* **2009**, *36*, 745–762. [CrossRef]
56. De Borger, B.; Proost, S.; Van Dender, K. Congestion and tax competition in a parallel network. *Eur. Econ. Rev.* **2005**, *49*, 2013–2040. [CrossRef]
57. Rothengatter, W. Evaluation of infrastructure investments in Germany. *Transp. Policy* **2000**, *7*, 17–25. [CrossRef]
58. Naess, P.; Nicolaisen, M.S.; Strand, A. Traffic forecasts ignoring induced demand: A shaky fundament for costs-benefit analyses. *Eur. J. Transp. Infrastruct. Res.* **2012**, *12*, 291–309. [CrossRef]
59. Bernacki, D.; Lis, C. Statistical estimation and prediction of Annual Average Daily Traffic (AADT) on the first/last mile road sections in the Port of Szczecin. *Ekon. Probl. Usług* **2017**, *128*, 67–80. [CrossRef]
60. Dunkerley, F.; Rohr, C.; Daly, A. *Road Traffic Demand Elasticities. A Rapid Evidence Assessment*; Final report; Rand Europe: Cambridge, UK, 2014.
61. Shen, S.; Fowkes, T.; Whiteing, T.; Johnson, D. Econometric modelling and forecasting of freight transport demand in Great Britain. In Proceedings of the European Transport Conference, Noordwijkerhout, The Netherlands, 5–7 October 2009.
62. Agnolucci, P.; Bonilla, D. UK freight demand elasticities and decoupling. *J. Transp. Econ. Policy* **2009**, *43*, 317–344.
63. Meersman, H.; Van de Voorde, E. The relationship between economic activity and freight transport. In *Developments in Transport Modelling. Lessons for the Freight Sector*; Ben Akiva, M., Meersman, H., Van de Voorde, E., Eds.; Emerald Publishing: Bingley, UK, 2013; pp. 15–43.
64. JASPERS. *Blue Book. Road Infrastructure*; Joint Assistance to Support Projects in European Regions—JASPERS: Warsaw, Poland, 2015.
65. CUPT. The Centre for EU Transport Projects. 2019. Available online: www.cupt.gov.pl (accessed on 13 June 2021).
66. European Investment Bank. *The Economic Appraisal of Investment Projects at the EIB*; European Investment Bank: Luxemburg, 2013.
67. Mańkowska, M.; Pluciński, M.; Kotowska, I. Biomass sea-based supply chains and the secondary ports in the era of decarbonization. *Energies* **2021**, *14*, 1796. [CrossRef]
68. CEFIC. *Guidelines for Measuring and Managing CO2 Emission from Freight Transport Operations*; CEFIC: Brussels, Belgium, 2011.
69. Mouter, N.; Dean, M.; Koopmans, C.; Vassallo, J.M. Comparing cost-benefit analysis and multi-criteria analysis. *Adv. Transp. Policy Plan.* **2020**, *6*, 225–254. [CrossRef]
70. Beria, P.; Maltese, I.; Mariotti, I. Multicriteria versus Cost Benefit Analysis: A comparative perspective in the assessment of sustainable mobility. *Eur. Transp. Res. Rev.* **2012**, *4*, 137–152. [CrossRef]

energies

MDPI

Article

A Comparison of Three Ridesharing Cost Savings Allocation Schemes Based on the Number of Acceptable Shared Rides

Fu-Shiung Hsieh

Department of Computer Science and Information Engineering, Chaoyang University of Technology, Taichung 413310, Taiwan; fshsieh@cyut.edu.tw

Abstract: Shared mobility based on cars refers to a transportation mode in which travelers/drivers share vehicles to reduce the cost of the journey, emissions, air pollution and parking demands. Cost savings provide a strong incentive for the shared mobility mode. As cost savings are due to cooperation of the stakeholders in shared mobility systems, they should be properly divided and allocated to relevant participants. Improper allocation of cost savings will lead to dissatisfaction of drivers/passengers and hinder acceptance of the shared mobility mode. In practice, several schemes based on proportional methods to allocate cost savings have been proposed in shared mobility systems. However, there is neither a guideline for selecting these proportional methods nor a comparative study on effectiveness of these proportional methods. Although shared mobility has attracted much attention in the research community, there is still a lack of study of the influence of cost saving allocation schemes on performance of shared mobility systems. Motivated by deficiencies of existing studies, this paper aims to compare three proportional cost savings allocation schemes by analyzing their performance in terms of the numbers of acceptable rides under different schemes. We focus on ridesharing based on cars in this study. The main contribution is to develop theory based on our analysis to characterize the performance under different schemes to provide a guideline for selecting these proportional methods. The theory developed is verified by conducting experiments based on real geographical data.

Keywords: shared mobility; cost; ridesharing; allocation

Citation: Hsieh, F.-S. A Comparison of Three Ridesharing Cost Savings Allocation Schemes Based on the Number of Acceptable Shared Rides. *Energies* **2021**, *14*, 6931. https://doi.org/10.3390/en14216931

Academic Editors: Elżbieta Macioszek, Anna Granà, Margarida Coelho, Paulo Fernandes and Grzegorz Karoń

Received: 26 August 2021
Accepted: 19 October 2021
Published: 21 October 2021

1. Introduction

Energy consumption and the associated impact on the environment due to urbanization and growth of population are two challenges directly linking to sustainability of cities that have attracted researchers' attention for decades. According to [1], the transportation sector accounts for a large proportion of global energy consumption and greenhouse gas (GHG) emissions. Fuel combustion in the transportation sector accounts for 24% of direct CO_2 emissions, which has significant impact on the environment. How to effectively minimize the impact on the environment by the transportation sector is critical to achieve sustainability of cities. Sharing economy has attracted much attention of practitioners, researchers, policy makers and individuals to improve efficiency and sustainability based on sharing of assets [2]. The transportation sector contributes to a large part of energy consumption and hence the study on green transportation and solutions has become an important research subject for the sustainable development of sharing economy [3]. In the past years, shared mobility emerges as a popular transportation mode that attracts the attention of researchers as well as practitioners in the transportation sector [4]. Shared mobility refers to a transportation mode in which travelers/drivers share vehicles to reduce the cost of the journey, energy consumption, emissions, air pollution and parking demands. It takes different forms in the world, including ridesharing, carpooling and bike sharing [5,6]. In this paper, we consider ridesharing systems based on cars. Shared mobility services enable users to access transportation services conveniently on an as-needed basis.

Thus, shared mobility services provide a flexible transportation option besides public transport. In addition to the advantage due to flexibility in terms of routing and timing, users of shared mobility services also can enjoy the benefits of cost savings due to the shared use of vehicles.

Cost savings provide a strong incentive for the shared mobility mode. Therefore, optimization of performance metrics relevant to cost savings in shared mobility systems has been extensively studied in the literature [7–9]. As the primary incentive for users to share a ride is cost savings, the way to allocate cost savings is an important factor to the adoption of ridesharing by the users. Allocation of a portion of cost savings to the service providers is also important for maintaining operation of ridesharing information systems. As cost savings are due to cooperation of the stakeholders in shared mobility systems based on cars, they should be properly divided and allocated to relevant participants. In the literature, several schemes to allocate costs/cost savings among players in a coalition of transportation systems have been studied in [10]. Due to computational complexity and the need for agile problem solvers [8], proportional methods have been adopted in practice to allocate costs/cost savings. For ridesharing systems, several schemes to allocate cost savings have been proposed based on proportional methods, e.g., [11–13]. However, there is no guideline to select the correct and proper scheme for allocation of cost savings due to the lack of a comparative study on these proportional schemes. Improper allocation of cost savings will discourage users and hinder acceptance of the shared mobility model. This is due to the lack of consideration of the satisfaction factor of drivers and passengers. For example, suppose all cost savings are allocated to the passengers. The drivers will not accept share rides with passengers. Therefore, how to allocate cost savings properly is an important issue. Although proportional methods have been proposed and studied in different problem domains [10], analysis of effectiveness of the three proportional allocation methods for ridesharing has not been carried out. Although shared mobility has attracted much attention in the research community, there is still a lack of study on the influence of cost savings allocation schemes on the performance of shared mobility systems. Motivated by deficiencies of existing studies on these issues, this paper aims to develop a framework to compare the three different cost savings allocation schemes proposed in [11–13] based on proportional methods and compare their performance in terms of the numbers of acceptable rides under different schemes. The contributions of this paper are three-fold. First, we develop theory to compare the performance of the three different cost savings allocation schemes. Second, we provide a guideline with rules for selecting a proper cost savings allocation scheme from the three cost savings allocation schemes for shared mobility systems. Third, policy makers may apply the guideline to increase the number of acceptable shared rides to enhance sustainability and service providers may apply the guideline to increase the number of acceptable shared rides and profits.

The decision support systems for shared mobility systems with cars typically include two stages—stage 1: determination of shared rides and stage 2: allocation of cost savings [13]. For the drivers and passengers, they must make decisions about whether they accept the recommended rides for ridesharing. In this paper, we assume that the drivers and passengers determine whether to accept shared rides based on the minimal expected rewarding rate. A ride recommended by the decision support systems for shared mobility systems with cars accepted by the driver and the passenger is called an acceptable ride. The stakeholders include drivers, passengers and the shared mobility service/information provider. Schemes to allocate cost savings to the stakeholders in shared mobility systems are defined based on these stakeholders. To achieve the goals of this paper, there are four tasks to complete: (1) identification of shared mobility decision model and definition of schemes to allocate cost savings to participants, (2) introduction of the concept of minimal expected rewarding rate of drivers and passengers and development of theory based on analysis of the influence of different cost savings allocation schemes on the number of acceptable shared rides, (3) verification of the developed theory by test cases and (4) a guideline for selecting the proper proportional scheme to allocate cost savings.

The structure of the remainder of this paper follows. In Section 2, we briefly review existing studies on shared mobility systems with cars. In Section 3, a shared mobility decision model and the three proportional schemes, including Fifty-Fifty Scheme, Local Proportional Scheme and Global Proportional Scheme, to allocate cost savings to the stakeholders in shared mobility systems are defined. We compare the Fifty-Fifty Scheme and Local Proportional Scheme in Section 4, compare the Local Proportional Scheme and Global Proportional Scheme in Section 5 and compare the Global Proportional Scheme and Fifty-Fifty Scheme in Section 6. We verify theoretical properties by examples in Section 7. In Section 8, we provide a discussion based on the results obtained and, to allocate cost savings, offer a guideline for selecting a proper proportional scheme from the three schemes studied in this paper. Finally, we conclude this paper in Section 9.

2. Literature Review

The widespread adoption of mobile communication devices and ICT to provide efficient access to transportation services has made shared mobility an important paradigm for cities [14]. Shared mobility may take different forms, including sharing of vehicles and sharing of rides [15]. The former allows cars, motorcycles, scooters, and bikes to be shared and the latter enables ridesharing and on-demand ride services for passenger rides. The benefits of shared mobility include significant reduction of total travel distance, trip cost, carbon emissions and congestion [16,17]. The advantages of ridesharing lead to several variants of transportation modes. Carpooling is an early variant of ridesharing in which employees in companies are encouraged to travel with colleagues for commuting to minimize the number of cars traveling to sites of the companies and travel cost of the employees [18]. Vanpooling is another variant of ridesharing transportation mode. In vanpooling, commuters drive to an intermediate location to take a van and ride together to the destination [19]. Generally speaking, ridesharing reduces the travel cost for people with similar itineraries and schedules by sharing vehicles [5]. In addition, ridesharing models with mixed passengers, parcels [20] or goods [21] were proposed. Recent trends in studies on ridesharing include social ridesharing, such as the one in presented in [22].

In this paper we focus on ridesharing systems with cars. Although shared mobility promises to achieve sustainability of cities, reduce energy consumption and provide an alternative way to meet passengers' transportation need without owning a car, it relies on the development of an efficient decision model and decision support tools to attain these goals. Actually, shared mobility poses several challenging research issues [7,8,23]. In the literature, optimization of shared mobility systems was discussed in [7,8]. The solution algorithms can be classified into exact methods and heuristic/metaheuristic methods. Exact methods are applied to find the optimal solutions for optimization of shared mobility systems [24]. Exact methods, however, can only be applied to small problems due to computational complexity. Therefore, many heuristic/metaheuristic methods have been proposed to solve optimization problems of shared mobility systems. For example, the genetic algorithm [25], tabu search [26], population-based metaheuristic algorithms [27], the differential evolution algorithm [28] and hybridization of metaheuristics [29] have been proposed to solve relevant optimization problems in shared mobility systems.

Beside optimization issues, efforts to realize shared mobility have developed. Mobility-as-a-service (MaaS) aims to provide users access to shared mobility services through a single friendly online interface from planning, booking to payment. Several cities have already implemented or are in the process of implementing MaaS trials or pilot projects. Although MaaS provides a potential alternative to private vehicle ownership, there are barriers and risks for adoption of MaaS in cities [30] such as the lack of appeal to public transport users and private vehicle users. Ridesharing is a potential transportation mode to reduce the number of cars, but it is still not widely adopted. Studies in [31,32] point out that savings of cost and time are the main incentives for ridesharing. For ridesharing drivers and passengers, a proper scheme to allocate cost savings among participants is important for adoption of ridesharing.

In the literature, the problem to allocate costs/cost savings among players in a coalition of transportation systems has been studied in [10]. Shapley value [33], nucleolus [34] and proportional methods [35] have been proposed in the literature to allocate costs/cost savings in cooperative game theory. Shapley value [33] and nucleolus [34] suffer from computational complexity problems [36,37]. Therefore, proportional methods [35] are often adopted in practice. There are many studies on application of proportional methods in different domains in the literature. For example, a pricing scheme is used in [38] is to maximize the total profit through exchanging transportation requests among collaborative carriers. In [39], proportional methods are used to allocate costs and emissions in a carrier's delivery network with multiple customers served by a single carrier. As transportation cost of low-value forest fuels, including trees, tree bark, branches, stumps and wood chips, accounts for a large part in logistic cost, several alternative ways applying proportional methods to lower logistic cost of forest fuels transport have been studied in [40].

In the context of ridesharing systems, several cost allocation methods have been developed. The problem to achieve fair cost allocation in ridesharing systems based on the nucleolus method has been studied in [41]. In [42], the problem to allocate passengers to drivers, charge passengers and create feasible schedules for drivers in a dynamic ridesharing scenario is addressed. Due to computational complexity and the need for agile algorithms in ridesharing systems, proportional methods have been widely used in practice. For example, the ones proposed in [11–13] are based on proportional methods. In [11], the authors consider a ridesharing problem in which each driver shares a ride with at most one rider. The cost allocation scheme proposed in [11] divides the cost savings of sharing a ride equally between the two matched participants (the driver and the rider) of the shared ride. Such cost allocation satisfies the property of Shapley value in the cooperative game theory. In [12], the authors considered ridesharing systems in which at most one pickup and delivery may take place during the trip and transfers are not allowed. In [12], it is assumed that costs are proportional to vehicle-miles driven. The cost savings allocated to the driver and the rider are proportionally to the lengths of their original trips. The cost allocation schemes in [11,12] are proposed for ridesharing systems in which each shared ride includes one driver and one rider. In [13], a ridesharing problem which allows multiple passengers to be transported by a driver is considered. A cost allocation scheme which divides the cost savings among three types of stakeholders in a ridesharing system, including ridesharing information provider, drivers and riders, is proposed. The cost savings allocation scheme proposed in [13] allocates a portion of the overall cost savings to the ridesharing information provider and the remaining cost savings to each ridesharing passenger and each driver, which is based on the weighting of their bid price with respect to overall bid price of all ridesharing passengers' and all drivers' bids. However, there still lacks a comparative study on whether the cost savings allocated by these allocation schemes are acceptable for drivers and passengers. That is, the cost savings allocation schemes will surely influence the number of acceptable shared rides for drivers and passengers. An interesting but unexplored research issue is to compare the number of acceptable rides that can be satisfied under the three proportional cost savings allocation schemes proposed in [11–13].

This paper does not focus on optimization of the ridesharing matching problems to minimize overall travel distance or maximize overall cost savings. In this paper, we will study effectiveness of three cost savings allocation schemes in terms of the number of acceptable shared rides that satisfy the minimal rewarding rate requirements of drivers and passengers. In the literature, different ridesharing matching methods have been proposed. To ensure fairness of comparison, the overall cost savings of a ridesharing system are obtained by the same ridesharing matching method. The cost savings allocation policies to be compared in this paper include those proposed in [11–13]. We compare the three cost savings allocation schemes by theoretical analysis. The properties and theorems established through analysis are verified by examples.

3. Decision Model and Cost Savings Allocation Schemes for Shared Mobility Systems

The decision support systems for shared mobility systems typically include two decision stages. A decision model for shared mobility systems is used in the first stage to find the shared rides. In the second stage, the solution obtained from the decision model is then used to divide the cost savings among the participants by applying different allocating schemes. To focus on the comparison of different schemes for allocating cost savings, the decision model in [13] based on maximization of cost savings will be used in this study. A list of notations used in this study is summarized in Table 1 for reference.

Table 1. Notations of symbols, variables and parameters.

Variable	Meaning
P	total potential passengers
D	total potential drivers
p	the index of a passenger, where $p \in \{1,2,3,\ldots,P\}$
d	the index of a driver, where $d \in \{1,2,3,\ldots,D\}$
k	the index of a location, $k \in \{1,2,\ldots,P\}$
J_d	total bids submitted by driver $d \in \{1,2,\ldots,D\}$
j	the index of the $j-th$ bid placed by a driver with $j \in \{1,2,\ldots,J_d\}$
BID_{dj}	$BID_{dj} = (q^1_{dj1}, q^1_{dj2}, q^1_{dj3}, \ldots, q^1_{djP}, q^2_{dj1}, q^2_{dj2}, q^2_{dj3}, \ldots, q^2_{djP}, o_{dj}, c_{dj})$, the $j-th$ bid of driver d, where q^1_{djk}: number of seats allocated at passenger k's pick-up location, q^2_{djk}: number of seats released at passenger k's drop-off location, o_{dj}: the cost when the driver travels alone, c_{dj}: the cost of the bid.
BID_p	$BID_p = (s^1_{p1}, s^1_{p2}, s^1_{p3}, \ldots, s^1_{pP}, s^2_{p1}, s^2_{p2}, s^2_{p3} \ldots, s^2_{pP}, f_p)$: the bid submitted by passenger p, where s^1_{pk}: number of seats requested for passenger k's pick-up location, s^2_{pk}: number of seats released at passenger k's drop-off location, f_p: the cost of passenger p without ridesharing.
x_{dj}	a decision variable indicating whether the $j-th$ bid of driver d is a winning bid ($x_{dj} = 1$) or not ($x_{dj} = 0$)
y_p	a binary decision variable indicating whether the bid of passenger p is a winning bid ($y_p = 1$) or not ($y_p = 0$)
N	the set of all players in a cooperative game, $N = \{1, \ldots, N\}$
Ω	the set Ω of all subsets of N
C	a characteristic function. $\Omega \to$ to assign the cost to each coalition S in Ω
β_n	the share value for a player $n \in N$, where $\sum_{n \in N} \beta_n = 1$
α	the share value for the information provider m
β^P_p	the share value for passenger $p \in P$
β^D_d	the share value for driver $d \in D$
Γ_{dj}	the set of passengers in the ride corresponding to the j-th bid submitted by driver d
$F(x,y)$	overall cost savings, $F(x,y) = \left(\sum_{p=1}^{P} y_p \left(f_p\right)\right) + \left(\sum_{d=1}^{D}\sum_{j=1}^{J_d} x_{dj} o_{dj}\right) - \left(\sum_{d=1}^{D}\sum_{j=1}^{J_d} x_{dj} c_{dj}\right)$
$F_{dj}(x,y)$	the cost savings of the j-th bid submitted by driver d, $F_{dj}(x,y) = \left[\left(\sum_{p \in \Gamma_{dj}} y_p f_p\right) + x_{dj} o_{dj} - \left(x_{dj} c_{dj}\right)\right]$

3.1. Decision Model

The decision support system for ridesharing makes decisions within a time period. Requests of drivers and passengers arriving before the start of the decision time period will be considered by the ridesharing decision support system. The decision support system follows the framework in [13] and the decision period is divided into two stages—stage 1: determination of shared rides and stage 2: allocation of cost savings. In the decision model used in stage 1, drivers and passengers submit bids [13]. Bids are generated by bid generation procedures to meet spatial, time and capacity requirements/constraints of drivers and passengers. The bid generation procedure for passengers generates bids based on the origins, destinations, earliest departure time, latest arrival time and number of seats requested by passengers. The bid generation procedure for drivers generates bids based on the origins, destinations, earliest departure time, latest arrival time, the number of seats for ridesharing and the maximum detour ratio specified by drivers. Therefore, the decision model can handle spatial, time and capacity requirements in ridesharing systems. In addition, the decision model is flexible and has been extended in [22] to deal with trust requirements in social ridesharing problems.

For the above decision model, suppose P passengers and D drivers submit bids to the systems. The bid of passenger p is $BID_p = (s^1_{p1}, s^1_{p2}, s^1_{p3}, \ldots, s^1_{pP}, s^2_{p1}, s^2_{p2}, s^2_{p3} \ldots, s^2_{pP}, f_p)$, where s^1_{pk} is the number of seats requested for passenger k's pick-up location, s^2_{pk} is the number of seats released at passenger k's drop-off location and f_p is the cost of passenger p without ridesharing. A driver may submit multiple bids, but only one bid can be accepted. A bid accepted by the system is called a winning bid. $BID_{dj} = (q^1_{dj1}, q^1_{dj2}, q^1_{dj3}, \ldots, q^1_{djP}, q^2_{dj1}, q^2_{dj2}, q^2_{dj3}, \ldots, q^2_{djP}, o_{dj}, c_{dj})$ denotes the $j-th$ bid of driver d, where q^1_{djk} is the number of seats allocated at passenger k's pick-up location, q^2_{djk} is the number of seats released at passenger k's drop-off location, o_{dj} is the cost when the driver travels alone and c_{dj} is the cost of the bid. It is assumed that the bids of passengers and drivers are generated by applying bid generation procedures developed in [13], taking into account the timing, capacity and spatial constraints of drivers and passengers. The bid generation procedures calculate the cost based on a function $Pf(d)$, where d is the travel distance. The decision model in [13] can handle the bids regardless of whether the function $Pf(d)$ used to calculate cost is linear or nonlinear.

The overall cost savings are the difference between the overall original cost (without ridesharing) and the overall cost after ridesharing. Original cost of passengers (without ridesharing) and original cost of drivers (without ridesharing) is $\left(\sum_{p=1}^{P} y_p f_p\right) + \left(\sum_{d=1}^{D}\sum_{j=1}^{J_d} x_{dj} o_{dj}\right)$. Cost of passengers and drivers with ridesharing is $\left(\sum_{d=1}^{D}\sum_{j=1}^{J_d} x_{dj} c_{dj}\right)$. The overall cost savings are the difference between these two terms. That is, the overall cost savings are $\left(\sum_{p=1}^{P} y_p f_p\right) + \left(\sum_{d=1}^{D}\sum_{j=1}^{J_d} x_{dj} o_{dj}\right) - \left(\sum_{d=1}^{D}\sum_{j=1}^{J_d} x_{dj} c_{dj}\right)$.

The decision problem aims to determine the winning bids submitted by the potential drivers and passengers to maximize the overall cost savings. Thus, the objective function is defined in (1) as follows:

$$F(x,y) = \left(\sum_{p=1}^{P} y_p f_p\right) + \left(\sum_{d=1}^{D}\sum_{j=1}^{J_d} x_{dj} o_{dj}\right) - \left(\sum_{d=1}^{D}\sum_{j=1}^{J_d} x_{dj} c_{dj}\right) \quad (1)$$

The problem to determine the winning bids is formulated as an integer programming problem subject to demand and supply constraints (2), (3), nonnegative cost savings constraint (4) and single winning bid constraints for drivers (5) as follows:

$$\max_{x,y} F(x,y)$$

$s.t.$

$$\sum_{d=1}^{D}\sum_{j=1}^{J_d} x_{dj} q^1_{djk} = y_p s^1_{pk} \ \forall \mathrm{p} \in \{1,2,\ldots,\mathrm{P}\} \ \forall \mathrm{k} \in \{1,2,\ldots,\mathrm{P}\} \quad (2)$$

$$\sum_{d=1}^{D}\sum_{j=1}^{J_d} x_{dj}q_{djk}^2 = y_p s_{pk}^2 \ \forall \mathrm{p} \in \{1,2,\ldots,\mathrm{P}\}\ \forall \mathrm{k} \in \{1,2,\ldots,\mathrm{P}\} \quad (3)$$

$$\sum_{p=1}^{P} y_p f_p + \sum_{d=1}^{D}\sum_{j=1}^{J_d} x_{dj}o_{dj} \geq \sum_{d=1}^{D}\sum_{j=1}^{J_d} x_{dj}c_{dj} \quad (4)$$

$$\sum_{j=1}^{J_d} x_{dj} \leq 1\ \forall d \in \{1,\ldots,D\} \quad (5)$$

$$x_{dj} \in \{0,1\}\ \forall d \in \{1,\ldots,D\}\ \forall j \in \{1,\ldots,J_d\}$$
$$y_p \in \{0,1\}\ \forall \mathrm{p} \in \{1,2,\ldots,\mathrm{P}\}$$

Due to computational complexity, only small instances of the above problem can be solved by applying exact algorithm due to exponential growth of solution space with respect to the problem size. In the literature, the problem to optimize cost savings in ridesharing systems has been extensively studied. Many approximate approaches can be applied to solve this problem, including evolutionary computation algorithms, bio-inspired evolutionary algorithms and metaheuristic approach. Therefore, many algorithms have been proposed to solve this problem, e.g., [13,28]. A comparison of these approaches appears in the literature [13]. The comparative study provides a guideline for selection of a solution methodology for solving the above cost savings optimization problem. As this paper focuses on comparison of cost allocation schemes, it is assumed that a solution algorithm is used to solve the cost savings optimization problem in the shared mobility model to ensure fairness in comparison of different cost allocation schemes. In this paper, the algorithm proposed in [13] is applied to determine the winning bids.

3.2. Three Cost Savings Allocation Schemes Based on Proportional Methods

Cost savings in ridesharing systems are due to collaboration of the participants in shared rides. Ridesharing can be considered as a type of collaborative transportation through which the participants, drivers and passengers in shared rides benefit from cost savings. Therefore, the participants in shared rides can be regarded as a coalition. The issue is to divide cost savings among participants in the coalition in collaborative transportation modes [10]. As cost savings are due to cooperation of participants in the coalition, they must be divided and allocated properly. Different ways to allocate cost savings can be developed for the shared mobility system. Several cost allocation methods have been proposed in the literature. Shared mobility systems can be modeled as a class of cooperative games in which participants represented by agents cooperate to benefit from cost savings. Consider a set $N = \{1, \ldots, N\}$ of all players in a cooperative game. The set N is called the "grand coalition." For the set Ω of all subsets of N, we define a characteristic function C: $\Omega \rightarrow R$ to assign the cost to each coalition S in Ω. A cost allocation vector is denoted as $z = (z_1, z_2, \ldots, z_n)$, which represents the value $z_n \in R$ assigned to each player n in N. A cost allocation vector is also called a preimputation. The characteristic function and cost allocation vector satisfy the "efficiency" property if $C(\Phi) = 0$ and $\sum_{n\in N} z_n = C(N)$. An allocation vector is said to be "rational" if there does not exist a subset S of players such that they would perceive less total cost than the total cost allocated to them by forming a coalition separately from the rest. Rationality of an allocation vector can be represented by the constraints $\sum_{n\in S} z_n \leq C(S)\ \forall S \in \Omega$. The set of cost allocation vectors or preimputations satisfying rationality is called the "core". An allocation in the core is stable as there is no subset S of the players such that its players would be better off by deviating from the coalition.

Several well-known methods have been proposed to allocate cost. These include Shapley value [33], nucleolus [34] and proportional methods [35]. Each cost allocation method has its pros and cons. Two appealing features of Shapley value allocation are efficiency and uniqueness. However, Shapley value allocation does not belong to the core in general. In addition, Shapley value allocation poses a computation complexity challenge [33]. Nucleolus is more complex than Shapley value although it always belongs to the core [34]. A simple way for cost allocation is proportional methods [35]. In the literature, several schemes to allocate cost savings in ridesharing systems were proposed in [11–13] based on the proportional methods. However, a further study on application of the proportional methods to shared mobility systems is needed to assess effectiveness of these methods. In this section, the schemes proposed in [11–13] to allocate cost savings in shared mobility systems are defined. These schemes to allocate cost savings are analyzed in the next section.

A proportional method allocates a share β_n of the total cost $C(N)$ to each player $n \in N$. That is, $z_n = \beta_n \times C(N)\ \forall n \in N$, where $\sum\limits_{n \in N} \beta_n = 1$. By assigning the share value β_n properly, we can obtain the schemes proposed in [11–13].

In shared mobility systems, the players are the stakeholders, including the information provider, a set of passengers, $\{1,2,3,\ldots P\}$ and a set of drivers, $\{1,2,3,\ldots,D\}$. The share values for the information provider, passenger $p \in P$ and driver $d \in D$ are denoted by α, β_p^P and β_d^D, respectively. Different cost savings allocation schemes can be defined by applying different rules to assign the share values α, β_p^P and β_p^D, where $\sum\limits_{p=1}^{P} \beta_p^P + \sum\limits_{d=1}^{D} \beta_d^D + \alpha = 1$. Instead of considering the original schemes proposed in [11–13] only, we generalize these schemes by taking into account allocation of cost savings to the information provider of shared mobility systems in addition to drivers and passengers. The three schemes proposed in [11–13] are variants of proportional methods defined by specifying the share values differently. For these three cost savings allocation schemes, it is assumed that the information provider is allocated $\alpha F(x,y)$ of cost savings. Let us define these three schemes formally.

For the convenience of discussions, we will call the schemes proposed in [11–13] "Fifty-Fifty Scheme", "Local Proportional Scheme" and "Global Proportional Scheme", respectively. We call the scheme proposed in [11] "Fifty-Fifty Scheme" as the cost savings of sharing a ride are equally divided between the two matched participants (the driver and the passenger) of each shared ride under this scheme. That is, fifty percent of cost savings of a shared ride is rewarded to the driver and the remaining fifty percent of cost savings is rewarded to the passenger. We call the scheme proposed in [12] "Local Proportional Scheme" as it allocates the cost savings of a ride to the driver and the passenger proportionally to the lengths of their trips. We call the scheme proposed in [13] "Global Proportional Scheme" as it allocates the cost savings to the drivers and passengers based on the weighting of individual bid price with respect to overall bid price of all ridesharing passengers' and drivers' bids. These three schemes are formally defined next. As the original schemes proposed in [11–13] do not consider the portion of cost savings allocated to information provider, the "Fifty-Fifty Scheme", "Local Proportional Scheme" and "Global Proportional Scheme" defined below extend the original schemes by taking into account allocation of cost savings to information provider. In each of the schemes, the cost savings are divided into three parts: the first part is allocated to the information provider, the second part is allocated to the drivers and the third part is allocated to the passengers.

The overall cost savings of a solution (x,y) is clearly defined in (1). Before defining the three schemes, we first distinguish the cost savings of a single ride from overall cost savings. For the ride corresponding to a winning bid, say the j-th bid submitted by driver d, let Γ_{dj} denote the set of passengers in the ride corresponding to the j-th bid submitted by driver d. The cost savings of the j-th bid submitted by driver d is defined in (6) as follows:

$$F_{dj}(x,y) = \left[\left(\sum_{p \in \Gamma_{dj}} y_p f_p \right) + x_{dj} o_{dj} - \left(x_{dj} c_{dj} \right) \right] \tag{6}$$

The three schemes are defined as follows.

In the Fifty-Fifty Scheme, which considers one driver and one passenger of a shared ride, fifty percent of cost savings of the ride is rewarded to the driver and the remaining fifty percent of cost savings is rewarded to the passenger.

Definition 1. *Fifty-Fifty Scheme*

(1) *This scheme allocates $\alpha F(x, y)$ to the information provider. The share value for the information provider is α.*
(2) *This scheme allocates $\sigma_p^P(1-\alpha)F_{dj}(x,y)$ to the ridesharing passenger $p \in P$ ($y_p = 1$) transported by driver d, where $\sigma_p^P = \frac{1}{2}$. The share value for passenger $p \in P(y_p = 1)$ is $\beta_p^P = \frac{y_p(1-\alpha)F_{dj}(x,y)}{2F(x,y)}$.*
(3) *This scheme allocates $\sigma_d^D(1-\alpha)F_{dj}(x,y)$ to each ridesharing driver $d \in D(x_{dj} = 1$ for some $j)$, where $\sigma_d^D = \frac{1}{2}$. The share value for the ridesharing driver $d \in D$ ($x_{dj} = 1$ for some j) is $\beta_d^D = \frac{\sum_{j=1}^{J_d} x_{dj}(1-\alpha)F_{dj}(x,y)}{2F(x,y)}$.*

Note that $\sum_{p=1}^{P} \beta_p^P + \sum_{d=1}^{D} \beta_d^D + \alpha = 1$.

The scheme proposed in [12] allocates cost savings to the driver and the passenger proportional to the lengths of their trips. Let us call it the Local Proportional Scheme. The Local Proportional Scheme allocates $\alpha F(x, y)$ to the information provider and allocates cost savings based on the weighting of the original bid price of each ridesharing passenger and driver in a shared ride with respect to overall bid price of the driver and the relevant passenger in the shared ride. The Local Proportional Scheme can be defined as follows:

Definition 2. *Local Proportional Scheme.*

(1) *This scheme allocates $\alpha F(x, y)$ to the information provider. The share value for the information provider is α.*
(2) *This scheme allocates $\sigma_p^P(1-\alpha)F_{dj}(x,y)$ to each ridesharing passenger $p \in P$ ($y_p = 1$) transported by driver d, where $\sigma_p^P = \frac{y_p f_p}{\left[\left(\sum_{p \in \Gamma_{dj}} y_p f_p\right) + x_{dj} c_{dj}\right]}$. The share value for the ridesharing passenger $p \in P$ ($y_p = 1$) transported by driver d is $\beta_p^P = \frac{\sigma_p^P y_p(1-\alpha)F_{dj}(x,y)}{F(x,y)}$.*
(3) *This scheme allocates $\sigma_d^D(1-\alpha)F_{dj}(x,y)$ to each ridesharing driver $d \in D$ ($x_{dj} = 1$ for some j), where $\sigma_d^D = \frac{x_{dj} c_{dj}}{\left[\left(\sum_{p \in \Gamma_{dj}} y_p f_p\right) + x_{dj} c_{dj}\right]}$.*

The share value for the ridesharing driver $d \in D$ ($x_{dj} = 1$ for some j) is

$$\beta_d^D = \frac{\sum_{j=1}^{J_d} x_{dj}\sigma_d^D\left(1-\alpha\right)F_{dj}(x,y)}{F(x,y)}.$$

Note that $\sum_{p=1}^{P} \beta_p^P + \sum_{d=1}^{D} \beta_d^D + \alpha = 1$.

The Global Proportional Scheme proposed in [13] allocates a portion of the overall cost savings to each participant according to the percentage of the participant's travel cost with respect to the overall travel cost. The Global Proportional Scheme is different from the Local Proportional Scheme defined previously. This scheme allocates $\alpha F(x, y)$ to the information provider and allocates the remaining cost savings to each ridesharing passenger and driver based on the weighting of their bid price with respect to overall bid price of all ridesharing passengers' and drivers' bids.

Definition 3. *Global Proportional Scheme*:

(1) *This scheme allocates $\alpha F(x, y)$ to the information provider. The share value for the information provider is α.*
(2) *This scheme allocates $\beta_p^P F(x, y)$ to the winning passenger $p \in P$ ($y_p = 1$), where the share value $\beta_p^P = \frac{(1-\alpha)y_p f_p}{\left[\left(\sum_{p=1}^{P} y_p f_p\right) + \left(\sum_{d=1}^{D}\sum_{j=1}^{J_d} x_{dj} c_{dj}\right)\right]}$.*
(3) *This scheme allocates $\beta_d^D F(x, y)$ to each winning driver $d \in D$ ($x_{dj} = 1$ for some j). The share value for driver $d \in D$ ($x_{dj} = 1$ for some j) is*

$$\beta_d^D = \frac{\sum_{j=1}^{J_d} (1-\alpha) x_{dj} c_{dj}}{\left[\left(\sum_{p=1}^{P} y_p f_p\right) + \left(\sum_{d=1}^{D}\sum_{j=1}^{J_d} x_{dj} c_{dj}\right)\right]}.$$

Note that $\sum_{p=1}^{P} \beta_p^P + \sum_{d=1}^{D} \beta_d^D + \alpha = 1$.

To compare different cost savings allocation schemes, it is important to study how many matched rides can satisfy cost savings expectations for relevant drivers and passengers. We define the minimal expected rewarding rate as follows to describe the minimal cost savings expectations of drivers and passengers.

Definition 4. *The minimal cost savings expectations for a ridesharing participant (either a driver or a passenger) are called the minimal expected rewarding rate, which is defined as the ratio of the cost savings to the travel cost of the participant. The minimal expected rewarding rate is denoted by r.*

In the real world, drivers or passengers may not be satisfied with the shared rides recommended by the shared mobility system due to low rewarding rate even if these rides can meet their transportation requirements. A driver or a passenger may choose either to accept or reject a ride found by the ride matching system. To assess effectiveness of different schemes to allocate cost savings, we distinguish acceptable shared rides from unacceptable ones.

Definition 5. *A shared ride is an acceptable ride if the rewarding rate for all the participants on the ride is greater than or equal to the minimal expected rewarding rate r. A shared ride is not an acceptable ride if the rewarding rate for any participant on the ride is less than the minimal expected rewarding rate.*

We will compare effectiveness of the Fifty-Fifty Scheme, Local Proportional Scheme and Global Proportional Scheme by analyzing the number of acceptable rides.

4. Comparison of the Fifty-Fifty Scheme and Local Proportional Scheme

In this section, we will compare the Fifty-Fifty Scheme and Local Proportional Scheme by analysis. Based on the analysis, a theorem will be stated to characterize the relation of these two schemes in terms of the number of acceptable rides.

A passenger or a driver will not accept to share a ride if any of their minimal cost savings expectations, i.e., the minimal expected rewarding rate, cannot be met. A shared ride is called an acceptable ride if the rewarding rate of the driver and the rewarding rate of the passenger are greater than or equal to the minimal expected rewarding rate. The comparison between cost savings allocation schemes is based on the number of acceptable shared rides. We first state two lemmas to pave the way for the establishment of a theorem that state the relation between the number of acceptable rides under the Fifty-Fifty Scheme and Local Proportional Scheme.

Lemma 1. *Suppose the original travel distance of the passenger $p \in \Gamma_{dj}$ is not greater than that of the driver d on a shared ride (with $x_{dj} = 1$ for some j). The rewarding rate for driver d (with $x_{dj} = 1$) under the Fifty-Fifty Scheme is less than or equal to that under the Local Proportional Scheme. The rewarding rate for passenger p in the shared ride is greater than or equal to the rewarding rate for driver d.*

Proof of Lemma 1. Under the Fifty-Fifty Scheme, the cost savings allocated to driver d in the ride corresponding to the j-th bid of driver d with $x_{dj} = 1$ is $\frac{(1-\alpha)F_{dj}(x,y)}{2}$. The rewarding rate for driver d is $\frac{(1-\alpha)F_{dj}(x,y)}{2c_{dj}}$. Under the Local Proportional Scheme, the cost savings allocated to driver d with $x_{dj} = 1$ is $\sigma_d^D(1-\alpha)F_{dj}(x,y) = \frac{c_{dj}(1-\alpha)F_{dj}(x,y)}{\left[\left(\sum_{p\in\Gamma_{dj}} y_p f_p\right)+x_{dj}c_{dj}\right]}$. The rewarding rate for driver d with $x_{dj} = 1$ is $\frac{(1-\alpha)F_{dj}(x,y)}{\left[\left(\sum_{p\in\Gamma_{dj}} y_p f_p\right)+x_{dj}c_{dj}\right]}$. In [12], it is assumed that there is at most one passenger transported by a driver; there is only one element in Γ_{dj}. Let p denote the passenger transported by driver d. Then $\left[\left(\sum_{p\in\Gamma_{dj}} y_p f_p\right)+x_{dj}c_{dj}\right] = y_p f_p + x_{dj}c_{dj} = f_p + c_{dj}$. As the original travel distance of the passenger $p \in \Gamma_{dj}$ is not greater than that of the driver in the shared ride, $f_p \leq c_{dj}$. Hence $f_p + c_{dj} \leq 2c_{dj}$.

Therefore, $\frac{(1-\alpha)F_{dj}(x,y)}{2c_{dj}} \leq \frac{(1-\alpha)F_{dj}(x,y)}{\left[\left(\sum_{p\in\Gamma_{dj}} y_p f_p\right)+x_{dj}c_{dj}\right]}$.

In this case, the rewarding rate for driver d with $x_{dj} = 1$ under the Fifty-Fifty Scheme is less than or equal to that under the Local Proportional Scheme.

Note that, as $f_p \leq c_{dj}$, $\frac{(1-\alpha)F_{dj}(x,y)}{2c_{dj}} \leq \frac{(1-\alpha)F_{dj}(x,y)}{2f_p}$.

Therefore, the rewarding rate for passenger p in the ride corresponding to the j-th bid of driver d with $x_{dj} = 1$ is greater than or equal to the rewarding rate for driver d. □

Lemma 2. *Suppose the original travel distance of passenger p is greater than that of the driver $d \in D$ in the shared ride corresponding to the j-th bid of driver d with $x_{dj} = 1$. The rewarding rate for passenger p under the Fifty-Fifty Scheme is less than that under the Local Proportional Scheme. The rewarding rate for driver d in the shared ride is greater than the rewarding rate for passenger p.*

Proof of Lemma 2. Under the Fifty-Fifty Scheme, the cost savings allocated to passenger p is $\frac{(1-\alpha)F_{dj}(x,y)}{2}$. The rewarding rate for passenger p is $\frac{(1-\alpha)F_{dj}(x,y)}{2f_p}$. Under the Local Proportional Scheme, the cost savings allocated to passenger p is $\sigma_p^P(1-\alpha)F_{dj}(x,y) = \frac{y_p f_p(1-\alpha)F_{dj}(x,y)}{\left[\left(\sum_{p\in\Gamma_{dj}} y_p f_p\right)+x_{dj}c_{dj}\right]}$.

The rewarding rate for passenger p is $\frac{y_p f_p(1-\alpha)F_{dj}(x,y)}{f_p\left[\left(\sum_{p\in\Gamma_{dj}} y_p f_p\right)+x_{dj}c_{dj}\right]} = \frac{y_p(1-\alpha)F_{dj}(x,y)}{\left[\left(\sum_{p\in\Gamma_{dj}} y_p f_p\right)+x_{dj}c_{dj}\right]}$.

In [12], it is assumed that there is at most one passenger transported by a driver; there is only one element in Γ_{dj}. Let p denote the passenger transported by driver d. Then $\left[\left(\sum_{p\in\Gamma_{dj}} y_p f_p\right) + x_{dj}c_{dj}\right] = y_p f_p + x_{dj}c_{dj} = f_p + c_{dj}$.

As $f_p > c_{dj}$, $2f_p > 2c_{dj} + f_p$.

Hence $\frac{(1-\alpha)F_{dj}(x,y)}{2f_p} < \frac{y_p(1-\alpha)F_{dj}(x,y)}{\left[\left(\sum_{p\in\Gamma_{dj}} y_p f_p\right)+x_{dj}c_{dj}\right]}$. In this case, the rewarding rate for passenger p under the Fifty-Fifty Scheme is less than that under the Local Proportional Scheme.

As $f_p > c_{dj}$, $\frac{(1-\alpha)F_{dj}(x,y)}{2c_{dj}} > \frac{(1-\alpha)F_{dj}(x,y)}{2f_p}$.

Therefore, the rewarding rate for driver d in the shared ride is greater than the rewarding rate for passenger p in the ride corresponding to the j-th bid of driver d with $x_{dj} = 1$. □

Theorem 1. *The number of acceptable shared rides under the Fifty-Fifty Scheme is not greater than the number of acceptable shared rides under the Local Proportional Scheme.*

Proof of Theorem 1. Under the Fifty-Fifty Scheme, the cost savings allocated to passenger p is $\frac{(1-\alpha)F_{dj}(x,y)}{2}$. The set of shared rides can be divided into two disjoint subsets: R_1 and R_2, where the original travel distance of the passenger is not greater than the original travel distance of the driver for each ride in R_1 and the original travel distance of the passenger is greater than the original travel distance of the driver for each ride in R_2. We first show that the number of acceptable shared rides in R_1 under the Fifty-Fifty Scheme is less than or equal to that under the Local Proportional Scheme. We then show the number of acceptable shared rides in R_2 under the Fifty-Fifty Scheme is also less than or equal to that under the Local Proportional Scheme.

(I) Whether a shared ride is acceptable is determined by the rewarding rate for the driver and the rewarding rate for the passenger in the ride. If either the rewarding rate for the driver or the rewarding rate for the passenger in the ride is less than the minimal expected rewarding rate r, the ride will not be accepted either by the driver or the passenger. The minimum of the rewarding rate for the driver and the rewarding rate for the passenger in a ride determines whether the ride is acceptable.

For a ride in R_1, as the travel distance of the passenger $p \in \Gamma_{dj}$ is not greater than that of the driver in a shared ride, according to Lemma 1, the rewarding rate for driver d with $x_{dj} = 1$ under the Fifty-Fifty Scheme is less than or equal to that under the Local Proportional Scheme and the rewarding rate for passenger p in the shared ride is greater than or equal to the rewarding rate for driver d. Thus, the minimum of the rewarding rate for the driver and the rewarding rate for the passenger in a ride in R_1 is the rewarding rate of the driver and is less than or equal to that under the Local Proportional Scheme. Therefore, if a ride in R_1 is accepted under the Fifty-Fifty Scheme, it must be accepted under the Local Proportional Scheme. As this holds for each ride in R_1, the number of acceptable shared rides in R_1 under the Fifty-Fifty Scheme is less than or equal to that under the Local Proportional Scheme.

(II) Next, we then show the number of acceptable shared rides in R_2 under the Fifty-Fifty Scheme is also less than or equal to that under the Local Proportional Scheme.

Whether a shared ride is acceptable is determined by the rewarding rate for the driver and the passenger in the ride. If either the rewarding rate for the driver or the passenger in the ride is less than the minimal expected rewarding rate r, the ride will not be accepted either by the driver or the passenger. The minimum of the rewarding rate for the driver and the rewarding rate for the passenger in a ride determine whether the ride is acceptable.

For a ride in R_2, as the travel distance of the passenger $p \in \Gamma_{dj}$ is greater than that of the driver in the shared ride, according to Lemma 2, the rewarding rate for passenger p under the Fifty-Fifty Scheme is less than that under the Local Proportional Scheme. The rewarding rate for driver d in the shared ride is greater than the rewarding rate for passenger p.

Thus, the minimum of the rewarding rate for the driver and the rewarding rate for the passenger in a ride in R_2 is the rewarding rate of the passenger and is less than that under the Local Proportional Scheme. Therefore, if a ride in R_2 is accepted under the Fifty-Fifty Scheme, it must be accepted under the Local Proportional Scheme. As this holds for each ride in R_2, the number of acceptable shared rides in R_2 under the Fifty-Fifty Scheme is less than or equal to that under the Local Proportional Scheme.

Based on the reasoning of (I), the number of acceptable shared rides in R_1 under the Fifty-Fifty Scheme is less than or equal to that under the Local Proportional Scheme. Based on the reasoning of (II), the number of acceptable shared rides in R_2 under the Fifty-Fifty Scheme is less than or equal to that under the Local Proportional Scheme. Therefore, the number of acceptable shared rides in $R_1 \cup R_2$ under the Fifty-Fifty Scheme is less than or equal to that under the Local Proportional Scheme. This completes the proof. □

5. Comparison of the Local Proportional Scheme and Global Proportional Scheme

In this section, we will compare the numbers of acceptable shared rides under the Local Proportional Scheme and Global Proportional Scheme. We present two theorems based on analysis. The first theorem in this section states that in case the rewarding rate under the Global Proportional Scheme is greater than or equal to the minimal expected rewarding rate, the number of acceptable shared rides under the Global Proportional Scheme will be greater than or equal to that under the Proportional Scheme. The next theorem in this section states that in case the rewarding rate under the Global Proportional Scheme is less than the minimal expected rewarding rate, the number of acceptable shared rides under the Local Proportional Scheme will be greater than or equal to that under the Global Proportional Scheme.

To characterize different cost savings allocation schemes, we define the concept of "all or nothing scheme" for cost savings allocation schemes.

Definition 6. *A cost savings allocation scheme is a local all or nothing scheme if the driver and passengers in the shared ride receive the same rewarding rate, but the driver and passengers in different shared ride may receive different rewarding rate under the scheme.*

Definition 7. *A cost savings allocation scheme is a global all or nothing scheme if all the drivers and passengers receive the same rewarding rate under the scheme.*

Based on these definitions, the Local Proportional Scheme proposed in [12] is a local all or nothing cost savings allocation scheme and the Global Proportional Scheme proposed in [13] is a global all or nothing scheme. In addition, the Fifty-Fifty Scheme is neither a local nor a global all or nothing cost savings allocation scheme as the driver and passenger in each ride may not receive the same rewarding rate under this scheme and all the drivers and passengers may not receive the same rewarding rate under the scheme.

Property 1. *Under a local all or nothing cost savings allocation scheme, the driver and passenger in each ride either both accept or reject the rewarding rate.*

Property 2. *Under a global all or nothing cost savings allocation scheme, all the drivers and passengers either accept or reject the rewarding rate.*

Before stating the main results in this section, we first introduce two lemmas as follows.

Lemma 3. *The minimal rewarding rate for passengers under Local Proportional Scheme is less than or equal to that under the Global Proportional Scheme.*

Proof of Lemma 3. Please refer Appendix A. □

Lemma 4. *The minimal rewarding rate for drivers under the Local Proportional Scheme is less than or equal to that under the Global Proportional Scheme.*

Proof of Lemma 4. Please refer Appendix B. □

Theorem 2. *If rewarding rate under the Global Proportional Scheme is greater than the minimal expected rewarding rate for all winning drivers and winning passengers, the number of acceptable shared rides under the Global Proportional Scheme is greater than or equal to that under the Local Proportional Scheme.*

Proof of Theorem 2. According to Property 2, the Global Proportional Scheme is an all or nothing cost savings allocation scheme. As the rewarding rate under the Global Proportional Scheme is greater than the minimal rewarding rate for all winning drivers and winning passengers, the number of acceptable shared rides under the Global Proportional Scheme is the number of all the shared rides for the solution (x, y). To show that the number of acceptable shared rides under the Global Proportional Scheme is greater than or equal to that of the Local Proportional Scheme, we show that there exists one or more rides in which either the rewarding rate of passengers or drivers is less than or equal to the Global Proportional Scheme. According to Lemma 3, the minimal rewarding rate for passengers under the Local Proportional Scheme is less than or equal to that of the Global Proportional Scheme. According to Lemma 4: The minimal rewarding rate for drivers under the Local Proportional Scheme is less than or equal to that of the Global Proportional Scheme. For the rides in which rewarding rate of passengers or drivers under the Local Proportional Scheme is less than or equal to Global Proportional Scheme, these rides may or may not be acceptable. However, these rides must be acceptable under the Global Proportional Scheme as the number of acceptable shared rides under the Global Proportional Scheme is all the shared rides for the solution (x, y). Therefore, the number of acceptable shared rides under the Global Proportional Scheme is greater than or equal to that of the Local Proportional Scheme. □

Before stating the next result, we first introduce the following lemma:

Lemma 5. *The maximal rewarding rate for passengers and drivers under the Local Proportional Scheme is greater than or equal to that of the Global Proportional Scheme.*

Proof of Lemma 5. Please refer Appendix C. □

Theorem 3. *If the rewarding rate is less than the minimal expected rewarding rate for all drivers and passengers under the Global Proportional Scheme, the number of acceptable shared rides under the Local Proportional Scheme is greater than or equal to that under the Global Proportional Scheme.*

Proof of Theorem 3. According to Property 2, the Global Proportional Scheme is an all or nothing cost savings allocation scheme. As the rewarding rate under the Global Proportional Scheme is less than the minimal rewarding rate for all winning drivers and winning passengers, the number of acceptable shared rides under the Global Proportional Scheme is zero for the solution (x, y). According to Lemma 5, the maximal rewarding rate for passengers and drivers under the Local Proportional Scheme is greater than or equal to that of the Global Proportional Scheme. Combining these facts, it follows that the number of acceptable shared rides of the Local Proportional Scheme is greater than or equal to that of the Global Proportional Scheme. □

6. Comparison of the Global Proportional Scheme and Fifty-Fifty Scheme

In this section, we will compare the numbers of acceptable shared rides under the Fifty-Fifty Scheme and Global Proportional Scheme under given minimal expected rewarding rate by analysis. Two theorems will be established to state the relation between the Global

Proportional Scheme and Fifty-Fifty Scheme in terms of the number of acceptable shared rides. One theorem characterizes the situation in which the number of acceptable shared rides under the Global Proportional Scheme dominates that under the Fifty-Fifty Scheme. The other theorem characterizes the situation in which the number of acceptable shared rides under the Global Proportional Scheme is dominated by that under the Fifty-Fifty Scheme.

Theorem 4. *If rewarding rate under the Global Proportional Scheme is greater than the minimal expected rewarding rate, the number of acceptable shared rides under the Global Proportional Scheme is greater than or equal to that of the Fifty-Fifty Scheme.*

Proof of Theorem 4. Under the condition that rewarding rate under the Global Proportional Scheme is greater than the minimal expected rewarding rate for all winning drivers and winning passengers, it follows from Theorem 2 that the number of acceptable shared rides under the Global Proportional Scheme is greater than or equal to that of the Local Proportional Scheme. According to Theorem 1, the number of acceptable shared rides of the Fifty-Fifty Scheme is not greater than that of the Local Proportional Scheme. Therefore, the number of acceptable shared rides under the Global Proportional Scheme is greater than or equal to that of the Fifty-Fifty Scheme under the condition that rewarding rate under the Global Proportional Scheme is greater than the minimal expected rewarding rate. □

Theorem 5. *If the rewarding rate under the Global Proportional Scheme is less than the minimal expected rewarding rate, the number of acceptable shared rides under the Fifty-Fifty Scheme is greater than or equal to that under the Global Proportional Scheme.*

Proof of Theorem 5. According to Property 2, the Global Proportional Scheme is an all or nothing cost savings allocation scheme. Under the condition that the rewarding rate under the Global Proportional Scheme is less than the minimal expected rewarding rate for all drivers and passengers, the number of acceptable shared rides of the Global Proportional Scheme is zero for the solution (x, y). As the number of acceptable shared rides of under the Fifty-Fifty Scheme must be non-negative, the number of acceptable shared rides under the Fifty-Fifty Scheme is greater than or equal to that under the Global Proportional Scheme.

To show that the number of acceptable rides may be greater than that under the Global Proportional Scheme, we characterize the number of acceptable shared rides under the Fifty-Fifty Scheme. We decompose the set of shared rides into two disjoint subsets: R_1 and R_2, where the original travel distance of the passenger is not greater than the original travel distance of the driver for each ride in R_1 and the original travel distance of the passenger is greater than the original travel distance of the driver for each ride in R_2. The set R_1 can be further decomposed into two disjoint subset, i.e., $R_1 = R_{11} \cup R_{12}$, where $R_{11} = \left\{\gamma_{dj} \middle| \gamma_{dj} \in R_1, \frac{(1-\alpha)F_{dj}(x,y)}{2c_{dj}} \geq r\right\}$ and $R_{12} = \left\{\gamma_{dj} \middle| \gamma_{dj} \in R_1, \frac{(1-\alpha)F_{dj}(x,y)}{2c_{dj}} < r\right\}$. That is, R_{11} is the set of shared rides in which the rewarding rate for the driver in each ride is greater than or equal to the minimal expected rewarding rate.

The set R_2 can be further decomposed into two disjoint subset, i.e., $R_2 = R_{21} \cup R_{22}$, where $R_{21} = \left\{\gamma_{dj} \middle| \gamma_{dj} \in R_2, \frac{(1-\alpha)F_{dj}(x,y)}{2f_p} \geq r\right\}$ and $R_{22} = \left\{\gamma_{dj} \middle| \gamma_{dj} \in R_2, \frac{(1-\alpha)F_{dj}(x,y)}{2f_p} < r\right\}$. That is, R_{21} is the set of shared rides in which the rewarding rate for the passenger in each ride is greater than or equal to the minimal expected rewarding rate.

We show that the number of acceptable shared rides under the Fifty-Fifty Scheme is $|R_{11} \cup R_{21}|$.

(i) We first show that all the rides in R_{12} are not acceptable rides and all the rides in R_{11} are acceptable rides.

As $R_{12} = \left\{\gamma_{dj} \middle| \gamma_{dj} \in R_1, \frac{(1-\alpha)F_{dj}(x,y)}{2c_{dj}} < r\right\}$, the rewarding rate for the driver in each ride in R_{12} is less than the minimal expected rewarding rate. Thus, all the shared rides in R_{12} are not acceptable. Next, we show that all the shared rides in $R_{11} = \left\{\gamma_{dj} \middle| \gamma_{dj} \in R_1, \frac{(1-\alpha)F_{dj}(x,y)}{2c_{dj}} \geq r\right\}$

are acceptable. To show this, we must show that the rewarding rate for the driver and the rewarding rate for the passenger in each ride are greater than or equal to the minimal expected rewarding rate. By definition, $R_{11} = \{\gamma_{dj} \big| \gamma_{dj} \in R_1, \frac{(1-\alpha)F_{dj}(x,y)}{2c_{dj}} \geq r\}$. The rewarding rate for the driver in each ride in R_{11} is obviously greater than or equal to the minimal expected rewarding rate r. As $R_{11} \subseteq R_1$ and the original travel distance of the passenger is not greater than the original travel distance of the driver for each ride in R_1, therefore $f_p \leq c_{dj}$. Therefore, $\frac{(1-\alpha)F_{dj}(x,y)}{2f_p} \geq \frac{(1-\alpha)F_{dj}(x,y)}{2c_{dj}}$. That is, the rewarding rate for the passenger in each ride in R_{11} is greater than or equal to the rewarding rate for driver d. Thus, all the shared rides in $R_{11} = \{\gamma_{dj} \big| \gamma_{dj} \in R_1, \frac{(1-\alpha)F_{dj}(x,y)}{2c_{dj}} \geq r\}$ are acceptable.

(ii) Similar to the proof of (i), we can show that all the rides in R_{22} are not acceptable rides and all the rides in R_{21} are acceptable rides.

Based on (i) and (ii), the set of acceptable shared rides under the Fifty-Fifty Scheme is $R_{11} \cup R_{21}$. Therefore, the number of acceptable rides under the Fifty-Fifty Scheme is $|R_{11} \cup R_{21}|$.

If $|R_{11} \cup R_{21}|$ is greater than zero, the number of acceptable shared rides under the Fifty-Fifty Scheme is greater than that under the Global Proportional Scheme.

This completes the proof. □

7. Results

The theoretical results presented in the previous sections will be verified by examples in this section. In this section, we conduct experiments by apply different schemes to allocate cost savings for several test cases. The test cases are generated based on a geographical area in Taiwan. For the given number of drivers and passengers in each test case, we create the test case data by randomly selecting the origins and destinations for drivers and passengers in Taichung City. The area of Taichung City is 2215 km^2. Eight test cases (Case 1 through Case 8) are generated to verify the theoretical results. Let us use one test case (Case 2) to illustrate a typical application scenario. For Case 2, there are three drivers and ten passengers. The origins and destinations generated for three drivers and ten passengers are shown in Table 2. The travel distance of the three drivers' routes fall within the range of 15 to 25 kilometers. The bids of passengers are shown in Table 3, where all the IDs of passengers are numbered from 1 to 10 and only s_{pk}^1 and s_{pk}^2 with nonzero values are shown. The bids of drivers are shown in Table 4, only q_{djk}^1 and q_{djk}^2 with nonzero values are shown. The results obtained by the decision model are displayed on a map in Figure 1, where the origin and destination of the route for driver i are denoted by Di+ and Di-, respectively, and the origin and destination of the route for passenger i are denoted by Pi+ and Pi-. Temporal data of itineraries are not shown in the tables to avoid occupying too much space in this paper. All test cases are randomly generated. The data of all test cases are available for download from the following link: https://drive.google.com/drive/folders/1No2d0lMwZz8zV5mF0yjfmKbSCCgON-38?usp=sharing (accessed on 20 August 2021).

The algorithm of the software used in the decision model is described in [13]. It is implemented in Java. We apply the algorithm to solve the ridesharing optimization problem to maximize the overall cost savings first and then apply the Fifty-Fifty Scheme, Local Proportional Scheme and Global Proportional Scheme to allocate cost savings. The cost savings optimization algorithm only determines the set of rides and relevant drivers/passengers for ridesharing. The rewarding rates due to cost savings allocated by the Fifty-Fifty Scheme, Local Proportional Scheme and Global Proportional Scheme for $\alpha = 0$ are summarized in Tables 5–8.

Table 2. Origins and destinations of participants.

Participant	Origin	Destination
Driver 1	24.23115, 120.57268	24.161486, 120.7313603
Driver 2	24.110586, 120.6839043	24.20659, 120.62886
Driver 3	24.06212, 120.69823	24.2223294, 120.690032
Passenger 1	24.17428, 120.64454	24.17337, 120.6714
Passenger 2	24.11314, 120.66166	24.1790507, 120.6657476
Passenger 3	24.09888, 120.70233	24.14285, 120.73275
Passenger 4	24.11637, 120.66388	24.13536, 120.70114
Passenger 5	24.14891, 120.66295	24.1142, 120.65267
Passenger 6	24.11666, 120.66595	24.152798, 120.6640863
Passenger 7	24.11308, 120.65914	24.13704, 120.67637
Passenger 8	24.114, 120.69139	24.118, 120.62749
Passenger 9	24.07937, 120.67911	24.14891, 120.66295
Passenger 10	24.16471, 120.66779	24.12877, 120.66223

Table 3. Bids submitted by passengers.

Passenger ID (p)	s_{pk}^1, s_{pk}^2	f_p
1	$s_{p1}^1 = s_{p1}^2 = 1$	8.4775
2	$s_{p2}^1 = s_{p2}^2 = 1$	23.1125
3	$s_{p3}^1 = s_{p3}^2 = 1$	18.002499
4	$s_{p4}^1 = s_{p4}^2 = 1$	13.3575
5	$s_{p5}^1 = s_{p5}^2 = 1$	11.47
6	$s_{p6}^1 = s_{p6}^2 = 1$	11.885
7	$s_{p7}^1 = s_{p7}^2 = 1$	9.8725
8	$s_{p8}^1 = s_{p8}^2 = 1$	22.4225
9	$s_{p9}^1 = s_{p9}^2 = 1$	24.3375
10	$s_{p10}^1 = s_{p10}^2 = 1$	12.9475

Table 4. Bid submitted by Driver 1.

Driver ID (d)	q_{djk}^1, q_{djk}^2	o_{d1}	c_{d1}
1	$q_{d11}^1 = q_{d11}^2 = 1$	59.1025	59.1025
2	$q_{d16}^1 = q_{d16}^2 = 1$	39.8075	39.8075
3	$q_{d19}^1 = q_{d19}^2 = 1$	59.0225	59.0225

For Test Case 1, there is only one shared ride and only one driver–passenger pair. The rewarding rates for the driver and the passenger are the same for the Local Proportional Scheme and Global Proportional Scheme in Test Case 1. The rewarding rates for the driver and the passenger are different for the Fifty-Fifty Scheme. For Test Case 2, the rewarding rates for all drivers and passengers are the same in the Global Proportional Scheme. The rewarding rates for the driver and passenger in a shared ride are the same under the Local Proportional Scheme. However, the rewarding rates for the driver and passenger in different rides are different under the Local Proportional Scheme. The rewarding rates for the driver and passenger in the same ride are different under the Fifty-Fifty Scheme.

Figure 1. A ridesharing scenario.

Table 5. Rewarding rate of Case 1 and Case 2 for $\alpha = 0$, where FF: Fifty-Fifty Scheme, LP: Local Proportional Scheme and GP: Global Proportional Scheme.

Case 1	Participant	FF	LP	GP	Case 2	Participant	FF	LP	GP
Ride 1	D1	0.072	0.12	0.12	Ride 1	D1	0.072	0.125	0.221
	P1	0.358	0.12	0.12		P1	0.5	0.125	0.221
					Ride 2	D2	0.149	0.23	0.221
						P2	0.5	0.23	0.221
					Ride 3	D3	0.206	0.292	0.221
						P3	0.5	0.292	0.221

Table 6. Rewarding rate of Case 3 and Case 4 for $\alpha = 0$.

Case 3	Participant	FF	LP	GP	Case 4	Participant	FF	LP	GP
Ride 1	D1	0.127	0.199	0.175	Ride 1	D1	0.162	0.245	0.22
	P1	0.461	0.199	0.175		P1	0.5	0.245	0.22
Ride 2	D2	0.064	0.103	0.175	Ride 2	D2	0.036	0.057	0.22
	P2	0.271	0.103	0.175		P2	0.139	0.057	0.22
Ride 3	D3	0.128	0.204	0.175	Ride 3	D3	0.295	0.371	0.22
	P3	0.5	0.204	0.175		P3	0.5	0.371	0.22

Table 7. Rewarding rate of Case 5 and Case 6 for $\alpha = 0$.

Case 5	Participant	FF	LP	GP	Case 6	Participant	FF	LP	GP
Ride 1	D1	0.194	0.279	0.217	Ride 1	D1	0.129	0.202	0.21
	P1	0.5	0.279	0.217		P1	0.457	0.202	0.21
Ride 2	D2	0.085	0.145	0.217	Ride 2	D2	0.072	0.126	0.21
	P2	0.5	0.145	0.217		P2	0.396	0.126	0.21
Ride 3	D3	0.217	0.234	0.217	Ride 3	D3	0.181	0.266	0.21
	P3	0.5	0.234	0.217		P3	0.5	0.266	0.21
					Ride 4	D4	0.183	0.268	0.21
						P4	0.5	0.268	0.21

Table 8. Rewarding rate of Case 7 and Case 8 for $\alpha = 0$.

Case 7	Participant	FF	LP	GP	Case 8	Participant	FF	LP	GP
Ride 1	D1	0.181	0.254	0.246	Ride 1	D1	0.163	0.233	0.21
	P1	0.422	0.254	0.246		P1	0.409	0.233	0.21
Ride 2	D2	0.168	0.251	0.246	Ride 2	D2	0.153	0.235	0.21
	P2	0.5	0.251	0.246		P2	0.5	0.235	0.21
Ride 3	D3	0.089	0.151	0.246	Ride 3	D3	0.204	0.29	0.21
	P3	0.5	0.151	0.246		P3	0.5	0.29	0.21
Ride 4	D4	0.096	0.161	0.246	Ride 4	D4	0.142	0.221	0.21
	P4	0.5	0.161	0.246		P4	0.5	0.221	0.21
Ride 5	D5	0.268	0.349	0.246	Ride 5	D5	0.09	0.152	0.21
	P5	0.5	0.349	0.246		P5	0.5	0.152	0.21
					Ride 6	D6	0.02	0.031	0.21
						P6	0.65	0.031	0.21
					Ride 7	D7	0.123	0.211	0.21
						P7	0.5	0.211	0.21

For Test Case 3 through Test Case 8, the rewarding rates for all drivers and passengers are the same under the Global Proportional Scheme. The rewarding rates for the driver and passenger in a shared ride are the same under the Local Proportional Scheme. However, the rewarding rates for the driver and passenger in different rides are different under the Local Proportional Scheme. The rewarding rates for the driver and passenger in the same ride are different under the Fifty-Fifty Scheme.

To compare the effectiveness of cost savings allocation schemes, we need to find the number of acceptable rides under the minimal rewarding rate, r, for drivers and passengers. Suppose $\alpha = 0$ and $r = 0.1$. In this case, a ride will be accepted as a successful shared ride only if the rewarding rates for the driver and the passenger are jointly greater than or equal to 0.1. For the results in Tables 5–8, the numbers of acceptable rides are summarized in Table 9 for $\alpha = 0$ and $r = 0.1$.

Table 9. Number of successful share rides for ($\alpha = 0$, $r = 0.1$) and ($\alpha = 0$, $r = 0.2$).

$\alpha = 0, r = 0.1$					$\alpha = 0, r = 0.2$				
Case	**Participant**	**FF**	**LP**	**GP**	**Case**	**Participant**	**FF**	**LP**	**GP**
1	D1	0	1	1	1	D1	0	0	0
2	D2	2	3	3	2	D2	1	2	3
3	D3	2	3	3	3	D3	0	1	0
4	D4	2	2	3	4	D4	1	2	3
5	D5	2	3	3	5	D5	0	2	3
6	D6	3	4	4	6	D6	0	3	4
7	D7	3	5	5	7	D7	1	3	5
8	D7	5	6	7	8	D7	1	6	7

According to Table 9, the numbers of acceptable rides under the Local Proportional Scheme are greater than or equal to those under the Fifty-Fifty Scheme for all test cases. These results are consistent with Theorem 1 from our previous analysis. The numbers of acceptable rides of the Global Proportional Scheme are greater than or equal to those under the Fifty-Fifty Scheme according to Table 9. These results are consistent with the theoretical result of Theorem 4 in this paper. Additionally, the numbers of acceptable rides of the Global Proportional Scheme are greater than or equal to those under the Local Proportional Scheme according to Table 9. These results are consistent with the theoretical result of Theorem 2.

To compare the performance of all test cases for $\alpha = 0$ and $r = 0.1$ under the three schemes, a bar chart based on the experimental results is shown in Figure 2a. It clearly indicates that the Global Proportional Scheme either performs as well as or the same as the other two schemes for all cases.

Figure 2. A comparison of the number of acceptable rides for different schemes under (**a**) $\alpha = 0$, $r = 0.1$ and (**b**) $\alpha = 0$, $r = 0.2$.

Suppose $\alpha = 0$ and $r = 0.2$. The numbers of acceptable rides are summarized in Table 9. According to Table 9, the numbers of acceptable rides of the Local Proportional Scheme are greater than or equal to those of the Fifty-Fifty Scheme for all test cases. These results are consistent with Theorem 1 from our previous analysis. For Case 2, Case 4, Case 5, Case 6, Case 7 and Case 8, the numbers of acceptable rides of the Global Proportional Scheme are greater than zero. Therefore, the numbers of acceptable rides of the Global Proportional Scheme are greater than or equal to those of the Fifty-Fifty Scheme according to Table 9. These results are consistent with the theoretical result of Theorem 4 in this paper. For Case 2, Case 4, Case 5, Case 6, Case 7 and Case 8, the numbers of acceptable rides of the Global Proportional Scheme are greater than zero. Therefore, the numbers of acceptable rides of the Global Proportional Scheme are greater than or equal to those of the Local Proportional Scheme according to Table 9. These results are consistent with the theoretical result of Theorem 2 in this paper.

For Case 1 and Case 3, the numbers of acceptable rides of the Global Proportional Scheme are zero. Therefore, the numbers of acceptable rides of the Global Proportional Scheme are less than or equal to those of the Fifty-Fifty Scheme according to Table 9. These results are consistent with the theoretical result of Theorem 5 developed in this paper. For Case 1 and Case 3, the numbers of acceptable rides of the Global Proportional Scheme are zero. Therefore, the numbers of acceptable rides of the Global Proportional Scheme are less than or equal to those of the Local Proportional Scheme according to Table 9. These results are consistent with the theoretical result of Theorem 3 in this paper.

To compare the performance of all test cases for $\alpha = 0$ and $r = 0.2$ under the three schemes, a bar chart based on the experimental results is shown in Figure 2b. It clearly indicates that the Global Proportional Scheme either performs as well as or the same as the other two schemes for Case 2, Case 4, Case 5, Case 6, Case 7 and Case 8 (Theorem 2 and Theorem 4). The Global Proportional Scheme, however, is not better than the other two schemes for Case 1 and Case 3 (Theorem 3 and Theorem 5).

Suppose $\alpha = 0.2$ and $r = 0.1$. In this case, a ride will be accepted as an acceptable shared ride only if the rewarding rates for the driver and the passenger are jointly greater than or equal to 0.1. For the results in Tables 10–13, the numbers of acceptable rides are summarized in Table 14 for $\alpha = 0.2$ and $r = 0.1$. According to Table 14, the numbers of acceptable rides under the Local Proportional Scheme are greater than or equal to those under the Fifty-Fifty Scheme for all test cases. These results are consistent with Theorem 1 from our previous analysis. The numbers of acceptable rides under the Global Proportional Scheme are greater than or equal to those under the Fifty-Fifty Scheme according to Table 14. These results are consistent with the theoretical result of Theorem 4 developed in this paper. Additionally, the numbers of acceptable rides under the Global Proportional Scheme are greater than or equal to those under the Local Proportional Scheme according to Table 14. These results are consistent with the theoretical result of Theorem 2.

Table 10. Rewarding rate of Case 1 and Case 2 for $\alpha = 0.2$.

Case 1	Participant	FF	LP	GP	Case 2	Participant	FF	LP	GP
Ride 1	D1	0.058	0.096	0.096	Ride 1	D1	0.057	0.1	0.176
	P1	0.286	0.096	0.096		P1	0.4	0.1	0.176
					Ride 2	D2	0.119	0.184	0.176
						P2	0.4	0.184	0.176
					Ride 3	D3	0.165	0.234	0.176
						P3	0.4	0.234	0.176

Table 11. Rewarding rate of Case 3 and Case 4 for $\alpha = 0.2$.

Case 3	Participant	FF	LP	GP	Case 4	Participant	FF	LP	GP
Ride 1	D1	0.102	0.159	0.14	Ride 1	D1	0.13	0.196	0.176
	P1	0.369	0.159	0.14		P1	0.4	0.196	0.176
Ride 2	D2	0.051	0.082	0.14	Ride 2	D2	0.029	0.045	0.176
	P2	0.217	0.082	0.14		P2	0.111	0.045	0.176
Ride 3	D3	0.102	0.163	0.14	Ride 3	D3	0.236	0.297	0.176
	P3	0.4	0.163	0.14		P3	0.4	0.297	0.176

Table 12. Rewarding rate of Case 5 and Case 6 for $\alpha = 0.2$.

Case 5	Participant	FF	LP	GP	Case 6	Participant	FF	LP	GP
Ride 1	D1	0.155	0.224	0.174	Ride 1	D1	0.103	0.161	0.168
	P1	0.4	0.224	0.174		P1	0.365	0.161	0.168
Ride 2	D2	0.068	0.116	0.174	Ride 2	D2	0.058	0.101	0.168
	P2	0.4	0.116	0.174		P2	0.397	0.101	0.168
Ride 3	D3	0.122	0.187	0.174	Ride 3	D3	0.145	0.213	0.168
	P3	0.4	0.187	0.174		P3	0.4	0.213	0.168
					Ride 4	D4	0.146	0.214	0.168
						P4	0.4	0.214	0.168

The numbers of acceptable rides are summarized in Table 14 for $\alpha = 0.2$ and $r = 0.2$. According to Table 14, the numbers of acceptable rides under the Local Proportional Scheme are greater than or equal to those under the Fifty-Fifty Scheme for all test cases. These results are consistent with Theorem 1 from our previous analysis.

For all cases, the numbers of acceptable rides under the Global Proportional Scheme are zero. Therefore, the numbers of acceptable rides under the Global Proportional Scheme are less than or equal to those under the Fifty-Fifty Scheme according to Table 14. These results are consistent with the theoretical result of Theorem 5 in this paper. For all cases, the numbers of acceptable rides under the Global Proportional Scheme are zero. Therefore, the numbers of acceptable rides under the Global Proportional Scheme are less than or equal to those under the Local Proportional Scheme according to Table 14. These results are consistent with the theoretical result of Theorem 3 in this paper.

Table 13. Rewarding rate of Case 7 and Case 8 for $\alpha = 0.2$.

Case 7	Participant	FF	LP	GP	Case 8	Participant	FF	LP	GP
Ride 1	D1	0.145	0.203	0.197	Ride 1	D1	0.131	0.187	0.152
	P1	0.337	0.203	0.197		P1	0.328	0.187	0.152
Ride 2	D2	0.134	0.201	0.197	Ride 2	D2	0.123	0.188	0.152
	P2	0.4	0.201	0.197		P2	0.4	0.188	0.152
Ride 3	D3	0.071	0.121	0.197	Ride 3	D3	0.163	0.232	0.152
	P3	0.4	0.121	0.197		P3	0.4	0.232	0.152
Ride 4	D4	0.077	0.129	0.197	Ride 4	D4	0.114	0.177	0.152
	P4	0.4	0.129	0.197		P4	0.4	0.177	0.152
Ride 5	D5	0.214	0.279	0.197	Ride 5	D5	0.072	0.122	0.152
	P5	0.4	0.279	0.197		P5	0.4	0.122	0.152
					Ride 6	D6	0.016	0.025	0.152
						P6	0.052	0.025	0.152
					Ride 7	D7	0.107	0.169	0.152
						P7	0.4	0.169	0.152

Table 14. Number of successful share rides for (a) $\alpha = 0.2$, $r = 0.1$ and (b) $\alpha = 0.2$, $r = 0.2$.

$\alpha = 0.2, r = 0.1$					$\alpha = 0.2, r = 0.2.$				
Case	Participant	FF	LP	GP	Case	Participant	FF	LP	0
1	D1	0	0	0	2	D2	0	1	0
2	D2	2	3	3	3	D3	0	0	0
3	D3	2	2	3	4	D4	1	1	0
4	D4	2	2	3	5	D5	0	1	0
5	D5	2	3	3	6	D6	0	2	0
6	D6	3	4	4	7	D7	1	3	0
7	D7	3	5	5	8	D7	0	1	0
8	D7	5	6	7					

To compare the performance of all test cases for $\alpha = 0.2$ and $r = 0.1$ under the three schemes, a bar chart based on the experimental results is shown in Figure 3b. It clearly indicates that the Global Proportional Scheme either performs as well as or the same as the other two schemes.

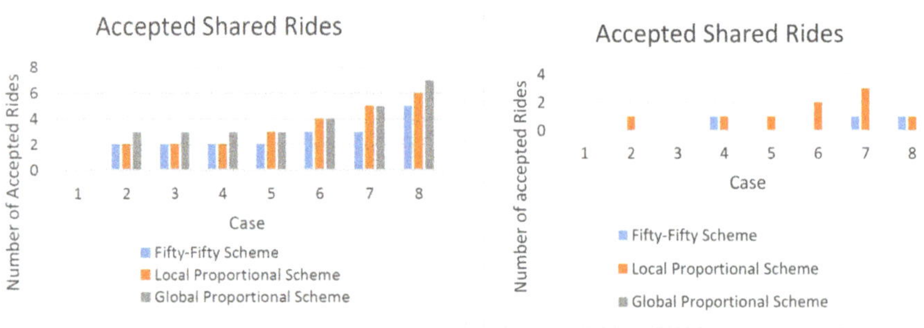

(a) (b)

Figure 3. A comparison of the number of acceptable rides for different schemes under (**a**) $\alpha = 0.2$, $r = 0.1$ and (**b**) $\alpha = 0.2$, $r = 0.2$.

To compare the performance of all test cases for $\alpha = 0.2$ and $r = 0.2$ under the three schemes, a bar chart based on the experimental results is shown in Figure 3. It clearly indicates that the Global Proportional Scheme is either the worst among all three schemes or not better than the other two schemes as the number of acceptable rides is zero for all cases (Theorems 3 and 5).

8. Discussion

The number of acceptable shared rides is an important performance index for assessing shared mobility systems. In a shared mobility system, drivers and passengers will share rides only if they can benefit from cost savings significantly. Depending on the scheme used to allocate cost savings, the benefits of individual participants vary. Adoption of different schemes to allocate cost savings will not only influence the cost savings for drivers and passengers but also the number of acceptable shared rides, which is closely related to the number of ridesharing passengers. An important issue is to study the influence of different schemes to allocate cost savings on the number of acceptable shared rides.

Development of a decision support system for shared mobility is usually broken down into two subproblems: (a) a ridesharing optimization subproblem to match potential drivers and passengers and (b) a cost savings allocation subproblem. In the literature, although the ridesharing optimization problem has been studied extensively and there are many schemes to allocate cost savings in ridesharing systems, the effectiveness of different schemes to allocate cost savings is less explored. In practice, simple rules have been used to solve the cost savings allocation problem based on proportional methods [11–13] to reduce computational complexity. Improper allocation of cost savings may lead to dissatisfaction of drivers and passengers and reduce the number of shared rides that can be accepted by the drivers and passengers. However, effectiveness of these proportional cost savings allocation schemes in a shared mobility system needs to be studied. The influence of applying different schemes to allocate cost savings on the number of acceptable shared rides is an important issue in the design of shared mobility systems.

In this study, we explore effectiveness of the three different proportional schemes proposed in [11–13] by theoretical analysis. Shared rides recommended by shared mobility systems are typically optimized for objective functions such as overall travel distance. Therefore, not all recommended shared rides are satisfactory or acceptable for drivers and passengers. The analysis in this paper characterizes several properties and conditions under which one proportional cost savings allocation scheme outperform another in terms of the number of rides acceptable by the drivers and passengers. These properties provide a foundation for selecting the proper proportional scheme to allocate cost savings. These properties are supported by experimental results of several test cases provided in this paper. Our theoretical analysis and properties are consistent with the experimental results. Although our analysis indicates that, among the three schemes analyzed in this study, none always dominates the other two under any situation, but a dominating property does exist under a specific situation. These properties provide the information about the influence of cost savings allocation schemes on the number of acceptable shared rides. If the minimal expected rewarding rate can be satisfied under the Global Proportional Scheme, then the Global Proportional Scheme is the best scheme among the three schemes studied in this paper. Otherwise, the Local Proportional Scheme is the best scheme among the three schemes. The Fifty-Fifty Scheme should not be used due to its poor performance.

The properties and results presented in this study are based on the assumption that drivers and passengers will not accept ridesharing in case the minimal expected rewarding rate cannot be attained. Several studies [31,32] indicate that cost savings are one important factor for ridesharing. Therefore, the results of these studies justify this assumption of this paper. The theory presented in this study does not hinge on the assumption that "travel cost is proportional to distance", which is used in [11,12]. The decision model adopted in this study consists of a practical general framework and procedures that consider spatial, time, capacity and detour ratio [13], and can consider even social factors [22]. The discussions above indicate that theory developed in this study is based on a problem setting in actual practice and also illustrate the validity and practicality of the theory developed in this study.

The results of this study generate interesting and important research questions: Do any other rules exist that are similar or dissimilar to the ones uncovered in this paper among other cost allocation schemes? Do any other rules exist among other cost allocation schemes and the three schemes studied in this paper? These open problems call for further studies.

9. Conclusions

Shared mobility is a transportation mode motivated by sharing vehicles to reduce the cost of journey, fuel consumption, emissions and air pollution. As cost savings are the primary incentive for the shared mobility mode, most studies on shared mobility focus on optimization of the travel distance to reduce cost. Due to computational complexity, the issue to allocate cost savings among stakeholders is often done by applying proportional methods such as the Fifty-Fifty Scheme, Local Proportional Scheme and Global Proportional Scheme, which are easy to implement in shared mobility systems. However, effectiveness of these proportional methods has not been analyzed and compared in the literature. In addition, there is no guideline on how to select the proper proportional method to allocate cost savings and how different proportional methods to allocate cost savings influence performance or acceptance of shared rides. This study provides a theoretical study on comparison of three different proportional methods in the literature for allocation of cost savings. Our study is based on the fact that potential drivers and passengers may reject the shared rides recommended by the shared mobility system if the rewarding rates are lower than their expectations. As the proportional method used to allocate cost savings among drivers and passengers directly influences the rewarding rates, the number of acceptable rides will be influenced by the method used to allocate cost savings. Therefore, we focus on comparing the number of acceptable rides under given minimal expected rewarding rate of the potential drivers and passengers. Through our analysis, we find several useful properties that can serve as a guideline for selecting a proper method from the Fifty-Fifty Scheme, Local Proportional Scheme and Global Proportional Scheme. These properties indicate that if the rewarding rate under the Global Proportional Scheme is greater than or equal to minimal expected rewarding rate, the Global Proportional Scheme is the best among the three proportional methods. Otherwise, the Fifty-Fifty Scheme and Local Proportional Scheme will be better than the Global Proportional Scheme. The properties established in this study also indicate that the Local Proportional Scheme is better or no worse than the Fifty-Fifty Scheme. These properties provide insights into which proportional method should be used to achieve more acceptable shared rides. Although the three schemes studied in this paper are often adopted in industry and academia, practitioners and researchers using these three schemes in ridesharing systems are not aware of their influence on acceptance of shared rides. The results of this study uncover some useful rules for selecting the three schemes to allocate cost savings. These results can be applied by policy makers and service providers to use the correct scheme to achieve more acceptable shared rides by following the rules: (1) the Fifty-Fifty Scheme should not be used, (2) choose the Global Proportional Scheme if the rewarding rate under the Global Proportional Scheme is greater than or equal to the minimal expected rewarding rate and (3) choose the Local Proportional Scheme if the rewarding rate under the Global Proportional Scheme is less the minimal expected rewarding rate. These rules can be easily implemented. An interesting future research direction will be to study and analyze effectiveness of other methods used for allocation of cost savings in the context of shared mobility systems to discover other useful rules to achieve more acceptable shared rides. Results call for further studies in this interesting and challenging research direction.

Funding: This research was supported in part by the Ministry of Science and Technology, Taiwan, under Grant MOST 110-2410-H-324-001.

Institutional Review Board Statement: Not applicable.

Informed Consent Statement: Not applicable.

Data Availability Statement: Data available in a publicly accessible repository described in the article.

Acknowledgments: The author would like to thank the anonymous reviewers and the editors for their comments and suggestions, which are invaluable for the author to improve presentation, clarity and quality of this paper.

Conflicts of Interest: The author declares no conflict of interest.

Appendix A

Proof of Lemma 3. To show the minimal rewarding rate for passengers under the Local Proportional Scheme is less than or equal to that of the Global Proportional Scheme, we must show that

$$\frac{(1-\alpha)y_pF(x,y)}{\left[\left(\sum\limits_{p=1}^{P} y_pf_p\right)+\left(\sum\limits_{d=1}^{D}\sum\limits_{j=1}^{J_d} x_{dj}c_{dj}\right)\right]} \geq \frac{y_p(1-\alpha)F_{d^*j^*}(x,y)}{\left[\left(\sum\limits_{p\in\Gamma_{d^*j^*}} y_pf_p\right)+x_{d^*j^*}c_{d^*j^*}\right]}.$$

Let d^*, j^* be the winning bid j^* of driver d^* such that $(d^*,j^*) = \arg\min\limits_{(d,j)} \frac{y_p(1-\alpha)F_{dj}(x,y)}{\left[\left(\sum\limits_{p\in\Gamma_{dj}} y_pf_p\right)+x_{dj}c_{dj}\right]}$.

Note that $\frac{y_p(1-\alpha)F_{dj}(x,y)}{\left[\left(\sum\limits_{p\in\Gamma_{dj}} y_pf_p\right)+x_{dj}c_{dj}\right]} \geq \frac{y_p(1-\alpha)F_{d^*j^*}(x,y)}{\left[\left(\sum\limits_{p\in\Gamma_{d^*j^*}} y_pf_p\right)+x_{d^*j^*}c_{d^*j^*}\right]}$.

$$\sum_{d=1}^{D}\sum_{j=1}^{J_d} y_p(1-\alpha)F_{dj}(x,y) \geq \sum_{d=1}^{D}\sum_{j=1}^{J_d}\left\{\frac{y_p(1-\alpha)F_{d^*j^*}(x,y)\left[\left(\sum\limits_{p\in\Gamma_{dj}} y_pf_p\right)+x_{dj}c_{dj}\right]}{\left[\left(\sum\limits_{p\in\Gamma_{d^*j^*}} y_pf_p\right)+x_{d^*j^*}c_{d^*j^*}\right]}\right\}$$

$$\frac{\sum\limits_{d=1}^{D}\sum\limits_{j=1}^{J_d} y_p(1-\alpha)F_{dj}(x,y)}{\left[\left(\sum\limits_{p=1}^{P} y_pf_p\right)+\left(\sum\limits_{d=1}^{D}\sum\limits_{j=1}^{J_d} x_{dj}c_{dj}\right)\right]} \geq \sum_{d=1}^{D}\sum_{j=1}^{J_d}\left\{\frac{y_p(1-\alpha)F_{d^*j^*}(x,y)\left[\left(\sum\limits_{p\in\Gamma_{dj}} y_pf_p\right)+x_{dj}c_{dj}\right]}{\left[\left(\sum\limits_{p\in\Gamma_{d^*j^*}} y_pf_p\right)+x_{d^*j^*}c_{d^*j^*}\right]\left[\left(\sum\limits_{p=1}^{P} y_pf_p\right)+\left(\sum\limits_{d=1}^{D}\sum\limits_{j=1}^{J_d} x_{dj}c_{dj}\right)\right]}\right\}$$

As $\sum\limits_{d=1}^{D}\sum\limits_{j=1}^{J_d} y_p(1-\alpha)F_{dj}(x,y) = y_p(1-\alpha)F(x,y)$,

$$\frac{y_p(1-\alpha)F(x,y)}{\left[\left(\sum\limits_{p=1}^{P} y_pf_p\right)+\left(\sum\limits_{d=1}^{D}\sum\limits_{j=1}^{J_d} x_{dj}c_{dj}\right)\right]} \geq \sum_{d=1}^{D}\sum_{j=1}^{J_d}\left\{\frac{y_p(1-\alpha)F_{d^*j^*}(x,y)\left[\left(\sum\limits_{p\in\Gamma_{dj}} y_pf_p\right)+x_{dj}c_{dj}\right]}{\left[\left(\sum\limits_{p\in\Gamma_{d^*j^*}} y_pf_p\right)+x_{d^*j^*}c_{d^*j^*}\right]\left[\left(\sum\limits_{p=1}^{P} y_pf_p\right)+\left(\sum\limits_{d=1}^{D}\sum\limits_{j=1}^{J_d} x_{dj}c_{dj}\right)\right]}\right\}$$

Note that

$$\sum_{d=1}^{D}\sum_{j=1}^{J_d}\left\{\frac{y_p(1-\alpha)F_{d^*j^*}(x,y)\left[\left(\sum_{p\in\Gamma_{dj}} y_pf_p\right)+x_{dj}c_{dj}\right]}{\left[\left(\sum_{p\in\Gamma_{d^*j^*}} y_pf_p\right)+x_{d^*j^*}c_{d^*j^*}\right]\left[\left(\sum_{p=1}^{P} y_pf_p\right)+\left(\sum_{d=1}^{D}\sum_{j=1}^{J_d} x_{dj}c_{dj}\right)\right]}\right\}=\frac{y_p(1-\alpha)F_{d^*j^*}(x,y)\sum_{d=1}^{D}\sum_{j=1}^{J_d}\left\{\left[\left(\sum_{p\in\Gamma_{dj}} y_pf_p\right)+x_{dj}c_{dj}\right]\right\}}{\left[\left(\sum_{p\in\Gamma_{d^*j^*}} y_pf_p\right)+x_{d^*j^*}c_{d^*j^*}\right]\left[\left(\sum_{p=1}^{P} y_pf_p\right)+\left(\sum_{d=1}^{D}\sum_{j=1}^{J_d} x_{dj}c_{dj}\right)\right]}$$

$$=\frac{y_p(1-\alpha)F_{d^*j^*}(x,y)\left[\left(\sum_{p=1}^{P} y_pf_p\right)+\left(\sum_{d=1}^{D}\sum_{j=1}^{J_d} x_{dj}c_{dj}\right)\right]}{\left[\left(\sum_{p\in\Gamma_{d^*j^*}} y_pf_p\right)+x_{d^*j^*}c_{d^*j^*}\right]\left[\left(\sum_{p=1}^{P} y_pf_p\right)+\left(\sum_{d=1}^{D}\sum_{j=1}^{J_d} x_{dj}c_{dj}\right)\right]}=\frac{y_p(1-\alpha)F_{d^*j^*}(x,y)}{\left[\left(\sum_{p\in\Gamma_{d^*j^*}} y_pf_p\right)+x_{d^*j^*}c_{d^*j^*}\right]}$$

Hence $\frac{y_p(1-\alpha)F(x,y)}{\left[\left(\sum_{p=1}^{P} y_pf_p\right)+\left(\sum_{d=1}^{D}\sum_{j=1}^{J_d} x_{dj}c_{dj}\right)\right]}\geq\frac{y_p(1-\alpha)F_{d^*j^*}(x,y)}{\left[\left(\sum_{p\in\Gamma_{d^*j^*}} y_pf_p\right)+x_{d^*j^*}c_{d^*j^*}\right]}$.

Therefore, the minimal rewarding rate for passengers under Local Proportional Scheme is less than or equal to that of Global Proportional Scheme. □

Appendix B

Proof of Lemma 4. We first show that the minimal rewarding rate for drivers under the Local Proportional Scheme is less than that of the Global Proportional Scheme. To show the minimal rewarding rate for drivers under the Local Proportional Scheme is less than or equal to that of the Global Proportional Scheme, we must show that

$$\frac{(1-\alpha)x_{dj}F(x,y)}{\left[\left(\sum_{p=1}^{P} y_pf_p\right)+\left(\sum_{d=1}^{D}\sum_{j=1}^{J_d} x_{dj}c_{dj}\right)\right]}\geq\frac{x_{d^*j^*}(1-\alpha)F_{d^*j^*}(x,y)}{\left[\left(\sum_{p\in\Gamma_{d^*j^*}} y_pf_p\right)+x_{d^*j^*}c_{d^*j^*}\right]}.$$

Let d^*, j^* be the winning bid j^* of driver d^* such that $(d^*, j^*) = \arg\min_{(d,j)}\frac{(1-\alpha)x_{dj}F_{dj}(x,y)}{\left[\left(\sum_{p\in\Gamma_{dj}} y_pf_p\right)+x'_{dj}c_{dj}\right]}$.

It follows that $\frac{(1-\alpha)x_{dj}F_{dj}(x,y)}{\left[\left(\sum_{p\in\Gamma_{dj}} y_pf_p\right)+x_{dj}c_{dj}\right]}\geq\frac{x_{d^*j^*}(1-\alpha)F_{d^*j^*}(x,y)}{\left[\left(\sum_{p\in\Gamma_{d^*j^*}} y_pf_p\right)+x_{d^*j^*}c_{d^*j^*}\right]}$.

$$(1-\alpha)x_{dj}F_{dj}(x,y)\geq\frac{x_{d^*j^*}(1-\alpha)F_{d^*j^*}(x,y)\left[\left(\sum_{p\in\Gamma_{dj}} y_pf_p\right)+x_{dj}c_{dj}\right]}{\left[\left(\sum_{p\in\Gamma_{d^*j^*}} y_pf_p\right)+x_{d^*j^*}c_{d^*j^*}\right]}$$

$$\sum_{d=1}^{D}\sum_{j=1}^{J_d}(1-\alpha)x_{dj}F_{dj}(x,y)\geq\frac{\sum_{d=1}^{D}\sum_{j=1}^{J_d}x_{d^*j^*}(1-\alpha)F_{d^*j^*}(x,y)\left[\left(\sum_{p\in\Gamma_{dj}} y_pf_p\right)+x_{dj}c_{dj}\right]}{\left[\left(\sum_{p\in\Gamma_{d^*j^*}} y_pf_p\right)+x_{d^*j^*}c_{d^*j^*}\right]}$$

$$\frac{\sum_{d=1}^{D}\sum_{j=1}^{J_d}(1-\alpha)x_{dj}F_{dj}(x,y)}{\left[\left(\sum_{p=1}^{P} y_pf_p\right)+\left(\sum_{d=1}^{D}\sum_{j=1}^{J_d} x_{dj}c_{dj}\right)\right]}\geq\frac{\sum_{d=1}^{D}\sum_{j=1}^{J_d}x_{d^*j^*}(1-\alpha)F_{d^*j^*}(x,y)\left[\left(\sum_{p\in\Gamma_{dj}} y_pf_p\right)+x_{dj}c_{dj}\right]}{\left[\left(\sum_{p\in\Gamma_{d^*j^*}} y_pf_p\right)+x_{d^*j^*}c_{d^*j^*}\right]\left[\left(\sum_{p=1}^{P} y_pf_p\right)+\left(\sum_{d=1}^{D}\sum_{j=1}^{J_d} x_{dj}c_{dj}\right)\right]}$$

As $\sum_{d=1}^{D}\sum_{j=1}^{J_d}(1-\alpha)x_{dj}F_{dj}(x,y)=(1-\alpha)F(x,y)$,

$$\frac{(1-\alpha)F(x,y)}{\left[\left(\sum_{p=1}^{P}y_pf_p\right)+\left(\sum_{d=1}^{D}\sum_{j=1}^{J_d}x_{dj}c_{dj}\right)\right]}\geq\frac{\sum_{d=1}^{D}\sum_{j=1}^{J_d}x_{d^*j^*}(1-\alpha)F_{d^*j^*}(x,y)\left[\left(\sum_{p\in\Gamma_{dj}}y_pf_p\right)+x_{dj}c_{dj}\right]}{\left[\left(\sum_{p\in\Gamma_{d^*j^*}}y_pf_p\right)+x_{d^*j^*}c_{d^*j^*}\right]\left[\left(\sum_{p=1}^{P}y_pf_p\right)+\left(\sum_{d=1}^{D}\sum_{j=1}^{J_d}x_{dj}c_{dj}\right)\right]}.$$

Note that

$$\begin{aligned}&\frac{\sum_{d=1}^{D}\sum_{j=1}^{J_d}x_{d^*j^*}(1-\alpha)F_{d^*j^*}(x,y)\left[\left(\sum_{p\in\Gamma_{dj}}y_pf_p\right)+x_{dj}c_{dj}\right]}{\left[\left(\sum_{p\in\Gamma_{d^*j^*}}y_pf_p\right)+x_{d^*j^*}c_{d^*j^*}\right]\left[\left(\sum_{p=1}^{P}y_pf_p\right)+\left(\sum_{d=1}^{D}\sum_{j=1}^{J_d}x_{dj}c_{dj}\right)\right]}=\frac{x_{d^*j^*}(1-\alpha)F_{d^*j^*}(x,y)\sum_{d=1}^{D}\sum_{j=1}^{J_d}\left[\left(\sum_{p\in\Gamma_{dj}}y_pf_p\right)+x_{dj}c_{dj}\right]}{\left[\left(\sum_{p\in\Gamma_{d^*j^*}}y_pf_p\right)+x_{d^*j^*}c_{d^*j^*}\right]\left[\left(\sum_{p=1}^{P}y_pf_p\right)+\left(\sum_{d=1}^{D}\sum_{j=1}^{J_d}x_{dj}c_{dj}\right)\right]}\\&=\frac{x_{d^*j^*}(1-\alpha)F_{d^*j^*}(x,y)\sum_{d=1}^{D}\sum_{j=1}^{J_d}\left[\left(\sum_{p\in\Gamma_{dj}}y_pf_p\right)+x_{dj}c_{dj}\right]}{\left[\left(\sum_{p\in\Gamma_{d^*j^*}}y_pf_p\right)+x_{d^*j^*}c_{d^*j^*}\right]\left[\left(\sum_{p=1}^{P}y_pf_p\right)+\left(\sum_{d=1}^{D}\sum_{j=1}^{J_d}x_{dj}c_{dj}\right)\right]}=\frac{x_{d^*j^*}(1-\alpha)F_{d^*j^*}(x,y)\left[\left(\sum_{p=1}^{P}y_pf_p\right)+\left(\sum_{d=1}^{D}\sum_{j=1}^{J_d}x_{dj}c_{dj}\right)\right]}{\left[\left(\sum_{p\in\Gamma_{d^*j^*}}y_pf_p\right)+x_{d^*j^*}c_{d^*j^*}\right]\left[\left(\sum_{p=1}^{P}y_pf_p\right)+\left(\sum_{d=1}^{D}\sum_{j=1}^{J_d}x_{dj}c_{dj}\right)\right]}\\&=\frac{x_{d^*j^*}(1-\alpha)F_{d^*j^*}(x,y)}{\left[\left(\sum_{p\in\Gamma_{d^*j^*}}y_pf_p\right)+x_{d^*j^*}c_{d^*j^*}\right]}\end{aligned}$$

Therefore the minimal rewarding rate for drivers under the Local Proportional Scheme is less than or equal to that of the Global Proportional Scheme. □

Appendix C

Proof of Lemma 5. Let r^* be the ride corresponding to the winning bid j^* of driver d^* such that $(d^*,j^*)=\arg\max_{(d,j)}\frac{y_p(1-\alpha)F_{dj}(x,y)}{\left[\left(\sum_{p\in\Gamma_{dj}}y_pf_p\right)+x_{dj}c_{dj}\right]}$.

To show the maximal rewarding rate for passengers under the Local Proportional Scheme is greater than or equal to that of the Global Proportional Scheme, we must show that $\frac{y_p(1-\alpha)F_{d^*j^*}(x,y)}{\left[\left(\sum_{p\in\Gamma_{d^*j^*}}y_pf_p\right)+x_{d^*j^*}c_{d^*j^*}\right]}\geq\frac{(1-\alpha)y_pF(x,y)}{\left[\left(\sum_{p=1}^{P}y_pf_p\right)+\left(\sum_{d=1}^{D}\sum_{j=1}^{J_d}x_{dj}c_{dj}\right)\right]}$.

Note that $\frac{y_p(1-\alpha)F_{dj}(x,y)}{\left[\left(\sum_{p\in\Gamma_{dj}}y_pf_p\right)+x_{dj}c_{dj}\right]}\leq\frac{y_p(1-\alpha)F_{d^*j^*}(x,y)}{\left[\left(\sum_{p\in\Gamma_{d^*j^*}}y_pf_p\right)+x_{d^*j^*}c_{d^*j^*}\right]}$

$$y_p(1-\alpha)F_{dj}(x,y)\leq\frac{y_p(1-\alpha)F_{d^*j^*}(x,y)\left[\left(\sum_{p\in\Gamma_{dj}}y_pf_p\right)+x_{dj}c_{dj}\right]}{\left[\left(\sum_{p\in\Gamma_{d^*j^*}}y_pf_p\right)+x_{d^*j^*}c_{d^*j^*}\right]}$$

$$\sum_{d=1}^{D}\sum_{j=1}^{J_d}y_p(1-\alpha)F_{dj}(x,y)\leq\sum_{d=1}^{D}\sum_{j=1}^{J_d}\left\{\frac{y_p(1-\alpha)F_{d^*j^*}(x,y)\left[\left(\sum_{p\in\Gamma_{dj}}y_pf_p\right)+x_{dj}c_{dj}\right]}{\left[\left(\sum_{p\in\Gamma_{d^*j^*}}y_pf_p\right)+x_{d^*j^*}c_{d^*j^*}\right]}\right\}$$

$$\frac{\sum_{d=1}^{D}\sum_{j=1}^{J_d}y_p(1-\alpha)F_{dj}(x,y)}{\left[\left(\sum_{p=1}^{P}y_pf_p\right)+\left(\sum_{d=1}^{D}\sum_{j=1}^{J_d}x_{dj}c_{dj}\right)\right]}\leq\sum_{d=1}^{D}\sum_{j=1}^{J_d}\left\{\frac{y_p(1-\alpha)F_{d^*j^*}(x,y)\left[\left(\sum_{p\in\Gamma_{dj}}y_pf_p\right)+x_{dj}c_{dj}\right]}{\left[\left(\sum_{p\in\Gamma_{d^*j^*}}y_pf_p\right)+x_{d^*j^*}c_{d^*j^*}\left[\left(\sum_{p=1}^{P}y_pf_p\right)+\left(\sum_{d=1}^{D}\sum_{j=1}^{J_d}x_{dj}c_{dj}\right)\right]\right]}\right\}$$

As $\sum_{d=1}^{D}\sum_{j=1}^{J_d} y_p(1-\alpha)F_{dj}(x,y) = y_p(1-\alpha)F(x,y)$,

$$\frac{y_p(1-\alpha)F(x,y)}{\left[\left(\sum_{p=1}^{P} y_p f_p\right)+\left(\sum_{d=1}^{D}\sum_{j=1}^{J_d} x_{dj}c_{dj}\right)\right]} \le \sum_{d=1}^{D}\sum_{j=1}^{J_d}\left\{\frac{y_p(1-\alpha)F_{d^*j^*}(x,y)\left[\left(\sum_{p\in\Gamma_{dj}} y_p f_p\right)+x_{dj}c_{dj}\right]}{\left[\left(\sum_{p\in\Gamma_{d^*j^*}} y_p f_p\right)+x_{d^*j^*}c_{d^*j^*}\right]\left[\left(\sum_{p=1}^{P} y_p f_p\right)+\left(\sum_{d=1}^{D}\sum_{j=1}^{J_d} x_{dj}c_{dj}\right)\right]}\right\}$$

Note that

$$\sum_{d=1}^{D}\sum_{j=1}^{J_d}\left\{\frac{y_p(1-\alpha)F_{d^*j^*}(x,y)\left[\left(\sum_{p\in\Gamma_{dj}} y_p f_p\right)+x_{dj}c_{dj}\right]}{\left[\left(\sum_{p\in\Gamma_{d^*j^*}} y_p f_p\right)+x_{d^*j^*}c_{d^*j^*}\right]\left[\left(\sum_{p=1}^{P} y_p f_p\right)+\left(\sum_{d=1}^{D}\sum_{j=1}^{J_d} x_{dj}c_{dj}\right)\right]}\right\}$$
$$= \frac{y_p(1-\alpha)F_{d^*j^*}(x,y)\sum_{d=1}^{D}\sum_{j=1}^{J_d}\left\{\left[\left(\sum_{p\in\Gamma_{dj}} y_p f_p\right)+x_{dj}c_{dj}\right]\right\}}{\left[\left(\sum_{p\in\Gamma_{d^*j^*}} y_p f_p\right)+x_{d^*j^*}c_{d^*j^*}\right]\left[\left(\sum_{p=1}^{P} y_p f_p\right)+\left(\sum_{d=1}^{D}\sum_{j=1}^{J_d} x_{dj}c_{dj}\right)\right]}$$
$$= \frac{y_p(1-\alpha)F_{d^*j^*}(x,y)\left[\left(\sum_{p=1}^{P} y_p f_p\right)+\left(\sum_{d=1}^{D}\sum_{j=1}^{J_d} x_{dj}c_{dj}\right)\right]}{\left[\left(\sum_{p\in\Gamma_{d^*j^*}} y_p f_p\right)+x_{d^*j^*}c_{d^*j^*}\right]\left[\left(\sum_{p=1}^{P} y_p f_p\right)+\left(\sum_{d=1}^{D}\sum_{j=1}^{J_d} x_{dj}c_{dj}\right)\right]}$$
$$= \frac{y_p(1-\alpha)F_{d^*j^*}(x,y)}{\left[\left(\sum_{p\in\Gamma_{d^*j^*}} y_p f_p\right)+x_{d^*j^*}c_{d^*j^*}\right]}$$
$$\frac{y_p(1-\alpha)F(x,y)}{\left[\left(\sum_{p=1}^{P} y_p f_p\right)+\left(\sum_{d=1}^{D}\sum_{j=1}^{J_d} x_{dj}c_{dj}\right)\right]} \le \frac{y_p(1-\alpha)F_{d^*j^*}(x,y)}{\left[\left(\sum_{p\in\Gamma_{d^*j^*}} y_p f_p\right)+x_{d^*j^*}c_{d^*j^*}\right]}$$

Therefore the maximal rewarding rate for passengers under the Local Proportional Scheme is greater than or equal to that of the Global Proportional Scheme.

As the rewarding rate for driver d^* is $\frac{y_p(1-\alpha)F_{d^*j^*}(x,y)}{\left[\left(\sum_{p\in\Gamma_{d^*j^*}} y_p f_p\right)+x_{d^*j^*}c_{d^*j^*}\right]}$ and $\frac{y_p(1-\alpha)F_{d^*j^*}(x,y)}{\left[\left(\sum_{p\in\Gamma_{d^*j^*}} y_p f_p\right)+x_{d^*j^*}c_{d^*j^*}\right]} \ge \frac{y_p(1-\alpha)F(x,y)}{\left[\left(\sum_{p=1}^{P} y_p f_p\right)+\left(\sum_{d=1}^{D}\sum_{j=1}^{J_d} x_{dj}c_{dj}\right)\right]}$, the maximal rewarding rate for the drivers under the Local Proportional Scheme is greater than or equal to that of the Global Proportional Scheme. □

References

1. The International Energy Agency (IEA). Tracking Report. Available online: https://www.iea.org/reports/tracking-transport-2020 (accessed on 20 August 2021).
2. Hossain, M. Sharing economy: A comprehensive literature review. *Int. J. Hosp. Manag.* **2020**, *87*, 102470. [CrossRef]
3. Klarin, A.; Suseno, Y. A state-of-the-art review of the sharing economy: Scientometric mapping of the scholarship. *J. Bus. Res.* **2021**, *126*, 250–262. [CrossRef]
4. Hyland, M.; Mahmassani, H.S. Operational benefits and challenges of shared-ride automated mobility-on-demand services. *Transp. Res. Part A: Policy Pr.* **2020**, *134*, 251–270. [CrossRef]
5. Furuhata, M.; Dessouky, M.; Ordóñez, F.; Brunet, M.; Wang, X.; Koenig, S. Ridesharing: The state-of-the-art and future direc-tions. *Transp. Res. Pt. B-Methodol.* **2013**, *57*, 28–46. [CrossRef]
6. Agatz, N.; Erera, A.; Savelsbergh, M.; Wang, X. Optimization for dynamic ride-sharing: A review. *Eur. J. Oper. Res.* **2012**, *223*, 295–303. [CrossRef]
7. Mourad, A.; Puchinger, J.; Chu, C. A survey of models and algorithms for optimizing shared mobility. *Transp. Res. Part B: Methodol.* **2019**, *123*, 323–346. [CrossRef]

8. Martins, L.D.C.; de la Torre, R.; Corlu, C.G.; Juan, A.A.; Masmoudi, M.A. Optimizing ride-sharing operations in smart sustainable cities: Challenges and the need for agile algorithms. *Comput. Ind. Eng.* **2021**, *153*, 107080. [CrossRef]
9. Santos, D.O.; Xavier, E.C. Taxi and Ride Sharing: A Dynamic Dial-a-Ride Problem with Money as an Incentive. *Expert Syst. Appl.* **2015**, *42*, 6728–6737. [CrossRef]
10. Guajardoa, M.; Ronnqvist, M. A review on cost allocation methods in collaborative transportation. *Int. Trans. Oper. Res.* **2016**, *23*, 371–392. [CrossRef]
11. Wang, X.; Agatz, N.; Erera, A. Stable Matching for Dynamic Ride-Sharing Systems. *Transp. Sci.* **2018**, *52*, 850–867. [CrossRef]
12. Agatz, N.A.H.; Erera, A.L.; Savelsbergh, M.W.P.; Wang, X. Dynamic ride-sharing: A simulation study in metro Atlanta. *Transp. Res. Part B: Methodol.* **2011**, *45*, 1450–1464. [CrossRef]
13. Hsieh, F.-S.; Zhan, F.-M.; Guo, Y.-H. A solution methodology for carpooling systems based on double auctions and cooperative coevolutionary particle swarms. *Appl. Intell.* **2018**, *49*, 741–763. [CrossRef]
14. Jin, S.T.; Kong, H.; Wu, R.; Sui, D.Z. Ridesourcing, the sharing economy, and the future of cities. *Cities* **2018**, *76*, 96–104. [CrossRef]
15. Shaheen, S.; Chan, N. Mobility and the sharing economy: Potential to facilitate the first- and last-mile public transit connec-tions. *Built Environ.* **2016**, *42*, 573–588. [CrossRef]
16. Lokhandwala, M.; Cai, H. Dynamic ride sharing using traditional taxis and shared autonomous taxis: A case study of NYC. *Transp. Res. Part C: Emerg. Technol.* **2018**, *97*, 45–60. [CrossRef]
17. Li, R.; Liu, Z.; Zhang, R. Studying the benefits of carpooling in an urban area using automatic vehicle identification data. *Transp. Res. Part C: Emerg. Technol.* **2018**, *93*, 367–380. [CrossRef]
18. Baldacci, R.; Maniezzo, V.; Mingozzi, A. An exact method for the car pooling problem based on Lagrangean column genera-tion. *Oper. Res.* **2004**, *52*, 422–439. [CrossRef]
19. Kaan, L.; Olinick, E.V. The Vanpool Assignment Problem: Optimization models and solution algorithms. *Comput. Ind. Eng.* **2013**, *66*, 24–40. [CrossRef]
20. Li, B.; Krushinsky, D.; Reijers, H.A.; Van Woensel, T. The Share-a-Ride Problem: People and parcels sharing taxis. *Eur. J. Oper. Res.* **2014**, *238*, 31–40. [CrossRef]
21. Masson, R.; Trentini, A.; Lehuédé, F.; Malhéné, N.; Péton, O.; Tlahig, H. Optimization of a city logistics transportation system with mixed passengers and goods. *EURO J. Transp. Logist.* **2017**, *6*, 81–109. [CrossRef]
22. Hsieh, F.-S. A Particle Swarm Optimization Algorithm to Meet Trust Requirements in Ridesharing Systems. In Proceedings of the 2020 11th IEEE Annual Ubiquitous Computing, Electronics & Mobile Communication Conference (UEMCON), New York, NY, USA, 28–31 October 2020; pp. 592–595.
23. Narayanan, S.; Chaniotakis, E.; Antoniou, C. Shared autonomous vehicle services: A comprehensive review. *Transp. Res. Part C: Emerg. Technol.* **2020**, *111*, 255–293. [CrossRef]
24. Hosni, H.; Naoum-Sawaya, J.; Artail, H. The shared-taxi problem: Formulation and solution methods. *Transp. Res. Part B: Methodol.* **2014**, *70*, 303–318. [CrossRef]
25. Schreieck, M.; Safetli, H.; Siddiqui, S.A.; Pflügler, C.; Wiesche, M.; Krcmar, H. A matching algorithm for dynamic rideshar-ing. *Transp. Res. Procedia* **2016**, *19*, 272–285. [CrossRef]
26. Cheikh-Graiet, S.B.; Dotoli, M.; Hammadi, S. A Tabu Search based metaheuristic for dynamic carpooling optimization. *Comput. Ind. Eng.* **2020**, *140*, 106217. [CrossRef]
27. Hsieh, F.S. A Comparative Study of Several Metaheuristic Algorithms to Optimize Monetary Incentive in Ridesharing Sys-tems. *ISPRS Int. J. Geo-Inf.* **2020**, *9*, 590. [CrossRef]
28. Hsieh, F.-S. Ridesharing based on a Discrete Self-adaptive Differential Evolution Algorithm. In Proceedings of the 2020 11th IEEE Annual Information Technology, Electronics and Mobile Communication Conference (IEMCON), Vancouver, BC, Canada, 4–7 November 2020; pp. 696–700.
29. Jung, J.; Jayakrishnan, R.; Park, J.Y. Dynamic Shared-Taxi Dispatch Algorithm with Hybrid-Simulated Annealing. *Comput. Civ. Infrastruct. Eng.* **2016**, *31*, 275–291. [CrossRef]
30. Butler, L.; Yigitcanlar, T.; Paz, A. Barriers and risks of Mobility-as-a-Service (MaaS) adoption in cities: A systematic review of the literature. *Cities* **2021**, *109*, 103036. [CrossRef]
31. Van Der Waerden, P.; Lem, A.; Schaefer, W. Investigation of Factors that Stimulate Car Drivers to Change from Car to Carpooling in City Center Oriented Work Trips. *Transp. Res. Procedia* **2015**, *10*, 335–344. [CrossRef]
32. Shaheen, S.A.; Chan, N.D.; Gaynor, T. Casual carpooling in the San Francisco Bay Area: Understanding user characteristics, behaviors, and motivations. *Transp. Policy* **2016**, *51*, 165–173. [CrossRef]
33. Shapley, L.S. A Value for N-Person Games. *A Value N-Person Games* **1952**, *28*, 307–317. [CrossRef]
34. Schmeidler, D. The Nucleolus of a Characteristic Function Game. *SIAM J. Appl. Math.* **1969**, *17*, 1163–1170. [CrossRef]
35. Kalai, E. Proportional solutions to bargaining situations: Intertemporal utility comparisons. *Econometrica* **1977**, *45*, 1623–1630. [CrossRef]
36. Fatima, S.S.; Wooldridge, M.; Jennings, N.R. A linear approximation method for the Shapley value. *Artif. Intell.* **2008**, *172*, 1673–1699. [CrossRef]
37. Perea, F.; Puerto, J. A heuristic procedure for computing the nucleolus. *Comput. Oper. Res.* **2019**, *112*, 104764. [CrossRef]
38. Berger, S.; Bierwirth, C. Solutions to the request reassignment problem in collaborative carrier networks. *Transp. Res. Part E: Logist. Transp. Rev.* **2010**, *46*, 627–638. [CrossRef]

39. Özener, O. Örsan Developing a Collaborative Planning Framework for Sustainable Transportation. *Math. Probl. Eng.* **2014**, *2014*, 1–14. [CrossRef]
40. Flisberg, P.; Frisk, M.; Röonnqvist, M.; Guajardo, M. Potential savings and cost allocations for forest fuel transportation in Sweden: A country-wide study. *Energy* **2015**, *85*, 353–365. [CrossRef]
41. Lu, W.; Quadrifoglio, L. Fair cost allocation for ridesharing services – modeling, mathematical programming and an algorithm to find the nucleolus. *Transp. Res. Part B: Methodol.* **2019**, *121*, 41–55. [CrossRef]
42. Zhang, C.; Xie, J.; Wu, F.; Gao, X.; Chen, G. Pricing and allocation algorithm designs in dynamic ridesharing system. *Theor. Comput. Sci.* **2020**, *803*, 94–104. [CrossRef]

energies

MDPI

Article

Evaluation of the Radar Speed Cameras and Panels Indicating the Vehicles' Speed as Traffic Calming Measures (TCM) in Short Length Urban Areas Located along Rural Roads

Heriberto Pérez-Acebo [1,*], Robert Ziolkowski [2] and Hernán Gonzalo-Orden [3]

[1] Mechanical Engineering Department, University of the Basque Country UPV/EHU, P° Rafael Moreno Pitxitxi, 2, 48013 Bilbao, Spain
[2] Faculty of Civil and Environmental Sciences, Bialystok University of Technology, Wiejska 45A, 15-351 Bialystok, Poland; robert.ziolkowski@pb.edu.pl
[3] Department of Civil Engineering, University of Burgos, c/Villadiego, s/n, 09001 Burgos, Spain; hgonzalo@ubu.es
* Correspondence: heriberto.perez@ehu.eus; Tel.: +34-94-601-7820

Abstract: Traffic calming measures (TCMs) are implemented in urban areas to reduce vehicles' speed and, generally speaking, results are obtained. However, speed is still a problem in rural roads crossing small villages without a bypass and with short-length urban areas, since drivers do not normally reduce their speed for that short segment. Hence, various TCM can be installed. It is necessary to maintain a calm area in these short segments to improve road safety, especially for pedestrian aiming to cross the road, and to save combustible by avoiding a constant increase-decrease of speed. Four villages were selected to evaluate the efficiency of radar speed cameras and panels indicating vehicle's speed. Results showed that the presence of radar speed cameras reduces the speed in the direction they can fine, but with a lower effect in the non-fining direction. Additionally, a positive effect was observed in the fining direction in other points, such as pedestrian crossings. Nevertheless, the effect does not last long and speed cameras may be considered as punctual measures. If the TCMs are placed far from the start of the village they are not respected. Hence, it is recommended to place them near the real start of the build-up area. Lastly, it was verified that longer urban areas make overall speed decrease. However, when drivers feel that they are arriving to the end of the urban area, due to the inexistence of buildings, they start speeding up.

Keywords: traffic calming measure; radar speed camera; urban area; road safety; rural roads; pedestrian; crosswalk

Citation: Pérez-Acebo, H.; Ziolkowski, R.; Gonzalo-Orden, H. Evaluation of the Radar Speed Cameras and Panels Indicating the Vehicles' Speed as Traffic Calming Measures (TCM) in Short Length Urban Areas Located along Rural Roads. *Energies* **2021**, *14*, 8146. https://doi.org/10.3390/en14238146

Academic Editors: Elżbieta Macioszek, Anna Granà, Margarida Coelho and Paulo Fernandes

Received: 10 November 2021
Accepted: 30 November 2021
Published: 5 December 2021

1. Introduction

Despite the great efforts that have been carried out to reduce the number of fatalities and injuries on road crashes, road safety continues to be a major problem around the world, even becoming the first cause of premature death [1,2]. Some figures could give an approximate idea about this reality. In the European Union (EU-27), there were 935,216 crashes resulting in injuries or death in 2019, which represents a low reduction of 3.9% from 2010, with 973,596. Some countries such as Spain and Romania registered an increase of 20% in the number of crashes during this period (2010–2019) [3]. Nevertheless, talking about fatalities, 22,700 people died in 2019, showing a decrease of 23.3% with regard to data from 2010, with 29,611 deaths. The only country with an increase of fatalities between 2010 and 2019 was the Netherlands [3]. In relative figures, the EU-27 reported a decrease from 67 fatalities per million inhabitants in 2010 to 51 in 2019 (−23%) but this positive trend in fatalities reduction has been distinctly flattened in last few years [3].

Nonetheless, it must be noted that one out of five (20.4%) of all road fatalities in the EU-27 are pedestrians, representing a higher proportion than other vulnerable road users (cyclist, 9%; mopeds, 3%; motorcycles 15%) [4,5]. Despite the fall in pedestrian fatalities from 2010 (5952) to 2018 (4763), implying a reduction of 20%, this figure is even lower to the global decrease of fatalities mentioned before (−23.3% for all the fatalities) and, hence, the proportion of pedestrians in total number of road fatalities remains constant (or even slightly higher). Individual data from Poland and Spain are shown in Table 1. As seen, both countries reported decreases in the number of fatalities, in pedestrian fatalities, and in the ratio fatalities per million inhabitants. However, the overall proportion of pedestrians killed in crashes is quite different. While it ranged from 31.6% to 27.3% in Poland, in Spain it is around 20%.

Table 1. Road safety data from Poland and Spain in 2010 and in 2019.

	Total Number of Road Crashes			Total Number of Fatalities			Fatalities Per Million			Pedestrian Fatalities		
Country	**2010**	**2019**	**Change (%)**	**2010**	**2019**	**Change (%)**	**2010**	**2019**	**Change (%)**	**2010 (% of the Total)**	**2019 (% of the Total)**	**Change (%)**
Poland	38,832	30,288	−22.0	3908	2909	−25.6	103	77	−25.2	1236 (31.6)	793 (27.3)	−35.8
Spain	85,503	104,080	+21.7	2479	1755	−29.2	53	37	−30.2	471 (19.0)	381 (21.7)	−19.1

Traditionally, road crashes are said to be multi-causal, which are generally grouped in factors related to driver, vehicle, and highway conditions [6–8], with variable proportions (Figure 1) [9].

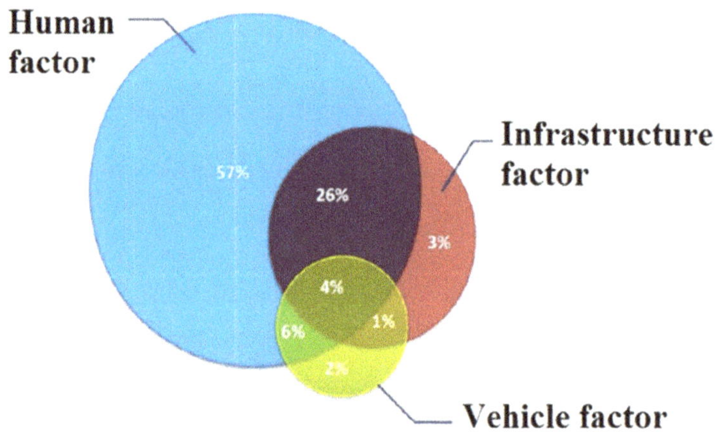

Figure 1. Interaction of the concurrent factors on road crashes.

Nevertheless, speed is said to be the key factor in serious and fatal crashes involving pedestrians, with a great influence on injury severity of pedestrians [5,10]. When vehicles travel below 30 km/h, the collisions between motorized vehicles and pedestrians are much less likely to happen, and if they occur, they do not normally result in fatality [10]. Correlations between the risk of pedestrian fatality and the speed impact have been established (Figure 2) [11].

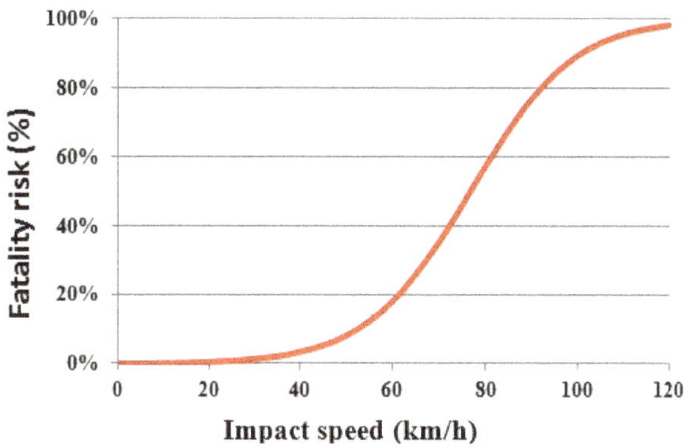

Figure 2. Risk of pedestrian fatality for different impact speeds (fatality rate is the percentage of pedestrians that are killed in a collision with a motorized vehicle).

In European countries, as expected, urban areas are the location where the majority of pedestrian fatalities occur (73%), increasing the relative percentage of the pedestrian death to the total deaths up to 38% (as previously said, pedestrian fatalities represent the 20% of the total). In Poland and in Spain, the proportion of pedestrians killed in urban areas represents 64.3% and 64.8%, respectively [3]. It has been reported that approximately two thirds of pedestrians involved in server road crashes were crossing the road [12]. Yue et al. [13] identified the main scenarios for fatal pedestrian crashes in Florida, indicating the percentage of occurrence:

- Vehicle going straight and pedestrian crossing the road (51%);
- Vehicle turning left and pedestrian crossing the road at the exit (17%) or the entry (3%) of the crossing;
- Vehicle turning right and pedestrian crossing the road at the exit (12%) or the entry (4%) of the crossing;
- Vehicle going straight and pedestrian in (3%) or adjacent to (6%) the road.

Similarly, Populer et al. [14] identified the most frequent scenarios for pedestrian crossing crashes and observed that the main contributing factors for signal-regulated crossing were non-compliance by the pedestrian (50%) and conflicting green-phases for pedestrians and turning motorists (25%). In non-signalized crossing, obstruction of view due to other traffic actors or parked vehicles was concluded to be the main factor [15]. Furthermore, the majority of accidents in Poland is reported in urban areas and driving speed remains the most important factor According to the International Transport Forum [16], unadjusted and excessive speed is the main cause of one third of all fatal road crashes in Hungary, Poland, and Lithuania. Equal conclusions emerged from official statistics published in Poland by the Police Headquarters [17].

With the aim of minimizing the vehicles' speed, traffic calming measures (TCM) have been introduced in urban areas [18–21]. They are usually defined as "*the combination of mainly physical measures that reduce the negative effect of motor vehicle use, alter driver behavior and improve conditions for non-motorized street users*" [22]. Their main goal is thus to reduce vehicle speed and volume in the area. Generally, TCMs are classified in four categories: vertical deflections (rumble strips, speed humps, speed cushions, raised crosswalk, raised intersection, etc.), horizontal deflections (chicane, gateway, raised median island, etc.); physical obstructions (semi and diagonal diverter, raised median in intersections, etc.); and sings and pavement markings (a meter) [23–34].

With regard to the environment, an important point to consider is fuel consumption and car emissions as a result of the driving speed and style in urban areas. Generally speaking, lower average speeds and higher number of stops are the main characteristics of driving in urban areas, implying both fuel consumption and pollutant emissions many times higher per vehicle·km. The effect of speed humps and speed tables on the pollutant emissions was analyzed by Obregón-Biosca [35]. Similarly, other authors evaluated the exhaust emissions and fuel consumption on Tempo-30-zoned streets [36,37]. However, it was reported that the speed reduction and fuel consumption and pollutant emissions is also dependent on other factors, such as the drivers' behavior and resulting from traffic jams [38], speed control schemes [39,40], shifting gears on the way through the TCMs [41], etc. For example, Wang et al. [42] indicated the acceleration should be incorporated, instead of mean speed of the vehicle, for estimating emission.

While in extensive urban areas the succession of TCMs leads to a situation wherein vehicles' speeds are relatively under control, a problem arises in the transition from interurban areas to urban areas. This problem is even greater when an interurban road goes through the urban area of a small village without by-passing it [43–46]. When the urban road segment is short, drivers tend not to reduce the speed adequately since the village is not their final destination, and they maintain the high speeds allowed in interurban areas. Hence, the risk for pedestrians is increased (Figure 2). Aiming to reduce the speed of vehicles when crossing these small villages, traffic calming measures are collocated at the entrance of villages to warn drivers that they are entering an urban area, even if it is short, and to force them to speed down.

The first traffic safety analyses on these rural roads crossing short urban areas without a by-pass were conducted in Denmark [47], and in Great Britain [48–50], and later in other countries [51–55]. Therefore, the introduction of traffic calming measures is recommended in many countries [56–63]. Usual solutions at the border between non-urban and urban areas are signs with speed limits, road humps, road markings, rumble devices, road narrowing measures, including chicanes, pinch-points or overrun areas, panels displaying vehicles' speed, radars (which could fine if the speed limit is exceeded), traffic lights turning red in case of exceeding speed limit, raised crosswalk, etc. [45–49,52–56,62–64]. Some studies have examined the effectiveness of these measures at this critical point [65,66]. Solowczuk and Kacprzak [45,46] analyzed the factors influencing the effectiveness of on-road chicanes in transition zones to villages subject to 70 and 50 km/h speed limits, and proposed one aggregate parameter combining various factors. Factors related to the surrounding landscape and visibility conditions were found to be key factors on speed reduction. Similarly, other authors examined the result of placing traffic lights that turn red if the speed limit is exceeded together with various displays of other measures in different small villages. It was observed that incorporating a panel indicating the vehicles' speed after or before the traffic light turning red in case of exceeding speed limit did not lead to variation in measured speeds [67,68]. Additionally, the point with the lowest speed was the traffic lights (since they could get a fine if police was controlling that the red light was controlled). However, after them, drivers speed up at the points where pedestrians could cross the road. Furthermore, it was shown that if the traffic lights turning red were placed with a pedestrian crossing, lower speeds were measured since pedestrians could appear and drivers tend to respect more the speed limit.

The aim of this paper is to evaluate the effectiveness of camera speed radars and panels indicting vehicles' speed at the entrance of small villages in rural roads, observing if drivers reduce their speed when passing next to the TCMs and further in the middle of the small village, where crosswalks are placed to connect the two parts of the small village divided by the road. Two small villages in Poland and two small villages in Spain located along the rural sections of roads were selected for the study. At the entrance of these villages radar speed cameras (which are supposed to fine) or panels indicating vehicles' speed are installed. If low speeds are obtained at the entrance and they are effectively maintained through the entire urban section, a safer and more energy-efficient driving will be developed.

2. Materials and Methods

The two small villages in Poland are Chodorówka Nowa and Suchowola and the two villages in Spain are Arrankudiaga and Arrespalditza/Respaldiza. The display of the traffic calming measures in each village is described individually.

2.1. Villages in Poland

Both analyzed villages in Poland are located on the National Road 8 (NR8), included in the international E-network as E67, from Helsinki (Finland) to Prague (Czech Republic). The NR8 road is the main international route in North-Eastern Poland, leading the traffic to the border with Lithuania and, hence, it is characterized with a high percentage of heavy traffic. An annual average daily traffic (AADT) on NR8 reaches almost 10,713 vehicles/day in 2019, implying 8% increase with regard to 2018. The average AADT for all international roads is almost 16,700 vehicles/day. On the other hand, the percentage of heavy traffic on NR8 (46%) is distinctly higher when compared to other international roads in Poland (30%).

The NR8 road crosses the village of Chodorówka Nowa in a length of approximately 790 m. In the middle part of the urban segment, there is a road intersection with a pedestrian crossing enabling to cross the NR8 road. At approximately 150 m before the crosswalk there is a radar speed camera. The existing elements in the urban segment, from the North to the South are as follows (Figure 3a):

- A sign with the village name (1);
- Radar speed camera, fining vehicles going southbound (2);
- Pedestrian crossing (3).

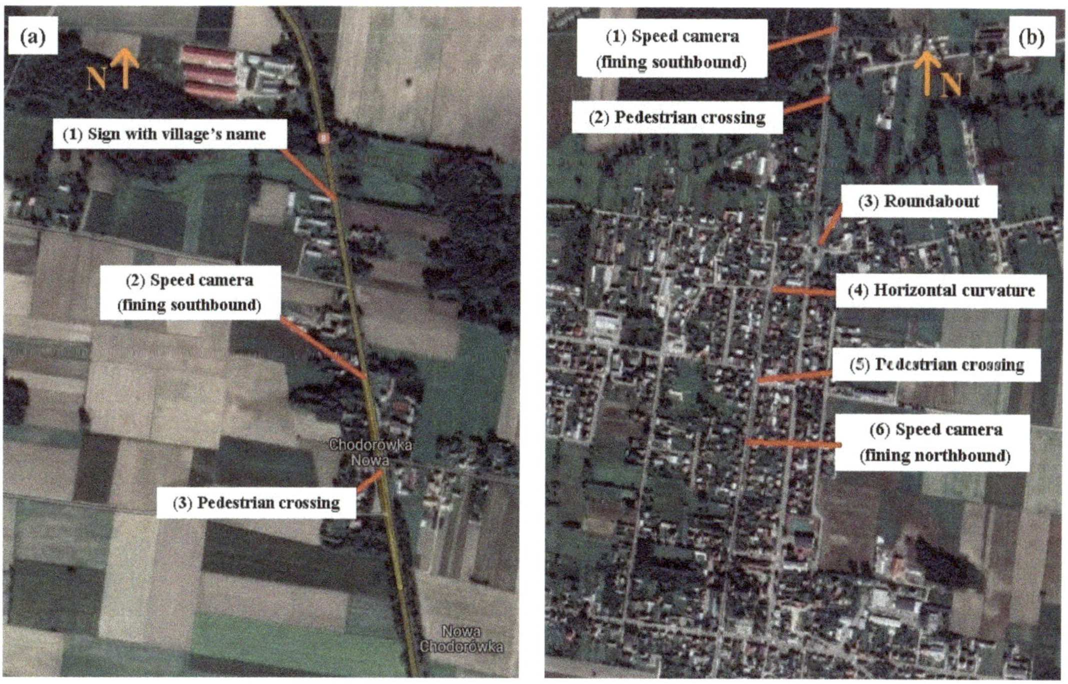

Figure 3. Locations of the traffic calming measures and other elements in (**a**) Chodorówka Nowa; (**b**) Suchowola.

Vehicles speeds were measured in both directions at the three points (the entrance/exit of the village, the speed camera, and the pedestrian crossing). Points are referred to as S1, S2, and S3, when measuring southbound, and N1, N2, and N3 northbound.

The second village investigated in Poland is Suchowola. The length of the urban segment is approximately 3500 m and there are speed cameras situated along the NR8 road and spaced at different distances. The traffic calming measures and important points in the urban segment (Figure 3b), from north to south are as follows:

- Speed camera (fining vehicles southbound) (1), 30 m south from the sign with the village name;
- Pedestrian crossing (2), 60 m away from the speed camera in point 1;
- Roundabout (3);
- Horizontal curvature (4);
- Pedestrian crossing (5), 210 m south from the horizontal curvature, which creates a physical deflection influencing vehicles' speed;
- Speed camera faced northbound (6), 120 m from the crosswalk in point 5.

Speed measurements were conducted in point 1 (speed camera), point 2 (pedestrian crossing), and point 5 (pedestrian crossing). They are called N1 or S1, N2 or S2, and N5 and S5, depending on the direction of the traffic (northbound or southbound), respectively.

2.2. Villages in Spain

The first village in Spain is Arrankudiaga, located in the province of Biscay. It is crossed by the road BI-625, which belongs to the basic network (orange network), the second level between the road network levels in the province [69]. The Regional Government of Biscay manages all the roads in the territory, even the freeways or national roads [70] so it means an important road. It has an AADT of 12,855 vehicles/day, with 8% of heavy vehicles [71]. The urban length is approximately 900 m and the sequence of traffic calming measures and elements in the urban area are (Figure 4), from north to south are:

- Signs and panels indicating that the speed limit is 50 km/h and the presence of a speed camera (1) and (2);
- Traffic lights with a pedestrian crossing and a pushbutton (3);
- Radar speed camera faced to control vehicles northbound (4);
- Traffic lights with pedestrian crossing and a pushbutton (5);
- Signs and panels indicating that the speed limit is 50 km/h and the presence of a speed camera (6) and (7).

Measurements were carried out at the pedestrian crossings with traffic lights and pushbuttons (3) and (5) and at the radar speed camera (4) northbound. Once again, points are referred to as N3, N4, and N5 or S3, and S5, depending on the controlled direction, northbound and southbound, respectively.

The second village in Spain is Arrespalditza/Respaldiza, in the province of Álava. The road A-624 crosses the village in a length of 415 m. The road belongs to the basic road network (the orange network), which is also the second level of the road networks in the province. The Regional Government of Alava also manages all the roads in the territory, even the freeways or national roads [70]. The road A-624 has an AADT of 3505 vehicles per day near Arrespalditza/Respaldiza, with 4% of heavy vehicles [72]. The traffic calming measures and the main elements on the urban segment (Figure 5) from north to south are as follows:

- A panel indicating the presence of traffic light (1);
- A panel indicating the speed of each vehicle (with a speed limit of 50 km/h) for vehicles going southbound (2);
- Traffic lights with a pedestrian cross walk and a pushbutton (3);
- Pedestrian crossing (4);
- Traffic lights warning about the presence of traffic lights (5);
- A panel indicating the speed of each vehicle (with a speed limit of 50 km/h) for vehicles going northbound (6).

Measures were conducted at the pedestrian crossing (4), and 25 m after the crosswalk with traffic lights with a pushing button at the point (3), southbound. This last point aimed to measure the speed not just at the TCM but at a certain distance. Additionally, the speed was measured in the places of both panels indicating the vehicles' speed (2) and (6). Points are referred to as N3, N4, and N6, or S2, S3, and S4, according to the direction of the vehicles, northbound or southbound, respectively.

Speed measures were obtained by fixed radars or by radar guns. A minimum of 200 vehicles were measured at each place, in each direction.

For the four villages, the following speed parameter values are provided: the maximum speed (V_{max}), the average speed (V_m), and the 85th percentile of the speed distribution (V_{85}), which is the speed at or below which 85 percent of the motorists drive on a given road. Additionally, the total number of vehicles controlled, the number of vehicles exceeding the speed limit (50 km/h) and the percentage of vehicles that exceeded the speed limit are also presented.

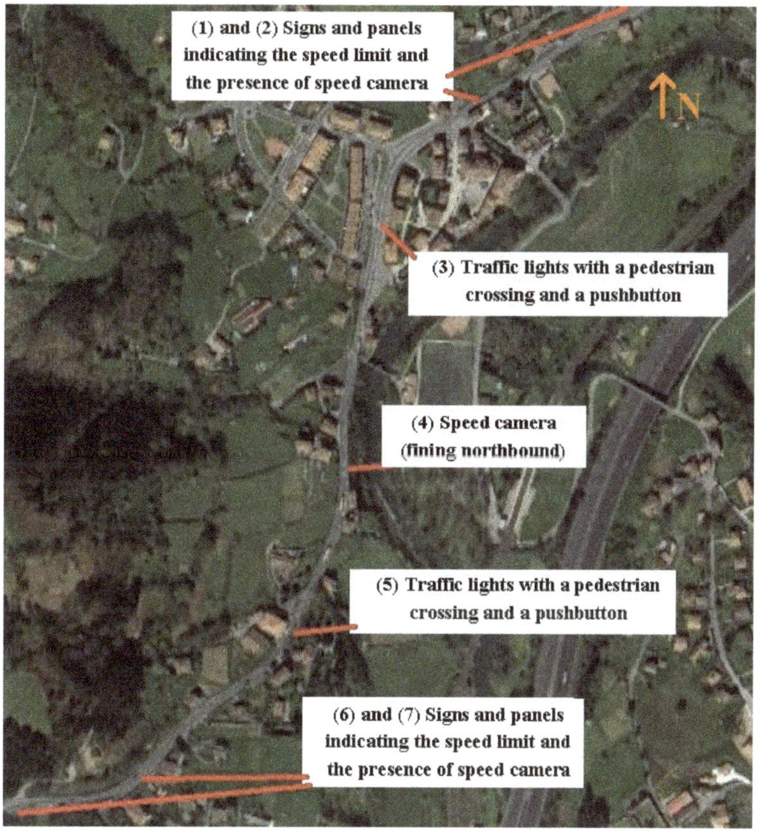

Figure 4. Locations of the traffic calming measures and other elements in Arrankudiaga (Spain).

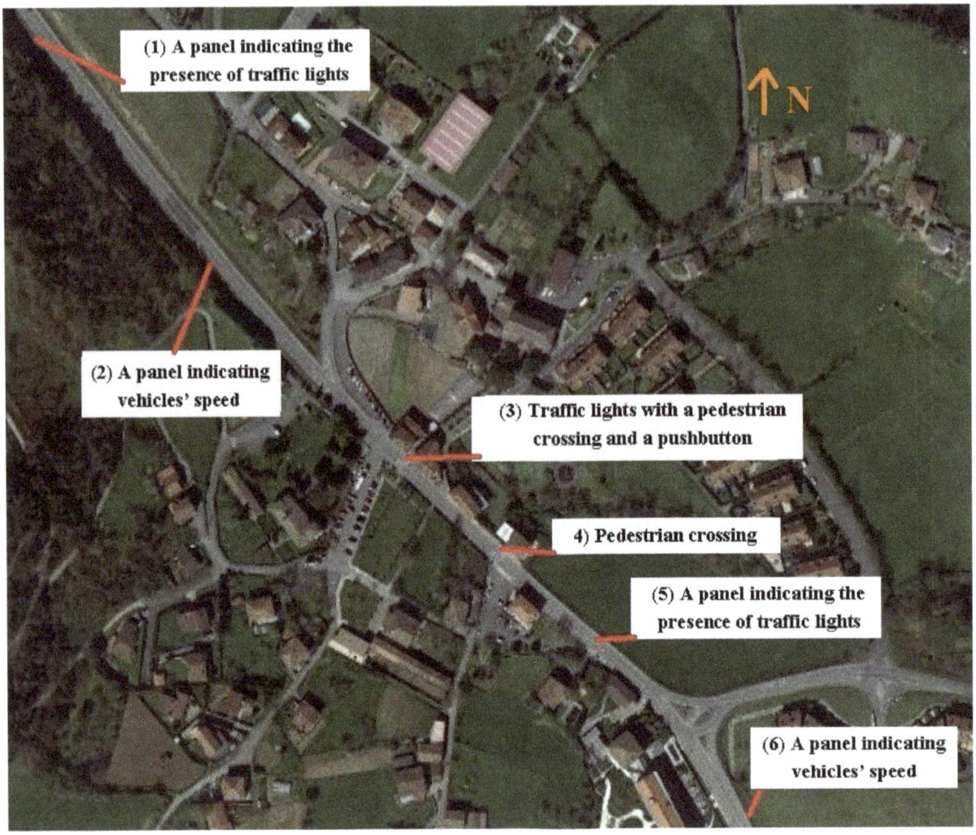

Figure 5. Locations of the traffic calming measures and other elements in Arrespalditza/Respaldiza (Spain).

3. Results and Discussion

More than 28,000 vehicles were controlled in all the locations. Results are presented individually.

3.1. Results in Chodorówka Nowa (Poland)

Table 2 presents the results for Chodorówka Nowa.

Table 2. Values of selected variables at control points in Chodorówka Nowa.

Points	S1 Entrance	S2 Radar	S3 Crosswalk	N3 Crosswalk	N2 Radar	N1 Exit
V_m (km/h)	64.5	48.6	54.4	51.7	49.5	64.6
V_{85} (km/h)	72	54	60	58	54	76
V_{max} (km/h)	96	64	76	84	65	99
Total number	200	200	200	200	200	200
Vehicles with v > 50 km/h (number)	196	69	166	111	101	178
Vehicles with v > 50 km/h (%)	98	34.5	83	55.5	50.5	89

As seen, southbound high speeds are registered at the entrance of the village, which is common, since drivers maintain the speed of an interurban road even if they are entering the area of 50 km/h speed limit. Then, in the next point, due to the presence of the radar speed camera, they speed down, reaching the lowest speeds in this controlled village.

However, 34.5% of the vehicles still exceeded the speed limit. In the final point, at the pedestrian crosswalk, speeds have increased so distinctly that the average went up over the limit. 83% of drivers violate the speed limit, even though they are still in an urban area at a place of potentially higher risk (pedestrian crossing) (Figure 6).

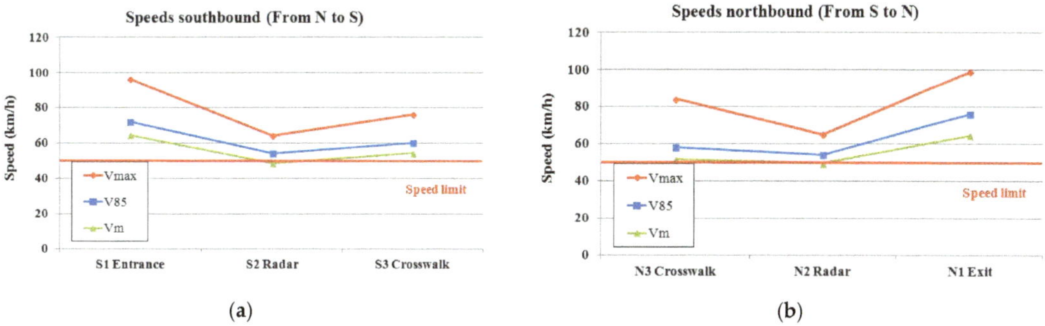

Figure 6. Speed at each controlled point in Chodorówka Nowa (Poland): (**a**) southbound, (**b**) northbound.

In the opposite direction, from south to north, lower speed values and number of violating drivers are registered. Drivers approaching the crosswalk seem to react correctly and reduce travel speed in advance. The visual effect of this is that only 55.5% exceed the speed limit comparing with 83% of speeding drivers in the opposite direction. Nonetheless, this value of exceeding drivers is still highly alarming. At the radar point, the average speed and 85th percentile are on a similar level to the values recorded in southbound direction. Despite the fact that the speed limit in force in urban areas is 50 km/h, it is common in Poland that speed cameras give drivers 10 km/h tolerance over the speed limit. This is partly due to technological limitations of the speed cameras and partly due to the limitations of the system and authority responsible for issuing the tickets. Unfavorably, this fact is generally known by drivers and they take risk of speeding over the limit, which is reflected in a high number of registered drivers' violations. Finally, at the exit of the village, which is approximately 400 m away from the speed radar, similar values to the entrance are registered.

Hence, it can be said that the radar makes drivers speed down, especially in the direction they could be fined, albeit not excessively as more than a third do not respect it, and at the crosswalk (the point more important to reduce the speed in terms of vulnerable road users' safety), the effect of the radar is not achieved. As soon as drivers pass the control point (speed radar) they accelerate regardless of the speed limit.

3.2. Results in Suchowola (Poland)

Results for Suchowola are listed in Table 3 and graphically presented in Figure 7.

Table 3. Values of selected variables at control points in Suchowola (Poland).

Points	S1 Radar	S2 Crosswalk	S5 Crosswalk	N5 Crosswalk	N2 Crosswalk	N1 Radar
V_m (km/h)	50.3	53.2	47.4	47.1	52.1	55.3
V_{85} (km/h)	54	59	53	54	58	63
V_{max} (km/h)	72	75	65	63	81	81
Total number	200	200	200	200	200	200
Vehicles with v > 50 km/h (number)	102	128	58	78	125	143
Vehicles with v > 50 km/h (%)	51	64	29	39	62.5	71.5

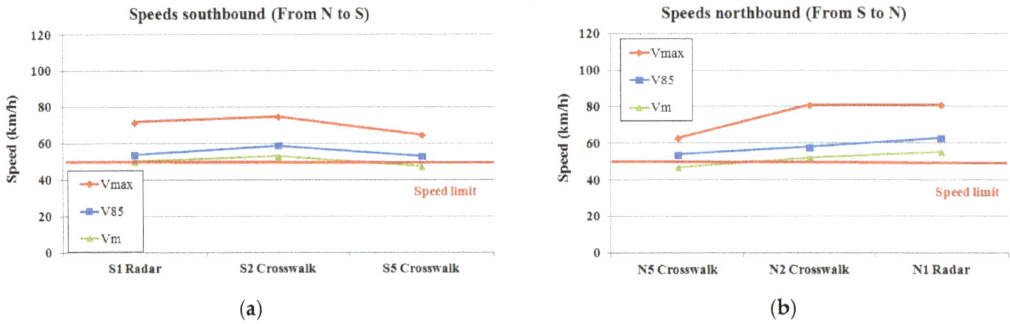

Figure 7. Speed at each controlled point in Suchowola (Poland): (**a**) southbound, (**b**) northbound.

As shown, in southbound, the first controlled point, S1, where a radar speed camera is installed, vehicles' speed decrease, and drivers adapt their values to the urban area. Although the average value is slightly over the speed limit, and hence, 51% of the vehicles do not respect the limit, but the V_{85} is very near to it, 54 km/h. It means, that approximately the majority of the vehicles adapted their speed due to the presence of the radar, although not totally conveniently. In the following control point, S2, a crosswalk, drivers speeded up and 64% exceeded the limit, meaning that the effect of the radar is very short as the pedestrian crossing is only 75 m away from the speed radar. Finally, at point S5, a crosswalk in the village located after a roundabout and a horizontal deflection the best values of the speed were obtained; only 29% of drivers travel over the speed limit and V_{85} is only 53 km/h. It clearly shows that in long urban areas speeds tend to be adapted to this area, getting better results since some vehicles turn in the intersections of the villages or new vehicles are incorporated to the main traffic flow. In the opposite direction, from South to North, very the same V_m and V_{85} values, in relation to S5, are registered in N5, due to the presence of a radar speed camera located 120 m before. In Suchowola, the speed values are almost the same in the same check points regardless the direction. In point N2, located at the end of the village, low-density housing in road vicinity makes drivers feel that the village has finished and even the presence of a crosswalk does not make speed down. The average speed is over the limit with 62.5% of drivers exceeding it. Finally, in N1, as the radar cannot fine vehicles northbound, higher values than in N2 are registered since drivers being forced to drive with a low speed along the whole village (3500 m) feel frustrated and trying to regain lost time accelerate earlier before the administrative border of the village is reached without worrying about the presence of the radar.

3.3. Results in Arrankudiaga (Spain)

Analyzed variables from the data registered in Arrankudiaga are shown in Table 4 and Figure 8.

Table 4. Values of selected variables at control points in Arrankudiaga (Spain).

Points	N5 Crosswalk	N4 Radar	N3 Crosswalk	S3 Crosswalk	S5 Crosswalk
V_m (km/h)	50.9	45.9	46.0	49.7	54.4
V_{85} (km/h)	59	51	54	59	63
V_{max} (km/h)	124	64	77	88	108
Total number	4047	794	4685	4615	4551
Vehicles with v > 50 km/h (number)	2187	141	1485	2332	3290
Vehicles with v > 50 km/h (%)	54.0	17.8	31.7	50.5	72.3

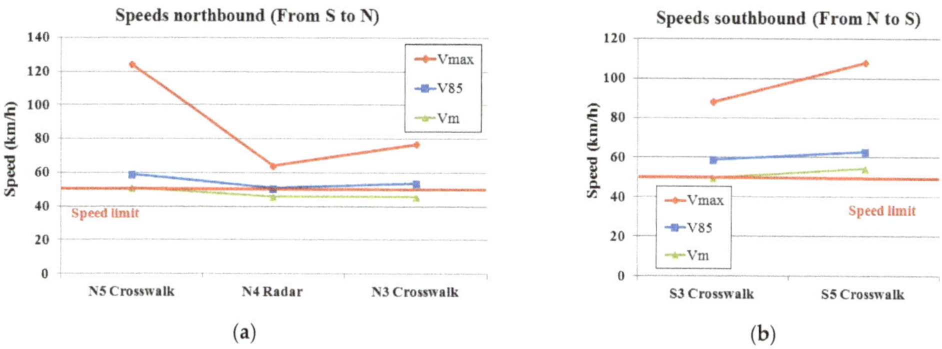

Figure 8. Speed at each controlled point in Arrankudiaga (Spain): (**a**) southbound, (**b**) northbound.

Observing the data, it can be deduced that the presence of a radar speed camera has a real influence on the vehicles' speed when comparing the values of the two directions. Northbound, the first crosswalk serves as a way to indicate that drivers are in an urban area. At the radar, in the direction to the north, the lowest speed values are registered, with the V_{85} near the speed limit (50 km/h), and only 17.8% of drivers exceed it. Later, at the following crosswalk (with traffic lights and a pushbutton), better speed values than in the previous one are obtained, meaning that the radar has an effect on drivers' behavior, making that only 31.7% of them drive with the speed over 50 km/h (Figure 8). Nevertheless, southbound results are not so satisfactory. Although there is a high AADT in the road, the traffic is local (no international traffic uses this road) and the habitual drivers know that the radar speed camera can only fine northbound. Hence, southbound drivers do not feel any possibility of infraction and higher values are recorded in both crosswalks, especially in the second one (S5), which is after the radar and near the exit of the village, with almost three quarts of the drivers exceeding the limit.

If compared with Polish data, fewer drivers exceeded the speed limit at the point of the radar. Spanish drivers also know that there is a tolerance between the speed limit and the real speed to get a fine, but it is observed that fewer drivers take the risk of exceeding the limit. The tolerance in radar speed camera measurement is a fact known in the entire European Union, but other factors can be attributed to the variable degrees of vehicles violating it such as a culture of driving fast, the observed probability of appearing a pedestrian according to the number of inhabitants in the village, etc.

3.4. Results in Arrespalditza/Respaldiza (Spain)

Analyzed variables from the data registered in Arrespalditza/Respalditza are displayed in Table 5.

Table 5. Values of selected variables at control points in Arrespalditza/Respaldiza (Spain).

Points	S2 Panel	S3 Crosswalk + Traffic Lights	S4 Crosswalk	N6 Panel	N4 Crosswalk	N3 Crosswalk + Traffic Lights
V_m (km/h)	68.6	48.5	43	63.9	46.7	52.1
V_{85} (km/h)	77	61	51	79	56	65
V_{max} (km/h)	112	112	92	117	91	113
Total number	200	1729	1483	211	1723	1951
Vehicles with v > 50 km/h (number)	196	690	255	168	539	1083
Vehicles with v > 50 km/h (%)	98	39.9	17.2	79.6	31.3	55.5

The first thing that can be observed from Table 5 is that the panels indicating the speed are not effective. They alert one about the presence of an urban area but they do not cause a speed reduction since drivers know that they cannot be fined at that point. They regard them as warning signals, but not as a punishing measure. Therefore, the majority of drivers do not respect the speed limit at these points (98% southbound and 79.6 northbound (Figure 9)). Later, at the pedestrian crossing, speed limit is more respected by motorists, with more than the half of vehicles respecting the speed limit, except for N3. The most appropriate drivers' behavior in terms of travel speed were obtained in point (3), at the location of the cross walks, where only 17.2% and 31.3% of the drivers do not respect the speed limit. Higher values were registered at 25 m from the traffic lights at the crosswalk with the pushing button, where the average speed slightly exceeds the limit of the urban area. However, V_{85} value, which is the usual variable in traffic engineering, is much above the limit.

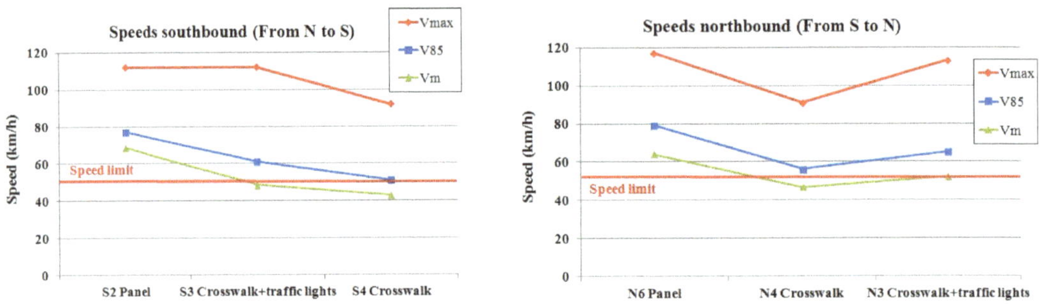

(a) (b)

Figure 9. Speed at each controlled point in Arrespalditza/Respaldiza (Spain): (**a**) southbound, (**b**) northbound.

3.5. Discussion

Very detailed equations cannot be extracted for some speed measurement in various villages since speed does not only depend on the existing traffic calming measures. Speed in short urban areas also depends on many other variables, such as, the exact location of the TCMs, the type of drivers (habitual or not), the geometry of the segment, the length of the short urban area, the usual speeds after and before the urban area, the building density of the village, the number of inhabitants, etc. Nevertheless, some general ideas are possible to be deduced from this study.

Firstly, the presence of radar speed cameras force drivers to speed down, mostly due to the risk of being fined if speed limit is exceeded. When comparing values in both directions, and the speed camera is only able to fine in one direction, better results (i.e., lower speeds) can be obtained in the direction that the radar can fine, especially if the speed camera is located in the outskirts of the built-up area. This can be clearly observed in Suchowola when values at S1 (fining) and N1 (non-fining) differ distinctly (10% difference). If the speed camera is located in a middle of a built-up area and drivers' speed is additionally affected by other factors (buildings around the road, road curvature or roundabout presence) the speeds in a certain distance are very similar regardless of the direction of the movement (S5 and N5). Another common finding for both villages in Poland is the short range of speed radar exposure. Drivers, being forced to slow down while passing the radar, start to accelerate, and their speed rises shortly after. This phenomenon is in accordance with the findings of previous researches [73,74]. At the place of the pedestrian crossings speed values depend on the direction of the movement if the speed camera is located within a reasonable distance; drivers approaching a speed camera reduce their speed more than those driving away. This is visible when comparing values in Chodorówka Nowa at the S3 (fining direction) with N3 (non-fining) directions. Furthermore, this effect of the presence of a radar speed camera and its orientation can be observed also in other elements. This is the case of Arrankudiaga, where northbound speeds were lower at the pedestrian crossing walks than southbound (N5 vs. S5 and N3 vs. S3). Hence, the positive effect of the radar speed camera on speed in that direction must be noted.

Secondly, placing traffic calming measures far from the real start of the village has low effect on vehicles' speed. This is perfectly observable in Arrespalditza/Respaldiza at points S2 and N6, where the panels indicating the speed, which include a speed limit of 50 km/h, are not effective, as drivers feel that the start of the village is not at that point and, hence, do not respect it. When compared these two panels, lower values are recorded in N6 than at S2, since N6 is placed nearer the entrance of the urban area. The effect on the closeness of the panel indicating the speed to the village has been previously investigated [68]. In that research it was shown that if the panel was further from the start of the village, higher speeds were registered. Additionally, something similar happens in Chodorówka Nowa, in N1. Although there is not a TCM at this point (it is just the panel with the village's name), and the speed limit of 50 km/h theoretically starts there, as drivers still do not really feel the presence of an urban area they regard the panel with the name as an informative one instead of a signal of a lower speed limit area. This additionally confirms the low effectiveness of administrative speed limitations (vertical signs) as a measure of speed management [75]. Consequently, with the aim of obtaining better results, TCMs at the borders between urban and non-urban areas should be placed as near as possible to the real urban area.

Thirdly, the effect of a radar speed camera is higher in the place of occurrence than at a pedestrian crossing located in some distance after the speed camera. The effect is highly disappointing comparing values of V_m and V_{85} between speed radar and the nearest pedestrian crossings. Drivers respect more a possible fine than a crosswalk, especially if no pedestrians are visible. Some points to observe this trend are S1 and S2 in Suchowola, and N4 and N3 in Arrankudiaga. In fact, radar speed camera must be regarded as a punctual TCM, which is generally respected by motorist (due to the risk of fine), but at that point, without achieving a real decrease of the speed in the subsequent segment [25,66]. In the measuring points after the radar speed cameras higher values are registered, confirming the idea of a punctual TCM. Furthermore, normally, lower speed values are measured some meters before the radar speed camera than some meters after it [66].

Lastly, if the urban segment is longer, lower values are obtained. The best example is the point S5 in Suchowola, where low speeds were measured, 970 m in urban area after S1 point, a roundabout and a horizontal deflection. The values in S5 are lower even than the ones in S1, where a radar speed camera is situated. The successive presence of measures affecting the travel speed (roundabouts, horizontal curvatures) distinctly influence on the speed. However, on the contrary, when drivers feel that the urban area is finishing, they speed up again, even if the real urban area is not finished. This can be regarded in Suchowola in points N2 and N1. The urban area is not finished "per se", but as there are not so many buildings in that area, motorists start adapting their speed to the new area, an interurban area.

4. Conclusions

With the aim of analyzing various traffic calming measures on rural roads that pass through short urban areas, four small villages without a by-pass road were selected and analyzed (two in Poland and two in Spain). Amongst analyzed traffic calming measures there were radar speed cameras and panels indicating vehicles' speed. Speed values were recorded at the place of the traffic calming measures and at the pedestrian crossings, which are the most important and sensitive places, as vulnerable road users will use them to go from one to the other side of the village.

Although the real effect of each TCM does not depend uniquely on the TCM itself, some general ideas can be obtained from the analysis of the results in these four villages.

If a radar speed camera can only fine in one direction, drivers on that direction would respect it more than in the other direction. Additionally, a positive effect was registered not only at the place of the radar speed camera. It also extended to some distance in front of and behind the camera. Values obtained at the crosswalks indicated that the presence of the speed camera in that direction directly influenced the motorists' speed behavior but that the impact diminishes with distance. It must be underlined that the effect of the speed camera does not last long, and drivers speed up after having crossed that critical point. Furthermore, it should be emphasized that the general effect of radar speed cameras is also partially mitigated by a tolerance for speeding drivers in the camera's settings. Continuous technological development should enable one to replace existing devices with new, more accurate ones, and the radar camera's speed limit thresholds should be narrowed and closer to the real limits.

Moreover, it was seen that if the TCMs are located far from the real start of the urban area, even if they impose a speed limit, they are not respected until drivers feel that the real urban area and the possibility of pedestrians appearing really started. Therefore, it is recommended to place TCMs near the real start of the built-up area.

Finally, longer urban areas make decrease the speed more, as the continuous succession of intersections have a real effect on speed. On the contrary, when motorists feel that the urban area is finished, they start speeding up to get usual speed of non-urban roads.

Author Contributions: All the authors have contributed to this work similarly. All authors have read and agreed to the published version of the manuscript.

Funding: This research was funded by GIRDER Ingenieros, S.L.P., grant number 2019.0478, by "Erasmus + programme—Call 2016—KA1—Mobility of Staff in higher education—Staff mobility for teaching and training activities", and by Project No WZ/WB-IIL/1/2020 and it was financially supported by Ministry of Science and Higher Education, Poland.

Institutional Review Board Statement: Not applicable.

Informed Consent Statement: Not applicable.

Data Availability Statement: The data presented in this study are available in the article.

Conflicts of Interest: The authors declare no conflict of interest.

References

1. Llopis-Castelló, D.; Findley, D.J. Influence of calibration factors on crash prediction on rural two-lane two-way roadway segments. *J. Transp. Eng. A Syst.* **2019**, *145*, 040190241–040190249. [CrossRef]
2. Shad, S.A.R.; Ahmad, N. Road infrastructure analysis with reference to traffic stream characteristics and accidents: An application of benchmarking based safety analysis and sustainable decision-making. *Appl. Sci.* **2019**, *9*, 2320. [CrossRef]
3. European Commission. *Annual Statistical Report on Road Safety in the EU 2020*; European Road Safety Observatory: Brussels, Belgium, 2021.
4. European Commission. *Facts and Figures—Pedestrian*; European Road Safety Observatory: Brussels, Belgium, 2020.
5. European Commission. *Road Safety Thematic Report—Pedestrians*; European Road Safety Observatory: Brussels, Belgium, 2021.
6. Hall, J.W.; Smith, K.L.; Titus-Glover, L.; Wambold, J.C.; Yager, T.J.; Rado, Z. *Guide for Pavement Friction*; Contractor's Final Report for NCHRP-Project 01-43; Transportation Research Board: Washington, DC, USA, 2009.
7. Chen, Y.; Li, Y.; King, M.; Shi, Q.; Wang, C.; Li, P. Identification methods of key contributing factors in crashes with high number of fatalities and injuries in China. *Traffic Inj. Prev.* **2016**, *17*, 878–883. [CrossRef]
8. Pérez-Acebo, H.; Gonzalo-Orden, H.; Rojí, E. Skid resistance prediction for new two-lane roads. *Proc. Inst. Civ. Eng. Transp.* **2019**, *172*, 264–273. [CrossRef]
9. Treat, J.R.; Tumbas, N.S.; McDonald, S.T.; Shinar, D.; Hume, R.D. *Tri-Level Study of the Causes of Traffic Accidents: Executive Summary*; Technical Report No. DOT HS-805 099; National Technical Information Services: Bloomington, IN, USA, 1979.
10. European Commission. *Road Safety Thematic Report—Speeding*; European Road Safety Observatory: Brussels, Belgium, 2021.
11. Rosén, E.; Sander, U. Pedestrian fatality risk as a function of car impact speed. *Accid. Anal. Prev.* **2009**, *41*, 536–542. [CrossRef] [PubMed]
12. Hesjesvol, I.; Hoye, A. Pedestrian Crossing Options. Handbook of Road Safety Measures. Available online: https://www.tshandbok.no/del-2/3-trafikkregulering/doc663/ (accessed on 15 February 2021).
13. Yue, L.; Abdel-Aty, M.; Wu, Y.; Zheng, O.; Yuan, J. In-depth approach for identifying crash causation patterns and its implications for pedestrian crash prevention. *J. Saf. Res.* **2020**, *73*, 119–132. [CrossRef]

14. Populer, M.; Chalanton, I.M. *Focant, N. Ongevallen met Voetgangers op of in de Buurt van Lichtengeregelde Voetgangersoversteekplaatsen*; VIAS Institute: Brussels, Belgium, 2018.
15. Dupriez, B.; Houdmont, A. *Gedetailleerde Analyse van Ongevallen in het Brussels Hoofdstedelijk Gewest (2000–2005)*; D/2009/0779/87; Brussels Mobility: Brussels, Belgium, 2009.
16. International Transport Forum—OECD. *Speed and Crash Risk. Research Report*; International Traffic Safety Data and Analyses Group: Paris, France, 2018.
17. The Polish General Police Headquarters. *Road Accidents in Poland in 2019*; Traffic Department: Warsaw, Poland, 2020.
18. Paszkowski, J.; Hermann, M.; Richter, M.; Szarata, A. Modelling the effects of traffic-calming introduction to volume-delay functions and traffic assignment. *Energies* **2021**, *14*, 3726. [CrossRef]
19. Ziolkowski, R.; Dziejma, Z. Investigations of the dynamic travel time information impact on drivers' route choice in an urban area—A case study based on the city of Bialystok. *Energies* **2021**, *14*, 1645. [CrossRef]
20. Solowczuk, A. Effect of traffic calming in a downtown district of Szczecin, Poland. *Energies* **2021**, *14*, 5838. [CrossRef]
21. Loprencipe, G.; Moretti, L.; Pantuso, A.; Banfi, E. Raised pedestrian crossing: Analysis of their characteristics on a road network and geometric sizing proposal. *Appl. Sci.* **2019**, *9*, 2844. [CrossRef]
22. Lockwood, I.M. ITE Traffic Calming Definitios. *ITE J.* **1997**, *67*, 22–24.
23. Ewing, R.H. *Traffic Calming: State of the Practice (FHWA-RD-99-135)*; Institute of Transportation Engineers (ITE) & Federal Highway Administration (FHWA): Washington, DC, USA, 1999.
24. Harvey, T. *A Review of Current Traffic Calming Techniques*; University of Leeds: Leeds, UK, 2013.
25. Gonzalo-Orden, H.; Rojo, M.; Pérez-Acebo, H.; Linares, A. Traffic calming measures and their effect on the variation of speed. *Transp. Res. Proc.* **2016**, *18*, 349–356. [CrossRef]
26. Kveladze, I.; Agerholm, N. Visual analysis of speed bumps using floating car dataset. *J. Locat. Based Serv.* **2018**, *12*, 119–139. [CrossRef]
27. Ziolkowski, R. Speed management efficacy on national road—Early experiences of sectional speed system functioning in Podlaskie Voivodship. *Transp. Probl.* **2018**, *13*, 5–12. [CrossRef]
28. Pérez-Sansalvador, J.C.; Lakouari, N.; García-Díaz, J. Pomares Hernández, S.E. The effect of speed humps in instantaneous traffic emissions. *Appl. Sci.* **2020**, *10*, 1592. [CrossRef]
29. Torres, J.; Cloutier, M.S.; Bergeron, J.; St-Denis, A. They installed a speed bump: Children's perception of traffic calming measures around elementary schools. *Child. Geogr.* **2020**, *18*, 477–789. [CrossRef]
30. Pérez-Acebo, H.; Ziolkowski, R.; Linares-Unamunzaga, A.; Gonzalo-Orden, H. A series of vertical deflections, a promising calming measure: Analysis and Recommendations for Spacing. *Appl. Sci.* **2020**, *10*, 3368. [CrossRef]
31. Jasiuniene, V.; Cygas, D. Analysis of older pedestrian accidents: A case study of Lithuania. *Balt. J. Road Bridge Eng.* **2020**, *15*, 147–160. [CrossRef]
32. Petru, J.; Krivda, V. The transport of oversized cargoes from the perspective of sustainable transport infrastructure in cities. *Sustainability* **2021**, *13*, 5524. [CrossRef]
33. Almoshaogeh, M.; Abdulrehman, R.; Haider, H.; Alharbi, F.; Jamal, A.; Alarifi, S.; Shafiquzzaman, M. Traffic accident risk assessment framework in Qassim, Saudi Arabia: Evaluating the impact of speed cameras. *Appl. Sci.* **2021**, *11*, 6682. [CrossRef]
34. Szagala, P.; Olszewski, P.; Czajewski, W.; Dabkowski, P. Active signage of pedestrian crossing as a tool in road safety measurements. *Sustainability* **2021**, *13*, 9405. [CrossRef]
35. Obregón-Biosca, S.A. Speed humps and speed tables: Externalities on vehicle speed, pollutant emissions and fuel consumption. *Results Eng.* **2020**, *5*, 100089. [CrossRef]
36. Panis, L.I.; Broekx, D.; Becks, C. Impact of 30 km/h Zone Introduction on Vehicle Exhaust Emissions in Urban Areas. In Proceedings of the European Transport Conference (ETC), Strasbourg, France, 18 September 2006.
37. Da Silva, F.N.; Custódio, R.A.L.; Martins, H. Low Emission Zone: Lisbon's Experience. *J. Traffic Logist. Eng.* **2014**, *2*, 133–139. [CrossRef]
38. De Vlieger, I.; De Keukeleere, D.; Kretzschmar, J. Environmental effects of driving behaviour and congestion related to passenger cars. *Atmos. Environ.* **2000**, *34*, 4649–4655. [CrossRef]
39. Liimatainen, H. Measures for Energy Efficient and Low Emission Private Mobility. In *Affordable and Clean Energy. Encyclopedia of the UN Sustainable Development Goals*; Filho, W.L., Azul, A.M., Brandli, L., Salvia, A.L., Wall, T., Eds.; Springer International Publishing: Cham, Switzerland, 2020; pp. 1–12. [CrossRef]
40. Tang, J.; McNabola, A.; Mistear, B. The potential impacts of different traffic management strategies on air pollution and public health for a more sustainable city: A modelling case study from Dublin, Ireland. *Sustain. Cities Soc.* **2020**, *60*, 102229. [CrossRef]
41. Beckx, C.; Panis, L.I.; De Vlieger, I.; Wets, G. Influence of Gear-Changing Behavior on Fuel Use and Vehicular Exhaust Emissions. In *Highway and Urban Environment. Alliance for Global Sustainability Bookseries*; Morrison, G.M., Rauch, S., Eds.; Springer: Dordrecht, The Netherlands, 2007; Volume 12. [CrossRef]
42. Wang, M.M.; Daamen, W.; Hoogendorrn, S.; Aren, B. Estimating acceleration, fuel consumption, and emissions from macroscopic traffic flow data. *Transp. Res. Rec.* **2011**, *2260*, 123–132. [CrossRef]

43. Abdi, A.; Rad, H.B.; Azimi, E. Simulation and Analysis of Traffic Flow for Traffic Calming. *Proc. Inst. Civ. Eng. Munic. Eng.* **2017**, *170*, 16–28. [CrossRef]
44. Mackie, A.M.; Ward, H.A.; Walker, R.T. *Urban Safety Project, Part 3: Overall Evaluation of Area Wide Schemes*; TRRL Report 263; Transport and Road Research Laboratory: Crowthorne, UK, 1990.
45. Sołowczuk, A.; Kacprzak, D. Identification of the determinants of the effectiveness of on-road chicanes in transition zones to villages subject to a 70 km/h speed limit. *Energies* **2020**, *13*, 5244. [CrossRef]
46. Solowczuk, A.B.; Kacprzak, D. Identification of the determinants of the effectiveness of on-road chicanes in the village transitition zones subject to a 50 km/h speed limit. *Energies* **2021**, *14*, 4002. [CrossRef]
47. *Urban Traffic Areas—Part 7—Speed Reducers*; Vejdirektoratet-Vejregeludvalget: Copenhagen, Denmark, 1991.
48. *Traffic Calming Guidelines*; Devon County Council Engineering & Planning Department: Devon, UK, 1992.
49. Sayer, I.A.; Parry, D.I. *Speed Control Using Chicanes—A Trial at TRL*; TRL Project Report PR 102; Transport Research Laboratory: Crowthorne, UK, 1994.
50. Sayer, I.A.; Parry, D.I.; Barker, J.K. *Traffic Calming—An Assessment of Selected On-Road Chicane Schemes TRL Report 313*; Transport Research Laboratory: Crowthorne, UK, 1998.
51. Crevier, C. *Les Aménagements en Modération de la Circulation, Étude et Applications*; École de Technologie Supérieure Université Du Québec: Montréal, QC, Canada, 2007.
52. Directives for the Design of Urban Roads. *RASt 06 Road and Transportation Research Association*; Working Group Highway Design FGSV: Köln, Germany, 2006.
53. Safe Road Design Manual. *Amendments to the WB Manual*; Transport Rehabilitation Project ID PO75207; Consulting Services for Safe Road Design: Loan, Sweden, 2011.
54. Hernández, E.; Abadía, X.; París, A.C. *Criterios de Movilidad ZONAS 30*; Fundación RACC: Barcelona, Spain, 2007.
55. *Guidelines for Traffic Calming*; City of Sparks, Public Works; Traffic Division: Reno, NV, USA, 2007.
56. Berger, W.J.; Linauer, M. *Speed Reduction at City Limits by Using Raised Traffic Islands*; Institut fuer Verkehrswesen (Institute for Transport Studies), Universitaet fuer Bodenkultur A-1190: Vienna, Austria, 1998.
57. Prato, C.G.; Rasmussen, T.K.; Kaplan, S. Risk Factors Associated with Crash Severity on Low-Volume Rural Roads in Denmark. *J. Transp. Saf. Secur.* **2014**, *6*, 1–20. [CrossRef]
58. Vahl, H.G.; Giskes, J. *Traffic Calming through Integrated Urban Planning*; Amarcande: Paris, France, 1990.
59. Seneci, F.; Avesani, F.; Bonomi, I. *Piani Particolareggiati per Mobilita' Ciclabile e Pedonale e Sicurezza Stradale*; Comune di Bassano del Grappa: Verona, Italy, 2012.
60. Sadeghi-Bazargani, H.; Saadati, M. Speed Management Strategies; A Systematic Review. *Bull. Emerg. Trauma* **2016**, *4*, 126–133. [PubMed]
61. González, D.D. Evaluación de las Zonas 30 en Europa y Definición de una Zona 30 Revisada. Ph.D. Thesis, Universitat Politècnica de Catalunya, Barcelona, Spain, 2012.
62. Bahar, G.B. *Guidelines for the Design and Application of Speed Humps*; Institute of Transportation Engineers: Washington, DC, USA, 2007.
63. Hallmark, S.L.; Peterson, E.; Fitzsimmons, E.; Hawkins, N.; Resler, J.; Welch, T. *Evaluation of Gateway and Low-Cost Traffic-Calming Treatments for Major Routes in Small Rural Communities*; Institute for Transportation, Iowa State University: Ames, IA, USA, 2007.
64. Ellebjerg, L. *Noise Control through Traffic Flow Measures—Effect and Benefits*; Raport 151; Danish Road Institute: Hedehusene, Denmark, 2007.
65. Daniels, S.; Martensen, H.; Schoeters, A.; Van de Berghe, W.; Papadimitrou, E.; Ziakopoulos, A.; Kaiser, S.; Aigner-Breuss, E.; Soteropoulos, A.; Wijnen, W.; et al. A systematic cost-benefit analysis of 29 road safety measures. *Accid. Anal. Prev.* **2019**, *133*, 105292. [CrossRef]
66. Gonzalo-Orden, H.; Pérez-Acebo, H.; Unamunzagas, A.L.; Arce, M.R. Effects of traffic calming measures in different urban areas. *Transp. Res. Proc.* **2018**, *33*, 83–90. [CrossRef]
67. Pérez-Acebo, H.; Otxoa-Muñoz, X.; Marquina-Llaguno, M.; Gonzalo-Orden, H. Analysis of the efficiency of traffic lights turning red in case of exceeding speed limit. *Ing. Investig.* **2021**, *41*, e86047. [CrossRef]
68. Pérez-Acebo, H.; Otxoa-Muñoz, X.; Marquina-Llaguno, M.; Gonzalo-Orden, H. Evaluation of the Efficiency of Traffic Lights Turning Red in Case of Exceeding Speed Limit with Previous Panels Indicating the Speed. In Proceedings of the XIV Congreso de Ingenieria del Transporte (CIT 2021), University of Burgos, Burgos, Spain, 6–8 July 2021; pp. 2987–2999. [CrossRef]
69. Hernández, H.; Alberdi, E.; Pérez-Acebo, H.; Álvarez, I.; García, M.J.; Eguía, E.; Fernández, K. Managing traffic data through clustering and radial basis functions. *Sustainability* **2021**, *13*, 2846. [CrossRef]
70. Pérez-Acebo, H.; Gonzalo-Orden, H.; Findley, D.J.; Rojí, E. Modeling the international roughness index performance on semi-rigid pavements in single carriageway roads. *Constr. Build. Mater.* **2021**, *272*, 121665. [CrossRef]
71. Diputación Foral de Bizkaia. *Evolución del Tráfico en las Carreteras de Bizkaia—Trafikoaren Bilakaera Bizkaiko Errepideetan 2018*; Departamento de Desarrollo Económico y Territorial: Bilbao, Spain, 2020.
72. Diputación Foral de Álava. *Estudio de Tráfico. 2018. Red de Carreteras del Territorio Histórico de Álava—2018. Trafiko Azterketa. Arabako Lurralde Historikoaren Errepide-Sarea*; Departamento de Infraestructuras Viarias y Movilidad: Vitoria-Gasteiz, Spain, 2020.

73. Ziółkowski, R. Speed Profile as a tool to estimate traffic calming measures efficiency. *J. Civ. Eng. Archit.* **2014**, *8*, 1585–1592.
74. Ziółkowski, R. Influence of Traffic Calming Measures on Drivers' Behaviour. In *9th International Conference Environmental Engineering: Selected Papers*; Vilnius Gediminas Technical University Press Technika: Vilnius, Lithuania, 2014. [CrossRef]
75. Ziółkowski, R. Effectiveness of Automatic Section Speed Control System Operating on National Roads in Poland. *Promet-Traffic Transp.* **2019**, *31*, 435–442. [CrossRef]

energies

Article

Synergy Effect of Factors Characterising Village Transition Zones on Speed Reduction

Alicja Sołowczuk * and Dominik Kacprzak

Department of Roads and Bridges, West Pomeranian University of Technology in Szczecin, al. Piastów 50, 71-311 Szczecin, Poland; dominik.kacprzak@zut.edu.pl
* Correspondence: alicja.solowczuk@zut.edu.pl; Tel.: +48-(91)-449-40-36

Abstract: There are various traffic calming measures that can be installed in village transition zones. So far, focus was placed on diversified use of pavement markings, amounts of horizontal deflection, shape of the installed chicanes or central islands, presence of gateway, etc., and their location along the transition zone. However, the combined effect of the different transition zone factors on speed reductions has been rarely studied so far. Authors put forward a hypothesis of there being some determinants, which in combination influence speed reduction. To corroborate the hypothesis on the combined impact of the transition zone features on speed reduction in the village transition zones and to validate the established relationships the authors conducted verification study in transition zones containing chicanes or central islands. To verify this hypothesis the authors studied twenty transition zones and managed to confirm the hypothesis at 95% confidence level. The authors used previously adopted binary methods, verified the previously defined factors and added a few new determinants. The contribution of this study is a further investigation of the synergy effect of various relevant factors and the findings can assist in planning new transition zones or suggest additional measures to achieve the desired speed reduction in existing transition zones.

Keywords: speed reduction; transition zone; chicane; horizontal deflection; central island; surrounding landscape; landscaping

Citation: Sołowczuk, A.; Kacprzak, D. Synergy Effect of Factors Characterising Village Transition Zones on Speed Reduction. *Energies* **2021**, *14*, 8474. https://doi.org/10.3390/en14248474

Academic Editors: Anna Granà, Margarida Coelho, Paulo Fernandes and Elżbieta Macioszek

Received: 11 November 2021
Accepted: 13 December 2021
Published: 15 December 2021

1. Introduction

Fast development of the automotive industry results in increased speeds of road traffic. Taking this into consideration, in 1990s design guidelines regarding the design of traffic calming measures were developed [1–3]. In the past, through roads usually crossed villages running through the main street on the way. After a few years, it was observed that speed limit zoning and proper traffic management were necessary on the through road sections within built-up areas [4,5] to inform drivers that the road environment had changed and that they were entering a built-up area and should adjust driving speed accordingly. Thus, the transition zone is a transitional section of the road between consecutive open and built-up areas, and it is important to obtain reduction of the speed of traffic along its length and, moreover, maintain this speed reducing effect on the way across the village [4].

To address the issues related to traffic calming in villages, it is necessary to analyse the evolution of the research related to this problem. Special methodology was applied to prepare the review of literature. The first mention of the need for traffic calming was made in the 1990s. At that time, design guidelines were mainly based on experimental studies carried out in several countries on specially adapted "artificial" test sites. The review of existing studies is based mainly on design guidelines, peer-reviewed scientific papers and reports of experimental studies published between 1990 and 2021 (Figure 1). The analysis has shown that the design guidelines were mainly developed in the last years of the 20th century and the first decade of the 21st century. In contrast, peer-reviewed scientific articles on traffic calming, as well as on traffic calming measures and their effective-ness in reducing speed and increasing traffic safety, were successively published between 2000 and 2020, with their greatest culmination in the period of 2010–2015. After 2020, few researchers have dealt with this issue.

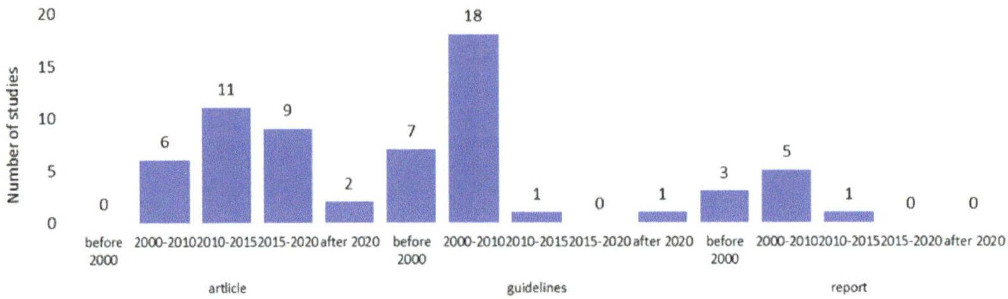

Figure 1. Progress of studies related to this issue.

Before the structural review of the literature, the authors searched the Scopus, Science Direct, Google Scholar and Google database in several phases of variously selected keywords, related to the analysed issue (Figure 2). Over a thousand publications have been found, mostly related to traffic calming. Therefore, a selection of keywords was necessary. After several combinations of keywords, a dozen or so design guidelines were finally selected, in which the principles for the design of horizontal and vertical traffic calming measures (TCM) were formulated. Several dozen peer-reviewed scientific articles and over a dozen reports from experimental studies have also been included. An overview of the final selection of literature based on different search phases and keywords is presented in Figure 2.

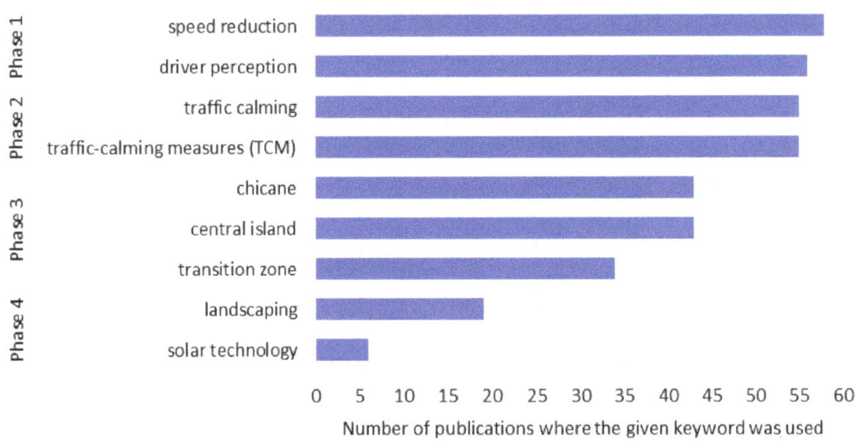

Figure 2. Number of found studies for particular keywords.

The main objective of traffic calming schemes implemented on through road sections crossing villages was to improve safety. This transpired for instance from studies [2,6,7] which demonstrated a possible direct relationship between road safety and reduced speed of traffic. The first traffic safety analyses on through roads crossing rural areas were conducted in the 1990s in Denmark [8], in the UK [1,9–11] and, later on, also in Canada [12], in Germany [13], in Sweden [14] and in the US [15–17]. Consequently, speed zoning and traffic calming schemes were recommended in many countries for through roads of lesser importance. An analysis of the traffic data from various countries compiled in Table 1 demonstrated that after implementation of traffic calming measures and various speed management strategies road crash rates were reduced on average by ca. 40–50% and 20% in rural areas and in village transition zones respectively.

Table 1. Compilation of the reductions in road crash rates on road sections through villages due to the use of traffic calming measures and various speed management strategies.

Rural Area		Transition Zone in Rural Area, after Implementation of Traffic Calming Measures	
Country and Source	**Reduction in Crash Rate**	**Country and Source**	**Reduction in Crash Rate**
UK [10]	54%	Ireland [18]	20%
Austria [19]	45%	Italy [20]	50%
Lithuania [21]	40–50%	Canada [22]	46%
Sweden [23]	47%	US [18]	20%
US [17]	40–50%	–	–
WHO [24]	44%	–	–

As mentioned before, traffic calming is relative to road traffic safety, which is directly associated with speed reduction. The obtained speed reduction in transition zones is further associated with reduction of noise, fuel consumption and air pollution, in the village transition zone and more importantly in the central zone (i.e., in the built-up area itself). In practice, traffic calming problems refer mainly to built-up areas. That said, part of the studies on the subject covered also the transition zones. In line with the research conclusions given in [4] there is still a wide array of problems referring to transition zones that need further investigation. Each location of a transition zone requires a specific approach to the aspects of landscape engineering, as they can be surrounded by a forest, open agricultural land, or land partially planted with greenery. Moreover, according to the conclusions presented in the guidelines [5,25], the easiest and the most economic traffic calming method was to implement proper traffic management measures (i.e., road signs informing drivers about the requirement to reduce speed) and various road markings. However according to [5,25] these measures turned out to be insufficiently effective. Thus, there arose a need for additional traffic calming measures. Table 2 shows the characteristics of the results of studies carried out in different countries with the use of various traffic calming schemes, involving mainly horizontal deflection.

Table 2. Considered parameters, area locations and results of previous research (supplemented based on [26,27]).

No.	Considered Parameters	Research Result	References
	Location of the study area: research track at the Transport Research Laboratory TRL		
1	Lane width, free view width, visual obstruction, length of stagger	Speed reduction results Δv are tabulated depending on of the parameters	[10]
	Location of the study area: residential road in city centre		
2	Length of stagger, narrow lane width, road width, free view width	Δv results are tabulated depending on of the parameters and type of vehicle	[2,13,28]
3	Chicanery schemes, lane width, different widths of the chicane island	Case study on a driving simulator, speed reduction results of speed v and speed reduction Δv graphically summarised	[29]
4	Various measures of traffic calming: raised pedestrian crossing, lane narrowing, speed cameras, warning signs, chicanes, middle islands	Δv results are given depending on the option of the tested traffic calming measures on different street classes	[21,30–32]
5	Horizontal deflection, length of stagger, the curves of the chicane	Recommended radii values are listed according to the speed limit on the street	[14]

Table 2. *Cont.*

No.	Considered Parameters	Research Result	References
		Location of the study area: transition zones	
6	Chicanes shape, narrowing lane width, horizontal deflection	Speed reduction results Δv are tabulated depending on the road class	[13]
7	Length of stagger l_v, horizontal deflection t_v	Δv results compiled according to the quotient l_v/t_v	[19]
8	Chicanes shape, horizontal deflection	Speed reduction results Δv are compiled depending on the v^{before} of a given road	[33]
9	Angle horizontal deflection and curb height at central islands gateways	Case study of different gateways—the analysis concerns: reducing speed, road accidents and driver distraction	[20]
10	Different types of gateways	Driving simulator tests to investigate the effect of various types of gateways on speed, standard deviation of longitudinal acceleration and deceleration (SDAD) and standard deviation of the lateral position (SDLP)	[34]
11	Different types of gateways: street trees, upgraded pavement treatments, median, lighting, signage and graphics, sculptures or public art	Driving simulator tests to investigate the effect of various types of gateways on speed v^{before} and v^{after}, spatial features of the surrounding environment, and driver behavioural factors	[35]
12	Designing Safe Road Systems (the design of the road and environment further to human) provides the background for study the effects of road design on driving behaviour.	The central theme is how design principles can reduce the probability of an error while driving.	[36]
13	Different types of gateways: trees, raised junction, lighting, signage, sculptures or public art	Driving simulator case study—research on the influence of various gateways on v^{before} and v^{after}	[15]
14	Four categories were investigated: geometric design (e.g., chicanes or central islands), traffic control devices (e.g., variable message signs or speed cameras), surface treatments (e.g., speed humps or transverse rumble strips) and roadside features (e.g., as gateways or landscaping).	Reduction Techniques for Rural High-to-Low Speed Transitions explore techniques for lowering traffic speeds in rural transition zones.	[25]
15	In transition zones with 70 km/h speed limit there were introduced three criteria pertaining to traffic management (5 factors), road condition (7 factors), landscape engineering and visibility (7 factors)	The authors introduced the binary system and one aggregated parameter being a sum of the analysed logical tautologies in all three criteria, i.e., of the confirmation of the presence of a given factor in the analysed transition zone	[26]
16	In transition zones with 50 km/h speed limit there were introduced three criteria pertaining to traffic management (7 factors), road condition (9 factors), landscape engineering and visibility (12 factors)	The authors introduced the binary system and one aggregated parameter being a sum of the analysed logical tautologies in all three criteria, i.e., of the confirmation of the presence of a given factor in the analysed transition zone	[27]

The review of the research status presented in Table 1 shows that various experimental studies on the reduction of speed of inbound traffic in transition zones have been conducted, but final conclusions regarding how to effectively obtain the expected inbound speed are yet to be made. It turned out that traffic management measures, such as a series of warning signs and road surface markings, though most relevant and easiest to apply, are insufficient when used alone [4,5,37]. Speed reduction is only one of the expected changes, and advising the drivers of an incoming change to the road environment and entering into a built-up area is a more important problem. This can be achieved through gradual application of various road geometry and traffic calming elements [38], in association with roadside landscape improvements [22], in order to make drivers slow down and keep the reduction on the way across the village. However, these changes should be implemented gradually and should provide clear information to drivers by means of conspicuous elements [26,27,38,39]. Spot traffic calming elements are effective for a short length only and the amount of speed reduction is hardly satisfactory, which has been confirmed by the results of studies presented in [29,33,35,39].

Still, one should remember that each transition zone is different and surrounded by different landscape elements, and that is why proposed design options should take into consideration these different aspects. The significance of considering the effect of the existing complexity of landscaped road scenes on speed reduction and on the driver's perception was confirmed presented in [22,40]. However, the findings of [22,38,41] confirmed that the most effective speed reduction schemes involve the use of variable geometry of the inbound lane characterised by appropriate amount of horizontal deflection, forcing drivers to change the driving path, slowing down as a result, that is installation of wide central islands or wide chicanes deflecting the inbound lane only. Furthermore, [25] found out that the path angle created by traffic deflection, especially in the case of central island, smaller than 3 degrees had no effect on speed reduction. This was confirmed also by [20], who additionally observed that actually the height of the kerb installed on the perimeter of the central island or chicane deflecting one lane had a greater impact on speed reduction than a small horizontal deflection.

Central islands or chicanes shifting the lane laterally installed on the carriageway result in the car changing its path in horizontal plane and therefore the driver must be warned of this change and get prepared for the manoeuvre [1,2,10,22,29,33]. Additionally, chicanes can differ in shape [19], size [10] and the applied soft landscaping [4,13,29]. An analysis of the installed chicanes and results of research on their impact on speed reduction presented in Table 2 demonstrated that the size of deflection of the vehicle path from a straight line and the associated path deflection angle are the two most relevant factors. However, this measure has only a localised effect [33,39].

The first results of experimental research on the effect of the size of the horizontal deflection on speed reduction were obtained on the TRL test track [10]. Other results were worked out based on in situ speed surveys conducted under natural conditions [26,27,29,33]. Nevertheless, in a majority of cases described in various studies the research data were obtained from traffic simulator tests [20,34,42–47] formulated conclusions from an analysis performed on the basis of driver survey data.

However, analysis of the research results given in the above publications indicates a big impact of the landscaped road scenes on final speed reduction results, which was particularly underlined in [5,22,40,48]. The same applies also to the combination of landscape features, road conditions and the implemented traffic management measures [26,27]. These basic relationships between the physical measures relating to the road geometry employed in traffic calming and integrated urban planning, taking into consideration the surrounding landscape features and village buildings were found already in the 1990s.

The review of the literature performed in Section 1 demonstrated that the existing design guidelines offer recommendations as to the effectiveness of specific traffic calming measures, and detailed recommendations regarding geometric parameters of traffic lane design in the UK [49], Denmark [2], and in Germany [13], yet there are no integrated design guidelines that consolidate the effect of the application of the above-mentioned solutions with consideration of the influence of the landscape surrounding the road on achieving the expected speed reduction of the entrance to the village.

According to [38] as regards village transition zones, it is most important to influence the driver's perception by facilitating his/her "reading of the road ahead", which sometimes can prove more effective than speed limit signs. Concepts such as "self-explaining" or "self-enforcing" road, were introduced in [4,38]. This is to be achieved through a visual design approach to explain the roadway function to drivers and thus achieve the desired speed of traffic. Similar conclusions on the need to consider in design guidelines different landscape elements present in village transition zones are presented also in [22,40,48]. The need to consider many elements of the roadside landscape was indicated also in [26,27], where the authors postulated three criteria characterizing transition zones on roads with 50 km/h and 70 km/h speed limits, including chicanes horizontally deflecting one traffic lane with various deflection angles.

In recent years, the development of photovoltaic technologies has also resulted in new studies conducted in transition zones, on the use of photovoltaic cells in traffic signs and road markings used in kerbs and on carriageway edges [50–52]. These are major factors associated with eyeball fixation, clear vision field and driver's perception [22,53].

The aim of this study is to confirm the hypothesis of synergistic effect of the combined influence of factors characterizing the surrounding of the transition zone to small villages on the expected speed reduction. The first study areas were analysed in [26] (70 km/h speed limit roads including chicanes) and in [27] (50 km/h speed limit roads including chicanes). In this paper, the authors studied a new area with a speed limit of 50 km/h and a central islands in the transition zones and presented new analyses. In addition, the authors performed a validation of the correlations presented in previous articles [26,27] and new correlations obtained in new study area. For conducted analysis the authors implemented a binary system of evaluation as applied in [26,27] to characterise the surroundings of the entry zone in three criteria, concerning traffic management (TM), road conditions (RC) and surrounding landscape and visibility (SLV), as shown in Section 3. As a result, statistically significant correlations were obtained confirming the effect of the combined influence of the determinants on speed reduction, thus fulfilling the purpose of the article. Sections 4 and 5 present the main conclusions and future research directions. Figure 3 below shows the structure of the article using the IMRAD scheme.

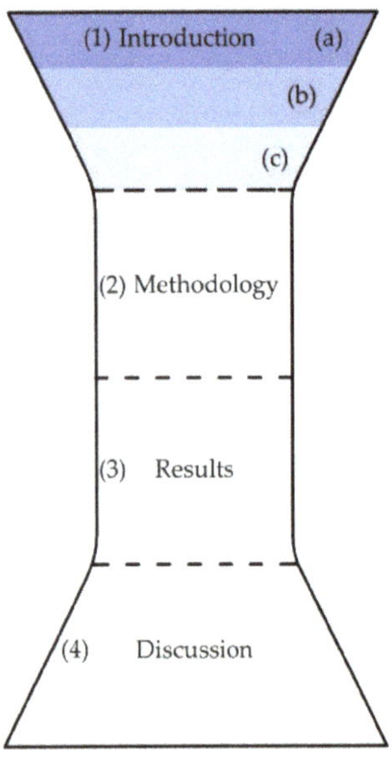

(a) The introduction discusses traffic calming, listing its primary objectives.

(b) The literature review has shown that the design principles of transition zones were formulated in design guidelines mainly between 1990 and 2010, while the culmination of scientific papers regarding this topic was between 2000 and 2010 (Figure 1). Later reduction in research intensity until 2020 has been observed.

(c) Based on the review of research results, the aim of the article was to confirm the synergy of the combined influence of factors characterizing the environment of the transition zone in small villages on the expected speed reduction.

(2) Following methodology has been used: data set from various source databases and measurements, defined research areas, statistical analyses, logic basics, binary method, aggregate parameter, statistical inference and regression analyses, validation.

(3) The results of the study, as well as the statistical analyses performed in the adopted research areas confirmed the obtained correlation, which was further confirmed in the validation process.

(4) The results of the analyses confirmed the research hypothesis of a synergistic effect of determinants on the expected speed reduction. As a result, a set of determinants characterizing transition zones was obtained, which brought novel knowledge. The developed research methodology can be used in other cases related to traffic calming.

Figure 3. IMRAD structure of the article.

2. Methodology

2.1. Study Area

Elaborating on the above problems, in this article, the authors undertook to validate the relationships presented in [26,27] and added one more study case with the transition zone located on a 50 km/h road with only a narrow central islands were installed, referred to as study area No. 3. The transition zones described in publication [26] are designated as study area No. 1 and in publication [27], as study area No. 2 (Figure 4). Figure 4 shows also the locations of four transition zones without chicanes, chosen for the purposes of comparison with the transition zones including chicanes and nine non-typical transition zones chosen for validation.

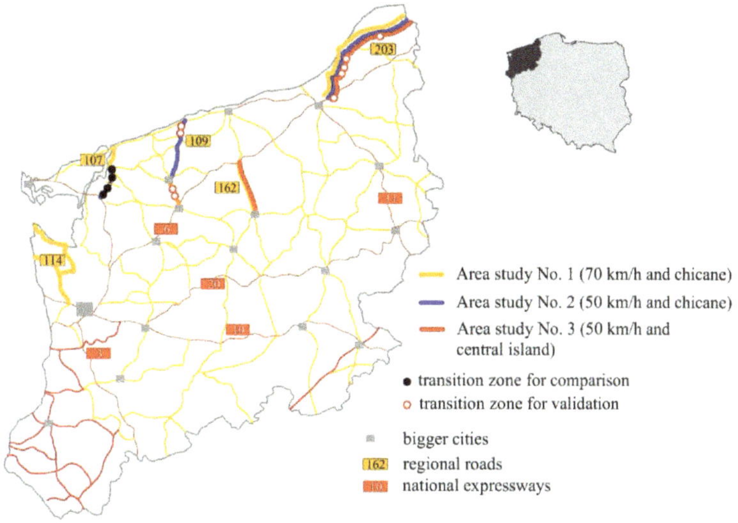

Figure 4. Map of the study area showing the locations of transition zones on regional roads in NW Poland. Source: own work of the authors.

Similar assumptions presented in previous articles [26,27] were also applied here. To validate the results presented in [26,27], in this study, for study areas No. 1 and No 2, transition zones were located on renovated provincial roads. In a village with a population of 35 to 1000 people, a short transition through the village (400 m to 2000 m) was also studied. The same research assumptions were made for the third study area. The transition zones selected for this study were located on roads of the same class with a roadway width of 6 or 7 m. When choosing the transition zones for validation purposes, 150 to 800 veh/h traffic volume was maintained, as in articles [26,27]. Mainly non-typical transition zones were chosen for validation of the relationships obtained in the study areas No. 1 and No. 2. With respect to the new study area No. 3 with central islands, the authors chose transition zones varying in terms of landscaping, roadside buildings and surrounding land (forest or farm fields). Additionally, two transition zones on each of the regional roads, with 50 km/h and 70 km/h speed limit with no chicanes in place were also chosen for analysis. This choice of additional transition zones was dictated by the intention to check whether the transition zone features (landscaping) in three criteria have a combined impact on speed reduction, with neither a median island nor a chicane in place.

2.2. Transition Zones Chosen for Validation Purposes

A few transition zones were chosen in each study area to confirm the demonstrated relationship between speed reduction and speed reduction ratio SRR on the one hand and the aggregate parameter on the other. In accordance with suggestions made in [22,38,40,48], as regards considering the transition zone features non-typical transition zones were chosen for validation for each of the respective study areas.

Taking into account the fact that on 70 km/h roads (study area No. 1) horizontal shifts in carriageway are applied in non-typical zones between villages, it is just these transition zones that were chosen for validation. Design details for on-road chicanes deflecting only one lane are given in the German guidelines [13]. Two zones without chicanes were chosen in addition.

In study area No. 2 three non-typical transition zones were chosen for validation, out of which one was located on a curvy road section with the view ahead additionally restricted by a rail bridge and two other zones were located in the same village in between built-up area composed of two distinct parts. Again, two zones with no chicanes installed were chosen additionally.

In study area No. 3 including central islands in place the following four non-typical transition zones were chosen: (A and B)—two zones, when the distance separating two villages was so small that they shared one transition zone with two opposite traffic directions (C)—one zone with buildings scattered loosely over the village area and the transition zone located amidst them (D) one zone, situated before a curvy road section, with view greatly restricted by a forest complex. Two additional zones, the same as in study area No. 2, without a central island were included in this group.

The analysis of the selected factors characterizing non-typical transition zones, located on 70 km/h roads demonstrated that four additional factors pertaining to specific conditions and presence of two staggered chicanes deflecting the traffic lanes in opposite directions should be considered. An example of such a non-typical transition zone is shown in Figure 5 below.

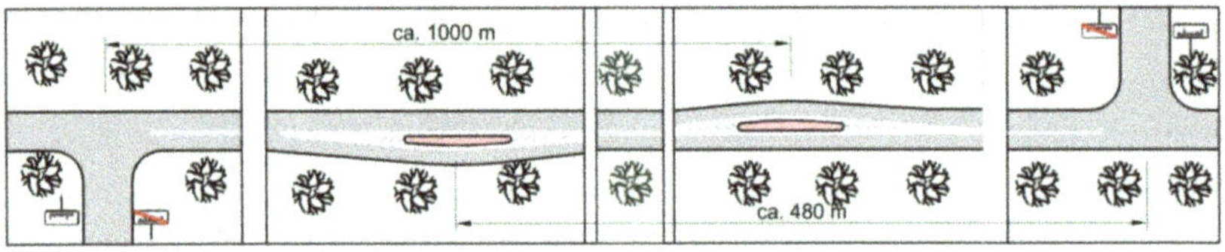

Figure 5. A non-typical transition zone with two staggered chicanes deflecting the traffic lanes in the opposite directions. Source: own work of the authors.

Additional factors in the road conditions criterion and quantification measures assumed for these factors were as follows:

- Distance between channelised intersection and the chicane (up to 500 m—1, above 500 m—0),
- Visibility of the splitter island on the channelised intersection to the driver passing through the chicane (yes—1; no—0),
- Presence of additional traffic calming measures (yes—1; no—0),
- Length of double solid line between the chicane and splitter island on channelised intersection (up to 250 m—1; greater distance or no double solid line at all—0).

2.3. Data Collection

Speed surveys were carried out in the transition zones under analysis, with simultaneous traffic counting before and after the chicane. It was assumed that speed surveys will be carried out under continuous traffic flow conditions with automatic traffic counting by means of a special device shown in Figure 6. The electronic software of the measuring device allowed for simultaneous recording of traffic volume data with breakdown of traffic by vehicle type and traffic direction and of speed data (under free and continuous flow conditions). Free traffic flow was understood in accordance with the definition given in [54], i.e., as a flow with 7 s. headway. Continuous flow was applied to all the vehicles travelling along a given road in a given time. However, for further analyses the authors chose only free-flow traffic situation for the sake of consistency of their analyses with studies by other researchers.

Based on preliminary measurements and taking into account the hourly traffic volumes in the analysed transition zones it was assumed that the survey would involve measuring the free speed of at least 100 passenger cars before and after the chicane. At sites without on-road chicanes or central islands vehicle speeds were measured before the E-17 "city/town/village" sign and after the D-42 "built-up area" sign respectively. Considering low traffic volumes and a small percentage share of heavy goods vehicles (3–8%), it was decided to measure their speeds only during the span of 100 passenger cars.

The speed measuring devices measured also the speeds of farming vehicles, yet they were not considered in further analyses.

(a)

(b)

Figure 6. The speed measuring device employed in traffic count and speed survey: (**a**) Schematic representation of the elevations of the sensors for the automatic traffic count and vehicle classification survey: C, D, E, F and G; (**b**) speed sensors A and B. Source: own work of the authors.

2.4. Research Methodology—Binary Method and Logical Tautologies

The analysis of the previously published results demonstrated lack of repeatability and homogeneity of results obtained in the tests by [10,15,19,20,33]. Sayer et al. [10] conducted the study on an "artificial" test track (Transport Research Laboratory TRL) and studied on a designated experimental roadway different shapes of chi-canes with different dimensions. The shape of the chicanes was made using horizontal and vertical elements. More than a dozen experienced drivers participated in the experiment. Many characteristic factors existing on the roads (i.e., green surroundings, open agricultural area, forest area, silhouette view of the village or single buildings, etc.) were omitted in the study. In other study [33] speed measurements were performed on existing roads with chicanes of various sizes. Unfortunately, other parameters of analysed roads were not given in the publication and there was no description of the development of the road surroundings. Beger & Linauer performed speed measurements in four transition zones with various chicanes [19]. Their results do not coincide with the results presented in [33], despite the fact that they were conducted under natural conditions on existing roads. Lantieri et al. [20] conducted speed measurements in 12 transition zones on an existing road with a length of 15 km and also obtained different results that do not coincide with the results described in [33], nor with the results reported in [19]. Dixon et al. [15] con-ducted experimental studies on various roads using traffic simulators and also the results from these studies did not match the results from the studies described above.

After analysing abovementioned information, the authors decided to analyse the results of other studies, which take into account other factors except physical effect of the chicane or central island on speed reduction in a short section of the transition zone. For instance, ref. [38] recommends considering in the transition zone design a few additional factors pertaining to the transition zone geometry, traffic calming treatments and presence of various road infrastructure elements such as bridges, culverts, traffic barriers.

While in [5,22,40,48] consideration of also the elements of the roadside landscape was further recommended. Still, it was only the guidelines [5] that proposed categorisation of certain road surrounding and landscape elements, which however referred only to built-up areas on the outskirts of towns, with pedestrian cross traffic, cycle traffic and public transport system stops.

Considering the above and taking into account the comparison of the results of other studies, summarised in Table 2, the authors adopted the research methodology presented in the flow chart below (Figure 7).

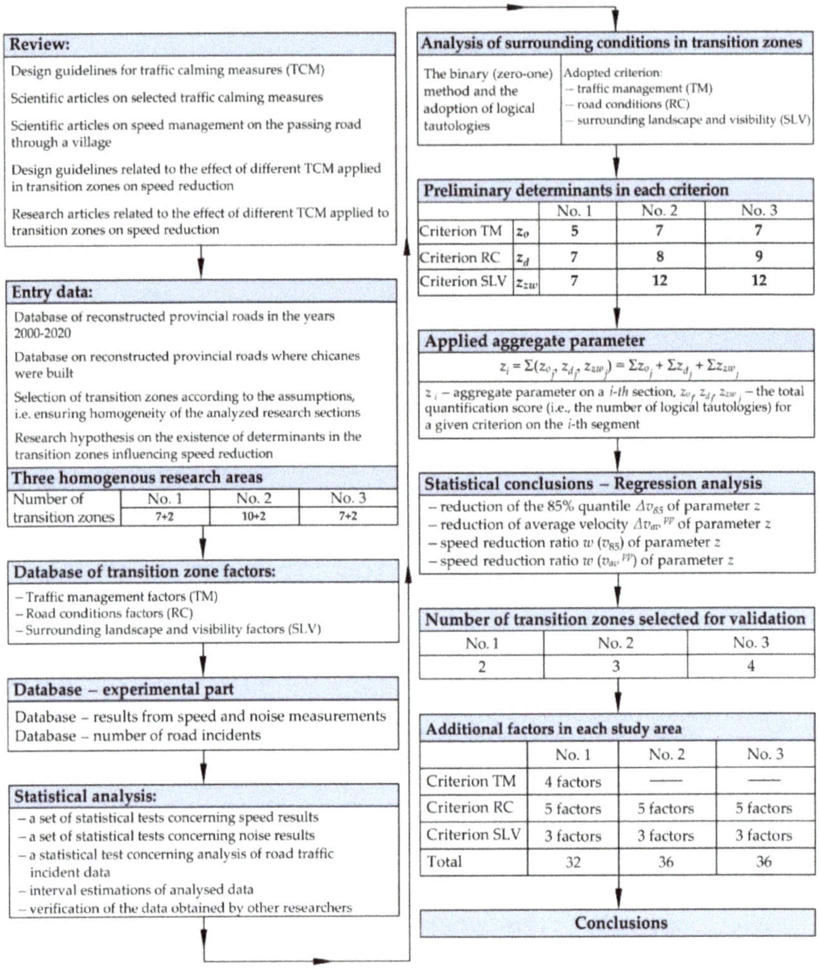

Figure 7. Flow chart of applied methodology. Source: own work of the authors.

Authors presented the results of research carried out in transition zones by taking into consideration the confirmation of the presence at the site of various factors characterising three adopted criteria by them related to: traffic management (TM), various factors related to road conditions (RC) and the surrounding landscape and visibility (SLV) [26,27]. To facilitate the evaluation of the varied influence of the selected factors on speed reduction, an aggregate parameter was introduced [26,27]. The aggregate parameter consisted in using the basics of logic and the binary method (zero-one coding) specified, for instance, in a publication by the [55]. In reference to the several dozen factors characterising the three adopted criteria, the authors proposed using logical tautologies, indicating whether a given factor is present or not present in the analysed transition zone [26,27]. The aggregate parameter was numerically equal to the sum of confirmed logical tautologies in the three adopted criteria, TM, RC and SLV. An analogical approach was proposed in this article to carry out validation of the relationships given by authors in [26,27] and compare the results obtained, which means that the binary method and logical tautologies were implemented also in this research.

Logical tautologies can be applied both for new and existing transition zones. They can also be subjected to further analysis with the aim of obtaining a practical indication on what should be improved in the transition zone to finally achieve the desired slowing effect.

2.5. Statistical Analysis

With regard to all data, it was checked if speed distribution in a given transition zone is a normal distribution, using the classic goodness-of-fit K–S test. In all cases it was confirmed that the distribution of speed populations obtained was normal and that further statistical analyses relating to them could be carried out. Different speeds are traditionally used in traffic calming studies when analysing the effectiveness of different traffic calming measures [4,38]. Based on the speed data obtained the authors assumed the following for further analyses: 85th percentile speed i.e., the speed at which or below which 85% of motorists drive on a given lane, v_{85}, mean free speed v_{av}, speed reductions Δv and SRR (calculated from the general formula SRR = $\Delta v / v^{before}$). Taking account of the SRR value in traffic calming analyses allows for comparing independent studies.

All the speed data were subjected to traditional statistical analyses. First the goodness-of-fit K–S test was performed in order to check the distribution of the investigated before and after speed populations (v^{before} and v^{after}) for normality (columns 2–3, 6–7, 10–11 in Table 3 and columns 5–6 in Table 4). As mentioned before, normal distribution was confirmed in all the speed populations under analysis.

Table 3. Results of standard statistical tests in additional transition zones accepted for validation (in relation to 85th percentile speed). Goodness-of-fit K–S test and Two–sample K–S test λ_α = 1.36 (α = 0.05), Median test $\chi_\alpha{}^2$ = 3.84 (α = 0.05).

No.	Study Area No. 1				Study Area No. 2				Study Area No. 3			
	$\lambda(v^{before})$	$\lambda(v^{after})$	λ	χ^2	$\lambda(v^{before})$	$\lambda(v^{after})$	λ	χ^2	$\lambda(v^{before})$	$\lambda(v^{after})$	λ	χ^2
1	*2*	*3*	*4*	*5*	*6*	*7*	*8*	*9*	*10*	*11*	*12*	*13*
Chosen for validation transition zones with chicanes deflecting one lane or central islands												
No. A	0.86	0.80	2.06	6.56	0.56	0.46	3.31	26.37	0.42	0.36	3.81	57.62
No. B	0.62	0.82	0.88	2.99	0.34	0.48	3.41	37.0	0.32	0.32	1.14	3.60
No. C	–	–	–	–	0.58	0.51	0.79	4.42	0.98	0.93	0.64	1.65
No. D	–	–	–	–	–	–	–	–	0.81	0.44	1.12	4.77
Chosen for validation transition zones without chicanes or central islands												
No. E	0.73	0.58	0.61	0.36	0.75	0.69	0.55	1.65	0.75	0.69	0.55	1.65
No. F	0.98	0.78	0.53	0.20	1.21	1.05	0.41	8.47	1.21	1.05	0.41	8.47

Table 4. Results of standard statistical tests on test in study area No. 3 with 50 km/h speed limit and central islands installed (in relation 85th percentile speed). Goodness-of-fit K–S test and Two–sample K–S test λ_α = 1.36 (α = 0.05), Median test $\chi_\alpha{}^2$ = 3.84 (α = 0.05).

No.	Survey Data			Statistical Test Results			
				Goodness-of-Fit K–S Test		*Two–Sample K–S Test*	*Median Test*
	$v_{85}{}^{before}$	$v_{85}{}^{after}$	Δv_{85}	λv^{before}	λv^{after}	λ	χ^2
1	*2*	*3*	*4*	*5*	*6*	*7*	*8*
No. 1	77.1	64.3	12.8	1.05	0.42	2.15	14.57
No. 2	76.5	65.5	11.0	0.38	0.67	2.20	7.03
No. 3	87.5	79.1	8.4	0.40	1.10	1.21	9.16
No. 4	82.0	75.6	6.4	0.95	0.75	2.03	2.41
No. 5	76.4	76.9	–0.5	0.41	0.54	1.08	1.76
No. 6	73.3	84.5	–11.2	0.34	0.43	1.46	5.64
No. 7	66.0	77.4	–11.4	0.25	0.75	1.52	15.45

The next statistical analysis involved checking: whether the "before" and "after" driving speed cumulative density functions belong to the same population or whether they are different, i.e., confirm the hypothesis of the desired slowing. The Two-Sample K–S test and median test were used for this purpose. Two-Sample *t*-Test was not used, as it can be applied only with positive scores of the *F*-test of equality of variances. Taking the above into consideration, the median test was used instead of the Two-Sample *t*-Test, this is in accordance with the principles of statistical analysis [56]. The results of the Two-Sample K–S test are given in Table 3—columns 4, 8, 12 and Table 4—column 7. The results of the median test are given in Table 3—columns 5, 9 and 13 and Table 4—column 8. In all the cases, when a positive Δv_{85} value was obtained both statistical tests produced positive results, which was a confirmation that the speeds measured before and after the chicane or central island were different, i.e., belonged to different populations. In the remaining cases the test results varied. The values of the 85th percentile "before" and "after" speed and of speed reduction are additionally given in Table 4 (columns 2, 3 and 4).

Based on the analyses made [26,27], authors adopted a research hypothesis that there exist determinants characterizing a transition zone in three criteria (TM, RC and SLV), whose combined energy has an effect on drivers' perception making them slow down. Only in very specific and rare situations can a single factor cause local speed reduction. In order to prove the above hypothesis or to validate the published relationships, the authors ensured homogeneity of the transition zones chosen, in accordance with the general theory of experimental research.

Taking the above into account, in study area No. 3 five transition zones with 1 m lateral shift and two transition zones with 1.5 m lateral shift were chosen for analysis. No additional factors characterizing a given transition zone were considered.

Figure 8 shows the accumulated relationships between speed reduction and its corresponding SRR value and the aggregate parameter for a given transition zone, in relation to the data from the publication [26,27] supplemented with new data from the study area No. 3.

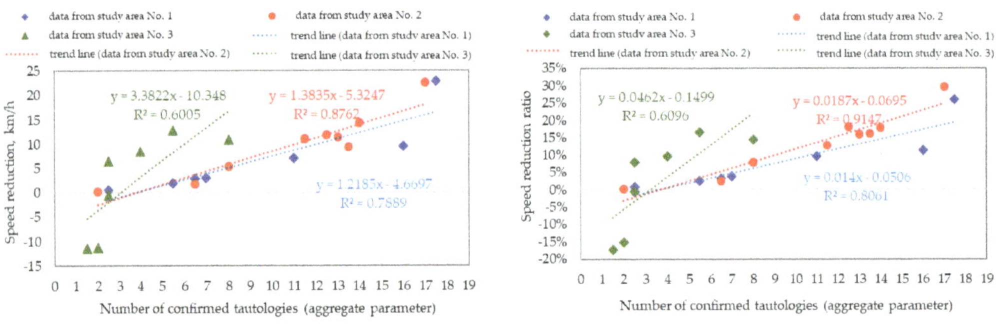

(a) (b)

Figure 8. Relationship between speed reduction and SRR on the one hand and aggregate parameter on the other in three analysed study areas. (Data from study areas No. 1 and No. 2—based on the data given in [26,27]): (**a**) speed reduction Δv_{85}; (**b**) speed reduction ratio SRR. Source: own work of the authors.

The analysis of the data presented in Figure 4, SRR in particular, justifies the division of the analysed transition zones into three study areas, depending on the posted speed limit and presence of a chicane or central island. The analysis of the value of determination coefficient R^2 also indicates that the adopted research hypothesis about the synergy of the combined impact of the proposed factors on the effectiveness of speed reduction is very likely to be true. The established functional relationships were particularly significant in the case of the study areas No. 1 and No. 2 in [26,27], in which the coefficient of determination covered 79% and 91% of the analysed data respectively. On the other hand, in the study area

No. 3 with lateral shift of only 1 to 1.5 m, significant statistical dependencies ($R^2 \approx 60\%$) were confirmed, yet no speed reduction but in fact its increase, was achieved in almost half of the transition zones, which, among other things, corroborates also the conclusions resulting from research [20,25,33] that a small deflection angle does not bring about slowing. However, it is possible that in conjunction with other factors even such a small deflection angle can achieve some speed reduction in specific cases.

3. Results

Considering the results of the statistical analyses presented in [26,27], in the next part of the analysis of the compared results regression analyses were carried out, including in each study area all the analysed transition zones, as presented in Figures 9–11 below. The regression analysis, determination model and determination factor are the same for all the analysed data and the purpose of colour coding, described in the legend is to facilitate distinguishing them in the following charts. The analysis of the established relationships demonstrated that all the data fall within the 95% prediction band, and almost all, with only small exceptions, fall within the 95% confidence interval. This testifies to robustness of the results obtained in the course of our research. The lines in the diagrams marked as 95% prediction bands are colloquially called the confidence limits for a single observation. This means that the data satisfy the basic principles of regression analysis, since they fall within this area according to [56]. High values of determination coefficient R^2 in study areas No. 1 and No. 2 indicate that speed reduction and SRR indeed depend on the combined impact of various factors operating in the transition zone, i.e., on the value of aggregate parameter proposed in [26,27]. The transition zones chosen for validation confirmed the established relationships since all but two of them fall within the 95% confidence interval in study area No. 3 (Figure 11).

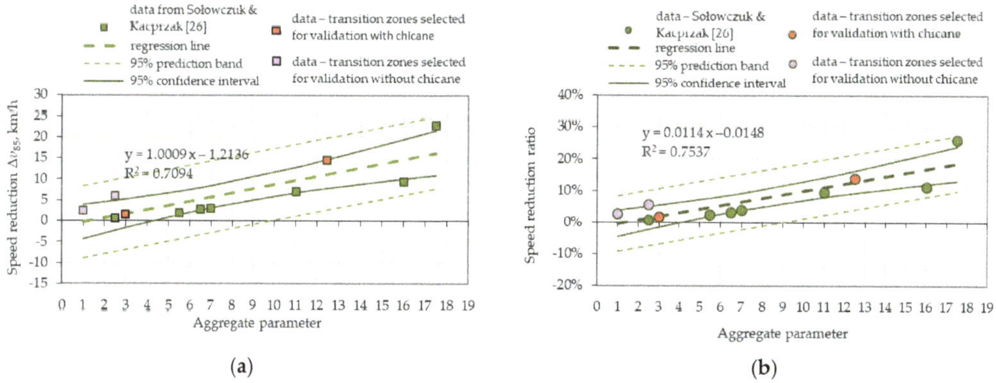

Figure 9. Linear regression function in the study area No. 1 defining the relationship between the aggregate parameter and speed reduction: (**a**) Speed reduction Δv_{85}; (**b**) Speed reduction ratio SRR. Source: own work of the authors.

The data presented in Figures 5–7 confirm also that the chosen study areas should remain unchanged. This is so, because other factors operate in transition zones on 70 km/h roads and other ones characterise the transition zones with 50 km/h speed limit. In study area No. 3 smaller values of determination coefficient R^2 were obtained, which is associated mainly with negative values of speed difference and a smaller number of logical tautologies, as well as very small horizontal deflection angles (1 m or 1.5 m). The presented results confirm also the conclusions given in [4,20,25,33], i.e., that central islands with a small horizontal deflection angle in transition zones are hardly effective as regards speed reduction.

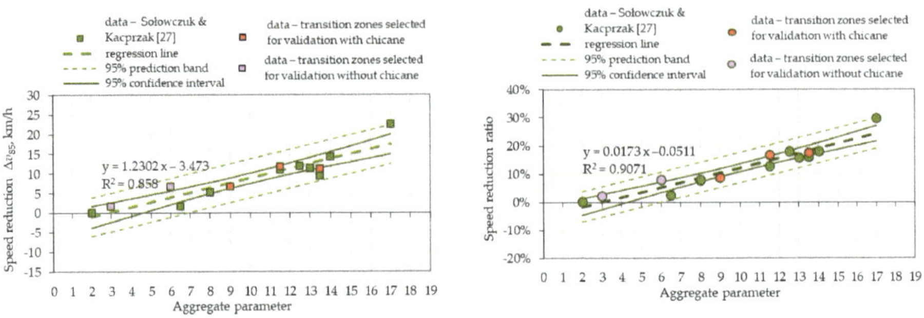

(a) (b)

Figure 10. Linear regression function in the study area No. 2 defining the relationship between the aggregate parameter and speed reduction: (**a**) speed reduction Δv_{85}; (**b**) speed reduction ratio SRR. Source: own work of the authors.

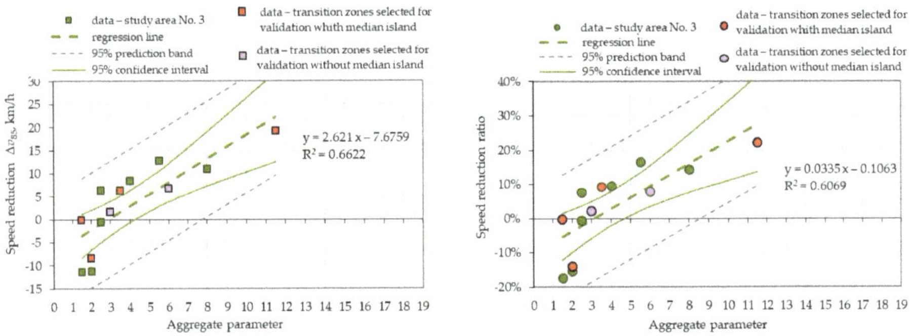

(a) (b)

Figure 11. Linear regression function in the study area No. 3 defining the relationship between the aggregate parameter and speed reduction: (**a**) speed reduction Δv_{85}; (**b**) speed reduction ratio SRR. Source: own work of the authors.

In order to obtain confirmation of the validity of the factors chosen in [26,27], correlation coefficients *R*, sums of the confirmed logical tautologies in each criterion and of the analysed speed distribution parameters in study area No. 3 and the coefficients of correlation *R* between the aggregated parameter and speed distribution parameters were compiled in Table 5 below. Bold characters denote correlation coefficients *R* greater than 0.7. Correlation coefficients *R* obtained in study area No. 3 confirm that although speed reductions may be in a significant way associated with the aggregate parameter, they are only slight, if any, due to a small horizontal deflection. Moreover, speed increases were even noted after passing the chicane.

Table 5. Coefficients of correlation *R* between the sum of logical tautologies in the respective criteria or the aggregate parameter and the selected speed distribution parameters (transition zones in study area No. 3 only).

The Adopted Sums of Logical Tautologies in the Three Criteria and Aggregate Parameter acc. to Sołowczuk and Kacprzak, (2020, 2021)	Speed Reduction		Speed Reduction Ratio	
	Δv_{85}	Δv_{av}	$w\ v_{85}$	$w\ v_{av}$
z_o	0.41	0.55	0.44	0.57
z_d	0.81	0.76	0.80	0.76
z_{zw}	0.61	0.47	0.62	0.49
z	0.80	0.74	0.82	0.75

Where: z_o—sum of confirmed logical tautologies for traffic management factors in the analysed transition zones; z_d—sum of confirmed logical tautologies for the traffic management factors in the analysed transition zones; z_{zw}—sum of confirmed logical tautologies for the traffic management factors in the analysed transition zones; z—aggregate parameter, being the sum of all confirmed logical tautologies in the analysed transition zones.

4. Discussion

Three criteria adopted in [26,27] and the factors characterizing the transition zone and its surroundings adopted preliminarily by authors were implemented in this article.

4.1. Traffic Management Criterion (TM)

In line with the research assumptions adopted in [26,27] under the TM criterion a preliminary analysis in the three study areas was performed, as presented in Figure 8. The factors are adopted preliminarily in study area No. 2 by authors [27] and implemented accordingly in the study area No. 3, with 50 km/h speed limit and central islands installed, are presented on the left side of Figure 8 against blue background. The choice of these factors was probably a direct result of the situations noted on site in the analysed transition zones. The central part of Figure 8 represents the quantification measures of the analysed logical tautologies adopted in [26,27]. The sums of logical tautologies in study areas No. 1 and No. 2 based on data given in [26,27], supplemented with the logical tautologies from the transition zones adopted for validation area shown on the right side of Figure 8. For better visual representation of the representative factors in this part of Figure 12, the sums of the noted logical tautologies greater than 4 are written in dark blue. Moreover, for better selection and ease of finding representative factors noted in the analysed transition zones in Figure 8, they are given in order from largest to smallest sum in respective study areas. An analysis of the data presented in Figure 12 indicates that the representative determinants in the study area No. 1 are different than the determinants in study areas No. 2 and No. 3, with a lower speed limit. Following a detailed analysis of the TM factors on 50 km/h roads [27] decided to:

- Use the view of residential buildings on the approach to chicane (line 6) in place of the village skyline (line 10),
- Use a different factor confirming the presence of a pedestrian crossing (line 4) in place of location of buildings after the road sign D-42 "city/town/village" (line 3),
- Use in the study area No. 2 location of the 40 km/h and 50 km/h speed limit signs (line 7) instead of the location of the road sign B-33 (70 km/h) in study area No. 1 (line 2).

		Adopted quantitative measures			Study area No.1	Study area No. 2	Study area No. 3
	Number of transition zones analyzed including transition zones used for validation:				9	13	11
No.	**Traffic management factors under analysis**	**1**	**0.5**	**0**	Σo_j		
1	Distance of the road infrastructure after D-42 sign	≤ 100 m		greater, none in sight	1.0	9.5	2.0
2	Location of B-33 sign (70 km/h) before the chicane	≤ 100 m		much further	7.0	-	-
3	Buildings after the D-42 sign	yes		no buildings	9.0	-	-
4	Pedestrian pavement after the D-42 sign	yes		no pedestrian pavement	-	4.0	0.0
5	Distance from the D-42 sign to the buildings located close to the road	≤ 100 m		much greater or no		0.0	0.0
6	Visible residential or farm buildings after the D-42 sign before the chicane	yes		no buildings	-	4.0	3.5
7	Location of B-33 sign (40-50 km/h) before the chicane	≤ 100 m	> 100 m	no sign	-	4.0	1.0
8	Location of additional B-33 sign (70 km/h) before the chicane	≤ 100 m		no sign	-	2.0	0.0
9	Location of the second B-33 sign (40 km/h) after the chicane	≤ 100 m		no sign	-	1.5	1.0
10	Visibility of the village skyline after the D-42 sign before the chicane	yes		no village skyline	1.0	-	-
	Max. number of tautologies in the analysed transition zones – max $\{o_j\}$				19.0	25.0	7.5

Figure 12. Comparison of initially adopted TM factors in the analysed study areas based on [26,27]. Legend: o_j—number of confirmed logical tautologies in all the analysed transition zones in a given study area; (-) indicates that the factor was excluded from consideration in a given study area.

Furthermore, noted in the study area No. 2 was an additional sign B-33 (70 km/h) in advance of the 50 km/h speed limit sign (line 8) and another B-33 40 km/h speed limit sign located after the chicane (line 9). In Figure 12 all the above replacement factors are presented against the blue background.

The analysis of the data presented in Figure 8 indicates that the most representative factors in study area No. 1 pertain to the location of sign B-33 (line 2), location of buildings after the sign D-42 (line 3) and the distance between these buildings and sign B-33 (line 5). While in study area No. 2 the most representative factors also pertain to the location of the B-33 sign (line 7), buildings in view after the D-42 sign (line 6) or pedestrian crossing (line 4) and the distance of the road infrastructure from the D-42 sign indicating the beginning of a built-up area (line 1). In the case of study area No. 3 only few factors were confirmed that disagreed with the representative factors in study areas No. 1 and No. 2. For the adopted TM factors, the following numbers of logical tautologies were confirmed in the respective study areas:

- 19 No. in the study area No. 1,
- 25 No. in the study area No. 2,
- Only ca. 8 No. in the study area No. 3.

4.2. Road Conditions Criterion (RC)

The factors of the RC criterion initially adopted in [26,27] are presented in Figure 13. The chronology of the description used in Figure 12 is similar to the description of the TM criterion in Figure 8. However, different font colours were used for the respective factors. The additional factors taken into account in study area No. 2 [27], in line with the actual road conditions are shown against green background. The additional factors taken into account by the authors in this article (Section 2.2), that characterised the non-typical transition zone selected for validation in the study area No. 1, are presented against orange background. Horizontal deflection represented by line 1 in Figure 13, turned out to be one of the most important factors, as corroborated for instance by results given in the study in [11,33]. In [26] only two measures relating to horizontal deflection in study area No. 1, since in the transition zones under analysis all the horizontal deflections fell in the range of 1–2 m. In the study area No. 2, authors adopted in [27] only two measures relating to horizontal deflection, since all the horizontal deflections under analysis fell in the range of 2–6 m. Conversely, in study area No. 3 hand horizontal deflections were only 1–1.5 m. Taking the above into consideration when comparing the three analysed study areas, the authors entered respectively three quantification measures (Figure 13), combining all the research assumptions, in order to determine robust factors relevant to speed reduction.

In the study area No. 1, the greatest number of logical tautologies was noted for the three following factors: the above mentioned horizontal deflection (line 1), presence of a horizontal curve after the chicane (line 7) and small distance between an access point and chicane (line 4). In study area No. 2, the two previously mentioned factors relating to the presence of horizontal curves in study area No. 1 (lines 2 and 9), were replaced by a single factor, combining the full characteristics of the horizontal curve (line 2), for which a large number of research tautologies was also noted. The other factors with a large number of confirmed logical tautologies were small distances between access points and junctions and the chicane (lines 3 and 4) and two factors characterizing the location of buildings in the village (lines 6 and 8).

Two transition zones with untypically placed chicanes (Figure 5) were also chosen for validation. In one of the additional zones the presence of all initially chosen factors and speed reduction of 14.4 km/h were noted, while in the other one only one logical tautology related to horizontal deflection and the associated speed reduction by 1.5 km/h were observed. Considering that schemes involving installation of chicanes horizontally deflecting one traffic lane are increasingly often employed on straight in plane road sections, the authors still recommend that these four additional factors should be also considered valid for non-typical transition zones, if the zones include staggered chicanes deflecting the lanes in the opposite directions.

No.	Road conditions factors under analysis	Adopted quantitative measures: 1	0.5	0	Σd_j No.1	No. 2	No. 3
	Number of transition zones analyzed including transition zones used for validation:				9	13	11
1	Horizontal deflection	2.15–6.0 m	1.5–2.15 m	0–1 m	6.5	9.5	0.0
2	Curved section after the chicane and the radius of the horizontal curve	yes, sharp curve, ≤ 250 m	smooth curve and large radius	lack of curve	–	8.5	1.5
3	Distance to the nearest intersection from the chicane	≤ 100 m	100-200 m	larger or no intersection	2.0	7.0	1.5
4	Distance to the nearest exit from the chicane	≤ 100 m	much greater	no exit	4.5	7.0	4.0
5	Limited visibility of the road surface after the chicane	yes ≤ 150 m	limited, ≈ 300 m	good visibility	3.0	6.0	1.5
6	Buildings or bridge/culvert on the inbound lane side on a horizontal curve after the chicanes	yes	on the opposite side ot the road	no buildinge or bridge/culvert	–	5.5	0.0
7	Curved section after the chicane	yes		no	5.0	–	–
8	Buildings located behind the pedestrian pavement after the chicane	a few buildings	single buildings	no buildings	2.0	4.0	1.5
9	Horizontal curve after the chicane, radius	sharp curve, ≤ 250 m radius	smooth curve and large radius	lack of curve	3.5	–	–
10	Chicane combined with pedestrian refuge	yes		no	–	2.0	1.5
11	Distance between the safety barrier end and the chicane	< 50 m	greater	no barier	–	0.5	1.0
12	Distance between channelized intersection and the chicane	≤ 500 m		greater than 500 m	1.0	–	–
13	Visibility of the splitter island on the channelized intersection	yes		no	1.0	–	–
14	Presence of additional traffic calming measures	yes		no	1.0	–	–
15	Length of double solid line between the chicane and splitter island on channelized intersection	≤ 250 m		greater than 250 m or none	1.0	–	–
	Max. number of tautologies in the analysed transition zones – max $\{d_j\}$				30.5	50.0	12.5

Figure 13. Comparison of initially adopted RC factors in the analysed study areas, based on [26,27]. Legend: d_j—number of confirmed logical tautologies in all the analysed transition zones in a given study area; (–) indicates that the factor was excluded from consideration in a given study area.

Keeping in mind the confirmed impact of gateways on slowing [20,22], which effect was neglected in [26,27], due to a lack of this treatment in the study areas No. 1 and No. 2, in the authors' opinion this factor should be included in the RC criterion. Taking into account the recommendations given in [20,22], according to which a gateway should be located past the chicane, in advance of the speed limit related to the built-up area and that the gateway height should be greater than the carriageway width these two factors and their corresponding quantification measures should be added to the RC criterion.

Additionally, based on the study results Lantieri et al. [20] claimed that high kerbing on the edges of chicanes or central islands is highly relevant to speed reduction. However, this measure in agricultural areas may be an obstacle to farm vehicles, as they may have to run over such high kerb with one wheel, which will lead to its premature deterioration. It should also be noted that the use of mountable kerbs results in passenger car wheels running over them freely and surfacing the shoulder with asphalt millings results in passenger and other car wheels running over the shoulder without slowing (Figures 14 and 15).

The strip of asphalt rubble "artificially" extends the travel lane sideways (Figure 14a), tempting the driver to negotiate the treatment without slowing. Figure 15a shows the existing design with high kerbing around the central island and along the carriageway edge. A missing element for the aggressive taper of 1:5 is hazard marker post to optically narrow the traffic corridor and thus make the drivers slow down. Taking the above into consideration, it is necessary to include an appropriate factor and its corresponding quantification measures into the factors under the RC criterion. However, considering the examples of shoulder, kerb or planting strip damage shown in Figures 14 and 15 and the conclusions of the research in [11,57] it is necessary to include a factor related to the use of bollards at the entry to and exit from the chicane or central island, that determine the travel trajectory and have a strong impact on driver's perception, which was presented in visualisations in [27]. Even with the overrun areas extending over the width of three natural paving stones, as shown in Figure 15b, damaged kerbs and wheel tracks at the exit from the village past the pedestrian refuge and at the entry to the village before the refuge are clearly visible, marked in Figure 15a by orange delineations. The arrow in Figure 15a indicates a central island situated ca. 150 m from a pedestrian refuge. Similar damage to the road shoulder, reaching even 0.7 m sideways of the carriageway edge was noted at the central island.

(a) (b)

Figure 14. Examples of poor design of horizontal deflection treatments related to the road surface construction: (**a**) Agricultural machine overrunning the shoulder surfaced with milled asphalt after passing the central island; (**b**) resulting road shoulder deterioration. Source: own work of the authors.

(a) (b)

Figure 15. Example of poor design in village transition zone: (**a**) example of high kerbing installed at the pedestrian refuge situated ca. 150 m from the central island; (**b**) grass strip deterioration caused by overrunning vehicle tyres, past the pedestrian refuge depicted in Figure 11a. Source: own work of the authors.

To sum up the discussion on transition zones in small villages, the problem transition zones on roads with paved shoulders still needs to be considered. An important result found in study area No. 2 was the analysis of the different speed reductions obtained for four one-sided chicanes with a 6 m horizontal deflection of the travel path, as described in detail in the article [27]. This large variation of the obtained speed reduction can be justified by the actual different factors in the three criteria (TM, RC, SLV), confirmed in situ in the analysed transition zones [27]. The conclusions drawn by Agkol et al. [29] has been tested in this study on a two-lane two-way road with roadway narrowing to one lane and the recommended use of two chicanes consecutively. However, considering that both lanes should be maintained on passing roads, the authors propose to add one more factor in the RD criterion and then apply one of the proposed solutions shown in Figure 16. The solution is to apply a bulb-outs island before the chicane and remove the paved shoulder after the chicane. However, no such studies have been carried out so far. This being so, in designing the proposed scheme the engineer can only refer to the principles of design of small roundabouts on roads with hard shoulders, removed before the roundabout. Thus, further research is needed, even with the use of road traffic simulation techniques.

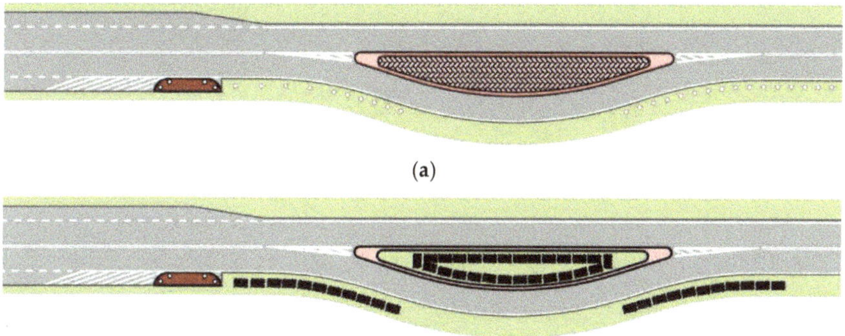

(a)

(b)

Figure 16. Chicane scheme, where the chicane is preceded by a bulb-out with removal of hard shoulders following the chicane: (**a**) with bollards; (**b**) with hedgerows. Source: own work of the authors.

Recapitulating the above observations, a compilation of the additional factors and their corresponding quantification measures, which should be taken into account in the RC criterion is given in Table 6.

Table 6. Additional factors and their corresponding quantification measures.

No.	Factors	Quantification Measures		
		Confirmed Logical Tautology (1)	Intermediate Value (0.5)	Unconfirmed Logical Tautology (0)
1	Gateway located past the chicane closer to the village	yes	–	no gateway
2	Gateway height	≥6 m	lower	no gateway
3	High kerbing around chicane or central island	yes	–	low or crossing type kerb
4	Use of bulb-outs before the chicane	yes	–	no bulb-outs
5	Removal of paved shoulders behind the chicane on a two-way two lane roads	yes	–	no

4.3. *Surrounding Landscape and Visibility Criterion (SLV)*

Surrounding landscape and visibility (SLV) is the last criterion adopted in [26,27]. The factors in the three study areas are presented graphically in Figure 17 similarly to the previous criteria except for different colour coding. In the study area No. 2 authors applied five factors in addition (lines 8–12) to account for different surroundings or approach conditions with 50 km/h speed limit [27]. In the study area No. 1, the greatest number of confirmed logical tautologies was obtained for the surrounding landscape (line 1) and restricted view of the road ahead (line 5). In the study area No. 2, the greatest number of confirmed logical tautologies was obtained for the first six SLV factors (lines 1–5). In the study area No. 3, the most important factors were the surrounding features (line 1) and a view of the nearby village buildings (line 6) and their distance from the central island (line 4). No more logical tautologies which could increase slowing were found in this study area. The additional five factors at 50 km/h speed limit were noted only in rare cases and concerned primarily the driver's perception and the associated driving speed on the approach to the chicane. The number of confirmed logical tautologies was small, yet whenever confirmed, they had a considerable effect on the achieved slowing.

				Study area	No. 1	No. 2	No. 3
				Number of transition zones analyzed including transition zones used for validation:	9	13	11
		Adopted quantitative measure					
No.	Surrounding landscape and visibility factors under analysis	1	0.5	0	Σzw_j		
1	Landscape elements in the surroundings of the chicane	open space and visible village skyline	groves or tree row and poorly visible village	forest and no buildings in view	4.0	9.0	5.5
2	Lateral obstructions visible after the chicane	visible		not visible	1.0	9.0	2.0
3	Closely spaced buildings at a close distance, visible on the way through the chicane	visible	poorly visible	not visible	3.5	8.0	2.5
4	Distance between the chicane and buildings in view situated close	< 100 m	ca. 200–250 m	greater > 350 m	2.5	8.0	4.0
5	Limited visibility the road ahead	yes	moderate	no visibility limited	4.0	6.5	1.5
6	Clearly visible single buildings or pedestrian pavements located close to the road	cleary visible		poorly visible or not visible	3.0	6.0	4.0
7	Village skyline visible from the approach to the chicane	< 300 m, closely spaced buildings	>300 m, vast built-up area	village skyline not visible	3.5	0.0	0.0
8	Behind the chicane an additional sign B-33 (40 km/h) is visible	visible		not visible	–	3.0	0.0
9	Visible buildings or handrails of a bridge behind chicane	visible		not visible	–	2.0	0.5
10	Traversable median after the chicane	visible		not visible	–	1.0	0.0
11	Barriers on the side of the road, separating them from the slope	yes		no barriers	–	1.0	0.0
12	Length of visible barriers at the approach to the chicane	300-500 m	shorter	no barriers	–	0.5	1.0
				Max. number of tautologies in the analysed transition zones – max $\{zw_j\}$	18.0	54.0	21.0

Figure 17. Comparison of initially adopted SLV factors in the analysed study areas based on the studies in [26,27]. Legend: zw_j—number of confirmed tautologies on all the analysed transition zones in a given study area; (–) excluded from consideration in a given study area.

Presence of vegetation on the chicane or central island is an important factor related to the SLV criterion that previously was omitted in [26,27], as this was not the case in the study areas No. 1 and No. 2. According to [11,15] any vertical obstruction installed on the chicane which obscures the forward view has a determining effect on slowing. Croundall in [58] postulated distinguishing two types of driver's eyeball movements: fixations and saccades. Fixations are movements made to focus on a given feature present in the visual scene and saccadea are ballistic jumps serving orientation in the space and shifting of focus from one point of interest to another. Figure 13 presents the share of driver's attention paid to the different elements of the road, i.e., summed fixation on a single element, taking into account saccadea. The above theories on allocation of visual attention were confirmed also by the studies of Werneke & Vollrath [59], who additionally related perception of road elements in the inner fringe with the speed of travel. Fixations of the driver's eyeballs and their orientation on features present in the inner fringe were also confirmed by Babkov [53] leading to the conclusion that the driver focuses the eyes primarily on the road centreline and its nearest surroundings (Figure 18), which confirms the relevance of soft landscaping in the chicane or central island to the slowing effect [15,20]. Dixon also made a distinction between lower and higher vegetation [15], which, when falling in the inner fringe, can hinder the view of the road ahead. Soft landscaping in chicanes is popularly used in the Netherlands with due attention paid in the guidelines [60] and report [4]. The conclusions from the studies by Babkov [53] and Chartier [43] are compiled in Figure 19 and an example of a chicane containing vegetation, located in the outskirts of a Dutch town is shown next to it (Figure 19).

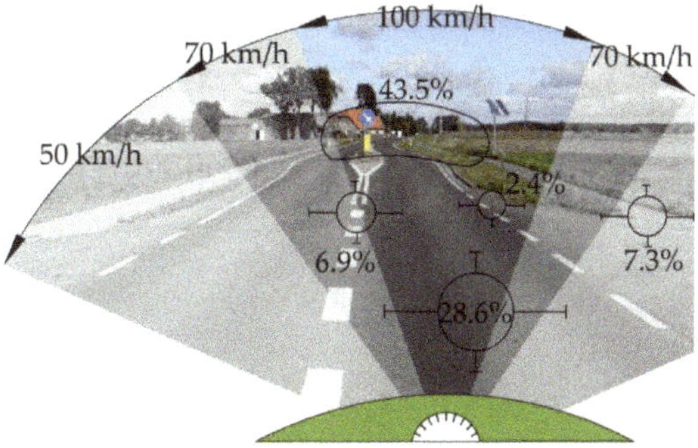

Figure 18. Conclusions of the studies by authors [26,27]. Source: own work of the authors.

Figure 19. Example of a chicane containing vegetation. Source: Google Earth [61].

Additional factors should be included in the SLV criterion based on the conclusions made by Jun et al. [50] and Jägerbrand et al. [51,52] in their studies on the use of solar-powered devices installed in the chicane or central island kerbs and on the roadway edge and their effect on the driver's perception and slowing down.

To sum up the above observations, additional factors to be included in the SLV criterion and the applicable quantification measures are compiled in Table 7.

Table 7. Additional factors and the applicable quantification measures.

No.	Factors	Quantification Measures		
		Confirmed Logical Tautology (1)	**Intermediate Value (0.5)**	**Unconfirmed Logical Tautology (0)**
1	Use of vegetation in the chicane or central island area	yes, shrubs obscuring the view	low shrubs	no vegetation
2	Bollards installed in the shoulder alongside the chicane or central island	yes	–	no bollards
3	Use of LED SolarCAP i SolarCAP Road solar-powered devices	yes	–	none

5. Conclusions

Altogether, ten factors have been proposed in the TM criterion, eighteen in the RC criterion, including the new factors and fifteen in the SLV criterion. From the data presented in Figures 4–6 it transpires that for a minimum speed reduction of 5 km/h at least 6–8 robust logical tautologies should be confirmed in the field.

In the case of existing transition zones with a chicane or central island, in which the desired slowing was not achieved, it would be advisable to apply a few factors which are likely to increase the speed reduction. Cheaper options, such as gateways, solar-powered devices fitted in the kerbs and on the roadway edge, barriers [10] (p. 11 Figure 7) or hedgerows [4] (p. 46 Appendix A), solar-powered lighting, etc., can be used in this case.

On the other hand, in the case of transition zones planned as part of the road renewal projects it is recommended to analyse the road surroundings as the first step (for example by examining satellite imagery of the approach section to the village) and on this basis locate a chicane or central island, considering various relevant factors, such as view of the village skyline or scattered buildings, presence of road infrastructure elements, etc. Next, the appropriate amount of horizontal deflection should be determined, taking into consideration the available width of the right-of-way. For transition zones located on a straight section with a narrow right-of-way it is advisable to analyse the possible horizontal deflection of the route with shifting the road centreline in order to fit in a minimum 3 m wide chicane deflecting one lane. Use of vegetation, solar-powered devices, barriers or hedgerows is also suggested for planned transition zone schemes to improve the effectiveness of speed reduction measures, in line with the conclusions of the studies of [11,15,20,22,50–52].

Author Contributions: Conceptualisation, A.S.; methodology, A.S.; validation, A.S., D.K.; formal analysis, A.S. and D.K.; investigation, D.K.; resources, D.K.; data curation, D.K.; writing—original draft preparation, A.S. and D.K.; writing—review and editing, A.S. and D.K.; visualisation, A.S. and D.K.; supervision, A.S.; project administration, D.K.; funding acquisition, D.K. All authors have read and agreed to the published version of the manuscript.

Funding: The research included in the paper are partly financed by Dean Grants of Faculty of Civil and Architecture Engineering West Pomeranian University of Technology in Szczecin: DKD Decision No. 517-02-033-6728/17.

Institutional Review Board Statement: Not applicable.

Informed Consent Statement: Not applicable.

Data Availability Statement: Not applicable.

Conflicts of Interest: The authors declare no conflict of interest.

References

1. *Traffic Calming Guidelines*; Devon County Council Engineering & Planning Department: Devon, UK, 1992.
2. *Urban Traffic Areas—Part 7—Speed Reducers*; Vejdirektoratet-Vejregeludvalget: Copenhagen, Denmark, 1991.
3. Vahl, H.G.; Giskes, J. *Traffic Calming through Integrated Urban Planning*; Amarcande: Paris, France, 1990.
4. Stamatiadis, N.; Kirk, A.; Cull, A.; Dahlem, A. *Transition Zone Design—Final Report*; Research Report Kentucky Transportation Center KTC—13-14/SPR431-12-1F; University of Kentucky: Lexington, MA, USA, 2013.
5. *Guidelines for Urban/Rural Speed Thresholds, RTS 15*; Land Transport Safety Authority: Wellington, New Zealand, 2002.
6. Poe, C.M.; Tarris, J.M.; Mason, J.M. *Relationship of Operating Speeds to Roadway Geometric Design Speeds*; Report FHWA-RD-96-024, PTI RR 9606; Pennsylvania State University: University Park, PA, USA; Federal Highway Administration: Washington, DC, USA, 1996.
7. Prato, C.G.; Rasmussen, T.K.; Kaplan, S. Risk Factors Associated with Crash Severity on Low-Volume Rural Roads in Denmark. *J. Transp. Safety Secur.* **2014**, *6*, 1–20. [CrossRef]
8. *Urban Traffic Areas, Booklet 3, Cross-sectional Profiles*; Danish Road Directorate—Road Regulation Committee DRD-RRC: Copenhagen, Denmark, 2000.
9. *Roads Development Guide*; Strathclyde Roads & South Ayrshire Council SAC: Ayrshire, UK, 1995.
10. Sayer, I.A.; Parry, D.I. *Speed Control Using Chicanes—A Trial At TRL*; TRL Project Report PR 102; Transport Research Laboratory: Crowthorne, UK, 1994.
11. Sayer, I.A.; Parry, D.I.; Barker, J.K. *Traffic Calming—An Assessment of Selected On-Road Chicane Schemes, Report 313*; Transport Research Laboratory TRL: Crowthorne, UK, 1998.
12. Crevier, C. *Les Aménagements en Modération de la Circulation, Étude et Applications*; École De Technologie Supérieure Université Du Québec: Montreal, Canada, 2007.
13. *Directives for the Design of Urban roads. RASt 06*; Road and Transportation Research Association; Working Group Highway Design; FGSV: Köln, Germany, 2006.

14. *Safe Road Design Manual, Amendments to the WB Manual*; Transport Rehabilitation Project ID PO75207; Swedish National Road Consulting AB (SweRoad): Solna, Sweden, 2011.
15. Dixon, K.; Hong Zhu, H.; Ogle, J.; Brooks, J.; Hein, C.; Aklluir, P.; Crisler, M. *Determining Effective Roadway Design Treatments for Transitioning from Rural Areas to Urban Areas on State Highways*; Final Report SPR 631, FHWA-OR-RD-09-02; Federal Highway Administration: Washington, DC, USA, 2008.
16. *Guidelines for Traffic Calming, Public Works*; Traffic Division; City of Sparks: Sparks, NV, USA, 2007.
17. *Streetscape Design Guidelines, Chapter 6*; City of Seattle Staff Directory: Seattle, DC, USA, 2020.
18. Curtis, L. *Traffic Calming of Towns and Villages on National Roads*; RS 472, Traffic Advisory Leaflet 1/05; Department for Transport: London, UK, 2008.
19. Berger, W.J.; Linauer, M. *Speed Reduction at City Limits by Using Raised Traffic Islands*; Institut fuer Verkehrswesen, Universitaet fuer Bodenkultur: Vienna, Austria, 1998.
20. Lantieri, C.; Lamperti, R.; Simone, A.; Costa, M.; Vignali, V.; Sangiorgi, C.; Dondi, G. Gateway design assessment in the transition from high to low speed areas. *Transp. Res. Part F Traffic Psychology Behav.* **2015**, *34*, 41–53. [CrossRef]
21. Jateikienė, L.; Andriejauskas, T.; Lingytė, I.; Jasiūnienė, V. Impact assessment of speed calming measures on road safety. *Transp. Res. Procedia* **2016**, *14*, 4228–4236. Available online: https://cyberleninka.org/article/n/1426267 (accessed on 18 March 2020). [CrossRef]
22. Mok, J.H.; Landphair, H.C.; Rosenblatt-Naderi, J. Landscape improvement impacts on roadside safety in Texas. *Landsc. Urban Plan.* **2006**, *78*, 263–274. [CrossRef]
23. Hummel, T.; Mackie, A.; Wells, P. *Traffic Calming Measures in Built-Up Areas*; Literature Review, PR/SE/622/02 Vägverket TR80 2002:15779; Swedish National Road Administration: Vägverket, Sweden, 2002. Available online: https://citeseerx.ist.psu.edu/viewdoc/download?doi=10.1.1.585.960&rep=rep1&type=pdf (accessed on 18 September 2021).
24. *Global Status Report on Road Safety 2018*; World Health Organization WHO: Geneva, Switzerland, 2018; Available online: http://apps.who.int/iris/bitstream/handle/10665/277370/WHO-NMH-NVI-18.20-eng.pdf?ua=1 (accessed on 18 June 2021).
25. Forbes, G.J. *Speed Reduction Techniques for Rural High-to-Low Speed Transitions. A Synthesis of Highway Practice*; TRB's National Cooperative Highway Research Program (NCHRP) Synthesis 412; Transportation Research Board: Washington, DC, USA, 2011; pp. 1–111. Available online: https://www.infrastructureusa.org/wp-content/uploads/2011/04/nchrp_syn_412.pdf (accessed on 18 June 2020).
26. Sołowczuk, A.; Kacprzak, D. Identification of the Determinants of the effectiveness of on-road chicanes in transition zones to villages subject to a 70 km/h speed limit. *Energies* **2020**, *13*, 5244. [CrossRef]
27. Sołowczuk, A.; Kacprzak, D. Identification of the determinants of the effectiveness of on-road chicanes in transition zones to villages subject to a 50 km/h speed limit. *Energies* **2021**, *14*, 4002. [CrossRef]
28. *Local Transport Note 01/07; Traffic Calming*; Department for Regional Development (Northern Ireland) DTDRD N-I: London, UK, 2007. Available online: https://assets.publishing.service.gov.uk/government/uploads/system/uploads/attachment_data/file/918429/ltn-1-07_Traffic-calming-guidance.pdf (accessed on 10 January 2020).
29. Akgol, K.; Gunay, B.; Aydin, M.M. Geometric optimisation of chicanes using driving simulator trajectory data. In *Proceedings of the Institution of Civil Engineers—Transport*; Transport ICE Publishing: London, UK, 2019; pp. 1–11. [CrossRef]
30. Distefano, N.; Leonardi, S. Effects of speed table, chicane and road narrowing on vehicle speeds in urban areas. In Proceedings of the VI International Symposium New Horizons 2017 of Transport and Communications, Sarajevo, Bosnia and Herzegovina, 17–18 November 2018; Available online: https://www.researchgate.net/publication/328738163_EFFECTS_OF_SPEED_TABLE_CHICANE_AND_ROAD_NARROWING_ON_VEHICLE_SPEEDS_IN_URBAN_AREAS (accessed on 10 January 2020).
31. Gonzalo-Ordena, H.; Pérez-Acebob, H.; Unamunzagaa, A.L.; Arcea, M.R. Effects of traffic calming measures in different urban areas. *Transp. Res. Procedia* **2018**, *33*, 83–90. [CrossRef]
32. Yilmaz, S.; Yildiz, N.D.; Irmak, M.A.; Yilmaz, H. Livable city: An approach to pedestrianization through tactical urbanism. *Alex. Eng. J.* **2014**, *58*, 251–259. [CrossRef]
33. *Wirksamkeit Geschwindigkeitsdämpfender Maßnahmen Außerorts; Dezernat Verkehrssicherheit und Verkehrstechnik*; Hessisches Landesamt für Straßen—Und Verkehrswesen: Hessen, Germany, 1997. Available online: https://mobil.hessen.de/sites/mobil.hessen.de/files/content-downloads/Wirksamkeit%20geschwindigkeitsd%C3%A4mpfender%20Ma%C3%9Fnahmen%20au%C3%9Ferorts.pdf (accessed on 10 January 2020).
34. Ariën, C.; Brijs, K.; Brijs, T.; Ceulemans, W.; Vanroelen, G.; Jongen, E.; Daniels, S.; Wets, G. Does the effect of traffic calming measures endure over time? A simulator study on the influence of gates. *Geogr. Psychol. Transp. Res. Part Traffic Psychol. Behav.* **2013**, *22*, 63–75. [CrossRef]
35. Afghari, A.P.; Haque, M.M.; Washington, S. Applying fractional split model to examine the effects of roadway geometric and traffic characteristics on speeding behaviour. *Bus. Med. Eng. Traffic Inj. Prev.* **2018**, *19*, 860–866. [CrossRef] [PubMed]
36. Theeuwes, J.; van der Horst, R.; Kuiken, M. *Designing Safe Road Systems: A Human Factors Perspective*, 1st ed.; Taylor & Francis Group: London, UK, 2012. [CrossRef]
37. *A Policy in Geometric Design of Highways and Streets*; American Association of State Highway Transportation Officials AASTO: Washington, DC, USA, 2004.

38. Lamm, R.; Cafiso, S.; La Cava, G.; Beck, A. To What Extent the Human Being Is So Far Regarded in Modern Highway Geometric Design—An International Review and a Personal Outlook, Compendium of Papers CD-ROM. In Proceedings of the 3rd International Symposium on Highway Geometric Design, Chicago, IL, USA, 29 June–1 July 2005.
39. Stamatiadis, N. *International Scanning Tour on Highway Geometric Design*; FHWA-PL-01-026; Federal Highway Administration: Washington, DC, USA, 2000.
40. Rosenblatt-Naderi, J. Landscape design in the clear zone: Effect of landscape variables on pedestrian health and driver safety. *Transp. Res. Rec.* **2003**, *1851*, 119–130. [CrossRef]
41. Andersson, P.K.; Kjemtrup, K. Byporte: Hvordan gøres Designet Trafiksikkert? pp. 26–28. Available online: https://www.trafitec.dk/sites/default/files/publications/byporte%20artikel.pdf (accessed on 10 January 2020).
42. Caliendo, C.; de Guglielmo, M.L. Road Transition zones between the rural and urban environment: Evaluation of speed and traffic performance using a microsimulation approach. *J. Transp. Eng.* **2012**, *139*, 295–305. [CrossRef]
43. Chartier, G. *Rural-Urban Transition Zones: Problems, Principles, and Practice*; Presentation to the ITE BC Interior; Canada's Community of Transportation Professionals: Kamloops, BC, Canada, 2009.
44. Lamberti, R.; Abate, D.; De Guglielmo, M.L.; Dell'Acqua, G.; Esposito, T.; Galante, F.; Mauriello, F.; Montella, A.; Pernetti, M. Perceptual measures and physical devices for traffic calming along a rural highway crossing a small urban community: Speed behavior evaluation in a driving simulator. In Proceedings of the Transportation Research Board 88th Annual Meeting, Washington DC, USA, 11–15 January 2009; pp. 1–15. Available online: https://trid.trb.org/view/882174 (accessed on 18 January 2020).
45. Yssaad, R. Analyse Psychophysique du Champ Visuel—Détection, Identification, Effet de groupement et Apprentissage Perceptif. Ph.D. Thesis, Université Lumière, Lyon, France, 28 September 2001. Available online: http://theses.univ-lyon2.fr/documents/lyon2/2001/yssaad_r#p=0&a=top (accessed on 18 July 2021).
46. Xu, T.; Sun, X.; Wang, W.; He, Y. Speed transition zone design based on driving simulator research. In Proceedings of the Tenth International Conference of Chinese Transportation Professionals (ICCTP), Beijing, China, 4–8 August 2010; Available online: https://ascelibrary.org/doi/abs/10.1061/41127%28382%2945 (accessed on 10 February 2020).
47. Wang, X.; Bie, Y.; Qiu, T.Z.; Niu, L. Effect of speed limits at speed transition zones. *Can. J. Civil Eng.* **2017**, *44*, 10–17. [CrossRef]
48. Stamatiadis, N.; Pigman, J.; Hartman, D. *Safety Consequences from Flexibility in Highway Design for Rural Communities*; Final Report NCHRP 15-22; TRB, National Research Council: Washington, DC, USA, 2004.
49. *Roads Development Guide*; Strathclyde Regional Council: East Ayrshire, UK, 1995.
50. Jun, W.; Ha, J.; Jeon, B.; Lee, J.; Jeong, H. LED traffic sign detection with the fast radial symmetric transform and symmetric shape detection. In Proceedings of the 2015 IEEE Intelligent Vehicles Symposium (IV), Seoul, Korea, 28 June–1 July 2015. [CrossRef]
51. Jägerbrand, A.K.; Johansson, M.; Laike, T. Speed responses to speed humps as affected by time of day and light conditions on a residential road with light-emitting diode (LED) road lighting. *Safety* **2018**, *4*, 10. [CrossRef]
52. Jägerbrand, A.K.; Sjöbergh, J. Speed responses of trucks to light and weather conditions. *J. Cogent Eng.* **2019**, *6*, 1685365. Available online: https://www.tandfonline.com/doi/full/10.1080/23311916.2019.1685365 (accessed on 10 February 2020). [CrossRef]
53. Babkov, V.F. *Road Design Parameters and the Safety of Traffic*; WKŁ: Warszawa, Poland, 1975.
54. Lamm, R.; Choueri, E.M.; Psarianos, B.P. A practical safety approach to highway geometric design, international case studies. In Proceedings of the International Symposium on Highway Geometric Design Practices, Boston, MA, USA, 30 August–1 September 1995.
55. *Logic, Chapter 1*; University of Victoria, Department of Mathematics and Statistics: Victoria, BC, Canada, 2015; Available online: https://www.math.uvic.ca/faculty/gmacgill/guide/Logic.pdf (accessed on 18 July 2021).
56. Taylor, J.R. *An Introduction to Error Analysis. The Study of Uncertainties in Physical Measurements*, 2nd ed.; University Science Books: Sausalito, CA, USA, 1997.
57. Yang, Q.; Overton, R.; Han, L.D.; Yan, X.; Richards, S.H. Driver behaviours on rural highways with and without curbs—A driving simulator based study. *Int. J. Inj. Control. Safety Promot.* **2013**, *21*, 115–126. [CrossRef] [PubMed]
58. Crundall, D.; Underwood, G. Visual attention while driving. In *Handbook of Traffic Psychology*; Porter, B.E., Ed.; Elsevier: Amsterdam, The Netherlands, 2011; pp. 137–148. [CrossRef]
59. Werneke, J.; Vollrath, M. What does the driver look at? The influence of intersection characteristics on attention allocation and driving behawior. *Accid. Anal. Prev.* **2012**, *45*, 610–619. [CrossRef] [PubMed]
60. van Schagen, I. (Ed.) *Traffic Calming Schemes, Opportunities and Implementation Strategies, R-2003-22*; SWOV Institute for Road Safety Research: Leidschendam, The Netherlands, 2003; Available online: https://www.gtkp.com/assets/uploads/20100110-011359-5067-R-2003-22.pdf (accessed on 18 July 2021).
61. Google Earth. Available online: http://www.earth.google.com (accessed on 15 July 2019).

energies

Article

Assessing the Environmental Performances of Urban Roundabouts Using the VSP Methodology and AIMSUN

Francesco Acuto [1], Margarida C. Coelho [2], Paulo Fernandes [2], Tullio Giuffrè [3], Elżbieta Macioszek [4] and Anna Granà [1,*]

1 Department of Engineering, University of Palermo, Viale delle Scienze ed 8, 90128 Palermo, Italy; francesco.acuto@unipa.it
2 Centre for Mechanical Technology and Automation (TEMA), Department of Mechanical Engineering, Campus Universitario de Santiago, University of Aveiro, 3810-193 Aveiro, Portugal; margarida.coelho@ua.pt (M.C.C.); paulo.fernandes@ua.pt (P.F.)
3 Faculty of Engineering and Architecture, University of Enna Kore, Viale delle Olimpiadi, 94100 Enna, Italy; tullio.giuffre@unikore.it
4 Department of Transport Systems, Traffic Engineering and Logistics, Faculty of Transport and Aviation Engineering, Silesian University of Technology, Krasinskiego 8 Street, 40-019 Katowice, Poland; elzbieta.macioszek@polsl.pl
* Correspondence: anna.grana@unipa.it; Tel.: +39-091-2389-9727

Citation: Acuto, F.; Coelho, M.C.; Fernandes, P.; Giuffrè, T.; Macioszek, E.; Granà, A. Assessing the Environmental Performances of Urban Roundabouts Using the VSP Methodology and AIMSUN. *Energies* **2022**, *15*, 1371. https://doi.org/10.3390/en15041371

Academic Editors: Yair Wiseman and Ivan Arsie

Received: 28 December 2021
Accepted: 10 February 2022
Published: 14 February 2022

Abstract: In line with globally shared environmental sustainability goals, the shift towards citizen-friendly mobility is changing the way people move through cities and road user behaviour. Building a sustainable road transport requires design knowledge to develop increasingly green road infrastructures and monitoring the environmental impacts from mobile crowdsourced data. In this view, the paper presents an empirically based methodology that integrates the vehicle-specific power (VSP) model and microscopic traffic simulation (AIMSUN) to estimate second-by-second vehicle emissions at urban roundabouts. The distributions of time spent in each VSP mode from instantaneous vehicle trajectory data gathered in the field via smartphone were the starting point of the analysis. The versatility of AIMSUN in calibrating the model parameters to better reflect the field-observed speed time trajectories and to enhance the estimation accuracy was assessed. The conversion of an existing roundabout within the sample into a turbo counterpart was also made as an attempt to confirm the reproducibility of the proposed procedure. The results shed light on new opportunities in the environmental performance evaluation of road units when changes in design or operation should be considered within traffic management strategies and highlighted the potential of the smart approach in collecting big amounts of data through digital communities.

Keywords: roundabout; vehicle-specific power; pollutant emission; microsimulation; road infrastructure

1. Introduction

Urban mobility is going through a time of great change and road user needs are changing with it [1]. Since the Paris Agreement, sustainable urban mobility has become a key concept to achieve the goals set with respect to climate change, economic growth and road safety [2,3]. Owing to their predominant role in promoting the sustainable mobility and transport, road infrastructure issues are of special concern. The new forms of personal micro-mobility on the market, and new technologies in autonomous driving call for road infrastructures suited to meet the industrial challenges posed by the decarbonization and digitalization of the urban transport system [4,5]. The question is now whether the (invisible) network of digital technology systems may really be conceived as the "infrastructure" of the (visible) networks of road infrastructures whose efficiency is increasingly linked to the use of Internet of Things (IoT)-based devices and Information and Communication Technologies (ICT) tools under normal or crises conditions [4]. Despite the perspective of a close link between the invisible network of digital technology in operational systems and the visible networks of road infrastructures, the introduction of the new technologies and services for smart roads and the penetration rate of modern vehicle technologies are still slow; at the same time, the adaptation of vehicle fleet to new emission rules set at national, European and international levels is still proceeding on the basis of the physiological replacement rate of old vehicles by new ones [6]. Nowadays, there is a great potential for the widespread development of computing platforms to support distributed mobile sensing due to the extensive employ of always-connected smartphones and mobile devices as a link between people and things also to monitor the health and environmental impacts of air pollution from vehicles and to analyse risky driving events and road infrastructure conditions [7,8]. Although safer and environmentally friendly management of urban mobility is recognized as crucial for contributing to a greener future, the impact of new engineering solutions which are, individually and in combination, at different levels of maturity is not entirely clear [9]. Hence a sustainable urban mobility strategy has to incorporate a well-balanced set of policies, planning methods and measures to transit to a carbon-neutral status [10,11].

For the reasons given, estimation of exhaust emissions of car engines is still an active field of research. Road-related air quality management has based the estimation of vehicular traffic emissions mainly on average speed-based approaches and instantaneous (modal) models [12,13]. The average speed-based approaches rely on emission functions derived from measurements of the emission rates over a wide range of driving patterns at different levels of speed. Since these approaches do not capture the differences in emissions estimates due to the changes in modal activity, their usual applications concern emission inventories for large geographical areas at the road network level [12]. In turn, the instantaneous (modal) models describe the emissive behaviour of vehicles at the microscale level by relating emission rates to vehicle operation over a short time interval; they usually use vehicle activity parameters to estimate second-by-second emission rates and vehicle fuel consumption [14,15]. The average emission rates derived from vehicles and integrated into commonly used emission models, however, have returned in some experiences great differences between the predicted and field-based measured emissions for individual vehicles [13,16]. Recent experiences in connected eco-driving on signalised arterial corridors have highlighted the advantages of real-time traffic sensing and cloud platforms for applications as well as information services for networked cars and infrastructure with the purpose of reducing emissions and fuel consumption; thus, the integrated use of communications technology can offset current (macroscale or mesoscale) emission models in capturing differences among deceleration, acceleration, idling and cruising modes on the same road unit [17]. Despite the extensive applications of mobile-source emissions models at the macroscopic level (e.g., the Computer Programme to calculate Emissions from Road Transport (COPERT) model [18]; the EMission FACtors (EMFAC) model [19,20], the Transport Emission Model for Line Sources (TREM) [21]), or mesoscopic level (e.g., Signalised and unsignalised Intersection Design and Research Aid software (Sidra Intersection) [22]; Mobile Source Emission Model [23]), the modal emissions models are sensitive to changes in vehicle speed and acceleration and are used to evaluate operational level projects for arterials and intersections [24]. In the case of the Comprehensive Modal Emissions Model (CMEM) [25] second-by-second engine-out emissions, fuel consumption, and tailpipe emissions of carbon monoxide (CO), carbon dioxide (CO_2), hydrocarbon (HC), and nitrogen oxides (NOx) can be predicted for different modal operations from driving cycle data of light-duty vehicles in different driving conditions. However, the emission model should be corrected to accurately calculate CO_2 emissions on urban roads for a traffic speed below 20 km/h [26]. In turn, the Vehicle Specific Power (VSP) methodology is able of capturing the dependence of emissions on speed and its changes, and the effect of roadway grade on engine power demand [14]. In this view, the EPA's MOtor Vehicle Emission Simulator (MOVES) estimates emissions for mobile sources at the national-level, county-level and project-level for criteria air pollutants, greenhouse gases and air toxics, and uses VSP and instantaneous speed as basic variables [27]. Other models have been developed for passenger cars with the purpose of linking real-world emissions to driving cycle characteristics at a microscale level and have been already employed for traffic management, control measures, air quality modelling; e.g., [28]. The use of Global Positioning System (GIS) technologies and on-board diagnostics devices in collecting second-by-second trajectory data give now great flexibility to calculate vehicle emissions in the real-world, since the emissions estimates can be allocated spatially [29,30]. Literature also informs on capabilities of microscopic traffic simulation models to produce emissions and fuel consumption data related to dynamic traffic-flow conditions on the road entity or network [31]. However, there is still not enough evidence to suggest that accurate vehicle speed and acceleration data (or their probability distributions) may be provided due to inadequate car-following equations and challenges with modelling the driver behaviour on multi-lane roadways [32]. Nevertheless, research efforts to improve both capability of microsimulation models to replicate realistic traffic behaviour or vehicle trajectory data, and traffic generation accuracy at the network or single node level, can be especially useful for managing traffic safety and environmental issues on urban roads because microsimulation allows the modelling of engineering solutions for road infrastructures and their traffic conditions, sometimes difficult to observe in the real world or not yet implemented in the field.

The Aim of the Paper

Given the role played by the environmental impact assessment in raising awareness of sustainability issues at the road design level, a research project was started having the general objective to develop a methodological approach which combines the employ of the VSP methodology and a microscopic traffic simulation model to estimate vehicle emissions at urban roundabouts.

In this view, the paper investigated the following research question: can fine-tuning adjustments of some vehicle attributes of a traffic model simulation be made to enable a better match of speed profiles experienced through roundabouts and to improve the emission estimations calculated by employing vehicle trajectory data? Trying to answer the above question simultaneously represented the primary objective of the research reported in this paper.

A sample of six roundabouts installed in the road network of Palermo City, Italy, was the starting point to collect vehicle trajectory data and to test the proposed approach also with reference to specific effects of the curvilinear design of the roundabouts on the emissive phenomenon. The novel aspect of the research is that data collection was inspired by a crowdsensing logic [7], where a "sentinel vehicle" here used as a test vehicle travels through the selected road entities to acquire vehicle trajectory data by using a smartphone installed on board. Thus, beyond being an economic data acquisition method, it allows that the collected vehicle trajectory data can be processed immediately to return the observed speed-time profiles and to obtain the vehicle acceleration and deceleration values.

To pursue the objectives outlined above, to start the calibration process of the microscopic traffic simulation model here used and to estimate emissions by applying the VSP methodology [14], speed-time profiles both gathered in the field and simulated in AIMSUN [33] were needed. The comparison between the second-by-second GPS trajectories collected by a smartphone app in a test vehicle and individual second-by-second vehicle speed profiles derived from AIMSUN called for model calibration. In relation to the research objectives defined above, the versatility of the micro-simulation model to explain emissions was explored. Another novel aspect was that the roundabout network models were calibrated using the mean values of the 85th or 95th percentiles of the accelerations and decelerations extracted from all the field observed trajectories regardless of driving direction through the roundabouts.

A further important objective also involved the conversion of an existing roundabout within the sample into a turbo counterpart; this was made as an attempt to confirm the applicability and reproducibility of the proposed procedure. Insights gained from this model validation also included how to assess and compare alternative designs from an environmental perspective.

Thus, the main considerations of the study concern the feasibility of the smart approach which integrated the use of real-world and simulated vehicle activity data to estimate emissions at urban roundabouts.

The paper is structured as follows. Section 2 presents a literature review on current models and methods for estimating emissions at roundabouts, while Section 3 introduces the sampled roundabouts, and traffic and trajectory data collection. Section 4 presents AIMSUN modelling, calibration and results; the last ones will be discussed in Section 5, while Section 6 gives some conclusive considerations.

2. Literature Review

The operating flexibility of roundabouts in urban settings where competing needs of safety, capacity and costs coexist, is due to the aptitude of the curvilinear design to reduce stops and delays compared to other forms of intersection control, and to moderate speeds and to reduce speed changes of vehicles [34]. Planning a roundabout should include potential trade-offs between design, operations, and safety issues based on an economic feasibility study showing that a roundabout may compare favourably with alternative design or control modes of intersection from a cost-benefit perspective [35]. However, a transition toward greener road infrastructures also needs tools to assess the impact of new road infrastructural projects from an environmental perspective; e.g., [36].

Literature reports some studies on the contribution of roundabouts to the emission phenomenon, most of them relate to the employ of existing models using data from a portable emissions monitor on a vehicle instrumented for calculating exhaust emissions; e.g., [16,37–40]. It should be noted that macroscale-level models are not suitable for testing objects on a microscale level such as roundabouts [16,38]; however, the employ of microscopic traffic simulation models for estimating emissions on roundabouts is still limited.

Hallmark et al. [37] assessed the emission impacts of roundabouts along uncongested corridors using a portable emissions monitor on an instrumented vehicle. Compared to other types of traffic control, roundabouts either as isolated units or multiple schemes increased the time spent in some modal states and the modal events at which emissions were correlated; thus, they did not always provide lower emissions than unsignalised or signalised intersections. Coelho et al. [38] proposed a VSP-based approach to assess emissions at roundabouts. They assessed the role of the entry and exit geometry on differences between the circulating and cruise speeds and found that high rates of acceleration to reach the cruise speed increased the amount of emissions as the values of conflicting traffic volumes increased. Salamati et al. [39] extended the employ of the VSP-based methodology to multi-lane roundabouts. Findings showed higher emission rates in the right lane where faster speeds and sharper acceleration or deceleration rates occurred at values less than 700 vph for the sum of entry and circulating flows; as demand increased, however, balanced between the entry lanes, longer stop-and-go cycles suffered by entering vehicles from the left lanes provided higher emission rates. However, the differences in emission estimates between the left- and right-lanes should be redefined with greater depth with reference to more homogeneous roundabout samples. Fernandes et al. [40] also employed the VSP distributions derived from speed-travel time profiles detected throughout an arterial road where roundabouts operated as isolated sites. Besides hotspots where high speeds caused high emissions, acceleration, and deceleration events, and then emissions were mainly affected by a high enough spacing between roundabouts rather than by the entry geometry. They highlighted the need for further study to better characterize the spatial distribution of emissions also in the cases of low or extremely high spacing between subsequent roundabouts. Discrete models were also developed by [24] to relate distinct speed profiles of vehicles driving through turbo or multilane roundabouts, and traffic conditions; however, there was the need to gather further data on different roundabout configurations to perform real-world testing for validation purposes. Jaworski et al. [16] proposed a methodology for creating an exhaust emission model for roundabouts based on emission data from the Portable Emissions Monitoring System (PEMS) for real driving cycles of various types of vehicles. They obtained results of emission calculations on roundabouts which may be used to introduce and to prepare new roundabout design guidelines concerning emission data. Guerrieri et al. [41] used the COPERT IV® software to assess environmental performances at roundabouts with right-turn bypasses under increasing entering traffic volumes. Although roundabouts may be a cost-effective solution, the macroscale level tools used did not allow us to investigate in greater depth the dependence of emissions on driver behaviour. To optimize roundabout modelling, Lakuari et al. [42] employed numerical simulation to predict CO_2 emissions. Repeated changes in vehicle speed in the ring resulted in greater CO_2 emission rates than at entry or exit lanes, whereas CO_2 emissions reduced where many circulating vehicles slowed down or stopped because of the vehicles entering the roundabout without a safe gap to the vehicles in the circulating lane. However, emissions caused by frequent stops-and-go in high traffic conditions could be reduced by using traffic lights at entries. Other experiences concerned the employ of a hierarchical Bayesian regression analysis to model speed profiles of different drivers at roundabouts which can be useful in determining individual or group-wise emissions estimates [43]. However, the above papers concerned the emission testing at roundabouts based on previously developed general emission models. Thus, given the specific characteristics of geometry and traffic on roundabouts, there is still a need to develop emission models for exhaust gases for roundabouts based on real data.

The literature also informs on a few studies that have studied the environmental effects on roundabouts using microscopic traffic simulation models. Ahn et al. [44] applied INTEGRATION and Verkehr In Städten – SIMulationsmodell (VISSIM) software [45] to generate speed-time profiles through a roundabout along a high-speed road, and employed microscopic energy and emission models to estimate emissions and vehicle fuel consumption. Despite a greater reduction in queue lengths and delays at roundabouts than other intersection control strategies under free-flowing traffic, vehicle emissions and fuel consumption increased as demand increased under unbalanced entering traffic flows, reaching higher levels than a two-way stop-controlled or a traffic light-controlled alternative. However, the case study only allowed to test the capability of the microscopic traffic simulation models to reproduce high-speed driving patterns as input for the employed emission models and stressed the usefulness of combining traffic microsimulation models with other emission models to assess design alternatives or traffic control strategies from the perspective of environmental sustainability. Recent studies used the VSP model to estimate emissions starting from on-field observed and simulated vehicle trajectories on a corridor level. In this regard, Anya et al. [46] have explored the potential for calibration of microsimulation models with a view to improve emissions estimates from simulated vehicle activity for mixed corridors of traffic signals and roundabouts. Adjusting the parameters allowed the model to capture emissions hotspots along the routes more accurately than under the model parameters with default values. Attempts to simulate emissions driving along one mixed roundabout/traffic light/stop-controlled intersections corridor were also made by [47] using microscopic traffic modelling and VSP methodology. Their study showed that roundabouts could achieve lower emissions, however depending on the pollutant, than the traffic light; they also searched for the optimal spacing for intersections along a corridor, but the results can only be referred to the case examined. Thus, more experience should be carried out both to generalize the results and to test the transferability of the proposed methodology into other environmental contexts. In other case studies, operational and environmental impacts of turbo-roundabouts were also analysed both as isolated intersections [48,49] and along corridors [50]. Improvements in the estimates of CO_2 emissions and fuel consumption on arterial roads have been achieved coupling dynamic micro traffic models with instantaneous emissions models; however, the large computation times especially for large-scale urban networks is a drawback of this kind of combination [51]. In turn, Stogios et al. [52] examined the effects of driving settings with Automated Vehicles (AVs) on greenhouse gas emissions (GHG) at urban corridors by combining traffic microsimulation and emissions modelling. The results highlighted that there is potential for up to 24 percent in GHG emission reduction when the effects of vehicle powertrain technology are included; no significant reductions of GHG emission were found when AVs alone characterized driving settings. Thus, the value of the study is to open interesting scenarios on autonomous driving impacts on traffic.

Based on the above, this paper would like to address some gaps in the prior studies that are related to the data collection, and to the employ of microscopic traffic simulation models for estimating emissions with reference to urban roundabouts. As introduced in Section 1, data collection here performed is inspired by a crowdsensing logic [7], where a "sentinel vehicle" used as a test vehicle travelled through the sampled roundabouts to acquire second-by-second vehicle trajectory data by using a smartphone installed on board.

Starting from the literature review as above reported, there is the clear need for more economical and user-friendly systems of collecting instantaneous vehicle trajectory data to make smart both the use of the available resources and then the subsequent data process and analysis. In this regard, it should be noted as a further novel aspect that the calibration process of the model parameters of AIMSUN used field data in order to ensure that emissions were predicted accurately from simulated trajectories through the sampled roundabouts.

The research also includes the following societal and scientific contributions:

(1) Societal—It allows to send the collected trajectory in a central management platform able of processing a considerable amount of data, converting them into digital information to estimate traffic emissions from the mobile source, and then, returning data to the community of users equipped with their smartphones to collectively share information, for instance, about hotspot emission locations at urban roundabouts;
(2) Scientific—It identifies certain parameters of driving behaviour using a traffic modelling approach aimed at accurately analysing and comparing road units (i.e., road segments or road intersections, and so on) when changes in design or operation are considered from an environmental point of view as the life cycle thinking approaches strongly require.

At last, the further objective of the conversion of an existing roundabout of the sample into a turbo counterpart as introduced in Section 1. represents as an attempt to test the reproducibility of the empirically-based methodology here proposed, that integrates the VSP model and AIMSUN to estimate vehicle emissions at urban roundabouts.

3. Materials and Methods

In accordance with the aim of the paper as introduced in Section 1, the field data collection and the VSP methodology will be introduced below, before the section on the infrastructure modelling in AIMSUN developed to generate the speed-time profiles for the sampled roundabouts and to perform the calibration of the model. The reasons behind the assumptions and choices made will also be described, as well as the obtained results and the conclusions that were then drawn also in view of future research developments. Figure 1 shows an overview of the proposed methodology.

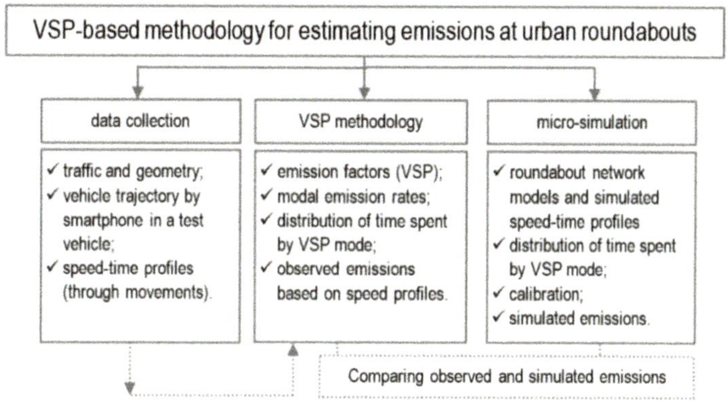

Figure 1. Overview of the proposed methodology.

3.1. Field Data Collection

A sample of six roundabouts installed in the road network of the City of Palermo, Italy, was selected for this pilot study. Although the examined roundabouts are sited in areas different from the urbanistic point of view, they were selected for the similar operating conditions observed in the field (i.e., no commercial area, similar entry traffic distribution between major and minor roads, analogy among the curvilinear paths experienced by the test vehicle in the two driving directions through each roundabout). Geometric dimensions were obtained from a combination of design plans and field measurements. Figure 2 shows the schematic drawings of the sampled sites, whereas Table 1 summarizes the main sites' geometric characteristics and details on the speed and traffic data. The geometry of entry (exit) lane, circulatory roadway width and deflection angles were consistent with the Italian standards on geometric design of interchanges and intersections [53]. The roundabouts were classified as conventional roundabouts for an outer diameter below 50 m, while large-diameter multilane roundabouts were recognized in the other cases. Despite the constraints of existing roadway alignments or buildings on roundabout design, and the large size of the outer diameter of some sampled roundabouts, speed control objectives could still be met. Entry design provided appropriate view angles to users and deflection angles on the whole comparable among the sites; however, the higher the deflection angle, especially over 41 degrees, the higher the reduction in approach speeds [34]. In the cases where the approach alignments are offset to the left of the roundabout centre, entry deflection and the exit radius on the same approach increased slightly, thus reducing the control of exit speeds and acceleration; e.g., Figure 2b,d (Roundabouts 2 and 4, respectively). Field observations confirmed that all the two-lane entries operated with a shared through-right lane in the right lane and a shared left-through lane in the left lane; both the inside lane and outside lane at entries often had sporadic queuing under traffic conditions observed during surveys. In this regard, the geometric alignment of the entry relative to the circulatory roadway encouraged drivers to handle the left-turns, right-turns, and through-movements under balanced entry lane-use patterns, and to avoid path overlap between adjacent lanes, thus reducing uncertainty in exiting from the circulatory roadway. Neither the spacing between subsequent intersections nor the destinations downstream of each roundabout influenced the lane choice behaviour at the entry so that each sampled roundabout operated as an isolated intersection. Since each roundabout is installed in a flat area where the grade is less than 2 percent, the effect of this parameter was neglected when the VSP methodology was applied (see next section).

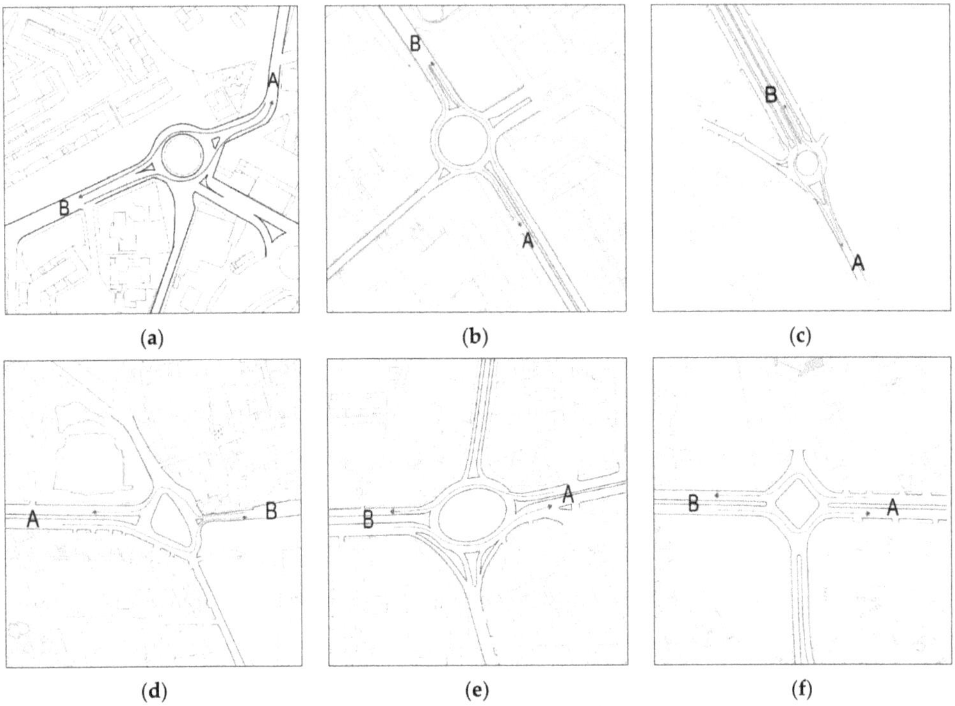

Figure 2. Schematic aerial view of Auto-Cad figures for the sampled roundabouts in the road network of Palermo City, Italy, each of them named as follows: (**a**) Roundabout 1; (**b**) Roundabout 2; (**c**) Roundabout 3; (**d**) Roundabout 4; (**e**) Roundabout 5; (**f**) Roundabout 6. Note: each image contains the origin or destination (A, B) of through movements travelled by the test vehicle in the driving directions from A to B and from B to A.

Table 1. Summary of geometric, kinematic and traffic data of the sampled roundabouts.

No	Entry (Exit)	Outer Diameter [m]	Entry (Exit) Lane Width [1] [m]	Ring Width [m]	Entry Traffic [2] [vph]	Conflicting Traffic [3] [vph]	Entry (Exit) Speed [4] [km/h]	Circulating Speed [km/h]
1	3 (4)	48.0	3.50 (3.50)	7.00	1576	1508	22.1 (30)	18.2
2	4 (4)	80.0	4.50 [5] (4.50)	8.00	3984	3196	25.9 (36)	23.1
3	4 (4)	50.0	3.50 (3.50)	9.00	2336	1904	23.3 (31)	19.9
4	4 (4)	60.0	4.75 [6] (4.75)	10.00	1306	1372	29.8 (42)	24.8
5	4 (4)	80.0	4.00 (5.00 [7])	9.00	3992	2971	25.1 (35)	23.1
6	4 (4)	80.0	5.00 (4.50)	10.00	988	972	30.0 (38)	25.5

[1] Entry width before widening the entry roadway or before the by-pass for right turns. [2] Average values of traffic volumes (in vehicles per hour—vph) from the right and left lanes videotaped during the morning peak period (7:00–8:30 a.m.) where the entry traffic distribution of 60-40 between major and minor roads was observed. [3] Average values of traffic volumes (all circulating lanes) observed for the morning peak period (7:00–8:30 a.m.). [4] Average values of speeds at the entry and exit lines. [5] 5.00 m (4.00 m) for the one-lane entry (exit) on minor street. [6] 3.50 m for the one-lane entry (exit) on minor street. [7] 4.50 m for the one-lane exit on minor street.

Field surveys were done to record traffic counts and vehicle trajectory data. Traffic flow data were videotaped on each entry and exit within the viewable area of each roundabout; this investigation was also integrated by manual counts simultaneously done by two operators to create a complete picture of traffic flows and driving behaviour where observations were difficult to be detected. The traffic flow data were recorded at each roundabout by driving direction, classified by manoeuvre and then computerized for processing as the input Origin-Destination (O-D) matrixes so as to allow for their subsequent simulation on each network model of roundabout built in AIMSUN (see Section 4). Hourly patterns of traffic flows often showed a marked similarity having only two peaks in the morning and the late afternoon without a significant peak at noon or in the late evening. Traffic flow measurements were done during the morning (7:00–8:30 a.m.) and afternoon peak hours (6:00–7:30 p.m.) on weekdays (Tuesday to Friday) from October 2018 to February 2019. In most cases the morning peak was reached over a longer time period than the afternoon peak whose duration suddenly dropped to its lowest point; thus, the peak hours observed during afternoons have not been considered in the subsequent analysis compared to data recorded during the morning peak hours. Table 1 also shows the average values of entering and circulating traffic volumes observed for the morning peak period (7:00–8:30 a.m.) at the sampled roundabouts. Since effects of pedestrian impedance on the vehicular entry capacity as a function of driver yielding behaviour were insignificant in the surveyed time periods, pedestrian flows were neglected in the subsequent analysis [54]; in turn, heavy traffic did not exceed 10 percent at each roundabout during surveys and it was also neglected. Data collection also covered the recording and retaining of vehicle trajectory data through each sampled roundabout during 7:00 to 8:30 a.m. peak periods on weekdays. According to the aim of the paper (see Section 1), reference was made only to the trajectory data of the test vehicle. Specifically, vehicle trajectory data were measured in sampling values per second at 1Hz frequency using the Speedometer GPS PRO for Android smartphone, installed on a diesel light-duty vehicle complying with the Euro IV emission standards and the specifications employed for deriving emissions rates for the VSP modes [55,56]. Second-by-second GPS trajectories experienced by the test vehicle entering each roundabout from the left lane covered about 15 km in 12 h of on-road surveys. Based on [24,57], the number of runs per roundabout counted 7 to 10 replications per site and travel direction for a total of 94 travel runs of through movements; it has been deemed adequate to have appropriate results from the data collection. Longitudinal acceleration (or deceleration) values have been determined from second-by-second GPS speed data collected in the field using the following equation:

$$a_{t_2} = \frac{v_{t_2} - v_{t_1}}{t_2 - t_1} \tag{1}$$

where a_{t_2} is the acceleration or deceleration (m/s^2) at time t_2 (s), while v_1 is the speed (m/s) at time t_1 (s), and v_2 is the speed (m/s) at time t_2 (s) [58]. According to [59], acceleration manoeuvre ended in the event that the increment in speed between two consecutive data points was less than 0.1 m/s^2 for the next 5 s. Similarly, deceleration values have been determined from the time onwards where its absolute values from Equation (1) were greater than or equal to 0.1 m/s^2 for 5 consecutive seconds [58,60]. The observed speed profiles of all the sampled roundabouts were divided into two subsets depending on the driving direction where the corresponding trajectories were experienced in the field (see directions AB and BA in Figure 2). An analogy among the curvilinear paths of the test vehicle in the two driving directions has been found. To test whether the direction AB was the same as BA, statistical tests were performed to compare the two corresponding subsets of data. The two-sample *t*-test was done to determine if the means of the two subsets of data were significantly different from each other. The two-tailed *F*-test was also done to answer the question whether the variances of the two samples were equal against the alternative that they were not (see Tables 2 and 3 for the summary statistics). Based on the *p*-values in the tables, it cannot conclude that a significant difference exists between the two driving directions in all the sampled roundabouts; in turn, the *F*-test results show that there is not enough evidence to reject the null hypothesis that the two sample variances are equal at the 0.05 significance level.

Table 2. Summary statistics for the distributions of key kinematic parameters collected in the field.

Parameter	Maximum Speed [m/s]	Maximum Acceleration [m/s^2]	Maximum Deceleration [m/s^2]
μ_{AB} [1] (s.e.)	14.459 (0.384)	1.770 (0.077)	2.498 (0.108)
μ_{BA} [1] (s.e.)	14.117 (0.357)	1.553 (0.074)	2.417 (0.111)
95% c.i. for difference in means	(−0.700, 1.385)	(0.004, 0.430)	(−0.227, 0.390)
$t_{0.05,92}$ value [2]	0.652	2.023	0.522
t-critical value	1.986	1.986	1.986
p-Value (α = 0.05)[3]	0.516	0.05	0.603
$F_{0.05,46,46}$ value [4]	1.158	1.102	0.946
F-critical value	1.796	1.796	1.796
F-probability	0.620	0.740	0.850

[1] μ_{AB} and μ_{BA} stand for the mean values of the samples of the observations of each parameter in AB and BA directions; [2] *t*-value is the result of the two-sample *t*-test done to compare the equality of the means (μ_{AB} and μ_{BA}) of samples from two populations with equal sample size: reject the null hypothesis that the two means are equal if $|t| > t_{1-\alpha/2,N}$ where $t_{1-\alpha/2,N}$ is the critical value of the *t* distribution with *N* degrees of freedom at the significance level α = 0.05); [3] α is the significance level; [4] *F*-value is the result of the two-tailed *F*-test done to answer the question whether two samples come from populations with equal variances (the hypothesis that the two variances were equal is rejected if $F > F_{\alpha/2,\ N_1-1,N_2-1}$, where $F_{\alpha/2,N_1-1,N_2-1}$ is the critical value of the *F* distribution with N_1*-1* and N_2*-1* degrees of freedom at the significance level of α = 0.05).

Table 3. Summary statistics for the distributions of the 85th or 95th percentile accelerations and decelerations observed in the field.

Parameter	85th Percentile Acceleration [m/s^2]	95th Percentile Acceleration [m/s^2]	85th Percentile Deceleration [m/s^2]	95th Percentile Deceleration [m/s^2]
μ_{AB} [1] (s.e.)	0.916 (0.030)	1.286 (0.039)	1.423 (0.082)	2.054 (0.089)
μ_{BA} [1] (s.e.)	0.868 (0.025)	1.195 (0.042)	1.251 (0.059)	1.882 (0.085)
95% c.i. for difference in means	(−0.031, 0.126)	(−0.024, 0.205)	(−0.022, 0.381)	(−0.073, 0.417)
$t_{0.05,92}$ value [2]	1.205	1.56	1.76	1.393
t-critical value	1.986	1.986	1.986	1.986
p-Value (α = 0.05) [3]	0.231	0.121	0.10	0.167
$F_{0.05,46,46}$ value [4]	1.362	1.18	1.76	1.081
F-critical value	1.796	1.796	1.796	1.796
F-probability	0.30	0.60	0.10	0.80

[1] μ_{AB} and μ_{BA} stand for the mean values of the samples of the observations of each parameter in AB and BA directions; [2] *t*-value is the result of the two-sample *t*-test done to compare the equality of the means (μ_{AB} and μ_{BA}) of samples from two populations with equal sample size: reject the null hypothesis that the two means are equal if $|t| > t_{1-\alpha/2,N}$ where $t_{1-\alpha/2,N}$ is the critical value of the *t* distribution with *N* degrees of freedom at the significance level of α = 0.05); [3] α is the significance level; [4] *F*-value is the result of the two-tailed *F*-test done to answer the question whether two samples come from populations with equal variances (the hypothesis that the two variances were equal is rejected if $F > F_{\alpha/2,N_1-1,N_2-1}$, where $F_{\alpha/2,N_1-1,N_2-1}$ is the critical value of the *F* distribution having N_1*-1* and N_2*-1* degrees of freedom at the significance level of α = 0.05).

3.2. Characterisation of the Speed-Time Profiles

The speed profiles collected in the field were classified in three main groups whose probability of occurrence was closely depending on the traffic level [24,38,61,62]: (1) speed profiles without stopping and yielding to circulating traffic, where the test vehicle entered the roundabout, negotiated the circulating area and accelerated back to cruise speed as it exited the roundabout; (2) speed profiles with one stop experienced by the test vehicle before finding a useful headway to enter the roundabout, accelerated to move along the ring and exited the roundabout; (3) speed profiles with multiple stops during queueing on the entry approach before the test vehicle entered the roundabout, faced the circulating traffic and then accelerated to exit. In this study, the speed profiles with or without one complete stop were considered, while those with multiple stopping on the entry approach were removed due to their low relative occurrence (less than 5% of the cases) under the observed traffic conditions ranging from free flow traffic to a saturation degree of 0.85 to 0.90. Characterization of the speed profiles obtained from processing GPS data collected for roundabouts in Palermo City, Italy, is reported in [62] to which we refer. During traffic surveys it was observed that the number of cases of the driver negotiating the roundabout without a stop decreased as the sum of entry and circulating flows (the total traffic flow) increased: for values of the total traffic flow below about 800 vph, the greater the probability of the driver that enters the roundabout without a stop; for higher values of the total traffic flow, the greater the probability for at least one stop. The result is consistent with what was detailed in [24]. Each speed profile extracted from the raw data was trimmed and adjusted as having about 60 data points distributed along the trajectory; thus, a direct and punctual comparison of the profiles has been done by referring to their order numbers. In order to have consistency among the speed profiles and then extract a contribution to the emissive phenomenon comparable among the sampled roundabouts, an average influence area of 250 m [24,62], corresponding to around three times the diameter was chosen for each roundabout; the influence area included the deceleration distance of the vehicle from the cruise speed to the entry line of the roundabout, and the acceleration distance of the exiting vehicle up to the section where the cruise speed is about to be reached again. A representative speed profile was chosen for each roundabout within the average influence area as introduced above. By way of example, Figure 3 shows all second-by-second speed-time profiles through Roundabout 2 in Figure 2b and Roundabout 4 in Figure 2d where the dashed black line identified the representative profile employed to compare the speed-time profiles then simulated in AIMSUM. There was a large number of profiles that seemed to be somewhere between the two forms identified above (i.e., speed profiles with or without a stop), especially in the case of atypical layout where not all the profiles fitted the identified shapes perfectly (see Figure 3b). Consistent with the objectives set above for this paper, one single sentinel vehicle interacted with other vehicles in traffic; the driving style was consistent with the characteristics of curvilinear design dominated by slight speed changes and intentional speed corrections by the driver to maintain his desired speed [63]. The vehicle adopted different initial speeds before entering due to differences in the upstream geometry or operational conditions at each sampled roundabout.

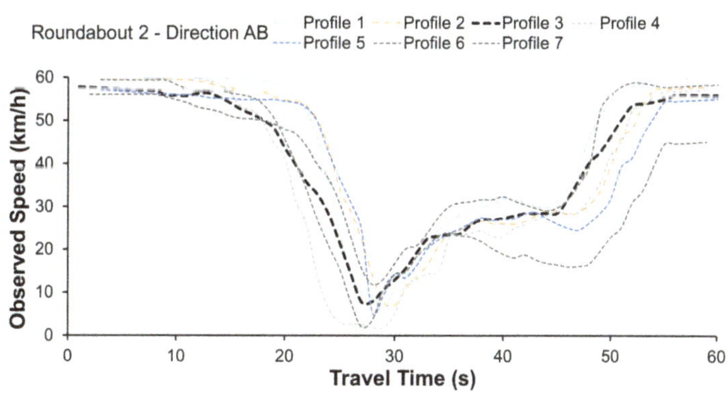

(a)

Figure 3. *Cont.*

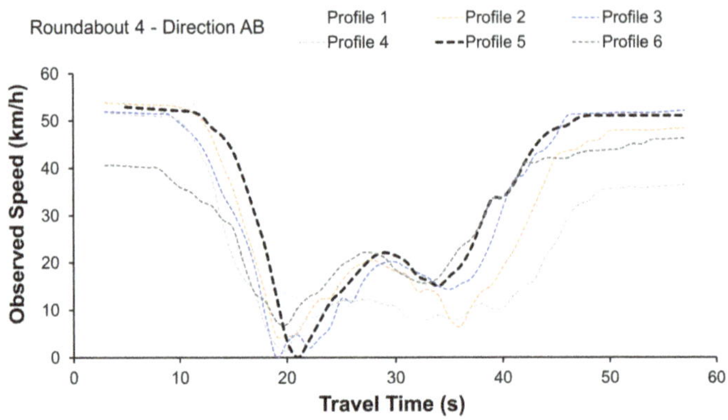

(b)

Figure 3. Second-by-second speed-time profiles of through movement observed in direction AB: (**a**) profile without stopping at Roundabout 2 in Figure 2b; (**b**) profile with one stop at Roundabout 4 in Figure 2d.

3.3. The VSP Methodology

In order to characterize vehicle driving profiles by using real-world data, the VSP methodology has been employed [14]. This methodology was selected since its suitability on emission estimation at roundabout level analysis [24,38,61]. Equation (2) expresses the instantaneous power that is generated by the engine and is used to overcome the rolling resistance F_r and aerodynamic drag F_a, and to increase the potential P_e and kinetic energies K_e of the vehicle as follows [14]:

$$VSP = \frac{\frac{d}{dt}\cdot(K_e + P_e) + F_r\cdot v + F_a\cdot v}{m} \quad (2)$$

where m and v are the mass, in metric tons, and the speed of the vehicle, in m/s, respectively. The US Environmental Protection Agency (USEPA) modal emissions model MOVES [27] employs a simplified form of Equation (2) where the VSP is calculated as a function of the vehicle's speed, acceleration, and grade:

$$VSP = \frac{1}{m}\cdot\left(A\cdot v + B\cdot v^2 + C\cdot v^3\right) + (a + g\cdot sin\theta)\cdot v \quad (3)$$

where VSP is expressed in kW/t; v is the instantaneous vehicle speed (m/s), a is the vehicle acceleration (m/s^2); A is the coefficient associated with tire rolling resistance (kW·s/m), and B is the coefficient associated with the mechanical rotating friction and higher-order rolling resistance losses both expressed (kW·s^2/m^2); C is the coefficient associated with the aerodynamic drag (kW s^3/m^3); m expresses the mass for the specific vehicle type in metric ton; g is the acceleration due to gravity; sin θ is the road grade. Thus, the VSP can be equal to zero under zero vehicle speed or idling, while the VSP is positive (or negative) during acceleration (or deceleration). Given the impact of various factors influencing VSP and the variability in instantaneous vehicle emissions on road segments of different types of roads, there is the need to promote the development of a binning process in order to reduce variability found in the cases examined [15].

A further simplified form of the VSP equation has been developed for a typical light passenger vehicle as follows [64]:

$$VSP = v\cdot[1.1\cdot a + 9.81\cdot sin(arctan(grade)) + 0.132] + 0.000302\cdot v^3 \quad (4)$$

where *grade* is the road grade (decimal fraction), while *VSP*, *v* and *a* are as introduced above. For light-duty vehicles, each second of driving is categorized into 14 modes that represent the different driving regimes [64]. While VSP modes 1 to 2 correspond to deceleration modes or downhill road, VSP mode 3 represents idling or low-speed situations. Finally, VSP modes 4 to 14 correspond to cruising, acceleration modes or uphill road sections [64]. Although a 14 mode VSP-based approach is not unique in its ability to predict tailpipe emissions, it allows us to simplify the design of a modelling system in comparison to other emission methods [65]. Each mode is associated with an emission factor of CO_2, CO, NOx, and HC that is fixed for a specific vehicle type (regulation class, fuel, model, year, mileage, or weight). Since a Light Passenger Diesel Vehicle (LPDV) was used as a test vehicle in this study, Table 4 depicts each interval of power requirements corresponding to each 14 VSP mode, and CO_2, CO, NOx, and HC emission factors by VSP mode for diesel powertrain [24,55]; the emission rates for light passenger gasoline vehicles and light commercial diesel vehicles can be found in [24]. Figure 4 shows the relative frequencies of time spent in each VSP mode for representative speed-time profiles experienced by the "sentry" test vehicle through the sampled roundabouts. The time percentages in both driving directions have been shown to be broadly consistent with each other both in the VSP modes 1 to 2 (deceleration) and 4 to 5 (acceleration); the vehicle spent a high proportion of time in VSP mode 4 or 5 as it exited the roundabout to reach the cruise speed, while the proportion of time sensibly appears to reduce from VSP mode 5 onward. In some cases, a low proportion of time still appears in 10 to 13 VSP modes denoting high acceleration events. Based on the speed-time profiles and the distribution of time spent in each VSP mode, emissions by source pollutant were estimated [24]:

$$E_{ij} = \sum_{k=1}^{N_k} F_{kj} \tag{5}$$

where E_{ij} is the total emission for the speed-time profile *i* and pollutant *j* (g); *k* is the label for second of travel (s); F_{kj} is the emission factor for each pollutant *j* in label for second *k* (g/s); N_k is the total seconds (s). Equation (5) was used to calculate second-by-second emission rates for each speed-time profile experienced by the test vehicle; the total emission by pollutant at each roundabout can be calculated as average of emissions for each pollutant and speed-time profile [62].

Table 4. The VSP modes, power requirements and mean emission rates by VSP mode for LPDV [1].

VSP Mode	Kw/ton [2]	Mean Modal Emission Rates (g/s)			
		CO_2	CO	NO_x,	HC
1	VSP < −2	0.21	0.00003	0.0013	0.00014
2	−2 ≤ VSP < 0	0.61	0.00007	0.0026	0.00011
3	0 ≤ VSP < 1	0.73	0.00014	0.0034	0.00011
4	1 ≤ VSP < 4	1.50	0.00025	0.0061	0.00017
5	4 ≤ VSP < 7	2.34	0.00029	0.0094	0.00020
6	7 ≤ VSP < 10	3.29	0.00069	0.0125	0.00023
7	10 ≤ VSP < 13	4.20	0.00058	0.0155	0.00024
8	13 ≤ VSP < 16	4.94	0.00064	0.0178	0.00023
9	16 ≤ VSP < 19	5.57	0.00061	0.0213	0.00024
10	19 ≤ VSP < 23	6.26	0.00101	0.0325	0.00028
11	23 ≤ VSP < 28	7.40	0.00115	0.0558	0.00037
12	28 ≤ VSP < 33	8.39	0.00096	0.0743	0.00042
13	33 ≤ VSP < 39	9.41	0.00077	0.1042	0.00040
14	VSP ≥ 39	10.48	0.00073	0.1459	0.00042

[1] LPDV stands for Light Passenger Diesel Vehicles; [2] as calculated by Equation (5).

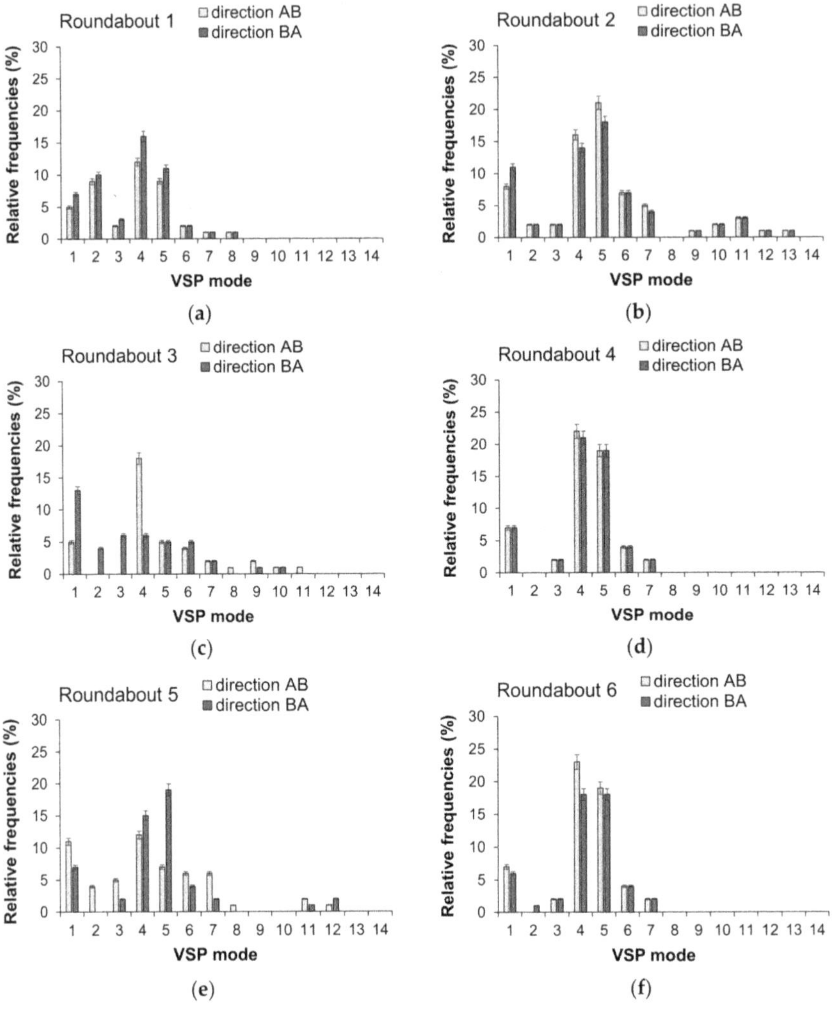

Figure 4. Relative frequencies of time spent in VSP modes at the sampled roundabouts in Figure 2 as follows: (**a**) Roundabout 1; (**b**) Roundabout 2; (**c**) Roundabout 3; (**d**) Roundabout 4; (**e**) Roundabout 5; (**f**) Roundabout 6. Note: direction AB means the through movement from A to B in Figure 2; direction BA means the through movement from B to A in Figure 2.

By way of example, Figure 5 shows the cumulative percentages of CO_2 emissions from the time spent in each VSP mode and the second-by-second emission rates for the speed profiles through the Roundabouts 2 and 4 in Figure 2 along the entire distance travelled on the field. The distributions of CO_2 emissions for the other roundabouts exhibited similar patterns. The relative increase in the percentage of CO_2 emissions with the distance from the entry resulted highest in acceleration and short stop-and-go events, the slope being the steepest. In this regard, Figure 5a shows that the test vehicle while interacting with the other vehicles in traffic experienced short stop-and-go and produced high percentages of total CO_2 emissions during the acceleration from zero to the cruise speed. Repeated speed changes in the circulatory roadway caused greater CO_2 emission rates than at entries. The relative increase of CO_2 emissions percentage with the distance was higher in the acceleration mode especially when the test vehicle got in the circulatory roadway with a minimum speed and started accelerating to reach the desired speed to exit (see Figure 5b). According to previous experiences on roundabouts in Palermo City, Italy [62] acceleration events in the circulating and exiting areas contributed to more than 25 percent of the emissions for a given speed profile.

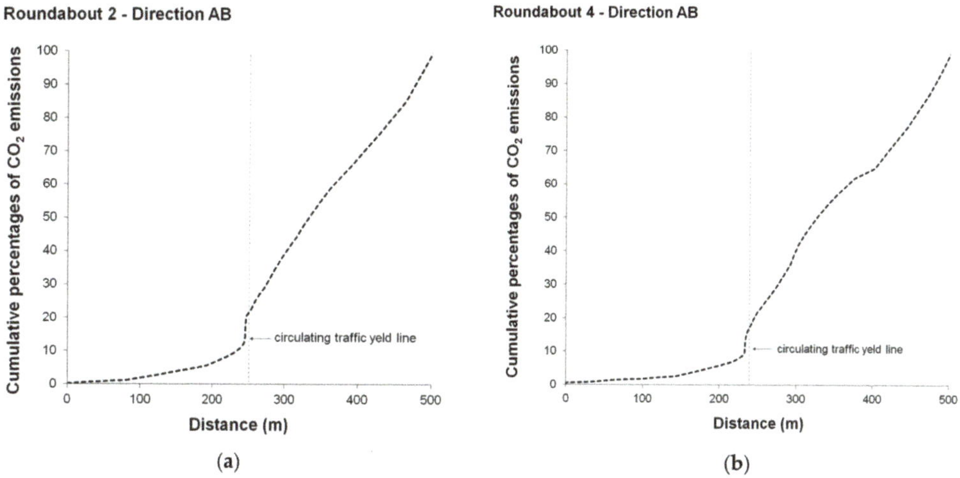

Figure 5. Spatial distribution of CO_2 emissions based on speed profiles for: (**a**) Roundabout 2 in Figure 2b and (**b**) Roundabout 4 in Figure 2d.

4. AIMSUN Modelling

The AIMSUN software was employed to model the sampled roundabouts and to investigate their contribution to the emission phenomenon [33]. Currently, it is known that AIMSUN allows to code the roundabout model among different roundabout types within its project window; this ensures simulated driving behaviour and lane usage corresponding as much as possible to the area under examination and real levels of traffic flow [66]. In the implementation stage of AIMSUN the roundabouts in Figure 2 were converted into the corresponding abstract network models, each of them represented by a system of links and connectors suited for microscopic simulation. Since roundabouts can have multiple turning movements, subpaths between each entry and exit were built to identify the roundabout turns and then extract the manoeuvres to be examined [33]. The urban speed limit of 50 km/h was set as the maximum speed at which each roundabout can be travelled under free-flow conditions. The traffic demand was given as a total O-D matrix of the traffic counts; thus, centroids were defined for each roundabout model. In order to build the traffic demand scenario and then feed each roundabout network model, two further matrices, one for cars and another for heavy vehicles were extracted from each O-D matrix. The vehicle attributes of AIMSUN were set by a class of vehicles and O-D pairs of through movements so as at extracting speed-time profiles of an individual vehicle quite similar to the test vehicle driven during surveys [33]. Traffic scenarios were built considering the same morning peak-hour data collected in the field. Each dynamic scenario was located into the 7:30–8:30 a.m. slot with a single replication of 5400 s and warm-up time of 1800 s. Once carried out the *i-th* replication in AIMSUN, imposing a resolution time scan per second, trajectory data useful for building the speed-time profiles were stored in the output database (with **.sqlite* extension); the usable information was referred to the previously selected O-D pair and to the class of vehicles which have travelled the roundabout in a given time interval [33]. The vehicle trajectory data from AIMSUN were aggregated based on 30 replications by through movement and driving direction on all the roundabout network models; they were plotted as a curve with speed on the Y-axis and travel time on the X-axis. By way of example, Figure 6 shows the observed speed-time profiles and those simulated under default parameters for Roundabout 1 (movement AB in Figure 2a) and Roundabout 3 (movement BA in Figure 2c). The comparison between the speed-time profiles experienced in the field and those simulated under default parameters of AIMSUN was made to figure out whether (or not) adjusting the model parameters was needed to better reflect the field data and then obtain greater accuracy in emission estimates.

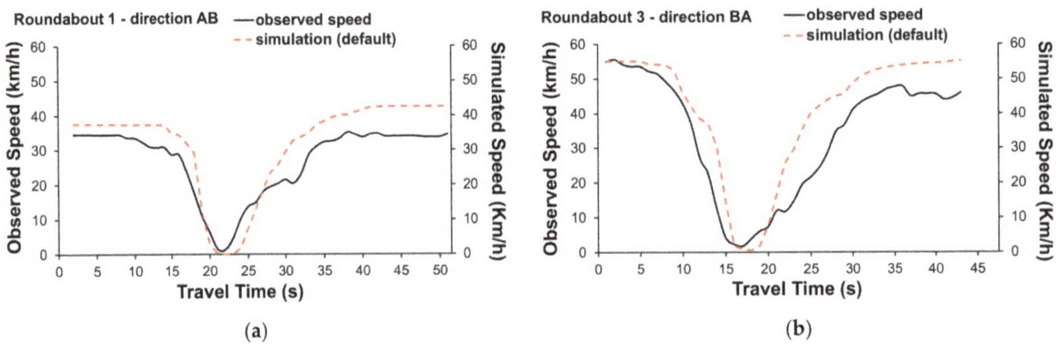

Figure 6. Observed vs. simulated speed-time profiles at: (**a**) Roundabout 1 (movement AB) in Figure 2a; (**b**) Roundabout 3 (movement BA) in Figure 2c.

4.1. AIMSUN Calibration

The comparison between the GPS trajectories collected in the field and speed-time profiles simulated in AIMSUN under default parameters required model calibration. According to the calibration procedures proposed by [46,67], a vehicle attribute calibration was made to enable the model to better match the field measurements and to estimate emissions with better accuracy. Two vehicle attributes of AIMSUN were selected: the maximum acceleration having a default value of 3 m/s^2, and normal deceleration having a default value of 4 m/s^2. A sensitivity analysis explored the effects of different combinations of model parameters on the modelled outputs so that they matched or were comparable with the real-world observations. Calibration under the mean values of the 85th or 95th percentiles of accelerations and decelerations extracted from all the field-observed trajectories travelled through the sampled roundabouts regardless of the driving direction best fitted the observed data (see Table 3); it also provided VSP distributions closer to field data than the distribution under default parameters. No distributions of 85th or 95th percentile desired speeds were defined since the maximum speeds from empirical data appropriately represented the maximum speeds desired by drivers under normal circumstances. By way of example, Figure 7a shows the comparison between the observed and simulated speed profiles for Roundabout 5 in Figure 2e under default parameters and calibration with 85th and 95th percentile parameters. Figure 7b shows, in turn, an example of scattergram analysis done to compare the observed versus simulated speeds under calibration with the 95th percentile parameters extracted from all the field-observed trajectories experienced by the test vehicle through the roundabouts regardless of driving direction. The same figure shows the regression lines of observed versus simulated speeds at the detection stations plotted along with the 95% prediction interval; the R^2 values and the fact that only a few points fall outside the confidence band in both graphs imply that the model could be accepted as significantly close to the reality. The results confirmed improvements in vehicle speed after calibration, while traffic volumes resulted slightly modified. Encouraging results under calibrated parameters were also obtained in terms of Geoffrey E. Havers' statistic (GEH) [31]. With reference to Figure 7b, the GEH values of 100% for each driving direction meant that the deviation of the simulated speed values under calibrated parameters with respect to the corresponding measurements resulted in less than 5 in 100% of the cases and the model could be accepted.

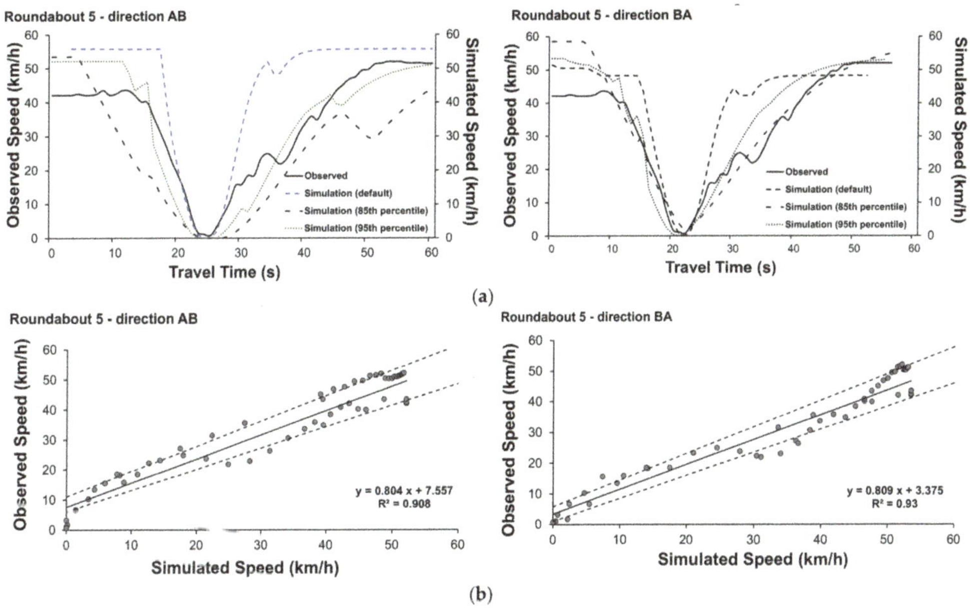

Figure 7. Observed vs. simulated data at Roundabout 5 (directions AB and BA in Figure 2e): (**a**) speed profiles; (**b**) regression line of observed vs. speeds simulated under calibration with 95th percentile parameters. Note: observed stands for observed data; simulation (default) stands for default data; simulation (85th percentile) or simulation (95th percentile) stand for simulation under parameters calibrated with the mean values of the 85th or 95th percentiles of the values of acceleration and deceleration observed in the field.

A two-sample *t*-test was done to see if the means (i.e., μ_{obs} and μ_{sim}) are equal for the two samples made by VSP distributions based on the observed and simulated speed-time profiles for the movements AB (or BA) through each roundabout; the null hypothesis that the two means are equal is rejected if $|t| > t_{1-\alpha/2,N}$, where $t_{1-\alpha/2,N}$ is the critical value of the *t* distribution with *N* degrees of freedom at the significance level of $\alpha = 0.05$. Table 5 shows the results of the *t*-test for the observed and simulated VSP distributions in the driving direction AB through each roundabout in Figure 2. Specifically, it can be deduced that there is no significant difference in the two distributions of the observed and simulated VSP values by driving direction; analogous results have been also obtained for the opposite direction BA.

Figure 8 shows, by way of example, the distributions of time spent in the VSP modes both under the default parameters of AIMSUN and the parameters calibrated with the mean values of the 85th or 95th percentiles of the accelerations and decelerations extracted from all trajectories experienced by the test vehicle through the roundabouts 2 and 4 in Figure 2. Figure 8a shows that the distribution of time spent in VSP modes under parameters calibrated with the mean values of the 95th percentiles of the acceleration and deceleration matched the VSP distribution from empirical data more closely than the distribution under default parameters or under calibration with the mean values of the 85th percentiles of the acceleration and deceleration. The simulated vehicle appears to spend a high proportion of time in VSP mode 4 under the default parameters; the proportion of time sensibly appears to reduce from VSP mode 5 onward, but it still appears in VSP modes 11 to

13 denoting high acceleration events. Under the default parameters of AIMSUN, simulated vehicle activity from roundabouts was not representative of real-world vehicle activity.

Table 5. The results of the two-sample *t*-test for the observed vs. simulated VSP distributions for the sampled roundabouts (direction AB in Figure 2).

VSP	μ_{obs} [1] (s.e.)	μ_{sim} [1] (s.e.)	95% c.i. for Difference in Means	$t_{(\alpha = 0.05)}$ (d.f) [2]	t-Critical	*p*-Value (α = 0.05) [3]
			Roundabout 1			
obs. vs. default	1.90 (0.80)	3.15 (1.48)	(−4.628, 2.135)	0.73 (55)	2.004	0.46
obs. vs. 85th percentile	0.98 (0.60)	1.80 (1.09)	(−3.313, 1.667)	0.66 (71)	1.993	0.51
obs. vs. 95th percentile	1.67 (0.77)	1.96 (1.24)	(−3.228, 2.654)	0.20 (55)	2.004	0.85
			Roundabout 2			
obs. vs. default	3.52 (3.06)	5.11 (2.69)	(−9.747, 6.571)	0.39 (58)	2.002	0.69
obs. vs. 85th percentile	3.56 (2.36)	2.54 (1.23)	(−2.340, 8.371)	1.13 (53)	2.005	0.27
obs. vs. 95th percentile	3.09 (2.02)	2.80 (1.30)	(−3.135, 7.366)	0.39 (64)	1.997	0.43
			Roundabout 3			
obs. vs. default	1.60 (1.75)	3.31 (2.57)	(−7.969, 4.557)	0.55 (49)	2.009	0.58
obs. vs. 85th percentile	1.52 (1.69)	2.40 (1.47)	(−5.372, 3.621)	0.39 (57)	2.002	0.69
obs. vs. 95th percentile	1.14 (1.48)	1.44 (1.58)	(−4.630, 4.019)	0.14 (68)	1.995	0.88
			Roundabout 4			
obs. vs. default	1.60 (1.57)	5.99 (2.66)	(−10.51, 1.942)	1.39 (46)	2.014	0.17
obs. vs. 85th percentile	1.75 (1.32)	0.82 (1.05)	(−4.288, 2.420)	0.55 (72)	1.993	0.58
obs. vs. 95th percentile	2.47 (0.85)	2.32 (1.09)	(−2.607, 2.899)	0.10 (88)	1.980	0.91
			Roundabout 5			
obs. vs. default	2.08 (1.21)	3.67 (3.33)	(−8.749, 5.573)	0.44 (39)	2.023	0.66
obs. vs. 85th percentile	2.75 (0.79)	1.73 (0.80)	(−1.220, 3.259)	0.90 (104)	1.983	0.37
obs. vs. 95th percentile	2.46 (0.96)	1.77 (1.71)	(−3.242, 4.606)	0.35 (68)	1.995	0.73
			Roundabout 6			
obs. vs. default	1.98 (1.21)	1.09 (1.58)	(−3.089, 4.881)	0.45 (74)	1.994	0.65
obs. vs. 85th percentile	2.01 (1.22)	0.22 (1.22)	(−1.722, 5.294)	1.01 (74)	1.992	0.31
obs. vs. 95th percentile	1.86 (1.28)	0.98 (1.67)	(−3.331, 5.080)	0.41 (70)	1.996	0.68

[1] μ_{obs} and μ_{sim} stand for the mean values of the samples of the observed vs. simulated VSP distributions for each roundabout in direction AB; [2] $|t|$ value of the two-sample *t*-test done to compare the equality of the means of samples of two populations with equal sample size; [3] α = 0.05 is the significance level.

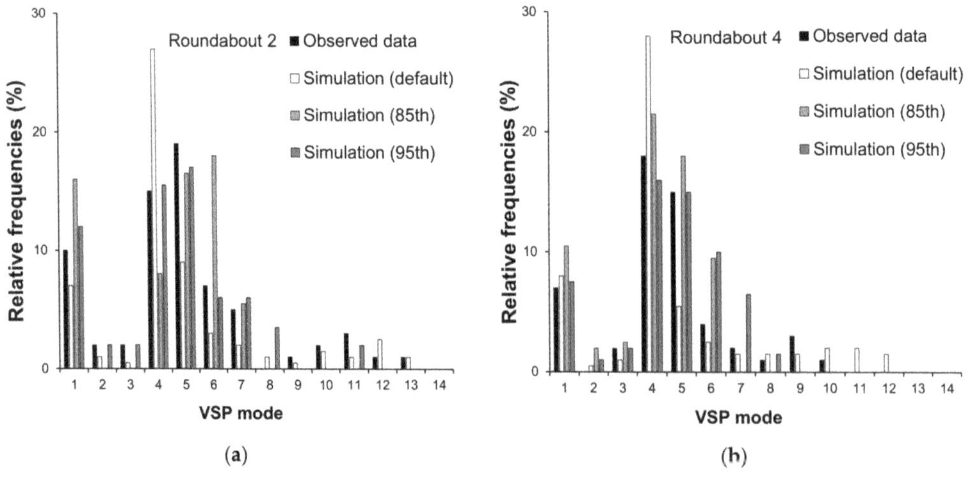

Figure 8. The relative frequencies of time spent in VSP modes under simulation with the default parameters and calibration with 85th or 95th percentile values of the selected vehicle attributes of

AIMSUN for: (**a**) Roundabout 2, movement AB, in Figure 2b; (**b**) Roundabout 4, movement AB, in Figure 2d. Note: simulation (default) stands for default data; simulation (85th) or simulation (95th) stand for simulation under parameters calibrated with the mean values of the 85th or 95th percentiles of the values of acceleration and deceleration observed in the field.

Since emissions are strongly associated with vehicle speed and acceleration, the process of high speed and aggressive acceleration can produce higher emissions than the process of braking [42]. Under the parameters calibrated with the mean values of 95th percentiles of acceleration and deceleration with reference to the sampled roundabouts, the time percentages were more realistic in the VSP modes 1 to 2 (deceleration), mode 3 (idle) and modes 4 to 7 (acceleration and cruising). Similar considerations can be drawn from Figure 8b where the more pronounced curvature of the trajectories, together with an atypical design of Roundabout 4, accentuated the effect of speed moderation; however, compared to scheme with a more typical design (see Figure 8a), greater variability of the relative frequencies by VSP mode was observed even in relation to short stop-and-go cycles on entry or in the acceleration phase.

According to Anya et al. [46], sampling from the 85th percentile distribution of accelerations and decelerations measured in the field limited the maximum values of the accelerations or decelerations obtainable in AIMSUN. Based on data observed in the field, 85 percent of the accelerations were below about 1.0 m/s^2 (that is the mean value of the distribution of the 85th percentile accelerations observed in the field); thus, there would be modelling no realistic scenario for a vehicle entering the simulated roundabout network model, if it could not achieve any instantaneous accelerations greater than 1.0 m/s^2 when travelling unconstrained below the speed limit. Among other things, the vehicles took a longer time to travel the circulatory roadway without significant changes in speed. Calibrating the model with 95th percentile values of the relevant parameters observed in the field at the existing roundabouts, in turn, tended to be more effective in producing VSP distributions enough consistent with what was drawn from the observed trajectory data. However, in this study, the absolute values of emissions were not of much concern given the potential for differences depending on the test vehicle characteristics; among the other things, the intent of this paper was not to derive definitive emission inventories but to explain the relative emission differences associated with various values of acceleration and deceleration.

4.2. Results

The speed-time profiles detected in the field by smartphone through the existing roundabouts and those simulated by AIMSUN were the starting point for estimating emissions of CO_2, CO, NOx and HC. Emissions for each source pollutant and representative speed profile by driving direction were calculated employing Equation (5) based on the time spent by the test vehicle in each VSP mode and the second-by-second emission rates in Table 4.

A two-sample t-test was performed for comparing average emissions between the observed and simulated data (default, 85th and 95th) by pollutant and by driving direction through the sampled roundabouts. Table 6 shows the results which confirmed that there is no significant difference between the emissions calculated using observed and simulated data in directions AB and BA through the roundabouts; the results also confirmed the need of calibrating the model to improve the accuracy of emission estimates.

Table 6. The results of the two-sample *t*-test for comparing average emissions between observed and simulated data (default, 85th and 95th) by pollutant and by driving direction AB and BA.

Pollutants	μ_{AB} [1] (s.e.)	μ_{BA} [1] (s.e.)	95% c.i. for Difference in Means	$t_{(\alpha = 0.05)}$ (d.f) [2]	t-Critical	*p*-Value ($\alpha = 0.05$) [3]
			CO_2			
default (AB)	66.54 (4.89)	69.41 (4.17)	(−17.20; 11.46)	0.45 (10)	2.228	0.66
default (BA)	76.85 (8.78)	74.64 (3.87)	(−19.19; 23.60)	0.23 (7)	2.364	0.82
85th percentile (AB)	66.54 (4.89)	64.27 (6.77)	(−16.36; 20.90)	0.27 (9)	2.262	0.80
85th percentile (BA)	76.85 (8.78)	65.99 (6.05)	(−12.92; 34.63)	1.01 (9)	2.262	0.34
95th percentile (AB)	66.54 (4.89)	71.26 (6.66)	(−23.14; 13.70)	0.57 (9)	2.262	0.60
95th percentile (BA)	76.85 (8.78)	67.80 (5.66)	(−14.24; 32.33)	0.86 (9)	2.262	0.42
CO						
default (AB)	0.011 (0.001)	0.012 (0.001)	(−0.005; 0.002)	0.76 (10)	2.228	0.46
default (BA)	0.010 (0.001)	0.011 (0.001)	(−0.003; 0.001)	0.84 (10)	2.228	0.42
85th percentile (AB)	0.011 (0.001)	0.010 (0.001)	(−0.003, 0.004)	0.44 (10)	2.228	0.67
85th percentile (BA)	0.010 (0.001)	0.010 (0.001)	(−0.002; 0.003)	0.36 (10)	2.228	0.73
95th percentile (AB)	0.011 (0.001)	0.012 (0.001)	(−0.004; 0.003)	0.23 (10)	2.262	0.82
95th percentile (BA)	0.010 (0.001)	0.010 (0.001)	(−0.002; 0.003)	0.15 (10)	2.262	0.88
NOx						
default (AB)	0.29 (0.03)	0.41 (0.04)	(−0.247, 0.009)	2.06 (9)	2.262	0.07
default (BA)	0.27 (0.04)	0.38 (0.04)	(−0.233, 0.013)	1.99 (10)	2.228	0.07
85th percentile (AB)	0.29 (0.03)	0.26 (0.03)	(−0.064, 0.133)	0.78 (10)	2.228	0.45
85th percentile (BA)	0.27 (0.04)	0.26 (0.03)	(−0.093; 0.114)	0.22 (9)	2.262	0.83
95th percentile (AB)	0.29 (0.03)	0.28 (0.04)	(−0.108, 0.117)	0.10 (10)	2.228	0.93
95th percentile (BA)	0.27 (0.04)	0.27 (0.03)	(−0.098; 0.109)	0.12 (9)	2.262	0.90
HC						
default (AB)	0.007 (0.001)	0.006 (0.001)	(−0.001; 0.003)	0.87 (8)	2.306	0.41
default (BA)	0.006 (0.001)	0.005 (0.001)	(−0.001, 0.002)	0.85 (8)	2.306	0.42
85th percentile (AB)	0.007 (0.001)	0.006 (0.001)	(−0.003; 0.003)	0.27 (10)	2.228	0.80
85th percentile (BA)	0.006 (0.001)	0.007 (0.001)	(−0.002; 0.0006)	1.23 (10)	2.228	0.25
95th percentile (AB)	0.007 (0.001)	0.006 (0.001)	(−0.002; 0.003)	0.46 (9)	2.262	0.65
95th percentile (BA)	0.006 (0.001)	0.006 (0.001)	(−0.001; 0.0012)	0.03 (9)	2.262	0.98

[1] μ_{AB} and μ_{BA} stand for the mean values of the samples of the observed and simulated emissions for each pollutant and driving directions AB and BA; [2] $|t|$ value of the two-sample *t*-test done to compare the equality of the means of samples of two populations with equal sample size; [3] $\alpha = 0.05$ is the significance level.

The calculated emissions were then averaged in the two driving directions AB and BA for each relevant speed-time profile as introduced before, to obtain the total emissions for source pollutant and roundabout (see Figure 9).

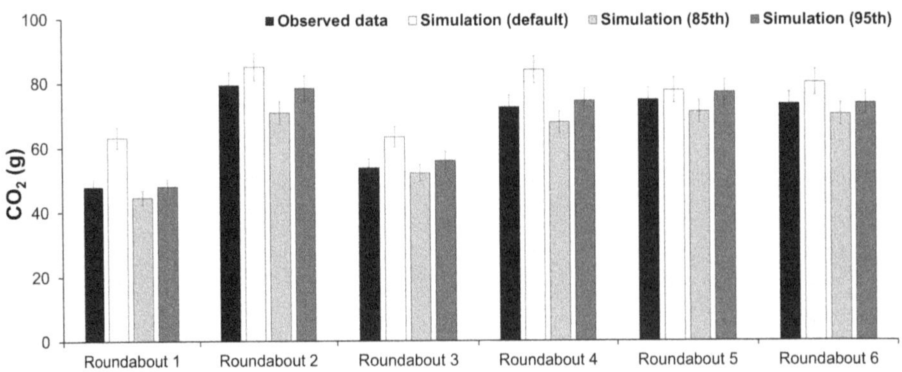

(a)

Figure 9. *Cont.*

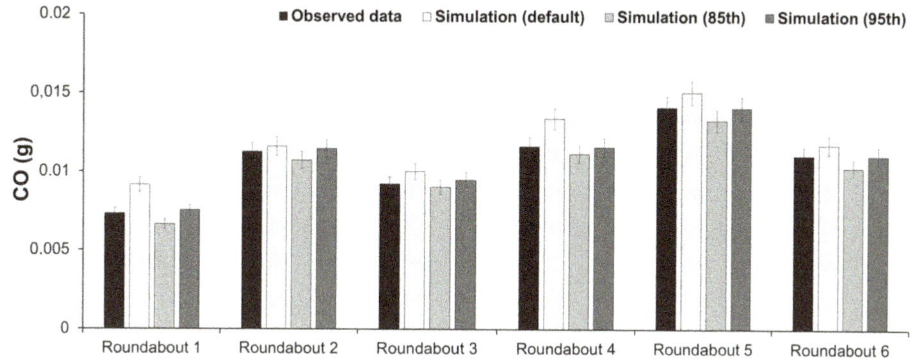

(b)

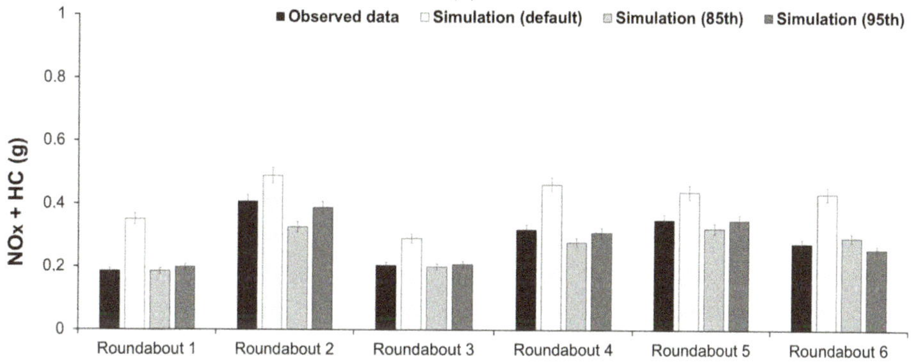

(c)

Figure 9. Observed versus simulated emissions at sampled roundabouts (through movements): (**a**) CO_2; (**b**) CO; (**c**) NOx + HC.

The results concerning CO_2 emissions in Figure 9a show that the percentage variations between the observed data and those simulated with the default parameters of AIMSUN were significantly higher than 5% (in the case of Roundabouts 1, 3 and 4 even higher than 15%); under simulation with the 85th percentiles of the acceleration and deceleration, the simulated values underestimated the observed ones and also showed percentage reductions greater than or equal to 5% (Roundabouts 1, 2, 4, 5). Emission values simulated with the 95th percentiles of the acceleration and deceleration were close to the observed ones (Roundabouts 1 and 6) and showed percentage increases less than 5%; only a percentage reduction of 1% resulted on Roundabout 2, however less than the reduction percentage found under 85th percentiles of the acceleration and deceleration (approximately equal to 11% on the same roundabout).

Concerning CO emissions in Figure 9b, the percentage variations between the observed data and those simulated with the default parameters resulted significantly higher than 5% (but higher than 15% on Roundabouts 1 and 4); under simulation with the 85th percentiles of the acceleration and deceleration, the percentage reductions between observed and simulated data were greater than or equal to 5% (intersections 1, 2, 5, 6), while under simulation with the 95th percentiles of the acceleration and deceleration, the simulated values met the observed data (Roundabouts 4, 5 and 6), with percentage increases less than 3% on the other roundabouts. In turn, Figure 9c shows the aggregated results for (NO_X + HC) emissions: under simulation with the default parameters of AIMSUN, the percentage increases between the observed and simulated data were significantly higher than 25%, while under simulation with the 85th percentile values of the acceleration and deceleration, the percentage reductions were greater than or equal to 5% (Roundabouts 2, 4, 5) or the percentage increase was around 7% for Roundabout 6. In turn, under simulation with the 95th percentiles of the acceleration and deceleration, the simulated values showed a certain variability by roundabout where they were close to the observed data or having slight percentage reductions (e.g., on Roundabouts 2 and 4), however less than under simulation with the 85th percentile values of the relevant parameters on the same roundabouts. In order to ensure the validity of the methodological approach proposed in this paper and to confirm its applicability and reproducibility, the conversion of the Roundabout 2 in Figure 2b into a turbo counterpart was conceptualized based on [68]. This roundabout was chosen as an example with reference to its typical design. Speed profiles were collected again at Roundabout 2 during the morning (7:00–8:00 a.m.) and afternoon peak hours (6:00–8:00 p.m.) on weekdays (Tuesday to Friday) in July 2020.

Traffic flow data were recorded by travel direction and classified by manoeuvre so that they can be identified in the roundabout network models built in AIMSUN. Speed profiles were simulated for the existing roundabout and the turbo roundabout based on the procedure developed by [69]. Table 7 shows the elements of the geometric design of the turbo roundabout here designed and used as a term of comparison; by way of example, Figure 10 shows the outcome of the comparison between the CO_2 emissions for the two roundabout layouts which will be discussed in the next section.

Table 7. Geometric conceptualization of the turbo roundabout.

Cross Section		[m]
Radius of the inside roadway, inner edge	R_1	15.00
Radius of the inside roadway, outer edge	R_2 [1]	20.40
Radius of the outside roadway, inner edge	R_3 [2]	20.70
Radius of the outside roadway, outer edge	R_4 [3]	25.95
Inner or outer edge line offset		0.45
Inside lane		4.35
Divider inner or outer line offset		0.20
Divider		0.30
Outside lane		4.25
Inside roadway width		5.50
Outside roadway width		5.25
Shift 1—inside to middle		5.15
Shift 2—middle to outside		4.95
Bias 1 for R_1 (Bias 2 for other radii)		2.58 (2.48)
Arc centre bias for R_1 (Bias 2 for other radii)		2.60 (2.45)

[1] $R_2 = R_1$ + inside roadway width-bias difference (differences match roadway widths); [2] $R_3 = R_2$ + divider width; [3] $R_4 = R_3$ + outside roadway width.

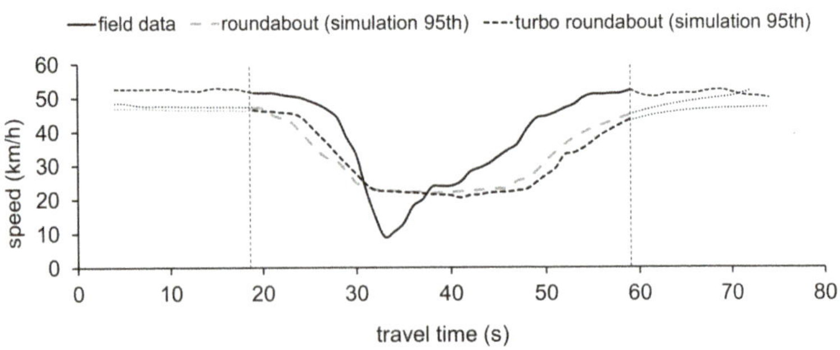

(**a**)

Figure 10. *Cont.*

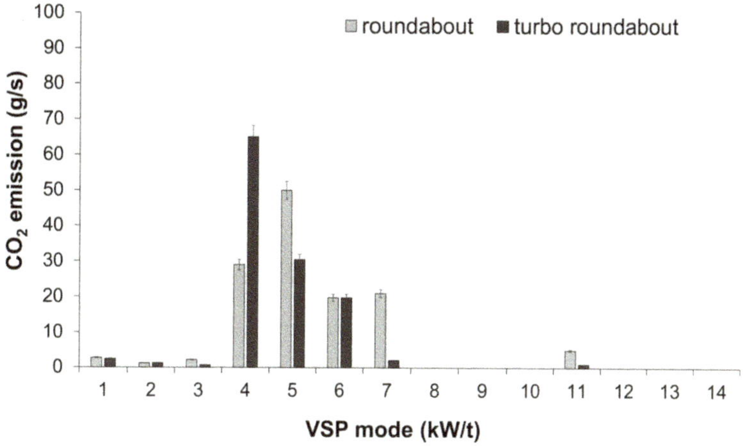

(b)

Figure 10. Turbo roundabout versus roundabout: (**a**) speed-time profiles; (**b**) simulated CO_2 emissions for through movements.

5. Discussion

The research is referred to a case study of six roundabouts installed in the road network of Palermo, Italy, where pre-existing conditions of the built environment may have constrained the entry and exit geometry, or approach alignment, and consequently driving behaviour through the roundabouts. Examination of the speed profiles as shown in Figure 3 highlighted that each type of situation had a quite distinctive shape but, not all the profiles completely fitted the typical shapes as literature presents [38]. There was a large group of profiles that appear to be somewhere "in between" the two typical shapes. After identification of the main characteristics of trajectories driven through the sampled roundabouts, the data were interpreted in terms of representative speed profiles provided by the test vehicle by roundabout to use them in the subsequent simulation, and in indirect estimation of emissions based on the VSP methodology. The comparison between individual GPS trajectories collected in the field and second-by-second speed profiles derived from AIMSUN as shown in Figure 6 called for model calibration. The speed-time profiles simulated under calibrated model parameters exhibited a better match with the second-by-second trajectories experienced in the field than those ones simulated under default values of AIMSUN. In turn, the comparison in Figure 9 showed that accuracy in emission estimates tended to be improved under calibration with the 95th percentiles of the acceleration and deceleration values extracted from vehicle trajectory data collected in the field. Consistent with literature [46], sampling from the 85th percentile distribution of accelerations and decelerations measured in the field limited the maximum values of the accelerations or decelerations obtainable in AIMSUN. In turn, calibrating the model parameters with 95th percentile values showed to be effective in producing VSP distributions more consistent with the VSP distributions derived from field-collected vehicle activity data (see Figure 8).

The results in Figure 9 also confirmed the versatility of AIMSUN in estimating emissions employing instantaneous speed-time profiles. Improvements in the predictive capacity of AIMSUN can be observed under calibration with the 95th percentiles of accelerations and decelerations; in this case, indeed, the simulated emissions resulted closer to the observed emissions with differences between the simulated values of CO_2 and CO emissions lower than other cases. Specifically, the results concerning CO_2 and CO emissions in Figure 9a,b show that the percentage variations between the observed data and those ones simulated with the default parameters of AIMSUN were significantly higher than 5%, and in some cases even higher than 15% (Roundabouts 1, 3 and 4 for CO_2 and Roundabouts 1 and 4 for CO). Under simulation with the 85th percentiles of the acceleration and deceleration, the percentage reductions between observed and simulated data were greater than or equal to 5% (Roundabouts 1, 2, 4, 5 in the case of CO_2; Roundabouts 1,2, 5,6 in the case of CO). Simulation with the 95th percentiles of the acceleration and deceleration provided simulated values close to those observed (Roundabouts 1, 2 and 6 in the case of CO_2 and Roundabouts 4, 5 and 6 in the case of CO) with increases from about 1 to 3%. The results for (NOx + HC) emissions, in turn, showed a certain variability by roundabout where simulations with the 95th percentiles of the acceleration and deceleration have returned both slight percentage increases and reductions between observed and simulated data, but, in the complex, smaller in absolute value than simulation under the 85th percentile values of the relevant parameters. However, the results may have been influenced by the selected test vehicle driven in the field, the geometric design of the sampled roundabouts not always corresponding to typical layouts, or the calibration process done to adjust those parameters which allowed the model to better match the observed emissions.

Having in mind the purpose to validate the methodological approach proposed in this paper and to confirm its applicability and reproducibility, the conversion of the roundabout 2 into a turbo counterpart was conceptualized [68]. For validation purposes, the values of the calibrated parameters were employed again to quantify the similarity between the observed and simulated trajectory data. In this regard, Figure 10a compares the simulated speed-time profiles under calibration with the 95th percentile values of the relevant parameters for both roundabouts within the influence area of 250 m identified by the vertical dashed lines; the speed profile from field data concerned the existing roundabout and was introduced just as a term of comparison. In turn, Figure 10b shows the simulated CO_2 emissions only through the existing roundabout and turbo roundabout. According to the literature [70], the results showed that higher time was spent in acceleration through the turbo roundabout than the roundabout as typical standard for turbo roundabouts. However, under the low traffic volume conditions surveyed in the field, the conversion of a two-lane roundabout to a turbo roundabout gave a comparable amount of emissions; thus, the two-lane roundabout still remains as the more appropriate layout in the context of installation under examination. In general, the turbo roundabout option still represents the best solution from a safety point of view when a two-lane roundabout instead of a single-lane roundabout should be designed unless a multi-lane roundabout remains the preferred option if a maximum output of capacity is expected [71].

6. Conclusions

The paper gives a contribution that covers the topic of estimating the vehicle emissions on urban roundabouts based on the integrated use of vehicle trajectory data collected in the field by a smartphone app, the VSP methodology, and AIMSUN.

Data collection was inspired by a crowdsensing logic, where a "sentinel vehicle" namely the test vehicle travelled through the sampled roundabouts to acquire vehicle trajectory data by using a smartphone installed on board. This system of collecting instantaneous vehicle trajectory data, beyond being economic and user-friendly, made smart both the use of the available resources and the subsequent data process and analysis. Thus, the collected vehicle trajectory data were processed immediately to return the observed speed-time profiles and to obtain the vehicle acceleration and deceleration values then used in the calibration process. In addition to this, two further aspects have been also met: (a) using measures of kinematic parameters from vehicle trajectories collected in the field in order to calibrate the modelling parameters of AIMSUN; (b) using individual vehicle trajectories simulated in AIMSUN under calibrated parameters in order to increase the accuracy in the emission estimation. In this view, since the paper is focused in demonstrating the versatility of AIMSUN in estimating emissions on roundabouts, the application of the VSP methodology can be considered as supplemental to the methods of calibration already used in practice. For validation purposes, the paper also introduced the conversion of an existing roundabout into a turbo roundabout giving insights into how to assess the impacts of alternative roundabout designs from an environmental perspective. In spite of the research efforts done to pursue the stated objectives, limitation of the implemented methodology cannot be denied and can be associated with:

(1) The selected test vehicle driven in the field;
(2) Interactions with pedestrians or cyclists;
(3) Variability in driving behaviour profiles experienced in the field;
(4) Comparison of emissions just for one diesel car;
(5) Roundabouts located in flat roads;
(6) The selected driving movements here considered.

Although the pilot study here reported reflects local conditions, it can be repeated in the simulation environment.

A few suggestions can be also outlined regarding future research:

- the chance to expand the roundabout sample in order to afford the general validation of the proposed methodology and to make a correlation between the prevailing geometric characteristics and the results obtained in terms of emissions; this is closely linked both to the use of more sentinel vehicles to better characterize the speed profiles experienced through the road units, given that the relative occurrence of each possible profile may be sensitive to the prevailing traffic levels or entry demand;
- the calibration process that should also include the "global parameters" or further "local parameters" of AIMSUN for their possible effects on the simulated vehicle activity, however in combination with the "vehicle attributes" that were here fine-tuned based on the percentile values extracted from the parameter distributions surveyed in the field;
- the transferability of the methodology should be tested with reference to other design alternatives to assess the environmental effects due to the conversion of an existing layout to another with similar space footprint, and to estimate the life-cycle costs of intersection design alternatives before the installation in the real world.

However, the study underlines the potential of new attitudes in the performance evaluation of road units in order to align infrastructural projects with the long-term ambitions about low-emission urban mobility. The proposed methodological approach has also proved to be friendly in collecting data via smartphone and in the subsequent data analysis and addressed novel opportunities to collect large-scale data through digital communities of users equipped with their smartphones to collectively share data and to derive some conclusions on any processes of common interest in order to optimize their current and future conditions of mobility.

Author Contributions: Conceptualization, A.G., T.G. and F.A.; methodology, F.A., M.C.C., and P.F.; software, F.A.; calibration and goal, F.A.; formal analysis, A.G., T.G. and F.A.; investigation, A.G., T.G. and F.A.; resources, A.G., T.G. and E.M.; data curation, F.A.; writing—original draft preparation, A.G., T.G. and F.A.; writing—review and editing, F.A., A.G., T.G., M.C.C., E.M. and P.F.; visualization, F.A., A.G., T.G., M.C.C., E.M. and P.F.; supervision, A.G., T.G., M.C.C., P.F. and E.M.; funding acquisition, A.G., and T.G. All authors have read and agreed to the published version of the manuscript.

Funding: This research was partially funded by CROWDSENSE project (POFESR2014-20_Sicily—Sicilian Region—PO FESR 2014–2020), Department of Engineering, University of Palermo, Italy. The authors also acknowledge the support of the projects: UIDB/00481/2020 and UIDP/00481/2020—Fundação para a Ciência e a Tecnologia; and CENTRO-01-0145-FEDER-022083—Centro Portugal Regional Operational Programme (Centro2020), under the PORTUGAL 2020 Partnership Agreement, through the European Regional Development Fund; DICA-VE (POCI-01-0145-FEDER-029463) and Driving2Driverless (POCI-01-0145-FEDER-031923) projects, co-funded by COMPETE2020, Portugal2020-Operational Program for Competitiveness and Internationalization (POCI) and European Union's ERDF (European Regional Development Fund); and "PAC Portugal AutoCluster for the Future" project, funded by PORTUGAL 2020 Partnership Agreement.

Institutional Review Board Statement: Not applicable.

Informed Consent Statement: Not applicable.

Data Availability Statement: Data are available after kind request to the corresponding author.

Acknowledgments: The authors would like to gratefully acknowledge the reviewers who provided helpful comments and insightful suggestions on the draft of the manuscript. Special thanks also go to the Organizing Committees of the 17th Scientific and Technical Conference TSTP 2021 and CIGOS 2021 where an earlier version of this work, as drawn from the PhD thesis [49], was presented in 2021.

Conflicts of Interest: The authors declare no conflict of interest that could influence the work reported in this paper.

References

1. Martins, V.; Anholon, R.; Quelhas, O. Sustainability Transportation Methods. In *Encyclopedia of Sustainability in Higher Education*; Filho, W.L., Ed.; Springer Nature: Cham, Switzerland, 2019; pp. 1–7.
2. The Paris Agreement. 2016 United Nations Framework Convention on Climate Change (UNFCCC). Available online: https://unfccc.int/sites/default/files/resource/parisagreement_publication.pdf (accessed on 20 November 2021).
3. Communication from the Commission: The European Green Deal. Available online: https://eur-lex.europa.eu/legal-content/EN/TXT/?qid=1588580774040&uri=CELEX:52019DC0640 (accessed on 20 November 2021).
4. Horan, T.; Rae Zimmerman, R. *Digital Infrastructures: Enabling Civil and Environmental Systems through Information Technology*; Networked Cities Series; Routledge: London, UK, 2004; 272p.
5. Litman, T. *Autonomous Vehicle Implementation Predictions: Implications for Transport Planning*; Victoria Transport Policy Institute: Victoria, BC, Canada, 2020.
6. Frey, H.C. Trends in on-road transportation energy and emissions. *J. Air Waste Manag. Assoc.* **2018**, *68*, 514–563. [CrossRef]
7. Abdelrahman, A.; El-Wakeel, A.S.; Noureldin, A.; Hassanein, H.S. Crowdsensing-Based Personalized Dynamic Route Planning for Smart Vehicles. *IEEE Netw.* **2020**, *34*, 216–223. [CrossRef]
8. Castignani, G.; Derrmann, T.; Frank, R.; Engel, T. Driver Behavior Profiling Using Smartphones: A Low-Cost Platform for Driver Monitoring. *IEEE Intell. Transp. Syst. Mag.* **2015**, *7*, 91–102. [CrossRef]
9. Rapa, M.; Gobbi, L.; Ruggieri, R. Environmental and Economic Sustainability of Electric Vehicles: Life Cycle Assessment and Life Cycle Costing Evaluation of Electricity Sources. *Energies* **2020**, *13*, 6292. [CrossRef]
10. Monzon, A.; Wang, Y. Toward Sustainable and Low Carbon Road Transportation: Policies, Tools, and Planning Methods. *Sustainability* **2019**, *11*, 1709. [CrossRef]
11. Hamadneh, J.; Esztergár-Kiss, D. The Influence of Introducing Autonomous Vehicles on Conventional Transport Modes and Travel Time. *Energies* **2021**, *14*, 4163. [CrossRef]
12. Lyu, P.; Wang, P.S.; Liu, Y.; Wang, Y. Review of the studies on emission evaluation approaches for operating vehicles. *J. Traffic Transp. Eng. (Engl. Ed.)* **2021**, *8*, 493–509. [CrossRef]
13. Fernandes, P.; Macedo, E.; Bahmankhah, B.; Tomas, R.F.; Bandeira, J.M.; Coelho, M.C. Are internally observable vehicle data good predictors of vehicle emissions? *Transp. Res. Part D Transp. Environ.* **2019**, *77*, 252–270. [CrossRef]
14. Jiménez-Palacios, J.L. Understanding and Quantifying Motor Vehicle Emissions with Vehicle Specific Power and TILDAS Remote Sensing. Ph.D. Thesis, Massachusetts Institute of Technology, Department of Mechanical Engineering, Cambridge, MA, USA, 1999.
15. Wyatt, D.W.; Li, H.; Tate, J.E. The impact of road grade on carbon dioxide (CO_2) emission of a passenger vehicle in real-world driving. *Transp. Res. Part D Transp. Environ.* **2014**, *32*, 160–170. [CrossRef]
16. Jaworski, A.; Mądziel, M.; Lejda, K. Creating an emission model based on portable emission measurement system for the purpose of a roundabout. *Environ. Sci. Pollut. Res. Int.* **2019**, *26*, 21641–21654. [CrossRef]
17. Oh, G.; Peng, H. Eco-driving at Signalized Intersections: What is Possible in the Real-World? In Proceedings of the 21st International Conference on Intelligent Transportation Systems (ITSC), Maui, HI, USA, 4–7 November 2018; pp. 3674–3679.
18. Ntziachristos, L.; Gkatzoflias, D.; Kouridis, C.; Samaras, Z. COPERT: A European Road Transport Emission Inventory Model. In *Information Technologies in Environmental Engineering. Environmental Science and Engineering*; Athanasiadis, I.N., Rizzoli, A.E., Mitkas, P.A., Gómez, J.M., Eds.; Springer: Berlin/Heidelberg, Germany, 2009; pp. 491–504.
19. EMFAC 2017 Volume I—User's Guide V1.0.2 Mobile Source Analysis Branch, Air Quality Planning & Science Division, California Air Resource Board Sacramento, CA 2018. Available online: https://www.arb.ca.gov/msei/downloads/emfac2017-volume-i-users-guide.pdf (accessed on 6 December 2021).
20. EMFAC-HK Vehicle Emission Calculation. Available online: https://www.epd.gov.hk/epd/english/environmentinhk/air/guide_ref/emfac-hk.html (accessed on 6 December 2021).
21. Tchepel, O.; Costa, A.M.; Amorim, J.H.; Miranda, A.; Borrego, C. Transport emission model and dispersion study for Lisbon air quality at local scale. In Proceedings of the 11th International Symposium, Transport and Air Pollution, Graz, Austria, 19–21 July 2002; Volume I, pp. 109–116.
22. Akçelik, R.; Smit, R.; Besley, M. Recalibration of a vehicle power model for fuel and emission estimation and its effect on assessment of alternative intersection treatments. In Proceedings of the 4th International Roundabout Conference, Seattle, WA, USA, 16–18 April 2014.

23. Transportation Research Board and National Research Council. *Modeling Mobile-Source Emissions*; The National Academies Press: Washington, DC, USA, 2000. [CrossRef]
24. Fernandes, P.; Pereira, S.R.; Bandeira, J.M.; Vasconcelos, L.; Bastos Silva, A.; Coelho, M.C. Driving around turbo-roundabouts vs. conventional roundabouts: Are there advantages regarding pollutant emissions? *Int. J. Sustain. Transp.* **2016**, *10*, 847–860. [CrossRef]
25. National Academies of Sciences, Engineering, and Medicine. *Predicting Air Quality Effects of Traffic-Flow Improvements: Final Report and User's Guide*; The National Academies Press: Washington, DC, USA, 2005. [CrossRef]
26. Ryu, B.Y.; Jung, H.J.; Bae, S.H. Development of a corrected average speed model for calculating carbon dioxide emissions per link unit on urban roads. *Transp. Res. Part D* **2015**, *34*, 245–254. [CrossRef]
27. US EPA. *Population and Activity of On-Road Vehicles in MOVES2014*; United States Environmental Protection Agency (US EPA): Washington, DC, USA, 2016.
28. Smit, R.; Smokers, R.; Schoen, E. VERSIT+ LD: Development of a new emission factor model for passenger cars linking real-world emissions to driving cycle characteristics. In Proceedings of the 14th Symposium Transport and Air Pollution, Graz, Austria, 1–3 June 2005; Volume 1, pp. 177–186.
29. Zeng, W.; Miwa, T.; Morikawa, T. Prediction of vehicle CO_2 emission and its application to eco-routing navigation. *Transp. Res. Part C* **2016**, *68*, 194–214. [CrossRef]
30. Hao, P.; Boriboonsomsin, K.; Wu, G.; Barth, M.J. Modal Activity-Based Stochastic Model for Estimating Vehicle Trajectories from Sparse Mobile Sensor Data. *IEEE Trans. Intell. Transp. Syst.* **2017**, *18*, 701–711. [CrossRef]
31. Barceló, J. *Fundamentals of Traffic Simulation*, 1st ed.; Springer: London, UK, 2010.
32. Rahimi, A.M.; Dulebenets, M.A.; Mazaheri, A. Evaluation of Microsimulation Models for Roadway Segments with Different Functional Classifications in Northern Iran. *Infrastructures* **2021**, *6*, 46. [CrossRef]
33. *AIMSUN, Version 8 Dynamic Simulator User Manual*; TSS-Transport Simulation Systems: Barcelona Spain, 2011.
34. National Academies of Sciences, Engineering, and Medicine. *Roundabouts: An Informational Guide*, 2nd ed.; The National Academies Press: Washington, DC, USA, 2010. [CrossRef]
35. Giuffrè, O.; Granà, A.; Giuffrè, T.; Acuto, F.; Lo Pinto, A. Life-Cycle Costing Decision-Making Methodology and Urban Intersection Design: Modelling and Analysis for a Circular City. In *Methods in Modern Urban Transportation Systems and Networks*; Macioszek, E., Sierpiński, G., Eds.; Springer Nature: Cham, Switzerland, 2021; pp. 59–86.
36. Parikka-Alhola, K.; Nissinen, A. Environmental impacts of transport as award criteria in public road construction procurement. *Int. J. Constr. Manag.* **2014**, *12*, 35–49. [CrossRef]
37. Hallmark, S.L.; Wang, B.; Mudgal, A.; Isebrands, H. On-Road Evaluation of Emission Impacts of Roundabouts. *Transp. Res. Rec.* **2011**, *2265*, 226–233. [CrossRef]
38. Coelho, M.C.; Farias, T.L.; Rouphail, N.M. Effect of roundabout operations on pollutant emissions. *Transp. Res. Part D Transp. Environ.* **2006**, *11*, 333–343. [CrossRef]
39. Salamati, K.; Coelho, M.; Fernandes, P.; Rouphail, N.M.; Frey, H.C.; Bandeira, J. Emission Estimation at Multilane Roundabouts: Effect of Movement and Approach Lane. *Transp. Res. Rec.* **2014**, *2389*, 12–21. [CrossRef]
40. Fernandes, P.; Salamati, K.; Rouphail, N.; Coelho, M. Identification of emission hotspots in roundabouts corridors. *Transp. Res. Part D Transp. Environ.* **2015**, *37*, 48–64. [CrossRef]
41. Guerrieri, M.; Corriere, F.; Rizzo, G.; Casto, B.L.; Scaccianoce, G. Improving the Sustainability of Transportation: Environmental and Functional Benefits of Right Turn By-Pass Lanes at Roundabouts. *Sustainability* **2015**, *7*, 5838–5856. [CrossRef]
42. Lakouari, N.; Oubram, O.; Bassam, A.; Pomares Hernandez, S.E.; Marzoug, R.; Ez-Zahraouy, H. Modeling and simulation of CO_2 emissions in roundabout intersection. *J. Comput. Sci.* **2020**, *40*, 101072. [CrossRef]
43. Mudgal, A.; Hallmark, S.; Carriquiry, A.; Gkritza, K. Driving behavior at a roundabout: A hierarchical Bayesian regression analysis. *Transp. Res. Part D Transp. Environ.* **2014**, *26*, 20–26. [CrossRef]
44. Ahn, K.; Kronprasert, N.; Rakha, H. Energy and Environmental Assessment of High-Speed Roundabouts. *Transp. Res. Rec.* **2009**, *2123*, 54–65. [CrossRef]
45. PTV. *VISSIM Version 5, User Manual*; Planung Transport Verkehr AG: Karlsruhe, Germany, 2008.
46. Anya, A.R.; Rouphail, N.M.; Frey, H.C.; Schroeder, B. Application of AIMSUN Microsimulation Model to Estimate Emissions on Signalized Arterial Corridors. *Transp. Res. Rec.* **2014**, *2428*, 75–86. [CrossRef]
47. Fernandes, P.; Coelho, M.C.; Rouphail, N.M. Assessing the impact of closely-spaced intersections on traffic operations and pollutant emissions on a corridor level. *Transp. Res. Part D Transp. Environ.* **2017**, *54*, 304–320. [CrossRef]
48. Mądziel, M.; Campisi, T.; Jaworski, A.; Kuszewski, H.; Woś, P. Assessing Vehicle Emissions from a Multi-Lane to Turbo Roundabout Conversion Using a Microsimulation Tool. *Energies* **2021**, *14*, 4399. [CrossRef]
49. Acuto, F. Integrating Vehicle specific Power Methodology and Microsimulation in Estimating emissions on Urban Roundabouts. Ph.D. Thesis, Department of Engineering, University of Palermo, Palermo, Italy, 2021.
50. Fernandes, P.; Rouphail, N.M.; Coelho, C.M. Turboroundabouts along Corridors: Analysis of Operational and Environmental Impacts. *Transp. Res. Rec. J. Transp. Res. Board* **2017**, *2627*, 46–56. [CrossRef]
51. Samaras, C.; Tsokolis, D.; Toffolo, S.; Magra, G.; Ntziachristos, L.; Samaras, Z. Improving fuel consumption and CO_2 emissions calculations in urban areas by coupling a dynamic micro traffic model with an instantaneous emissions model. *Transp. Res. Part D Transp. Environ.* **2018**, *65*, 772–783. [CrossRef]

52. Stogios, C.; Saleh, M.; Ganji, A.; Tu, R.; Xu, J.; Roorda, M.J.; Hatzopoulou, M. Determining the Effects of Automated Vehicle Driving Behavior on Vehicle Emissions and Performance of an Urban Corridor. In Proceedings of the TRB 97th Annual Meeting, Washington, DC, USA, 7–11 January 2018.
53. Ministero delle Infrastrutture e dei Trasporti [Ministry of Infrastructure and Transport]. Available online: http://www.mit.gov.it/mit/mop_all.php?p_id=13799 (accessed on 18 December 2021).
54. Schroeder, B.J.; Rouphail, N.; Salamati, K.; Bugg, Z. Effect of Pedestrian Impedance on Vehicular Capacity at Multilane Roundabouts with Consideration of Crossing Treatments. *Transp. Res. Rec. J. Transp. Res. Board* **2012**, *2312*, 14–24. [CrossRef]
55. Coelho, M.C.; Frey, C.H.; Rouphail, N.M.; Zhai, H.; Pelkmans, L. Assessing methods for comparing emissions from gasoline and diesel light-duty vehicles based on microscale measurements. *Transp. Res. Part D* **2009**, *14*, 91–99. [CrossRef]
56. Zhai, Z.; Song, G.; Lu, H.; He, W.; Yu, L. Validation of temporal and spatial consistency of facility- and speed-specific vehicle-specific power distributions for emission estimation: A case study in Beijing, China. *J. Air Waste Manag. Assoc.* **2017**, *67*, 949–957. [CrossRef] [PubMed]
57. Li, S.; Zhu, K.; van Gelder, B.; Nagle, J.; Tuttle, C. Reconsideration of sample size requirements for field traffic data collection with global positioning system devices. *Transp. Res. Rec.* **2002**, *1804*, 17–22. [CrossRef]
58. Bokare, P.S.; Maurya, A.K. Acceleration-Deceleration Behaviour of Various Vehicle Types. *Transp. Res. Procedia* **2017**, *25*, 4733–4749. [CrossRef]
59. Wang, J.; Dixon, K.; Li, H.; Ogle, J. Normal acceleration behavior of passenger vehicles starting from rest at all-way stop-controlled intersections. *Transp. Res. Rec.* **2004**, *1883*, 158–166. [CrossRef]
60. Wang, J.; Dixon, K.; Li, H.; Ogle, J. Normal deceleration behavior of passenger vehicles at stop sign controlled intersections evaluated within-vehicle global positioning system data. *Transp. Res. Rec.* **2005**, *1937*, 120–127. [CrossRef]
61. Fernandes, P.; Tomás, R.; Acuto, F.; Pascale, A.; Bahmankhah, B.; Guarnaccia, C.; Granà, A.; Coelho, M.C. Impacts of roundabouts in suburban areas on congestion-specific vehicle speed profiles, pollutant and noise emissions: An empirical analysis. *Sustain. Cities Soc.* **2020**, *62*, 102386. [CrossRef]
62. Giuffrè, O.; Granà, A.; Giuffrè, T.; Acuto, F.; Tumminello, M.L. Estimating pollutant emissions based on speed profiles at urban roundabouts: A pilot study. In *Smart and Green Solutions for Transport Systems, Advances in Intelligent Systems and Computing*; Sierpiński, G., Ed.; Springer: Heidelberg, Germany, 2020; Volume 1091, pp. 184–200.
63. Jurecki, R.S.; Stanczyk, T.L. A Methodology for Evaluating Driving Styles in Various Road Conditions. *Energies* **2021**, *14*, 3570. [CrossRef]
64. US EPA. *Methodology for Developing Modal Emission Rates for EPA's Multi-Scale Motor Vehicle & Equipment Emission System, Report number EPA420-R-02-027*; United States Environmental Protection Agency (US EPA): Washington, DC, USA, 2002.
65. Frey, H.C.; Unal, A.; Chen, J.J.; Song, L. Evaluation and Recommendation of a Modal Method for Modeling Vehicle Emissions. Corpus ID: 292981. 2003. Available online: https://www3.epa.gov/ttnchie1/conference/ei12/mobile/frey.pdf (accessed on 20 November 2021).
66. Giuffrè, O.; Granà, A.; Tumminello, M.L.; Sferlazza, A. Capacity-based calculation of passenger car equivalents using traffic simulation at double-lane roundabouts. *Simul. Model. Pract. Theory* **2018**, *81*, 1–30. [CrossRef]
67. Acuto, F.; Giuffrè, T.; Granà, A. Environmental Performance Evaluation at Urban Roundabouts. In *Emerging Technologies and Applications for Green Infrastructure*; Ha-Minh, C., Tang, A.M., Bui, T.Q., Vu, X.H., Huynh, D.V.K., Eds.; Lecture Notes in Civil Engineering; Springer: Singapore, 2022; Volume 203, pp. 1613–1621.
68. Royal Haskoning DHV. Roundabouts—Application and Design. A Practical Manual. Available online: https://nmfv.dk/wp-content/uploads/2012/06/RDC_Netherlands.pdf (accessed on 20 November 2021).
69. Acuto, F.; Giuffrè, T.; Granà, A. Environmental Performance Assessment of Urban Roundabouts. In *Intelligent Solutions for Cities and Mobility of the Future*; Lecture Notes in Networks and Systems; Sierpiński, G., Ed.; Springer: Cham, Switzerland, 2022; Volume 352, pp. 27–45.
70. Vasconcelos, L.; Silva, A.B.; Seco, Á.M.; Fernandes, P.; Coelho, M.C. Turboroundabouts: Multicriterion Assessment of Intersection Capacity, Safety, and Emissions. *Transp. Res. Rec. J. Transp. Res. Board* **2014**, *2402*, 28–37. [CrossRef]
71. Granà, A.; Giuffrè, T.; Macioszek, E.; Acuto, F. Estimation of Passenger Car Equivalents for Two-Lane and Turbo Roundabouts Using AIMSUN. *Front. Built Environ.* **2020**, *6*, 86. [CrossRef]

MDPI
St. Alban-Anlage 66
4052 Basel
Switzerland
Tel. +41 61 683 77 34
Fax +41 61 302 89 18
www.mdpi.com

Energies Editorial Office
E-mail: energies@mdpi.com
www.mdpi.com/journal/energies

www.ingramcontent.com/pod-product-compliance
Ingram Content Group UK Ltd.
Pitfield, Milton Keynes, MK11 3LW, UK
UKHW060122010826
14089UKWH00006B/104

9 783036 555232